GIScience and Geo-environmental Modelling

Series Editors

The "GIScience and Geo-environmental Modelling" book series seeks to publish a broad portfolio of scientific books addressing the interface between geography and the environment. The aim of the book series is to present geospatial technology approaches to data mining techniques, data analytics, modeling, risk assessment, visualization, and management strategies. The series includes peer-reviewed monographs, edited volumes, textbooks, and conference proceedings. The focus of Geo-environmental is on geospatial modelling in the frontier area of earth-environment related fields, such as urban and peri-urban environmental issues, ecology, hydrology, basin management, geohazards, estuarine-ecology, groundwater, agriculture, climate change, land-water, and forest resources, and related topics.

Geo-environmental modelling techniques have enjoyed an overwhelming interest in recent decades among the earth environmental and social sciences research communities for their powerful ability to solve and understand various complex problems and develop novel approaches toward sustainable earth and human society. Geo-environmental modelling using data mining, machine learning, and cloud computing technology is focused on spatiotemporal data analysis and modeling for sustainability in our environment.

Jayanta Das · Somenath Halder
Editors

Advancement of GI-Science and Sustainable Agriculture

A Multi-dimensional Approach

Springer

Editors
Jayanta Das
Department of Geography
Rampurhat College
Rampurhat, West Bengal, India

Somenath Halder
Department of Geography
Kaliachak College
Kaliachak, West Bengal, India

ISSN 2730-7506 ISSN 2730-7514 (electronic)
GIScience and Geo-environmental Modelling
ISBN 978-3-031-36824-0 ISBN 978-3-031-36825-7 (eBook)
https://doi.org/10.1007/978-3-031-36825-7

This Springer imprint is published by the registered company Springer Nature Switzerland AG
The registered company address is: Gewerbestrasse 11, 6330 Cham, Switzerland

Dedicated to scholars and planners

Foreword

It is an immense pleasure to pen a foreword for *Advancement in GIScience and Sustainable Agriculture*, which covers multi-dimensional topics of current relevance. The book is edited by two young scholars, Dr. Jayanta Das and Dr. Somenath Halder, from the fields of humanities and social sciences, geography, regional studies, and planning. The book is published by Springer and is dedicated to agricultural scientists, regional scientists, and policymakers.

Today the applications, as well as adaptation of methodological knowledge of GI-Science, have been observed across the disciplines. Thus, the approach of the said branch has been transformed into a multi-disciplinary milieu. Contemporarily, the core discipline of *agriculture* is encouragingly drifting toward sustainable science and policy studies. Nevertheless, it is important to note that agriculture is inherently interlinked with various branches and sub-branches, such as food production, food security, poverty eradication, agro-economics, soil science, pre- and post-disaster management, spatial assessment of agricultural scenarios, agro-mechanics, and many other relevant facts and facets like climate change, meteorological drought, agro-based pollution and pollutants, and so forth.

Because of this, classical agricultural studies have evolved as sustainable agriculture, with advantages compared to the former as a liberal adaptation of advanced methods like GIScience and other cutting-edge technologies. The current environmental challenges, such as frequent floods and drought, climate change phenomenon (CCP), deteriorating soil health, food crisis, supply-chain failure, etc., are compelling agricultural scientists, social scientists, and policymakers to enhance the periphery of the discipline. Meanwhile, the ever-increasing concentration and density of people in a specific geo-entity have also generated immense pressure on multi-cropping, soil exhaustion, and over-dependency of the workforce in the agro-sector, especially in the Global South. From its origin, human activity like agriculture has been in ever-expansion mode throughout the world; in due course, the afresh types and forms of agricultural activities also increase new challenges to human society. Here the present concern to the scientists and policymakers is to handle the emerging challenges as the focal points and balance the environment, economy, and demography. This present volume may throw light on this regard. In the near future, when a major part of the world is urbanized, the current academic contributions may also help manage the transformation in agriculture and reframed policy implementation. It is the primary economic activity, i.e., agriculture, across underdeveloped and developing countries that are proven to be a prime supporting way for millions of households, ensuring the country's economic growth, environmental sustainability, demographic stability, and poverty alleviation for a peaceful life.

The present volume is a collection of 21 research studies from across the trans-disciplinary communities of the globe covering topics ranging from changing nature of land and water usage, advanced technology and competitive farming, shifting cultivation, landslide and micro-level agro-loss, geopolitical urban networking, MCDM model and crop yielding, road network and agro-planning, and use of nanoparticles in sustainable agriculture. The themes identified and incorporated in this edited book are vital components of the current agricultural development, and the issues addressed are of utmost significance in view of the present-day crisis management for sustainable agriculture. I want to convey my appreciation to the editors and applaud them on attempting to carry this volume for readers of diverse bands of disciplines in the areas of GIScience and sustainable agriculture.

The case-study approach used in the chapters, with instances across the various corners of the globe covering diverse geo-environment, brings the insights of the authors and editors to bear upon the regional pattern of interconnection between GIScience and sustainable agriculture and other linked issues of the present era. I hope this edited piece will be widely acclaimed by agricultural scientists, regional scientists, policymakers, and scholars of allied disciplines on the emerging issues of sustainable agriculture.

Dr. Ranjan Roy
Professor and Former Head of the Department
Department of Geography and Applied Geography
University of North Bengal
Darjeeling, West Bengal, India

Preface

Contemporarily, we are living in the age where rocket-rising innovations leave a marked question on our served dishes, from breakfast to dinner, where the main agenda lies within the tag-line *Advancement of GI-Science and Sustainable Agriculture*. The current century also faces numerous problematic issues like food shortages, food insecurity, malnutrition, malfunctioning food-chain-delivery-systems, conflicts on policy proposal and ground-level implementation of food security and agricultural policies. Consecutively, sustainable agriculture may provide the desired results affording with the food crisis and declining agro-productivity, post-pandemic food security, zonation and mapping technique viewing food crisis, biotechnology and sustainable agricultural, scaling hunger indices, health hazard and food crisis, changing climate and food availability, consumer load and fertilizer usage, growing demand and increasing usage of harmful chemical in agro-fields, scope of sustainable agricultural potentiality (SAP) modeling, impact of pandemic on sustainable agriculture, using waste water as non-sustainable agricultural practice, applying geospatial techniques on extreme weather susceptibility and agro-production, soil erosion and poor agricultural production, questioning shifting cultivation on the issue of sustainability, meteorological drought and irrigational gaps, occupational mobility and loss of agricultural heritage, GI-Science and sustainable agro-management, community preparedness in food-crisis management, climate change declining sustainable agro-production, and many more.

Without any doubt, most of the entire world suffers from several natural and human-induced problems; among them, food crisis and unsustainable agriculture throws major challenge to human existence. Contrastingly, if the modern technology and means, with advanced monitoring and calibration methodology and policy guidance, can help, it will certainly reduce half of the world's problem and ensure future survival issue of human society. The present 'multi-dimensional approach' also can minimize the other partially linked problems, like climate change and food shortage, livelihood crisis and environmental refugee, international trade balance, global food supply-chain interruption, ever-expanding gap between rich and poor, and so on. Therefore, nurturing the knowledge in proper way on the application of GI-Science for agriculturally sustainable society and also their monitoring and management can curtail the gap between science, policy, and the ground-level scenario concerned.

The book comprises two major parts: (Part I) Sustainable Agriculture and Applied GI-Science, and (Part II) Agro-ecology, People and GI-Science. The incorporated 21 chapters are attempting to capture the diverse and complex trends of usage of GI-Science in sustainable agriculture and/or agricultural sustainability. The aforesaid phenomena are induced to project through case studies and discussions from different pockets of the developing world.

Under the Part I, there are eleven chapters. Chapter 1 broadly discloses and rebuilds the nexus between sustainable agriculture and GI-Science with a most recognized systematic literature review, across the countries of the globe. The subsequent chapter, Chap. 2, with the help of meta-data, is attempted to highlight the research academia regarding the application of GIS technique for resolving the management issue related to present-day agricultural crisis. While, Chap. 3 with an advanced technique, like hybrid multi-criteria decision making, is trying to pursue sustainability analysis of different protected cultivation structure. Chapter 4 attempts to foresee agricultural drought monitoring by applying remote sensing indices, in combination format. The said chapter applied DEMETAL method as a cutting-edge technique for drought monitoring agenda. The next chapter, Chap. 5, tries to explore contemporary trends of meteorological variables and its effect on agriculture in agriculturally dominant region. Subsequently, Chap. 6 keeps the prime focus on MCDM model and assesses the site suitability in the case of potato cultivation. With the help of geospatial approach (especially RS-GIS based model), Chap. 7 targets the agenda like projected suitability analysis of an industrially demanded crop (finger millet) through intra-regional assessment, micro- to macroregion. Chapter 8 seeks to find out the yield estimation of crops like rice and potato, based on vegetation indices in micro-region. The said chapter also incorporates Sentinel 2B satellite imagery data for geospatial and geostatistical breakthrough. The next chapter, Chap. 9, unveils the potential land suitability analysis of an important cash crop in a marginalized region, adopting MCDM model, whereas Chap. 10 approaches the integration of IT-OT for sustainable and competitive farming. On the contrary, Chap. 11 discusses how for mitigating of abiotic stress in crop plants the application of biogenic nanomaterial(s) is viable as a blueprint. This work may throw a new light in biogenic research arena.

Under the Part II, there are ten chapters, which are as follows. In view of exploring pro-poor employment potential Chap. 12 discusses the value chain analysis of sericulture in Bangladesh. The said chapter also put valuable recommendations for the sake of sustainable agriculture. In the next chapter, Chap. 13, in search of sustainable development in an underdeveloped district, a significant agricultural specific crop suitability modeling is worked out by using geospatial techniques. The above chapter outcome is also helpful for eradicating the phenomenon of unsuitability of maize cultivation in other micro-regions. In Chap. 14, with the help of advanced geostatistical technique, a critical analysis has been pursued for problematic scenario like declining groundwater level and its influence on irrigation and agro-production. Subsequently, Chap. 15 explores the effect of shifting cultivation and changing land use on a watershed in extreme northeastern part of India. Afterward, Chap. 16 seeks to find out the inter-relation between road network analysis with the sustainability of agriculture and the development. The mentioned chapter purposively induced GIS-based road network with graph theory to explore the targeted objectives. Simultaneously, Chap. 17 keeps its focus upon the impacts of vulnerable landslides, in landslide-prone areas, on agricultural sustainability; here geospatial mapping played a major analytical role. Chapter 18 investigates how the changing mode of agricultural land use controls on the farmers' livelihood assets. Subsequently, Chap. 19 assesses the phenomenon like drought upon the sustaining agricultural productivity in peri-urban region. The very chapter used more advanced technique of analysis, with the help of meteorological data and indicators like SPI and SPEI, to reach its prime goal. The next chapter is quite dissimilar one. In Chap. 20, an attempt has been made to understand the utility of print media for dissemination of insect pest management (IPM) tactics to rice farmers. Lastly, in Chap. 21, the authors have tried to explore the incongruous potentiality of faulty wastewater usage versus agricultural sustainability in the eastern part of mega-city like Kolkata (India). Ultimately, this volume should be measured as a significant development work in terms of its ground to laboratory expedition, methodological, and axiological rigor in expediting integrative sciences toward sustainable and resilient agricultural potentiality in the true sense of the term.

Birbhum, West Bengal, India — Jayanta Das
Malda, West Bengal, India — Somenath Halder

Acknowledgements

We would like to express our deep gratitude to all the contributing authors for their valuable time, effort, and innovative research ideas presented in the individual chapters. We would also express our sincere thanks to the series editors and publishing editor of 'GIScience and Geo-environmental Modeling' for their kind patience and bigheartedness to undertake the editorial work for such diverse topics on *Advancement of GI-Science and Sustainable Agriculture*. All the included chapters are peer-reviewed, and the authors have developed and modified their research in response to the reviewers' suggestions. We would like to express our heartfelt thanks to all of the anonymous reviewers who made their tough effort to submit comments and recommendations on the chapters. We are deeply indebted to Prof. Ranjan Roy (Department of Geography and Applied Geography, University of North Bengal) for agreeing to write the foreword and to continuously keep patience in revising it as we kept adding new chapters and/or changing chapter titles in the book. We would also like to express our special thanks to Prof. Deepak Kumar Mandal, HoD, Department of Geography and Applied Geography, the University of North Bengal, for his generous attitude and encouragement regarding the preparation of this book. Dr. Jayanta Das cordially indebt to his Ph.D. Supervisor Dr. Sudip Kumar Bhattarcharya, Retd. Associate Professor, Department of Geography and Applied Geography, University of North Bengal. He shows his gratitude to Dr. Asish Banerjee, President of Rampurhat College, and also Deputy Speaker of West Bengal Legislative Assembly, and Dr. Buddhadeb Mukherjee, Teacher-in-Charge, for their encouragement and support. He also expresses his due acknowledgment to the Department of Geography, Rampurhat College, for providing infrastructure facilities. Dr. Das also shows indebt to his colleagues and students of Department of Geography, Rampurhat College. Dr. Somenath Halder expresses his gratitude to Dr. Nazibar Rahaman, Principal, Kaliachak College, for continuous support and encouragement. Dr. Halder whole-heartedly indebted to his Ph.D. Supervisor Prof. Malay Mukhopadhyaya, (Retd.) Professor, Department of Geography, Visva-Bharati. He also thanks all of his colleagues, friends, well-wishers, and students from whom he got inspired a lot to take such a project in hand. Laterally, he expresses his due acknowledgment to the Department of Geography, Kaliachak College, for receiving right ambience and infrastructure. Finally, we also thank the Springer Management team for accommodating us and providing us such a platform for this book for its successful completion. It was a pleasure to work with them throughout the process.

Birbhum, West Bengal, India — Jayanta Das
Malda, West Bengal, India — Somenath Halder

About This Book

This book on *Advancement of GI-Science and Sustainable Agriculture* describes the contributing aspects of contemporary developments related to sustainable agricultural resources, assessment of sustainable agriculture in developing nations, and cutting-edge technology applied in solving region-based problems. Recently, almost the entire world suffers from several natural and human-induced problems; among them, food crisis and unsustainable agriculture throw major challenge to human society. Contrastingly, if the modern technology and means, with advanced monitoring and calibration methodology and policy guidance, can help, it will certainly reduce half of the world's problem and ensure future survival issue of human society. In addition, this approach also can minimize the other partially linked problems, like climate change and food shortage, livelihood crisis and environmental refugee, international trade balance, global food supply-chain interruption, ever-expanding gap between rich and poor, and so on. Therefore, nurturing the knowledge in proper way on the application of GI-Science for agriculturally sustainable society and also their monitoring and management can curtail the gap between science, policy, and the ground-level scenario concerned. This book is primarily confining within strategies for extraction and processing of bioactive compounds from agro-industrial wastes, emphasizes the therapeutic potential of bioactive constituents from food and agro-wastes, and presents novel drug delivery systems for enhancing the therapeutic efficacy of bioactive constituents. Comparatively, the present edited volume includes the broad agenda of usage of GI-Science in addressing agricultural sustainability for better monitoring and management. Moreover, this book has dealt with dissimilar aspects not only with statistical analysis but also with geospatial technologies for problem-oriented forecasting and policy suggestions, along with geospatial analysis. Therefore, the present volume is considerably unalike and can serve the multi-disciplinary fields of knowledge.

Key Features

- Throws limelight on the updated knowledge on sustainable agriculture contained with the advanced GI-Science application in India and abroad.
- Covers diverse topics including natural, anthropogenic, and environmental issues.

- Presents the role of technology, methods, and modeling used to facilitate controlling measures to unsustainable agriculture.
- Provides advanced methodologies for sustainable agriculture practices using MCDM, hybrid model, crop suitability model proposition, IT-OT integration, food-chain-supply discrepancies, climate change, and over-exploitation of agricultural resources, agricultural practices on problematic terrain and landscape, as well as relevant issues to ensure the SDGs and resilient society.
- International and cross-disciplinary contributions by scholars recognized globally.

Contents

Part I Sustainable Agriculture and Applied GI-Science

1 **Nexus Between GIScience and Sustainable Agriculture** 3
Sanjoy Saha, Jayanta Das, and Somenath Halder

2 **Application of GIS in Agricultural Crisis Management** 13
Sanjoy Saha

3 **Suitability Assessment of Different Protected Cultivation Structures Using Hybrid Multi-Criteria Decision Analysis Technique** 25
Debaditya Gupta, K. N. Tiwari, D. T. Santosh, and Subha M. Roy

4 **Spatio-Temporal Agricultural Drought Monitoring Using Remote Sensing Indices** 41
Syed Sadath Ali, Koyel Mukherjee, Papia Kundu, and Piu Saha

5 **Recent Trends of Meteorological Variables and Impacts on Agriculture in Northwest Bangladesh** 59
J. M. Adeeb Salman Chowdhury, Md. Abdul Khalek, and Md. Kamruzzaman

6 **Application of RS-GIS-Based Multi-Criteria Decision-Making Model (MCDM) on Site Suitability Analysis for Potato Cultivation in Jalpaiguri District, West Bengal, India** 81
Indrajit Poddar, Amiya Basak, Jiarul Alam, Jayanta Das, and Asraful Alam

7 **Comparative Assessment of Projected Suitability of Finger Millet Crops in Tamil Nadu and Parambikulam Aliyar Basin Using ECOCROP Model: A Geospatial Approach** 99
P. Dhanya, T. Sankar, and V. Geethalakshmi

8 **Vegetation Indices-Based Rice and Potato Yield Estimation Through Sentinel 2B Satellite Imagery** 113
Chiranjit Singha and Kishore C. Swain

9 A Multi-Criteria Decision-Making Approach for Land Suitability Assessment for Tea Cultivation in Hilly Aizawl District in Mizoram, India ... 135
Jonmenjoy Barman and Partha Das

10 Integration of Information Technology and Operation Technology in Agriculture Toward Sustainable and Competitive Farming ... 149
Jitranjan Sahoo, Manoranjan Dash, and Preeti Y. Shadangi

11 Green Synthesis and Application of Biogenic Nanomaterials as a Blueprint in Mitigation of Abiotic Stress in Crop Plants: A Conceptual Review ... 155
Saswati Bhattacharya and Jayita Saha

Part II Agro-ecology, People and GI-Science

12 Employment Potential of Sericulture for Underprivileged Section: Assessment of Value Chain Analysis in Bangladesh ... 185
Md. Kamruzzaman, Md. Abdullah Al Mamun, Jayanta Das, Kamruzzaman, and G. M. Monirul Alam

13 Development of Objective-Based Multi-criteria Decision-Making Approach in Crop Suitability Assessment for Maize Production Using GIS ... 199
Rajib Mitra, Amit Sarkar, Golap Hossain, Dipesh Roy, Goutam Mandal, Jayanta Das, and Deepak Kumar Mandal

14 Declining Groundwater Level and Its Impact on Irrigation and Agro-production ... 217
Shekhar Singh, Dheeraj Mohan Gururani, Anil Kumar, Yogendra Kumar, Manoj Singh Bohra, and Priyanka Mehta

15 Impact of Shifting Cultivation and Changing Land Use on the Hydrology of Iril Watershed, Manipur ... 225
Rebati Sinam

16 GIS-Based Road Network Connectivity Assessment and Its Impact on Agricultural Characteristics Using Graph Theory: A Block-Level Study in the Hill Area of Darjeeling District, West Bengal ... 243
Surajit Paul, Debasish Roy, and Bipul Chandra Sarkar

17 Landslide and Its Impact on Agriculture in Kottiyoor Panchayath, Kannur District, Kerala ... 257
R. Nirmala

18 Agricultural Land Use Change and Its Impact on the Farmers' Livelihood Assets of Maldah District, West Bengal, India ... 265
Tapash Mandal and Snehasish Saha

19 Assessment of Drought in Meteorological Data Using SPI and SPEI Indicators for Sustaining Agricultural Productivity in the Agra Division of Uttar Pradesh, India 285
Shekhar Singh, Anil Kumar, and Sonali Kumara

20 Understanding the Utility of Print Media for Dissemination of Insect Pest Management Tactics to Rice Farmers at Hooghly, West Bengal, India 305
Eureka Mondal and Kaushik Chakraborty

21 Faulty Waste Water Usage Versus Agricultural Sustainability: An Assessment of East Kolkata Wetlands 323
Divyadyuti Banerjee, Sweta Sinha, and Kathakali Bandopadhyay

Index 339

Editors and Contributors

About the Editors

Dr. Jayanta Das is an Assistant Professor at the Department of Geography in Rampurhat College, University of Burdwan, West Bengal, India. He has completed his post-graduate and Ph.D. degrees from the Department of Geography and Applied Geography, University of North Bengal, India. Dr. Das's broad area of Ph.D. thesis covers the theme of agricultural geography and GI-Science. His research interest includes agricultural modeling and sustainable management studies, groundwater, flood, drought analysis, climate change, watershed management, hydrological modeling, water quality, geospatial data analysis, data mining, and GIS applications with more than 15 academic years of experience. Dr. Jayanta Das has published more than 30 scholarly articles in peer-reviewed journals, focusing mainly on climate change, agricultural suitability analysis, natural and man-made hazards analysis, risk management, and spatial data analysis. Recently, Dr. Das has published an edited book entitled *Monitoring and Managing Multi-hazards: A Multidisciplinary Approach* jointly with Dr. Sudip Kumar Bhattacharya from Springer Nature.

Dr. Somenath Halder is an Assistant Professor at the Department of Geography in Kaliachak College, University of Gour Banga, West Bengal, India. He has completed his PG from Department of Geography and Applied Geography, University of North Bengal, India, and Ph.D. from Department of Geography, Visva-Bharati, India. Dr. Halder's field of interest includes multi-dimensional data modeling connecting to socio-political agenda, conflicting issues with laws and ecology, community practices, crisis management, spatio-temporal analysis, environmental management, policy science, climate change, watershed management, hydrological modeling, geospatial data analysis, and GIS applications with more than 14 academic years of experience. Dr. Halder has published more than 40 research articles in peer-reviewed journals. Meanwhile, he performs his academic endeavor as peer-reviewer in number of Scopus and Web of Science indexed prestigious journals like *Modeling Earth System and Environment, Journal of Environment, Development and Sustainability, Internal Journal of Geoheritage and Parks, GeoJournal, South Asian Survey, Research in Globalization, Current Psychology, Sage Open*, and many more.

Contributors

Md. Abdul Khalek Department of Statistics, University of Rajshahi, Rajshahi, Bangladesh

J. M. Adeeb Salman Chowdhury Institute of Bangladesh Studies, University of Rajshahi, Rajshahi, Bangladesh

Md. Abdullah Al Mamun Institute of Bangladesh Studies, University of Rajshahi, Rajshahi, Bangladesh

Asraful Alam Department of Geography, Rampurhat College, Rampurhat, Birbhum, India

G. M. Monirul Alam Department of Agribusiness, Bangabandhu Sheikh Mujibur Rahman Agricultural University, Gazipur, Bangladesh

Jiarul Alam Department of Geography and Applied Geography, University of North Bengal, Darjeeling, India

Syed Sadath Ali Civil Engineering Department, Ballari Institute of Technology and Management, Ballari, Karnataka, India

Kathakali Bandopadhyay Vidyasagar University, Kadamtala, Howrah, West Bengal, India

Divyadyuti Banerjee Jadavpur University, Makardaha, Howrah, West Bengal, India

Jonmenjoy Barman Department of Geography and RM, Mizoram University, Aizawl, India

Amiya Basak Department of Geography and Applied Geography, University of North Bengal, Darjeeling, India

Saswati Bhattacharya Department of Botany, Dr. A.P.J. Abdul Kalam Government College, New Town, Rajarhat, India

Manoj Singh Bohra Department of Soil and Water Conservation Engineering, G.B. Pant University of Agriculture and Technology, Pantnagar, Udham Singh Nagar, Uttarakhand, India

Kaushik Chakraborty Department of Zoology, Raiganj University, Raiganj, Uttar Dinajpur, West Bengal, India

Jayanta Das Department of Geography, Rampurhat College, Rampurhat, Birbhum, West Bengal, India

Partha Das Department of Geography, A.B.N. Seal College, Cooch Behar, India

Manoranjan Dash Faculty of Management Sciences, Siksha O Anusandhan (Deemed to be University), Bhubaneswar, India

P. Dhanya Directorate of Crop Management, Tamil Nadu Agriculture University, Coimbatore, Tamil Nadu, India

V. Geethalakshmi Director Crop Management, Tamil Nadu Agriculture University, Coimbatore, Tamil Nadu, India

Debaditya Gupta School of Agro and Rural Technology, Indian Institute of Technology, Guwahati, Assam, India

Dheeraj Mohan Gururani Department of Civil Engineering, Institute of Meerut, MIET College, Meerut, Uttar Pradesh, India

Somenath Halder Department of Geography, Kaliachak College, Malda, West Bengal, India

Golap Hossain Department of Geography and Applied Geography, University of North Bengal, Siliguri, West Bengal, India

Md. Kamruzzaman Institute of Bangladesh Studies, University of Rajshahi, Rajshahi, Bangladesh

Kamruzzaman Social Science Research Council, Ministry of Planning, Government Republic of Bangladesh, Dhaka, Bangladesh

Anil Kumar Department of Soil and Water Conservation Engineering, G.B. Pant University of Agriculture and Technology, Pantnagar, Udham Singh Nagar, Uttarakhand, India

Yogendra Kumar Department of Irrigation and Drainage Engineering, G. B. Pant University of Agriculture and Technology, Pantnagar, Udham Singh Nagar, Uttarakhand, India

Sonali Kumara Department of Soil Conservation and Watershed Development, Office of Project Director, Watersheds, Balangir, Odisha, India

Papia Kundu Department of Geography and Applied Geography, University of North Bengal, Siliguri, West Bengal, India

Deepak Kumar Mandal Department of Geography and Applied Geography, University of North Bengal, Siliguri, West Bengal, India

Goutam Mandal Department of Geography and Applied Geography, University of North Bengal, Siliguri, West Bengal, India

Tapash Mandal Department of Geography and Applied Geography, University of North Bengal, PO. North Bengal University, Darjeeling, India

Priyanka Mehta Department of Irrigation and Drainage Engineering, G.B. Pant University of Agriculture and Technology, Pantnagar, Udham Singh Nagar, Uttarakhand, India

Rajib Mitra Department of Geography and Applied Geography, University of North Bengal, Siliguri, West Bengal, India

Eureka Mondal Department of Zoology, Raiganj University, Raiganj, Uttar Dinajpur, West Bengal, India

Koyel Mukherjee Department of Geography, Rampurhat College, Rampurhat, West Bengal, India

R. Nirmala Department of Marine Geology, Mangalore University, Konaje, India

Surajit Paul Department of Geography and Applied Geography, University of North Bengal, Siliguri, West Bengal, India

Indrajit Poddar Department of Geography and Applied Geography, University of North Bengal, Darjeeling, India

Debasish Roy Department of Geography and Applied Geography, University of North Bengal, Siliguri, West Bengal, India

Dipesh Roy Department of Geography and Applied Geography, University of North Bengal, Siliguri, West Bengal, India

Subha M. Roy Faculty of Agricultural Sciences, Institute of Applied Sciences and Humanities, GLA University, Uttar Pradesh, Mathura, India

Jayita Saha Department of Botany, Rabindra Mahavidyalaya, Champadanga, Hooghly, West Bengal, India

Piu Saha Department of Geography and Applied Geography, University of North Bengal, Siliguri, West Bengal, India

Sanjoy Saha Department of Geography, Kaliachak College, Malda, West Bengal, India

Snehasish Saha Department of Geography and Applied Geography, University of North Bengal, PO. North Bengal University, Darjeeling, India

Jitranjan Sahoo Faculty of Management Sciences, Siksha O Anusandhan (Deemed to be University), Bhubaneswar, India

T. Sankar Agro Climate Research Centre, Directorate of Crop Management, Tamil Nadu Agriculture University, Coimbatore, Tamil Nadu, India

D. T. Santosh Centre of Smart Agriculture, Centurion University of Technology and Management, Paralakhemundi, Odisha, India

Amit Sarkar Department of Geography and Applied Geography, University of North Bengal, Siliguri, West Bengal, India

Bipul Chandra Sarkar Ananda Chandra College, University of North Bengal, Siliguri, West Bengal, India

Preeti Y. Shadangi Faculty of Management Sciences, Siksha O Anusandhan (Deemed to be University), Bhubaneswar, India

Rebati Sinam Centre for the Study of Regional Development, School of Social Sciences, Jawaharlal Nehru University, New Delhi, India

Chiranjit Singha Department of Agricultural Engineering, Institute of Agriculture, Sriniketan, West Bengal, India

Shekhar Singh Department of Soil and Water Conservation Engineering, G.B. Pant University of Agriculture and Technology, Pantnagar, Udham Singh Nagar, Uttarakhand, India

Sweta Sinha Mothijheel, West Bengal, India

Kishore C. Swain Department of Agricultural Engineering, Institute of Agriculture, Sriniketan, West Bengal, India

K. N. Tiwari Agricultural and Food Engineering Department, Indian Institute of Technology, Kharagpur, West Bengal, India

Abbreviations

ABA	Abscisic acid
AGRL	Agricultural land-generic
AHP	Analytic hierarchy process
AIC	Akaike information criteria
APX	Ascorbate peroxidase
ARIMA	Autoregressive integrated moving average
AUROC	Area under the receiver operating characteristic curve
Aus	A rice crop coinciding with late dry and early monsoon season
BBS	Bangladesh Bureau of Statistics
BIC	Bayesian information criteria
BNP	Biogenic nanoparticles
BOA	Bottom of atmosphere
Boro (Season)	Dry season rice, grown from December to April
BSDB	Bangladesh Sericulture Development Board
BSRTI	Bangladesh Sericulture Research and Training Institute
BU	Built-up land
cADPR	Cyclic adenosine diphosphate ribose
CAT	Catalase
CBOs	Community-based organizations
CF	Competitive farming
CF	Contact farmer
CGWB	Central Groundwater Board
CL	Cropland
CN	Curve number
CO_2	Carbon dioxide
CoCoSo	Combined compromise solution
CRITIC	Criteria Importance Through Intercriteria Correlation
CZM	Coastal zone management
DD	Drainage density
DEM	Digital elevation model
DF	Dense forest
DH	Dead heart
DMAI	De Martonne Aridity Index
E	Elevation
ECZ	Efficient cropping zone

EDS	Energy-dispersive spectroscopy
EDU	Ethyldiurea
EKW	East Kolkata Wetlands
EPA-SWMM	Environmental Protection Agency-Storm Water Management Model
ESA	European Space Agency
ET	Evapotranspiration
ETL	Economic threshold level
EWM	Entropy weighting method
FAHP	Fuzzy analytic hierarchy process
FAO	Food and Agriculture Organization
FAO-PM	Food and Agriculture Organization Penman–Monteith Method
FAPAR	Fraction of absorbed photosynthetically active radiation
FCC	False color composite
FGD	Focus group discussions
FMEI	Farm Magazine Effectiveness Index
FR	Frequency ratio
FRI	Flood Risk Index
FRSE	Forest-evergreen
FRST	Forest-mixed
FSC	Full supply capacity
FSI	Flood Susceptibility Index
FSZ	Flood susceptibility zonation
FTIR	Fourier transformed infrared spectroscopy
FVI	Flood Vulnerability Index
G	Geomorphology
GCPs	Ground control points
GDEM	Geographical digital elevation model
GDZ	Groundwater depletion zonation
Ge	Geology
GEV	Generalized extreme value
GIS	Geographic Information System
GIScience	Geographical Information Science
GLEAMS	Global Livestock Environmental Assessment Model
GNS	Green nanosilica
GSI	Geological Survey of India
HAP	Hydroxyapatite
HEC-HMS	Hydrologic Engineering Center–Hydrologic Modeling System
HM	Heavy metals
HYV	High-yield variety
IDW	Inverse distance weighing
IDW	Inverse distance weightage
IMD	Indian Meteorological Department
IoT	Internet of Things
IP3	Inositol 1,4,5, triphosphate

IPCC	Intergovernmental Panel on Climate Change
IPM	Integrated pest management
ISRO	Indian Space Research Organization
IT	Information technology
IT-OT	Information technology-operation technology
IUCN	International Union for Conservation of Nature
K	Potassium
KII	Key-informant interviews
LANDSAT	Land-use satellite
LAT	Latitude
LONG	Longitude
LPDAAC	Land Processes Distributed Active Archive Center
LSA	Land suitability assessment
LST	Land surface temperature
LUE	Light use efficiency
LULC	Land use land cover
MCDA	Multi-criteria decision analysis
MCDM	Multi-criteria decision making
mNDWI	Modified Normalized Difference Water Index
N	Nitrogen
N1	Currently not suitable
N2	Permanently not suitable
NASA	The National Aeronautics and Space Administration
NCF	Non-contact farmer
NDVI	Normalized Difference Vegetation Index
NEM	Northeast monsoon
NP	Nanoparticles
NSE	Nash–Sutcliffe efficiency
O_2	Oxygen
OC	Organic carbon
OLI	Operational Land Imager
OT	Operational technology
P	Phosphorus
PAB	Parambikulam Aliyar Basin
PBIAS	Percent bias
PCI	Precipitation Concentration Index
PETC	Photosynthetic electron transport chain
PM	Penman–Monteith
POX	Peroxidase
PPO	Polyphenol oxidase
PS-I	Photosystem-I
R	Rainfall
RCMs	Regional climate models
RCP	Representative concentration pathways
RegCM	Regional climate model
RMSE	Root mean square error
ROS	Reactive oxygen species

RS	Remote sensing
RSI	Relative Spread Index
RSNOs	R-S nitrotiols
RSR	Ratio of root mean square error to standard deviation
RYI	Relative Yield Index
S	Slope
S1	High suitability
S2	Moderate suitability
S3	Low suitability
SAVI	Soil Adjusted Vegetation Index
SL	Scrubland, shrubs, and mixed forest
SLR	Sea level rise
SMI	Soil Moisture Index
SOD	Superoxide dismutase
SOI	Survey of India
SPSS	Statistical Package for Social Sciences
SRTM	Shuttle Radar Topography Mission
SSA	Site suitability analysis
SVM	Support vector machine
SWAT	Soil and Water Assessment Tool
SWM	Southwest monsoon
T	Temperature
TCI	Temperature Condition Index
TIRS	Thermal infrared sensor
TWI	Topographic Wetness Index
UVB	Ultraviolet B
VCI	Vegetation Condition Index
VHI	Vegetation Health Index
VIs	Vegetation indices
Viz	Videlicet
WASPAS	Weight aggregated sum product assessment
WH	White head
WHO	World Health Organization
WMO	World Meteorological Organization
WPM	Weighted product model
WQI	Water Quality Index
WSM	Weighted sum model
XDR	X-ray diffraction
YSB	Yellow stem borer

Part I

Sustainable Agriculture and Applied GI-Science

1 Nexus Between GIScience and Sustainable Agriculture

Sanjoy Saha, Jayanta Das, and Somenath Halder

Abstract

Human civilization is rapidly developing in the age of advanced science and technology. Intact, the acute crisis for food, clothing, and shelter has been resolved to a large extent, but the spatial disparity of overall development is still a problem for humanity. Highly technology-based development is only sometimes friendly to the environment. The alternative thought for environmentally unfriendly development is crucial for civilization. In this context, the philosophy of sustainable development is highly appreciated. Also, the concept of balanced development is very much relevant in agriculture. Agricultural scientists are keenly interested in seeking sustainable ways to grow agrarian production. The development strategies for agriculture also appreciate the philosophy of balanced development. Contemporarily, data science has been developed notably due to the innovation of technologies in GIScience. Indeed, detailed geospatial data about the different aspects of agriculture has improved the research in agricultural science and its development-related studies. In the last few decades, many research works on sustainable agriculture have been published in which GIS and RS technologies have been used remarkably. Many studies about sustainable development in agriculture have also been performed using geospatial data appreciably by researchers worldwide. Despite the remarkable tendency of the current researchers to build different models suitable for balanced development in agriculture using geospatial data, a gap is needed in linking GIScience and sustainable agriculture. Thus, this chapter attempts to analyze why and how a strong linkage can be made between GIScience and sustainable agriculture. The strong connection between GIScience and research about the development of agriculture using geospatial data will enhance the progress.

Keywords

Geospatial science · Sustainable agriculture · GIScience · Balanced development · Geographical information system · Agrotourism · Agroclimatic zone

1.1 Introduction

The thought of balanced development is an unquestionable philosophy to the environmentalist, humanists, and also for ordinary people of

S. Saha (✉) · S. Halder
Department of Geography, Kaliachak College, Malda, West Bengal, India
e-mail: geosanjoysaha1408@gmail.com

J. Das
Department of Geography, Rampurhat College, Birbhum, West Bengal, India

J. Das and S. Halder (eds.), *Advancement of GI-Science and Sustainable Agriculture*, GIScience and Geo-environmental Modelling, https://doi.org/10.1007/978-3-031-36825-7_1

the world (Fang et al. 2023; Stables and Scott 1999; Tyburski 2008). The present civilization is keenly interested in searching the strategies for sustainable development. There are many hurdles to implementing the design for balanced growth in a real-world situation. Especially in the developing world, weak economic background and a massive need for food and shelter to sustain a large population compel a nation to continue the environmentally unfriendly but cheap production methods in agriculture and industry (Omisore 2018). To some extent, developed countries continue to use modern technologies to maintain their living standards compared to developing nations (Gollakota et al. 2020). Thus, the necessity of rational pathways to attain sustainable development is the main thrust of the researchers who think about balanced product (Zhang et al. 2022). While making a good strategy for properly managing some systems or events, a clear understanding of the systems or events is necessary. Knowledge of phenomena depends on detailed information (Boal and Schultz 2007). When discussing an activity like agriculture which depends on the geo-environmental condition, we need much geographical information for better management.

In this connection, geographic information science (GIScience) is a reliable discipline for efficiently collecting, analyzing, and representing real-world data. The prime goal of GIScience is to seek the best ways to expand geographical information system (GIS)-related data, software, and practice professionally (Ricker et al. 2020). Indeed, the current technologies and research fields of geographical information systems (GIS), cartographic techniques, remote sensing, photogrammetry, and geomatics are included in GIScience (Goodchild 2010). GIScience uses GIS to analyze and visualize spatial data in a concrete form to understand the real-world situation of geo-environmental phenomena (Jiang and Yao 2010). In earth sciences like Geomorphology, Geology, Climatology, and many more, the application of GIScience has revolutionized the research arena. As an important sector of the economy, agriculture also needs beneficial strategies for its development in a sustainable way (Jhariya et al. 2021). Thus, modern agriculture research requires detailed geospatial data for crisis management in real-world situations. In this context, incorporating GIScience is too relevant in advanced agricultural research (Hussain et al. 2021). In recent decades, many researchers have used GIS to analyze geospatial data about agriculture-related geo-environmental conditions. However, there is an urgent need to make a strong interlinkage between GIScience and research to attain a balanced development in the agricultural sector. This chapter attempts to understand the nature and status of recent research articles using GIScience to acquire and analyze geospatial data. In this chapter, our objective is to track how to connect GIScience and sustainable agriculture effectively.

1.2 Data Source and Methodology

This chapter is an analytical review work. Thus, the sources of data of the chapter and the published research articles in agricultural research specially applied GIScience to obtain spatial data. A systematic study of the research papers has analyzed the status of the application of GIScience in agricultural research. For re-establishing a substantial nexus between GIScience and sustainable agriculture and synthesizing the different published research works about the strategy of improvement in the farm sector, it has been explored practical ways to make a torch-bearing interpretation.

1.3 What is GI-Science?

GIScience is a discipline that uses the geographical information system (GIS) to understand the world (Clarke and Gaydos 1998; Lü et al. 2019). Goodchild (2010) defined *GIScience* as 'the science behind the systems, concerned with the fundamental questions raised by GIS and allied technologies'. GIScience is the tank of knowledge that assists GIS in performing its works more and more effectively (Fuhrmann et al. 2008; Kraak and Ormeling 2020). The continuous research in GIScience seeks effective, efficient, and rapid processes to acquire and visualize geospatial data to assist the research activities in the different fields of geography (Andrienko et al. 2007; Phung 2022). Therefore, a systematic scientific discipline works for searching technology-based systems to collect geospatial data intelligently to assist the geospatial analysis of the earth's phenomena. In the present age of machine learning, the demand for data science has been increasing rapidly in academic and industrial research. Consequently, an urgent necessity feels by the researchers to make a strong relationship between data science and GIScience (Arribas-Bel and Reades 2018; Zaman et al. 2021).

1.4 Why Sustainable Agriculture is Needed?

The societies of the modern world face problems in agriculture in two dimensions. The first one is the eco-environmental threats due excessive use of biotechnologically generated seeds and chemical fertilizers (Chen 2020; Ingrao et al. 2018). The other is the spatial inequality of the growth and development of agriculture worldwide, increasing economic disparity in human society (Señoret et al. 2022). A tremendous increase in agricultural production is obligatory to sustain the livelihoods of a large population in developing nations (Awazi et al. 2022). Generally, developing countries use high-yield variety (HYV) seeds, chemical fertilizers, and groundwater for irrigation to fulfill the ever-increasing demand for food and beverages of citizens (Kumar et al. 2021a, b). Consequently, environmental deterioration is joint in these nations (York and McGee 2016). Developing countries and developed nations face the problems of environmentally unfriendly agricultural practices to some extent (Hussain et al. 2022). Therefore, sustainable development in agriculture is urgently needed to shake human well-being and restore ecological balance.

1.5 Results and Discussion

1.5.1 Applicability of GIScience for Sustainable Agriculture

Balanced development in agriculture requires those strategies which can keep the ecological environment intact as much as possible on the one hand. On the other hand, increase agricultural production in such a way that the needs of all people for food, clothing, and shelter can be ideally met without hampering the future generation's needs (Keijzers 2002). Exploring strategies for suitable agriculture is a tough challenge for world agricultural science and social science practitioners (Braun et al. 2022; Meyfroidt et al. 2022). To meet this challenge, a proper understanding of geo-environmental factors that influence the nature and growth of agriculture in a spatial context is necessary. Thus, detailed and reliable updated geospatial data is required for the researchers to analyze agricultural conditions to assist sustainable agriculture. GIScience is the most dependable discipline for accumulating geospatial data in this context.

1.5.2 Present Scenario of the Application of GIScience in Agricultural Research

Indeed, in many agricultural research works, RS and GIS have been appreciably used to understand the agricultural crises concerning the physical environment and socio-economic aspects that influence crop production and trading. In the recent few decades, many research articles have been published about the agroclimatic zone, soil suitability classification, prediction of climatic disasters that hamper agricultural production, changing cropping patterns due to climatic change, and prediction for crop production, and so on, in which RS and GIS technology used for collecting, analyzing, and visualizing the geospatial data. An analysis of such works with examples has been included.

The summary of this analysis is displayed in tabular form (Table 1.1). Examples of recent works about sustainable development in agriculture using remote sensing and geographical information show that researchers are kingly interested in applying the technologies for collecting geospatial data. *Nevertheless*, in the research articles, more effort is needed to link sustainable agriculture and GIScience in the real world. Thus, there is necessary to take the initiative to connect sustainable agriculture and GIScience.

1.5.3 Need for Strong Connection Between Sustainable Agriculture and GIScience

GIScience, as an applied discipline, attempts to explore systematic and efficient methods and tools for extracting geospatial data quickly and reliably with the help of intelligent technologies (Machiwal et al. 2018; Teixeira et al. 2021). The researchers and professionals use these geospatial data according to the motives of the research. Thus, the purpose of applying this science depends on the thought of the users, which might be sustainable or not (Devillers et al. 2007; Vermeulen-Miltz et al. 2022). By analyzing the geospatial data related to agriculture, a researcher can explore the strategy (s) for the growth of agriculture that might be sustainable. Balanced development in agriculture aims to preserve the ecological balance and achieve an equilibrium condition in human welfare (Mensah 2019). So, the use of GIScience while seeking a strategy for agricultural development should always follow the philosophy of sustainability. The researchers in the field of agricultural development will likely aim to use the geospatial data provided by GIScience to implement sustainable methods and strategies for its development (Tian et al. 2022). Even though research works on agricultural development using the GIS and RS technologies, a gap is found while linking GIScience and sustainable agriculture. Therefore, there is a need to make a bridge linking sustainable agriculture and GIScience. Now, the debate is why sustainable agriculture and GIScience are strongly liked. The answer to the question is that more and more advanced strategies are needed for the complete implementation of the concept of balanced development in agriculture. The innovation of new techniques for sustainable agriculture requires more detailed geospatial data to be obtainable in GIScience. The need for precise geospatial in agricultural research will inspire the discovery of novel technologies in the GIScience. With the innovation of new technologies in GIScience, progress in agriculture will be enhanced. Therefore, nexus between sustainable agriculture and GIScience is necessary for researchers searching for balanced agricultural development strategies.

1.5.4 Pathways for Linking Sustainable Agriculture and GIScience

One of our objectives in this analytical chapter is to explore ways of connecting sustainable agriculture and GIScience. Recent studies on agricultural development rely on geospatial data to a large extent, and the use of GIS and RS is widespread almost in every agricultural research (Rao 2007; Tian et al. 2022). Thus, GIScience is accustomed by the researchers performing studies about agricultural development (Ghute et al. 2022). Now, the researchers must use the geospatial data obtained from the platform of GIScience more carefully from the sustainability angle. Researchers must seek a more efficient strategy for achieving sustainable agriculture in a real-world situation using the ground-level data collected by field surveys and geospatial data accrued by intelligent technologies innovating in GIScience. Scientists should focus on developing new technologies to assist sustainable agriculture. In this way, a substantial nexus can be established between GIScience and sustainable agriculture. However, Fig. 1.1 gives a descriptive idea about the overall links between the subject matter of 'GIScience' and 'sustainable agriculture'.

Table 1.1 Example of the works on sustainable agriculture using remote sensing and geographical information

Examples of research work on sustainable agriculture (Title)	Author(s)	Empirical insights into the work	Material and method used	Theory used in the work	Gaps in research agenda	Type of technology used form GIScience	Relevance for real-world applications
Is there a need for a more sustainable agriculture?	Gomiero, et al. (2011)	Artificial intensification of agricultural production causes environmental deterioration	An analysis based on literatures review	Agriculture to be in more sustainable pathways	Only a theoretical approach is used	There is no space for using RS and GIS technologies	Relevant for building a conceptual model
Modeling of alternative crops suitability to tobacco based on analytical hierarchy process in Dinhata subdivision of Koch Bihar district, West Bengal	Das et al. (2017)	Alternative crop suitability to tobacco toward the FAO-FCTC treaty	Analytical hierarchy process (AHP)	Climate and soil suitability traditional crops exist to replace tobacco cultivation	Cost–benefit analysis for alternative cultivation has not been considered	RS-GIS has been applied intensively	Relevance toward sustainable crop suitability
Modeling and assessment of irrigation water quality index using in semi-arid region for sustainable agriculture	El Behairy, et al. (2021)	Rational use of water for irrigation reduces the risk of environmental hazards	Inverse distance-weighted algorithms and the model builder function	Sustainable management of water resource	Generalized model has not been built for optimal use of water resource	RS and GIS technology were used efficiently	Applicable in areas suffers from water shortage
Assessment of spatial variability and mapping of soil properties for sustainable agricultural production using geographic information system techniques (GIS)	Denton, et al. (2017)	Selection of crop for cultivation is profitable by judging the suitability of the soil	Soil suitability classification model is used	Soil suitability classification is suitable for profitable farming	Feasibility for using the model not assessed	Geospatial data have been analyzed with smart technology	Can be applicable in a region when it is feasible in the socio-economic condition of the concerned region
Intelligent analysis of precision marketing of green agricultural products based on big data and GIS	Tian et al. (2022)	Decision for green farming is prosperous while understanding the market's status	Modern information technology is used for managing big data	Market analysis is necessary for the success of farming	Cost and profit analysis for green cultivation have not been considered	GIS and RS have been used efficiently	Universal application of this concept needs more analysis

(continued)

Table 1.1 (continued)

Examples of research work on sustainable agriculture (Title)	Author(s)	Empirical insights into the work	Material and method used	Theory used in the work	Gaps in research agenda	Type of technology used form GIScience	Relevance for real-world applications
Developing spatial model to assess agro-ecological zones for sustainable agriculture development in MENA region: Case study Northern Western Coast, Egypt	Amin et al. (2022)	Sustainable agricultural practice needs the identification of agro-ecological environments of a region	NDVI and land use/covers maps prepared by using Sentinel-2 images	Climate and ecological setup are the most important determining factors of agriculture	Other factors of agricultural development have not been considered	GIS and RS data have been used intensively	Relevance for the application in a real-world situation
Research on the use of aerial scanning and GIS in the design of sustainable agricultural production extension works in an agritourist farm in Romania	Călina, et al. (2022)	The integration of agrotourism and sustainable agriculture is a strategy for the balanced development	Spatial analysis of agricultural aspects	Integrated development is a good way for sustainable agriculture	Management strategy for agrotourism and sustainable agriculture has not been included	Areal Scanning and GIS have been used	Relevance for integrated development
Sustainable agricultural development: a micro-level GIS-based study on women's perceptions of environmental protection and entrepreneurship in Japan and Bangladesh	Ahamed et al. (2021)	Perception of the environment from an eco-friendly approach leads to the way of sustainable development	Multi-criteria decision analysis (MCDA)	Perception about the environment is the first step to the sustainable development	Approach of the decision-makers about sustainable development has not included	GIS has been applied intensively	The thought is relevant for sustainable development
Nano-fertilizer use for sustainable agriculture: Advantages and limitations	Zulfiqar, et al. (2019)	Environmentally friendly input in agriculture is less risky	Direct field observation method used	Eco-friendly agriculture is better for the present and future	Micro-level analysis has not included	GIScience is not used for getting geospatial data	Relevant for medium and large farming
Profitable and viable alternative to tobacco crop in Dinhata Subdivision of Koch Behar district, West Bengal	Das and Bhattacharya (2016)	Successful strategies require that address the availability of alternative crops to promote the transition away from tobacco cultivation	Direct field observation method and cost–benefit analysis	Profitable sustainable crops exist to replace tobacco cultivation	Social factors were not considered	There is no space for using RS and GIS technologies	Relevance for FAO-FCTC treaty toward alternative crops

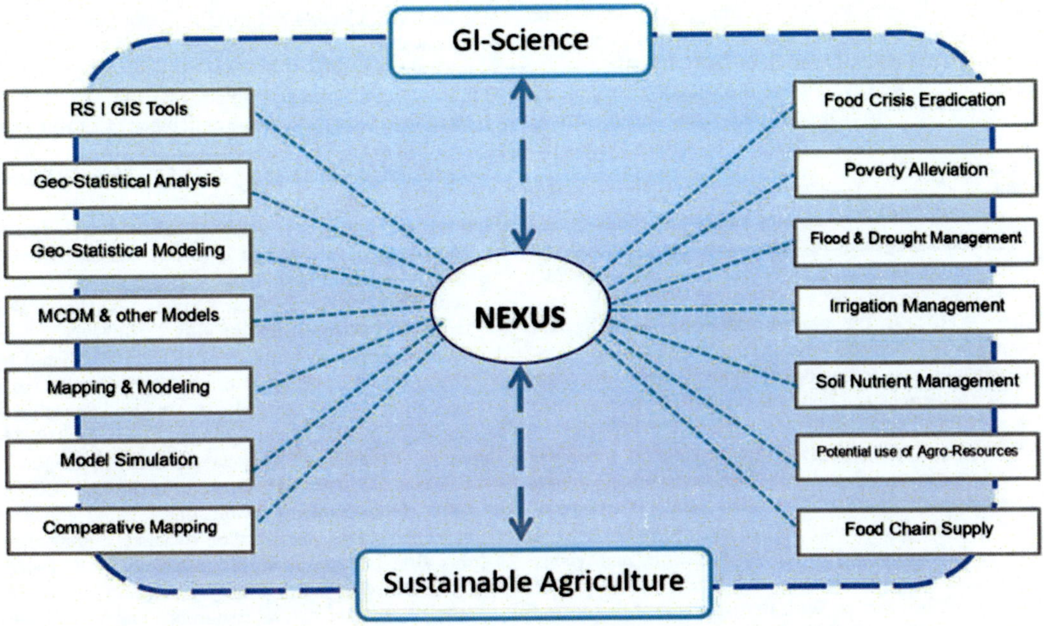

Fig. 1.1 Simplified linkages between 'GIScience' and 'sustainable agriculture'

1.6 Conclusions

Using geospatial data in agricultural science research and studies about agricultural development has opened a new horizon of scientific thinking worldwide. In this context, GIScience has induced a revolutionized scenario in the innovation of technologies that assist in extracting geospatial data. The researchers use this geospatial data to investigate the agriculture crisis related to physical or socio-economic phenomena. Sustainable agriculture is relevant to ecologists, humanists, and ordinary people worldwide. Consequently, searching for realistic ways to implement a wholly balanced development in agriculture is a challenge for the decision-makers of any nation. In this connection, GIScience is a reliable discipline that can assist in assessing agricultural aspects in the spatiotemporal context. So, there is a need for a perfect linkage between sustainable agriculture and GIScience. Researchers in farm development and GIScience must take responsibility for the relationship above to achieve balanced agricultural development and progress in GIScience, respectively.

References

Ahamed T, Noguchi R, Muhsin N, AyuPurnamasari R, Islam M, Tasnim F, Akmam W (2021) Sustainable agricultural development: a micro-level GIS-based study on women's perceptions of environmental protection and entrepreneurship in Japan and Bangladesh. GeoJournal 86(5):2071–2103

Amin ME, Mohamed ES, Belal AA, Jalhoum ME, Abdellatif MA, Nady D, Mahmoud AG et al (2022) Developing spatial model to assess agro-ecological zones for sustainable agriculture development in MENA region: case study Northern Western Coast, Egypt. Egypt J Remote Sens Space Sci 25(1):301–311

Andrienko G, Andrienko N, Jankowski P, Keim D, Kraak MJ, MacEachren A, Wrobel S (2007) Geovisual analytics for spatial decision support: setting the research agenda. Int J Geogr Inf Sci 21(8):839–857

Arribas-Bel D, Reades J (2018) Geography and computers: past, present, and future. Geogr Compass 12(10): e12403

Awazi NP, Quandt A, Kimengsi JN (2022) Endogenous livelihood assets and climate change resilience in the Mezam Highlands of Cameroon. GeoJournal 88(3):1–18

Boal KB, Schultz PL (2007) Storytelling, time, and evolution: the role of strategic leadership in complex adaptive systems. Leadersh Q 18(4):411–428

Braun CL, Bitsch V, Häring AM (2022) Behind the scenes of a learning agri-food value chain: lessons from action research. Agric Hum Values 39(1):119–134

Călina J, Călina A, Iancu T, Vangu GM (2022) Research on the use of aerial scanning and GIS in the design of sustainable agricultural production extension works in an agrotourist farm in Romania. Sustainability 14(21):14219

Chen W (2020) The transformation of the utilization mode of elements and resources with the agro-scientific and technological innovations at the core. In: Challenges and opportunities for Chinese agriculture. Palgrave Macmillan, Singapore, pp 125–182

Clarke KC, Gaydos LJ (1998) Loose-coupling a cellular automaton model and GIS: long-term urban growth prediction for San Francisco and Washington/Baltimore. Int J Geogr Inf Sci 12(7):699–714

Das J, Bhattacharya SK (2016) Profitable and viable alternative to tobacco crop in Dinhata Subdivision of Koch Behar district, West Bengal, India. J Landscape Syst Ecol Stud 39(2):70–88

Das J, Gayen A, Saha S, Bhattacharya SK (2017) Modelling of alternative crops suitability to tobacco based on Analytical hierarchy process in Dinhata subdivision of Koch Bihar district, West Bengal. Model Earth Syst Environ 3(4):1571–1587. https://doi.org/10.1007/s40808-017-0392-y

Denton OA, Aduramigba-Modupe VO, Ojo AO, Adeoyolanu OD, Are KS, Adelana AO, Oke AO et al (2017) Assessment of spatial variability and mapping of soil properties for sustainable agricultural production using geographic information system techniques (GIS). Cogent Food Agric 3(1):1279366

Devillers R, Bédard Y, Jeansoulin R, Moulin B (2007) Towards spatial data quality information analysis tools for experts assessing the fitness for use of spatial data. Int J Geogr Inf Sci 21(3):261–282

El Behairy RA, El Baroudy AA, Ibrahim MM, Kheir A, Shokr MS (2021) Modelling and assessment of irrigation water quality index using GIS in semi-arid region for sustainable agriculture. Water Air Soil Pollut 232(9):1–19

Fang WT, Hassan AA, LePage BA (2023) Research methods for environmental education. In: The living environmental education. Springer, Singapore, pp 49–90

Fuhrmann S, MacEachren A, Cai G (2008) Geoinformation technologies to support collaborative emergency management. In: Digital government. Springer, Boston, MA, pp 395–420

Ghute BB, Shaikh M, Halder B (2022) Impact assessment of natural and anthropogenic activities using remote sensing and GIS techniques in the Upper Purna River basin, Maharashtra, India. Model Earth Syst Environ 9:1–16.

Gollakota AR, Gautam S, Shu CM (2020) Inconsistencies of e-waste management in developing nations–facts and plausible solutions. J Environ Manage 261:110234

Gomiero T, Pimentel D, Paoletti MG (2011) Is there a need for a more sustainable agriculture? Crit Rev Plant Sci 30(1–2):6–23

Goodchild M (2010) Twenty years of progress: GIScience in 2010. J Spat Inf Sci 1:3–20

Hussain M, Tayyab M, Zhang J, Shah AA, Ullah K, Mehmood U, Al-Shaibah B (2021) GIS-based multi-criteria approach for flood vulnerability assessment and mapping in district Shangla: Khyber Pakhtunkhwa, Pakistan. Sustainability 13(6):3126

Hussain S, Amin A, Mubeen M, Khaliq T, Shahid M, Hammad HM, Nasim W et al (2022) Climate smart agriculture (CSA) technologies. In: Building climate resilience in agriculture. Springer, Cham, pp 319–338

Ingrao C, Bacenetti J, Bezama A, Blok V, Goglio P, Koukios EG, Huisingh D et al (2018) The potential roles of bio-economy in the transition to equitable, sustainable, post fossil-carbon societies: findings from this virtual special issue. J Clean Prod 204:471–488

Jhariya MK, Meena RS, Banerjee A (2021) Ecological intensification of natural resources towards sustainable productive system. In: Ecological intensification of natural resources for sustainable agriculture. Springer, Singapore, pp 1–28

Jiang B, Yao X (eds.) (2010) Geospatial analysis and modelling of urban structure and dynamics, vol 99. Springer Science & Business Media

Keijzers G (2002) The transition to the sustainable enterprise. J Clean Prod 10(4):349–359

Kraak MJ, Ormeling F (2020) Cartography: visualization of geospatial data. CRC Press

Kumar A, Pramanik M, Chaudhary S, Negi MS (2021a) Land evaluation for sustainable development of Himalayan agriculture using RS-GIS in conjunction with analytic hierarchy process and frequency ratio. J Saudi Soc Agric Sci 20(1):1–17

Kumar R, Saurabh K, Mishra JS, Mondal S, Hans H, Sundaram PK, Sinha NK et al (2021b) Impact of conservation agriculture on soil properties and crop productivity under rice-fallow ecology in Eastern India. In: Conservation agriculture: a sustainable approach for soil health and food security. Springer, Singapore, pp 165–193

Lü G, Batty M, Strobl J, Lin H, Zhu AX, Chen M (2019) Reflections and speculations on the progress in Geographic Information Systems (GIS): a geographic perspective. Int J Geogr Inf Sci 33(2):346–367

Machiwal D, Cloutier V, Güler C, Kazakis N (2018) A review of GIS-integrated statistical techniques for groundwater quality evaluation and protection. Environ Earth Sci 77(19):1–30

Mensah J (2019) Sustainable development: meaning, history, principles, pillars, and implications for human action: literature review. Cogent Soc Sci 5(1):1653531

Meyfroidt P, de Bremond A, Ryan CM, Archer E, Aspinall R, Chhabra A, ZuErmgassen EK et al (2022)

Ten facts about land systems for sustainability. Proc Natl Acad Sci 119(7):e2109217118
Omisore AG (2018) Attaining sustainable development goals in Sub-Saharan Africa; the need to address environmental challenges. Environ Dev 25:138–145
Phung Duc T (2022) Improving data acquisition processes for geospatial building information applications. Master's thesis, Aalto University
Rao NH (2007) A framework for implementing information and communication technologies in agricultural development in India. Technol Forecast Soc Chang 74(4):491–518
Ricker BA, Rickles PR, Fagg GA, Haklay ME (2020) Tool, toolmaker, and scientist: case study experiences using GIS in interdisciplinary research. Cartogr Geogr Inf Sci 47(4):350–366
Señoret A, Ramirez MI, Rehner J (2022) Employment and sustainability: the relation between precarious work and spatial inequality in the neoliberal city. World Dev 153:105840
Stables A, Scott W (1999) Environmental education and the discourses of humanist modernity: redefining critical environmental literacy. Educ Philos Theory 31(2):145–155
Teixeira H, Magalhães A, Ramalho A, Pina MDF, Gonçalves H (2021) Indoor environments and Geographical Information Systems: a systematic literature review. SAGE Open 11(4):21582440211050380
Tian T, Zhang Y, Mei Y (2022) Intelligent analysis of precision marketing of green agricultural products based on big data and GIS. Earth Sci Inf 15(1): 1–12
Tyburski W (2008) Origin and development of ecological philosophy and environmental ethics and their impact on the idea of sustainable development. Sustain Dev 16(2):100–108
Vermeulen-Miltz E, Clifford-Holmes JK, Scharler UM, Lombard AT (2022) A system dynamics model to support marine spatial planning in Algoa Bay, South Africa. Environ Model Softw 160(4):105601
York R, McGee JA (2016) Understanding the Jevons paradox. Environ Sociol 2(1):77–87
Zaman U, Zahid H, Habibullah MS, Din BH (2021) Adoption of big data analytics (BDA) technologies in disaster management: a decomposed theory of planned behavior (DTPB) approach. Cogent Bus Manag 8 (1):1880253
Zhang J, Lin H, Li S, Yang E, Ding Y, Bai Y, Zhou Y (2022) Accurate gas extraction (AGE) under the dual-carbon background: green low-carbon development pathway and prospect. J Cleaner Prod 377:134372
Zulfiqar F, Navarro M, Ashraf M, Akram NA, Munné-Bosch S (2019) Nanofertilizer use for sustainable agriculture: advantages and limitations. Plant Sci 289:110270

Application of GIS in Agricultural Crisis Management

2

Sanjoy Saha

Abstract

The application of scientific knowledge to resolve the real-world problems faced by the people of society is essential for any scientific discovery or invention. In handling any problem or crisis, whether man-made or naturally induced, a proper investigation is needed to understand the actual scenario of the matter. Based on collected information about the targeted problem from the real world, an analysis can do in a meaningful way. After that, anybody can decide on beneficial strategies to resolve the targeted issue. In this context, geographical information system (GIS) is a widely used software technology for spatial mapping data about different phenomena in nature and human society. In human society, agriculture is one of the most important economic activities adopted by human beings to sustain their livelihoods. Agriculture is an environmentally influenced and artificially controlled activity that faces several environmental and man-created problems. To overcome the said problems in the agricultural sector, visualization of the real situation in the spatiotemporal context in a more precise way GIS is a powerful and suitable technology for the geographers and planners of the different development authorities in a nation. Based on accrued real-word data, a detailed analysis can be done and hence can decide better strategies to mitigate the problems related to agricultural activities. Again, GIS helps clear visualization of strategies in spatial context and assists in implementing the strategies in more fruitful ways. In the age of globalization, agriculture is considered a high-tech industry due to commercializing agricultural productions and agro-based industries. GIS is being used to estimate and identify the potential agricultural regions for more improvement in the agricultural sector. The selection of crops for production in more profitable ways is necessary to meet the challenge of the global market by a nation. On a national level, the government can estimate the net sown area, probable production of particular crops, and amount of loss in the agricultural area due to any disaster by mapping the spatial situation through GIS. On the basis of the estimation, government can take effective measures to resolve the crisis faced by the farmers. So, applying GIS to resolve the crisis in agriculture is beyond question.

Keywords

Spatial data · Agriculture sector · Globalization · Geographical information system · Agricultural region · Commercialization

S. Saha (✉)
Department of Geography, Kaliachak College, Malda, West Bengal, India
e-mail: geosanjoysaha1408@gmail.com

J. Das and S. Halder (eds.), *Advancement of GI-Science and Sustainable Agriculture*, GIScience and Geo-environmental Modelling, https://doi.org/10.1007/978-3-031-36825-7_2

2.1 Introduction

In the age of globalization in every sector, whether economic, social, political, health, or hygiene, efficient and systematic management is required to face the challenges in the global market (Hove-Sibanda et al. 2021; van Tulder and van Mil 2022). In this context, the old and traditional management and monitoring systems need to be updated in the arena of the digital management system. The intelligent management system involves applying advanced technologies to assess the exact scenario of activities performed in the spatiotemporal context to operate them rationally and effectively (Jimenez et al. 2020). As a decision maker, a human being is a principal agent for making a fruitful plan to enhance the development in a sector, but man requires effective techniques for understanding the cause-and-effect relationship of the events (Dall'Acqua 2021). The spatiotemporal analysis of data related to different natural and human activities on the digital platform relies on remote sensing (RS) and geographic information systems (GIS) appreciably (Souissi et al. 2022). GIS is a mapping tool that uses data extracted or collected manually and remotely sensed data through the remote sensing system. GIS can represent data more accurately and vividly and minimizes human labour. It can also store the information for further activity in future (Baban 2022). Utilizing this mapping tool, man can decide about certain phenomena in real situations based on the analysis of actual spatiotemporal conditions (Hassan et al. 2019). As an important sector of the economy, agriculture needs an efficient management system to promote its growth in a sustainable way to face the global market and sustain the environment in an eco-friendly manner (Khan et al. 2021; Sarkar 2013). Proper agricultural production management requires several quantitative analyzes to identify agricultural regions based on climate, soil fertility, crop efficiency, crop diversification, specialization of crops, cropping pattern, and prediction of climatic vulnerability which influences the production of crops and risk management. As mentioned above, correctly understanding the facts needs reliable data and good presentations (Behera and France 2016). In this context, colossal data is required in numerical and digital forms, which can be extracted from remotely sensed information gathered by a satellite's sensor or manually collected data through field observations. It is a complex and time-consuming task for the human being to handle or manage the mentioned extracted information manually.

Moreover, analyzing and visualizing this information are also a monumental task. Geographical information systems can resolve these difficulties more systematically and conveniently in a shorter time than a conventional manual system (Kirilenko 2022). Presently, agricultural sector faces a crisis for its management in two aspects. One aspect is changing the nature of production due to climate change; another is frequently changing global market challenges (Böhm et al. 2022). Thus, the chapter aims to analyze how innovative technology can assist researchers and decision-makers in making efficient management systems to resolve the crisis in agriculture. It is also an objective of this chapter that what will be a new dimension for researchers seeking better models for the future of the agricultural management sector for the present generation.

2.2 Data and Methods

This chapter deals with a proper understanding of the management system to tackle the crisis faced by the agricultural sector in the context of 'global climatic change' (GCC) and challenges in the global agricultural market. Therefore, it requires a detailed literature review of the published research articles relevant to the management of the farming sector by using GIS to analyze and visualize agriculture-related contemporary issues. This analytical chapter has expressed the applicability of GIS in the management system in modern agricultural activities for combating the crisis faced by this sector based on secondary information. This chapter attempts to analyze and synthesize the ideas and models explained by numerous researchers concerned with applying GIS in agriculture's crisis management issues. Exploring the several thoughts of the researchers in this field, a layout for making rational policies to manage the crisis faced in the agricultural sector by applying the GIS as an efficient tool has been recommended.

2.3 Relevance of Agricultural Crisis Management in the Context of Global Climate Change and Challenges of Global Agricultural Market

The contemporary world agricultural sector faces problems in two major dimensions, i.e. changes in environmental conditions due to global climatic change and many challenges to compete in the global agricultural market. In this chapter, the term agriculture confers the cultivation of plants keenly sensitive to a region's climate and soil. Even slight changes in climatic variables can alter the area's quality and quantity of crop production (Dhanya et al. 2022; Patra 2022). Presently, livelihoods of the agrarian society have been threatened by global climatic change, which is responsible for the alteration of cropping patterns of a region (Müller et al. 2011; Negi et al. 2017; Ngigi and Muange 2022). Adaptation to the climate, whether virgin or altered temporally, is necessary to sustain the people's livelihoods properly (Dhodho 2022; Hernández-Delgado 2015). Adapting to the climate requires a suitable strategy to be introduced by correctly understanding climatic phenomena prevailing in the geo-environmental region (Karl et al. 2010; Murshed et al. 2022).

Even though similar climatic conditions in a region, different types of crops might be cultivated splendidly, and a farmer should have to decide to select a particular crop for cultivation for commercial benefit and understand the market situation (Das and Bhattacharya 2016). Nowadays, cultivators are facing challenges in the market for selling the produced crops not only in the domestic market but also in the global market remarkably (Manser 2022). Indeed, intelligent cultivation management makes it easier to meet the challenges in domestic and international agricultural markets (Coulibably et al. 2022; Sinha and Dhanalakshmi 2022). So, an intelligent agricultural practice needs a proper understanding of suitable climatic conditions and the status of the current market (Taylor 2018). Besides these, the farm sectors face climatic disaster, which affects the production of crops and the agrarian economy to a large extent—so forecasting climatic disasters and preparedness to combat the situation necessary for efficient cultivation management (Shiferaw et al. 2014; Sziroczak et al. 2022). Thus, modern agricultural activities need an efficient management system. In this connection, GIS can help manage the environmental condition rationally and environmentally sustainable. Therefore, an analytical study about the application of GIS for resolving problems in the agricultural sector is relevant in the age of globalization (Sgroi 2022).

2.4 Discussions

This chapter needs to discuss the nature of the agricultural sector's crisis and how it can be resolved through an efficient management system applying modern technic like GIS. It is also incorporated in this chapter that is how researchers study the application of GIS to resolve the crisis in agriculture in real-world situations.

2.4.1 What is Agricultural Crisis?

As a sector of the economy, agriculture plays a vital role in strengthening the economic situation of a nation indeed (Denton 2002). However, like other sectors of the economy, agriculture also experiences other crises induced by nature and marketing policy and challenges in the global market (Stephens et al. 2022; Timmer 1988). Climatic disasters and especially the change in cropping patterns due to global climatic change have imposed a new dimension on the management of the agricultural sector (Baldos and Hertel 2015; Schipper and Pelling 2006). In addition, facing challenges in the global market due to the introduction of frequently changing marketing policy in the global market is a difficult task for cultivators (Das and Bhattacharya 2018; Saraf et al. 2022). Consequently, growth and development in the agricultural sector are only possible by making a solid linkage between the contemporary situation and an effective management system. Innovation of new strategies for managing agricultural activities in fruitful ways is a mandatory task for the policy markets. The intensive observations and analysis of the agriculture crisis can help make a better policy to mitigate the problem in this sector (Darnhofer et al. 2010; Knickel et al. 2018). In this context, GIS considers an innovative tool by researchers.

2.4.2 Why Management in Agriculture is Necessary to Meet Global Climatic Change (GCC) and Challenges in Global Market (GM)

Changes in climatic variables in a region due to GCC have influenced cropping patterns (Cooper et al. 2008; Fatima et al. 2022; Yin et al. 2022). Thus, farmers are compelled to make a wise decision in selecting the crops suited for production in response to the region's altered climatic variables (Kom et al. 2020). Not only the changes in climatic variables but also the climatic vulnerability due to climatic disasters hinder the production of crops (Nascimento et al. 2022). Global climatic change (GCC) leads to an increase in average temperature, duration of seasons, amount of precipitation, abnormality in the drought and flood phenomena, soil moisture, and nature of the ecosystem in the global and regional context (Gaur and Goyal 2022; Wuebbles and Hayhoe 2004). Such altered geo-environmental conditions influence a region's nature and amount of crop production.

It should note that altered climatic conditions positively and negatively influence crop production (Chaudhry and Sidhu, 2022; Rosenzweig and Parry 1994). Due to the changes in the duration and timing of season in a (particular) region the overall agricultural functions altered automatically in relation with the activities like sowing, growing, and harvesting times of (each) particular seasonal crops (Sakamoto et al. 2006; Southworth et al. 2000). Thus, an updated crop management system requires sustaining agrarian livelihoods concerning global climatic change (GCC).

Another significant crisis that the agricultural sector has to face on the commercial platform is the frequent change in the status of the global market regarding the trading of produced crops. As per the demand of the international market, the agricultural sector must follow an intelligent management system (Fan et al. 2021; Jayne et al. 2014). It is exciting that crop production has increased appreciably due to biogenetic advancements (Cho 1992). But massive production of crops creates a new problem, i.e. marketing produced crops in the domestic and international market at a profitable price (Dar et al. 2022). Marketing produced crops efficiently and systemically is tough for cultivators (Guillemin 2022). Here, an intelligent management system must face the challenges of trading crops.

2.4.3 Application of GIS in Agriculture Crisis Management (ACM)

The monitoring and management of agriculture from a local to global scale make convenient and reliable by GIS. Researchers, planners, and decision-makers rely on GIS to make good decisions and plan for agriculture's growth and development (Newburn et al. 2005). GIS can precisely visualize the spatiotemporal data from the updated digital database and continuous spatiotemporal data through remote sensing techniques (Gabriele et al. 2022; Pettit 2005). Proper spatiotemporal data analysis on the GIS platform helps search specific crop cultivation sites optimally. Management of agriculture-related activities and agricultural inputs like application of irrigation, the information of manure as required for crop nutrition, and understanding climatic conditions suitable for sowing, growing, and harvesting crops has become more effective through the application of GIS (Reddy et al. 2018; Wilson 1999). When the farmers function as economic men, the primary aim is to maximize profit and reduce production loss. In this connection, farmers need information about market trends and how they can compete with the producers of crops from other regions. They also need to know how the yield of the crops can be increased and weather predictions to reduce the risk of loss due to hostile weather conditions. Presently, remote sensing (RS) provides geospatial data about the phenomena, as mentioned earlier, and GIS gives the visualized information based on geospatial analysis of the same (Dogan and Gokovali 2012; Yousefi and Razdari 2015). For a sustainable agricultural practice, the researchers commonly use agro-climatic classification by quantitative analysis based on the time series rainfall data. Agro-climatic regions based on climatic data, the effectiveness of rainfall distribution for crop production, crop growth and yield prediction, disaster risk management, land suitability classification, and so many agricultures management-related mapping used by researchers worldwide (Table 2.1).

2.5 Policy Direction

The application of GIS for resolving the agricultural crisis has opened a new horizon for researchers and is essential for the development of the agrarian sector of a nation. Researchers need more analysis of present agricultural research models for better work more applicable to resolving the agrarian crisis. Modified as well as innovative models are to search by the researchers. Consecutively, any future researcher must pursue further academic endeavours based on the reliable database collected by smart technologies. Thus, GIS has a bright future for researchers seeking beneficial strategies to resolve the crisis in agriculture. In the meantime, through Fig. 2.1, *it can be easily understandable the rationality of using GIS in the case of agricultural crisis management.

Table 2.1 Examples of contemporary research works on agricultural crisis management using GIS technology

Sl. No	Purpose of activity-related agricultural practice	Name of the model/method review	Examples of related works	Researchers
1	Agro-climatic and agroecological zonation	Global agroecological zone model (matrix and cluster method)	Use of agro-climatic zones to upscale simulated crop yield potential	van Wart et al. (2013)
2	Crop growing zones identification	Multi-criteria climatic classification (MCC) system	Using historical weather data and a novel season temperature index to classification winegrape growing zones in Australia	Zhang et al, (2023)
3	Flood risk management for crop production	GIS-based multi-criteria decision-making (MCDM) and analytical hierarchy process (AHP)	Potential food-prone area identification and mapping using GIS-based multi-criteria decision-making and analytical hierarchy process in Dega Damot district, northwestern Ethiopia	Negese et al. (2022)

(continued)

Table 2.1 (continued)

Sl. No	Purpose of activity-related agricultural practice	Name of the model/method review	Examples of related works	Researchers
4	Agroecological logical zonation	Variability and trend analysis	Spatiotemporal rainfall and temperature variability in Suha watershed, Upper Blue Nile Basin, Northwest Ethiopia	Alemayehu, et al. (2022)
5	Availability of soil moisture for cropping	Multiscale extrapolative learning algorithm (MELA) framework	Multiscale extrapolative learning algorithm for predictive soil moisture modelling and applications	Chakraborty et al. (2023)
6	Early prediction of crop yield for strategic plan	Semi-empirical light use efficiency (LUE) model	Comparative performance of semi-empirical-based remote sensing and crop simulation model for cotton yield prediction	Prasad et al. (2022)
7	Suitable input for better crop production	Crop simulation model	The role of crop simulation modelling in managing fertilizer use in maize production systems in Northern Ghana	MacCarthy et al. (2022)
8	Crop performance level estimation	Crop simulation model	Crop yield estimation at gram panchayat scale by integrating field, weather, and satellite data with crop simulation models	Milesi, and Kukunuri (2022)
9	Weather forecasting for seasonal crop management	Crop simulation model	Season-specific management strategies for rainfed soybean in the South American Pampas based on a seasonal precipitation forecast	Rizzo et al. (2022)
10	Crop health assessment	Normalized difference vegetation index (NDVI)	Sensitivity of normalized difference vegetation index (NDVI) to land surface temperature, soil moisture and precipitation over district Gautam Buddh Nagar, UP, India	Sharma et al. (2022)
11	Strategies for enhancing crop yield by assessing availability of water in different seasons	Cumulative crop drought index (CDI)	Subfield maize yield prediction improves when in-season crop water deficit is included in remote sensing imagery-based model	Shuai and Basso (2022)
11	Management of crop yield	Data assimilation approach	Assimilation of remote sensing data into crop growth model for yield estimation: a case study from India	Gumma et al. (2022)
12	Risk assessment for crop production in terms of climatic disaster	Drought frequency estimation	System structure-based drought disaster risk assessment sensing and field experiment data	Cui et al. (2022)

(continued)

Table 2.1 (continued)

Sl. No	Purpose of activity-related agricultural practice	Name of the model/method review	Examples of related works	Researchers
13	Land suitability assessment for cultivation	Fuzzy analytical hierarchy model	Assessment of agricultural land suitability using GIS and fuzzy analytical hierarchy process approach in Ranchi District, India	Sengupta et al. (2022)
14	Collection information about soil properties needed for plant growth and yield	Geophysical method	Electrical geophysical method and GIS in agricultural crop productivity in a typical sedimentary environment	Ozegin and Salufu (2022)
15	Strategies for smart agriculture	Agricultural suitability model based on hybrid fuzzy logic and AHP model	Coupling geographical information system integrated fuzzy logic -analytical hierarchy process with global and machine learning-based sensitivity analysis for agricultural mapping	Talukdar et al. (2022)
16	Strategies for replacement tobacco crop	GIS-based multi-criteria decision-making (MCDM) and analytical hierarchy process (AHP)	Modelling of alternative crops suitability to tobacco based on analytical hierarchy process in Dinhata subdivision of Koch Bihar district, West Bengal	Das et al. (2017)

Source Compiled by authors

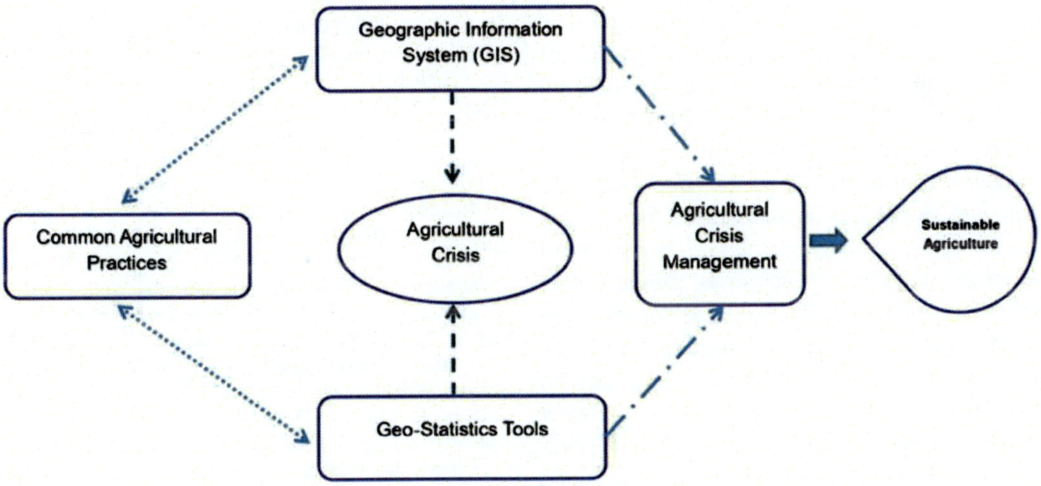

Fig. 2.1 Layout describing the usage of GIS for targeted agricultural sustainability

2.6 Conclusions

Modern agriculture needs an intelligent management system to meet the modified climatic situation induced by global climate change and face global market challenges. For this purpose, agriculture management strategies must be decided and adopted with a proper understanding of the real-world situation. Therefore, to resolve the crisis in agriculture in terms of the contemporary situation, necessary actions must be performed by the decision-makers of any region concerned to the agriculture development authority. Such activities include preparing detailed agricultural field land use maps from micro- to macro-scale based on updated data collected from the field observation and remotely sensed information using GIS. Therefore, there is a serious need to prepare land suitability maps of every region following the standard 'land suitability classification scheme' method. Side-by-side, a standard model should be followed for preparing the agro-climatic and agroecological maps of any targeted part. Probable environmental disaster-prone area zonation is required for every region. Soil fertility maps are prepared precisely based on the updated data for selecting suitable crop production. Prediction about the crop yield is to be made the continuously based standard model for estimating the loss and profit of the cultivators. Regular weather forecasting systems must be introduced and accessible to the cultivators. The local and global market status must be updated and reachable to the farmers. GIS experts must be employed at the village-level administrative unit to prepare agriculture-related information-based map making; agriculture-related research work should be furnished to the micro-unit of farming. Every region has to adopt an intelligent agricultural management system to face the present crisis in agriculture in terms of climate and as well for market management. GIS is an efficient helping tool for researchers to do the mentioned activities.

References

Alemayehu ZY, Minale AS, Legesse SA (2022) Spatiotemporal rainfall and temperature variability in Suha watershed, Upper Blue Nile Basin, Northwest Ethiopia. Environ Monit Assess 194(8):1–25

Baban SM (2022)The suitability of satellite remote sensing and GIS technologies for mapping, monitoring and managing water resources in the Middle East. In: Satellite monitoring of water resources in the Middle East. Springer, Cham, pp 29–47

Baldos ULC, Hertel TW (2015) The role of international trade in managing food security risks from climate change. Food Secur 7(2):275–290

Behera UK, France J (2016) Integrated farming systems and the livelihood security of small and marginal farmers in India and other developing countries. Adv Agron 138:235–282

Böhm S, Carrington M, Cornelius N, de Bruin B, Greenwood M, Hassan L, Shaw D et al (2022) Ethics at the centre of global and local challenges: thoughts on the future of business ethics. J Bus Ethics 180 (3):835–861

Chakraborty D, Başağaoğlu H, Alian S, Mirchi A, Moriasi DN, Starks PJ, Verser JA (2023) Multiscale extrapolative learning algorithm for predictive soil moisture modeling& applications. Expert Syst Appl 213:119056

Chaudhry S, Sidhu GPS (2022) Climate change regulated abiotic stress mechanisms in plants: a comprehensive review. Plant Cell Rep 41(1):1–31

Cho CY (1992) Feeding systems for rainbow trout and other salmonids with reference to current estimates of energy and protein requirements. Aquaculture 100(1–3):107–123

Cooper PJ, Dimes J, Rao KPC, Shapiro B, Shiferaw B, Twomlow S (2008) Coping better with current climatic variability in the rain-fed farming systems of sub-Saharan Africa: an essential first step in adapting to future climate change? Agr Ecosyst Environ 126(1–2):24–35

Couliably S, Kamsu-Foguem B, Kamissoko D, Traore D (2022) Deep learning for precision agriculture: a bibliometric analysis. Intell Syst Appl 16:200102

Cui Y, Tang H, Jin J, Zhou Y, Jiang S, Chen M (2022) System structure–based drought disaster risk assessment using remote sensing and field experiment data. Remote Sens 14(22):5700

Dall'Acqua L (2021) Artificial intelligence, human intelligence, and orientism management framework: a way to improve managerial mindset and manage emergency. In: Computational thinking for problem solving and managerial mindset training. IGI Global, pp 1–30

Dar ZA, Khan A, Lone AA, Bahar FA, Baghel S, Lone BA, Sultan A et al (2022) Specialty food crops: an alternate way for increasing farmers' income. In: Secondary agriculture. Springer, Cham, pp 59–68

Darnhofer I, Bellon S, Dedieu B, Milestad R (2010) Adaptiveness to enhance the sustainability of farming systems: a review. Agron Sustain Dev 30(3):545–555

Das J, Bhattacharya SK (2016) Profitable and viable alternative to tobacco crop in Dinhata Subdivision of Koch Behar district, West Bengal, Indian. J Landscape Syst Ecol Stud 39(2):70–88

Das J, Gayen A, Saha S, Bhattacharya SK (2017) Modelling of alternative crops suitability to tobacco based on Analytical hierarchy process in Dinhata subdivision of Koch Bihar district, West Bengal. Model Earth Syst Environ 3(4):1571–1587. https://doi.org/10.1007/s40808-017-0392-y

Das J, Bhattacharya SK (2018) Tobacco cultivation: time to ban its production in Dinhata subdivision of Koch Bihar district, West Bengal. Res Dir 6(1):1–10

Denton F (2002) Climate change vulnerability, impacts, and adaptation: why does gender matter? Gend Dev 10(2):10–20

Dhanya P, Geethalakshmi V, Ramanathan S, Senthilraja K, Sreeraj P, Pradipa C, Sathyamoorthy NK et al (2022) Impacts and climate change adaptation of agrometeorological services among the maize farmers of west Tamil Nadu. AgriEng 4(4):1030–1053

Dhodho CG (2022) Displacement and livelihood vulnerability among the Tonga Women of Binga, 1958–1980. In: Tonga livelihoods in Rural Zimbabwe. Routledge, pp 115–130

Dogan B, Gokovali U (2012) Geographical indications: the aspects of rural development and marketing through the traditional products. Procedia Soc Behav Sci 62:761–765

Fan S, Teng P, Chew P, Smith G, Copeland L (2021) Food system resilience and COVID-19–lessons from the Asian experience. Glob Food Sec 28:100501

Fatima Z, Naz S, Iqbal P, Khan A, Ullah H, Abbas G, Ahmad S et al (2022) Field crops and climate change. In: Building climate resilience in agriculture. Springer, Cham, pp 83–94

Gabriele M, Brumana R, Previtali M, Cazzani A (2022) A combined GIS and remote sensing approach for monitoring climate change-related land degradation to support landscape preservation and planning tools: the Basilicata case study. Appl Geomatics, pp 1–36

Gaur M K, Goyal RK (2022) Thar desert of India: application of geospatial technology for extreme weather events. In: Hydro-meteorological extremes and disasters. Springer, Singapore, pp 279–300

Guillemin P (2022) Food (in) justice and social inequalities in vegetable and market garden production in Normandy, France. Rev Agric Food Environ Stud 103:1–25

Gumma MK, Kadiyala MDM, Panjala P, Ray SS, Akuraju VR, Dubey S, Whitbread AM et al (2022) Assimilation of remote sensing data into crop growth model for yield estimation: a case study from India. J Indian Soc Remote Sens 50(2):257–270

Hernández-Delgado EA (2015) The emerging threats of climate change on tropical coastal ecosystem services, public health, local economies and livelihood sustainability of small islands: cumulative impacts and synergies. Mar Pollut Bull 101(1):5–28

Hove-Sibanda P, Motshidisi M, Igwe PA (2021)Supply chain risks, technological and digital challenges facing grocery retailers in South Africa. J Enterp Communities People Places Glob Econ 15(2):228–245

Jayne TS, Sturgess C, Kopicki R, SitkoN (2014) Agricultural commodity exchanges and the development of grain markets and trade in Africa: a review of recent experience. In: Food security collaborative working papers 188568. Department of Agricultural, Food, and Resource Economics, Michigan State University

Jimenez AF, Cardenas PF, Canales A, Jimenez F, Portacio A (2020) A survey on intelligent agents and multi-agents for irrigation scheduling. Comput Electron Agric 176:105474

Karl TR, Diamond HJ, Bojinski S, Butler JH, Dolman H, Haeberli W, Zillman J et al (2010) Observation needs for climate information, prediction and application: capabilities of existing and future observing systems. Procedia Environ Sci 1:192–205

Khan N, Jhariya MK, Raj A, Banerjee A, Meena RS (2021) Eco-designing for sustainability. In: Ecological intensification of natural resources for sustainable agriculture. Springer, Singapore, pp 565–595

Kirilenko AP (2022) Geographic Information System (GIS). In: Applied data science in tourism. Springer, Cham, pp 513–526

Knickel K, Redman M, Darnhofer I, Ashkenazy A, Chebach TC, Šūmane S, Rogge E et al (2018). Between aspirations and reality: making farming, food systems and rural areas more resilient, sustainable and equitable. J Rural Stud 59:197–210

Kom Z, Nethengwe NS, Mpandeli NS, & Chikoore H (2020) Determinants of small-scale farmers' choice and adaptive strategies in response to climatic shocks in Vhembe District, South Africa. GeoJournal 87:1–24

MacCarthy DS, Adiku SG, Kamara AY, Freduah BS, Kugbe JX (2022) The role of crop simulation modeling in managing fertilizer use in maize production systems in northern Ghana. In: Enhancing agricultural research and precision management for subsistence farming by integrating system models with experiments. Wiley, pp 48–68

Manser GM (2022) Systematizing authenticity and codifying values: the role of values, standards, and governance at farmers markets. J Rural Stud 96:154–166

Milesi C, Kukunuri M (2022) Crop yield estimation at gram panchayat scale by integrating field, weather and satellite data with crop simulation models. J Indian Soc Remote Sens 50(2):239–255

Müller C, Cramer W, Hare WL, Lotze-Campen H (2011) Climate change risks for African agriculture. Proc Natl Acad Sci 108(11):4313–4315

Murshed S, Griffin AL, Islam MA, Wang XH, Paull D (2022) Assessing multi-climate-hazard threat in the coastal region of Bangladesh by combining influential environmental and anthropogenic factors. Prog Disaster Sci 16:100261

Nascimento N, Lazaro LLB, AmaralMH (2022) How can the water-energy-food nexus approach contribute to enhancing the resilience of Amazonian cities to climate change? In: Water-energy-food nexus and climate change in cities. Springer, Cham, pp 77–92

Negese A, Worku D, Shitaye A, Getnet H (2022) Potential flood-prone area identification and mapping using GIS-based multi-criteria decision-making and analytical hierarchy process in DegaDamot district, northwestern Ethiopia. Appl Water Sci 12(12):1–21

Negi VS, Maikhuri RK, Pharswan D, Thakur S, Dhyani PP (2017) Climate change impact in the Western Himalaya: people's perception and adaptive strategies. J Mt Sci 14(2):403–416

Newburn D, Reed S, Berck P, Merenlender A (2005) Economics and land-use change in prioritizing private land conservation. Conserv Biol 19(5):1411–1420

Ngigi MW, Muange EN (2022) Access to climate information services and climate-smart agriculture in Kenya: a gender-based analysis. Clim Change 174 (3):1–23

Ozegin KO, Salufu SO (2022) Electrical geophysical method and GIS in agricultural crop productivity in a typical sedimentary environment. NRIAG J Astron Geophys 11(1):69–80

Patra S (2022) Uncertainty and chaos in farming: an interpretation on production, livelihood and income. Doctoral dissertation, Bidhan Chandra Krishi Viswavidyalaya

Pettit CJ (2005) Use of a collaborative GIS-based planning-support system to assist in formulating a sustainable-development scenario for Hervey Bay, Australia. Environ Plann B Plann Des 32(4):523–545

Prasad NR, Patel NR, Danodia A, Manjunath KR (2022) Comparative performance of semi-empirical based remote sensing and crop simulation model for cotton yield prediction. Model Earth Syst Environ 8 (2):1733–1747

Reddy GP, Ramamurthy V, Singh SK (2018) Integrated remote sensing, GIS, and GPS applications in agricultural land use planning. In: Geospatial technologies in land resources mapping, monitoring and management. Springer, Cham, pp 489–515

Rizzo G, Mazzilli SR, Ernst O, Baethgen WE, Berger AG (2022) Season-specific management strategies for rainfed soybean in the South American Pampas based on a seasonal precipitation forecast. Agric Syst 196:103331

Rosenzweig C, Parry ML (1994) Potential impact of climate change on world food supply. Nature 367 (6459):133–138

Sakamoto T, vanNguyen N, Ohno H, Ishitsuka N, Yokozawa M (2006) Spatio–temporal distribution of rice phenology and cropping systems in the Mekong Delta with special reference to the seasonal water flow of the Mekong and Bassac rivers. Remote Sens Environ 100(1):1–16

Saraf SA, Ali J, Bahar FA, Sheraz Mahdi S (2022) Marketing of agricultural produce in India: problems and prospects. In Secondary agriculture. Springer, Cham, pp 85–95

Sarkar AN (2013) Promoting eco-innovations to leverage sustainable development of eco-industry and green growth. Eur J Sustain Dev 2(1):171–171

Schipper L, Pelling M (2006) Disaster risk, climate change and international development: scope for, and challenges to, integration. Disasters 30(1):19–38

Sengupta S, Mohinuddin S, Arif M, Sengupta B, Zhang W (2022) Assessment of agricultural land suitability using GIS and Fuzzy Analytical Hierarchy Process approach in Ranchi District, India. Geocarto Int 37(3):1–34

Sgroi F (2022) Cooperation and innovation in Italian agribusiness between theoretical analysis and empirical evidence. J Agric Food Res 10:100406

Sharma M, Bangotra P, Gautam AS, Gautam S (2022) Sensitivity of normalized difference vegetation index (NDVI) to land surface temperature, soil moisture and precipitation over district Gautam Buddh Nagar, UP, India. Stoch Env Res Risk Assess 36(6):1779–1789

Shiferaw B, Tesfaye K, Kassie M, Abate T, Prasanna BM, Menkir A (2014) Managing vulnerability to drought and enhancing livelihood resilience in sub-Saharan Africa: technological, institutional and policy options. Weather Clim Extremes 3:67–79

Shuai G, Basso B (2022) Subfield maize yield prediction improves when in-season crop water deficit is included in remote sensing imagery-based models. Remote Sens Environ 272:112938

Sinha BB, Dhanalakshmi R (2022) Recent advancements and challenges of internet of things in smart agriculture: a survey. Futur Gener Comput Syst 126:169–184

Souissi D, Souie A, Sebei A, Mahfoudhi R, Zghibi A, Zouhri L, Ghanmi M et al (2022) Flood hazard mapping and assessment using Fuzzy Analytic Hierarchy process and GIS techniques in Takelsa, Northeast Tunisia. Arab J Geosci 15(16):1–24

Southworth J, Randolph JC, Habeck M, Doering OC, Pfeifer RA, Rao DG, Johnston JJ (2000) Consequences of future climate change and changing climate variability on maize yields in the midwestern United States. Agr Ecosyst Environ 82(1–3):139–158

Stephens EC, Timsina J, Snow V, Martin G, van Wijk M, Klerkx L, Reidsma P(2022) The immediate impact of the first waves of the global COVID-19 pandemic on agricultural systems worldwide: reflections on the COVID-19 special issue for agricultural systems. Agric Syst 201:103436

Sziroczak D, Rohacs D, Rohacs J (2022) Review of using small UAV based meteorological measurements for road weather management. Prog Aerosp Sci 134: 100859

Talukdar S, Naikoo MW, Mallick J, Praveen B, Sharma P, Islam ARMT, Rahman A et al (2022) Coupling geographic information system integrated fuzzy logic-analytical hierarchy process with global and machine learning based sensitivity analysis for agricultural suitability mapping. Agric Syst 196: 103343

Taylor M (2018) Climate-smart agriculture: what is it good for? J Peasant Stud 45(1):89–107

Timmer CP (1988) The agricultural transformation. Handb Dev Econ 1:275–331

van Tulder R, van Mil E (2022) Principles of sustainable business: frameworks for corporate action on the SDGs. Taylor & Francis

Van Wart J, van Bussel LG, Wolf J, Licker R, Grassini P, Nelson A, Cassman KG et al (2013) Use of agro-climatic zones to upscale simulated crop yield potential. Field Crop Res 143:44–55

Wilson JP (1999) Local, national, and global applications of GIS in agriculture. In: Geographical information systems: principles, techniques, management, and applications, pp 981–998

Wuebbles DJ, Hayhoe K (2004) Climate change projections for the United States Midwest. Mitig Adapt Strat Glob Change 9(4):335–363

Yin L, Tao F, Chen Y, Wang Y (2022) Reducing agriculture irrigation water consumption through reshaping cropping systems across China. Agric for Meteorol 312:108707

Yousefi MR, Razdari AM (2015) Application of GIS and GPS in precision agriculture (a review). Int J.adv Biol Biom Res 3(1):7–9

Zadeh AH, Zolbanin HM, Sharda R, Delen D (2019) Social media for nowcasting flu activity: spatio-temporal big data analysis. Inf Syst Front 21(4):743–760

Zhang P, Howell K, Li Y, Li L, Wang J, Eckard R, Barlow EWR (2023) Using historical weather data and a novel season temperature index to classify wine-grape growing zones in Australia. Sci Hortic 307: 111516

Check for updates

Suitability Assessment of Different Protected Cultivation Structures Using Hybrid Multi-Criteria Decision Analysis Technique

3

Debaditya Gupta, K. N. Tiwari, D. T. Santosh, and Subha M. Roy

Abstract

Protected cultivation structures are used to protect crops from adverse weather conditions and for off-season cultivation. The selection of protected cultivation structures mainly emphasizes minimizing the crop water requirement, i.e., reducing evapotranspiration and creating maximum hindrance against rainfall and insect infestation. In the present chapter, reference evapotranspiration (ET_0) is determined using the FAO-Penman–Monteith (FAO-PM) model and sensitivity analysis of ET_0 for the open field condition and four other protected cultivation structures, namely polyhouse, shadow hall, shade net house, and polytunnel. Sensitivity analysis is crucial to recognize the impact of variance in climatic variables on the change in ET_0. Suitability assessment of the structures was done based on Multi-Criteria Decision Analysis (MCDA) technique in which three different methods, viz., Criteria Importance Through Intercriteria Correlation (CRITIC), Analytic Hierarchy Process (AHP), and Fuzzy Analytic Hierarchy Process (FAHP) technique ensemble to obtain the final weights. The ultimate selection ranking of the structures is determined about the Weight Aggregated Sum Product Assessment (WASPAS) technique. The results found that the average value of ET_0 is the highest and the lowest for open field and polyhouse, respectively. The sensitivity study showed that solar radiation is the utmost sensitive input for ET_0, followed by mean temperature and wind speed, indicating the least sensitive for open conditions and other protected cultivation structures. The values of sensitivity coefficients are in a higher range for solar radiation and mean temperature, least for wind speed, and negative for related humidity. The MCDA results show that the shadowed hall is the most suitable structure, whereas the open condition is the least appropriate structure.

D. Gupta (✉)
School of Agro and Rural Technology, Indian Institute of Technology, Guwahati, Assam 781039, India
e-mail: debadityagupta24@gmail.com

K. N. Tiwari
Agricultural and Food Engineering Department, Indian Institute of Technology, Kharagpur, West Bengal 721302, India
e-mail: kamlesh@agfe.iitkgp.ac.in

D. T. Santosh
Centre of Smart Agriculture, Centurion University of Technology and Management, Paralakhemundi, Odisha 752050, India
e-mail: santosh.dt@cutm.ac.in

S. M. Roy
Faculty of Agricultural Sciences, Institute of Applied Sciences and Humanities, GLA University, Uttar Pradesh, Mathura 281406, India
e-mail: subhamanash@gmail.com

J. Das and S. Halder (eds.), *Advancement of GI-Science and Sustainable Agriculture*,
GIScience and Geo-environmental Modelling, https://doi.org/10.1007/978-3-031-36825-7_3

Keywords

Reference evapotranspiration · Sensitivity analysis · MCDA · CRITIC · AHP · FAHP · WASPAS

3.1 Introduction

Protected cultivation is a modern agricultural practice used for maximum crop production under a controlled or partially controlled environment (Singh et al. 2016). Protected cultivation structures create a suitable atmosphere for better yield and growth of crops. Various types of covers, cladding materials, shade, and insect-proof nets result in variations in interior weather conditions (microclimate) compared to those in the field with the open condition. The greenhouse cladding materials differ in the incidence of solar radiation's magnitude, affecting the greenhouse's microclimate. Diffused solar radiation infiltrates into deeper portions of plants' canopy than direct solar radiation; thus, the diffused film is preferable to use as cladding material (Hemming et al. 2008). The microclimate of the shade net house structure remains cooler than the ambient because of a shade net, which helps reduce incident solar radiation. The microclimate inside the polytunnel remains hot, as the polytunnel is deprived of ventilation. Therefore, the inside heat storage becomes more significant than greenhouse and shade net houses.

Evapotranspiration (ET) represents the significant contributions of the water cycle, which represents two hydrological processes, namely evaporation and transpiration. It is a complex process, and it depends upon different factors, namely temperature, wind speed, solar radiation, relative humidity, etc. It plays an important role in irrigation scheduling and crop production. The FAO-Penman–Monteith (FAO-PM) model assimilates climatic factors, namely temperature, solar radiation, and wind speed which can be altered by variations in climatic conditions (Irmak et al. 2006). In this manner, it is essential to identify the sensitivity of ET_0_FAO-PM to the change in these climatic factors. The sensitivity study may be a process that permits progressing understanding, not as it were on the connections between climatic conditions and ET_0 inconstancy but too on the distinguishing proof of the prevailing climatic factors in assessing evapotranspiration (Gong et al. 2006; Irmak et al. 2006; Mosaedi et al. 2017). A sensitivity study is required to depict the relative importance of different climate factors related to the computation of ET_0. Therefore, results from sensitivity studies play an essential part in determining the change in ET_0 for a known change in one of the independent climatic variables (Ndiaye et al. 2017).

Estévez et al. (2009) reported the sensitivity of ET_0 using the FAO-PM method with different weather parameters in southern Spain, where the climate is semi-arid. They observed that relative humidity, temperature, and radiation were the main influential weather parameters affecting ET_0. Tabari and Talaee (2014) analyzed the sensitivity of ET_0 using FAO-PM with temperature and sunshine hour under humid, semi-arid hot, semi-arid cold, and arid in Iran. They showed that temperature is more important than a sunshine hour in arid and semi-arid climates.

Multi-Criteria Decision Analysis (MCDA) is a popular method in energy decision-making because of its ability to deal with complex decision processes with multiple views and preferences, different sources of uncertainty, and distinguished time frames. The MCDA is performed in different agricultural water management (Banihabib and Shabestari 2017; Gómez-Limón et al. 2007), water resources planning and management (Geng and Wardlaw 2013; Zhu et al. 2019), groundwater potential zone mapping (Agarwal et al. 2013; Kumar et al. 2014; Machiwal et al. 2011), groundwater depletion mapping (Basak et al. 2021; Mandal et al. 2022), flood prediction mapping (Mitra et al. 2022; Mitra and Das 2022; Osman and Das 2023), and several other engineering design applications and project management.

Several studies have been done for the sensitivity study of ET_0_FAO-PM to the change of climatic variables of open field conditions (Ambas and Baltas 2012; Djaman et al. 2016;

Liang et al. 2008; Tabari and Talaee 2014). However, no literature on the sensitivity study of ET_0 under different protected cultivation structures is available. Therefore, the present study was conducted with the following objectives: (i) to estimate ET_0 applying the FAO-PM model and to analyze the sensitivity of ET_0 calculated by FAO-PM equations for open field and in four other protected cultivation structures (polyhouse, polytunnel, shade net house, and shadow hall); (ii) to calculate the sensitivity coefficients for FAO-PM equations for the structures mentioned above, and (iii) to assess the suitability of different protected cultivation structures using MCDA.

3.2 Materials and Methods

3.2.1 Study Area

The experiment was conducted at the research field of the Agricultural and Food Engineering (AgFE) Department, IIT Kharagpur, having 22° 18′ N latitude 87°19′ E longitude and an elevation of 52 m above mean sea level (Fig. 3.1). The climatic environment is sub-humid with an average rainfall of 1390 mm.

The standard FAO-PM Eq. (3.1) for calculating ET_0 is given by Allen et al. (1998):

$$ET_0 = \frac{0.408\Delta(R_n - G) + \frac{900\gamma u_2}{T_{mean} + 273}(e_s - e_a)}{\Delta + \gamma(1 + 0.34u_2)} \quad (3.1)$$

where ET_0 = reference evapotranspiration, R_n = net radiation, G = soil heat flux, u_2 = wind speed at 2 m height, $(e_s - e_a)$ = vapor pressure deficit, Δ = slope of vapor pressure versus temperature graph, γ = psychometric constant, and T_{mean} = mean temperature

Soil heat flux (G) is very small as compared to net radiation (R_n), and therefore, for a daily period, G is considered as 0 (Allen et al. 1998).

Vapor pressure deficit is computed according to Allen et al. (1998)

$e_s - e_a$ = vapor pressure deficit (kPa).

where

$$e_s = \frac{e^0(T_{max}) + e^0(T_{min})}{2} \quad (3.2)$$

e_s = mean saturated vapor pressure

$$e^0(T_{max}) = 0.6108\exp\left[\frac{17.27T_{max}}{237.3 + T_{max}}\right] \quad (3.3)$$

$$e^0(T_{min}) = 0.6108\exp\left[\frac{17.27T_{min}}{237.3 + T_{min}}\right] \quad (3.4)$$

$e^0(T_{max})$ and $e^0(T_{min})$ are saturated vapor pressure at T_{max} and T_{min}, respectively.

e_a = actual vapor pressure that is derived from relative humidity data as given in Allen et al. (1998)

$$e_a = \frac{RH}{100}\left[\frac{e^0(T_{max}) + e^0(T_{min})}{2}\right] \quad (3.5)$$

Δ = slope of vapor pressure versus temperature graph (kPa °C^{-1}) that is computed as in Allen et al. (1998)

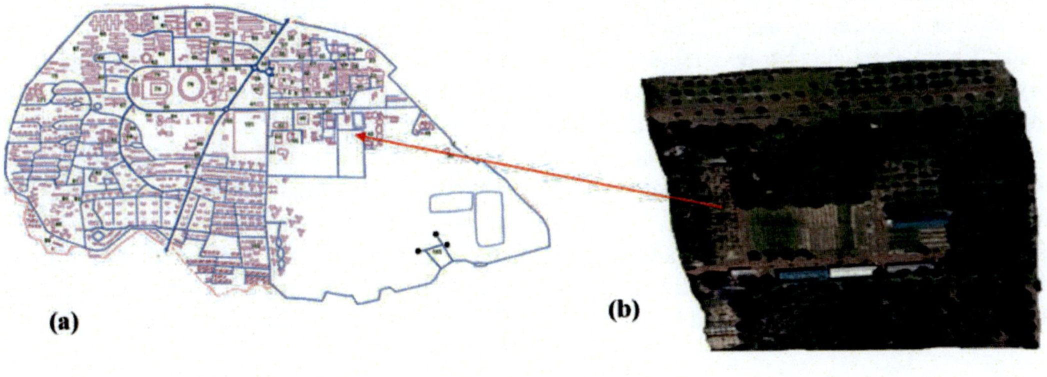

Fig. 3.1 Study area; a campus map of IIT Kharagpur, **b** PFDC field map

$$\Delta = \frac{4098\left[0.6108\exp\left(\frac{17.27T_{\text{mean}}}{T_{\text{mean}}+237.3}\right)\right]}{(T_{\text{mean}}+237.3)^2} \tag{3.6}$$

γ is computed following Allen et al. (1998); the expression of the psychometric constant γ is given as

$$\gamma = 0.665 \times 10^{-3}P \tag{3.7}$$

where

$$P = 101.3\left(\frac{(T_{\text{mean}}+273) - 0.0065z}{(T_{\text{mean}}+273)}\right)^{5.26} \tag{3.8}$$

where

T_{mean} = mean temperature.

z = elevation from mean sea level (m).

Mean temperature is calculated as given by Allen et al. (1998)

$$T_{\text{mean}} = \frac{T_{\max} + T_{\min}}{2} \tag{3.9}$$

where $T_{\max}$ and $T_{\min}$ are maximum and minimum temperature (°C), respectively.

The climatic variables for ET_0 measurement using FAO 56 Penman–Monteith methods, namely maximum temperature, minimum temperature, wind speed, solar radiation, and relative humidity data, were collected from the AgFE Department, IIT Kharagpur. The maximum and minimum temperatures were measured using maximum–minimum thermometers installed in the greenhouse. Wind speed was measured by using instantaneous pocket weather stations inside the greenhouses. Solar radiation inside the greenhouses was measured by using a hand-held Lux meter. Relative humidity has been measured using dry bulb–wet bulb thermometers installed inside the greenhouses. The outside climatic data were also obtained by the same instruments installed in the PFDC. Solar radiation (W/m^2) is converted from Lux unit by the factor of 1 W/m^2 = 100 lx. Relative humidity from the dry bulb and wet bulb thermometer was computed by the following formula (http://www.1728.org/relhum.htm):

$$\text{RH} = \left[\frac{e_w - N(1+0.00115T_w)(T_d - T_w)}{e_d}\right] \times 100 \tag{3.10}$$

$$e_d = 6.112\exp\left(\frac{17.502T_d}{240.97+T_d}\right) \tag{3.11}$$

$$e_w = 6.112\exp\left(\frac{17.502T_w}{240.97+T_w}\right) \tag{3.12}$$

T_d and T_w are dry bulb and wet bulb temperatures, respectively. $N = 0.6687451584$.

The sensitivity study is done to identify the variation in ET_0 with the variation of the input climatic factors, namely average temperature, wind speed, relative humidity, and solar radiation.

If y is expressed as a function of x_1, x_2, ……, x_n

$$y = f(x_1, x_2, \ldots, x_n) \tag{3.13}$$

Then sensitivity coefficient of y for variable x_i can be expressed as

$$Cx_i = \frac{\partial y}{\partial x_i} \tag{3.14}$$

for $i = 1, 2, 3, \ldots, n$

The climatic variables are obtained as mean temperature (T_{mean}), solar radiation (R_n), wind speed (u_2), and relative humidity (RH), and y is the ET_0.

This is such that

$$CT_{\text{mean}} = \frac{\partial ET_0}{\partial T_{\text{mean}}}, \; Cu_2 = \frac{\partial ET_0}{\partial u_2}, \; \text{CRH} = \frac{\partial ET_0}{\partial \text{RH}}, \; CR_n = \frac{\partial ET_0}{\partial R_n} \tag{3.15}$$

where CT_{mean}, Cu_2, CRH, and CR_n are sensitivity coefficients of ET_0 with respect to T_{mean}, u_2, RH, and R_n, respectively.

Since ET_0 is the factor of maximum and minimum temperature, wind speed, relative humidity, and solar radiation, the sensitivity coefficients of ET_0 for various climatic factors are achieved by taking the partial derivative of ET_0 concerning the climatic factors.

For this chapter, the sensitivity study of ET_0 determined by the FAO-PM method is done on an open field and four other structures, viz., polyhouse (plan 14 m × 6 m), polytunnel (plan 20 m × 5 m), shade net house (plan 20 m × 5 m), and shadow hall (plan 20 m × 5 m). Daily climatic variables, viz., maximum and temperature, relative humidity, solar radiation, and wind speed for the open condition and all four protected structures are taken from January 1, 2018, to December 31, 2018 (Fig. 3.2).

3.2.2 Multi-Criteria Decision Analysis (MCDA)

MCDA for selection of suitable conditions mainly depends on three major criteria, viz., climatic factors except for rainfall, rainfall (*R*), and insect infestation (*I*). The criteria (*C*) is selected based on the four sub-criteria: average temperature, net radiation, wind speed, and relative humidity. The selection process is given in Fig. 3.3.

3.2.3 Weight Determination for Climatic Factors

Weight is determined by using the sensitivity range matrix of the climatic factors, namely mean temperature (T_{mean}), net radiation (R_n), wind speed (u_2), and relative humidity (RH), using the Criteria Importance Through Intercriteria Correlation (CRITIC) method (Triantaphyllou 2000), Analytic Hierarchy Process (AHP) method (Mitra et al., 2022; Saaty 2008), and Fuzzy Analytic Hierarchy Process (FAHP) method (Pehlivan et al. 2017). The sensitivity range matrix is generated based on the sensitivity study of the FAO-PM method concerning the climatic variables (sub-criteria). The pairwise comparison

Fig. 3.2 Images of protected cultivation structures; **a** polytunnel, **b** polyhouse, **c** shade net house, and **d** shadow hall

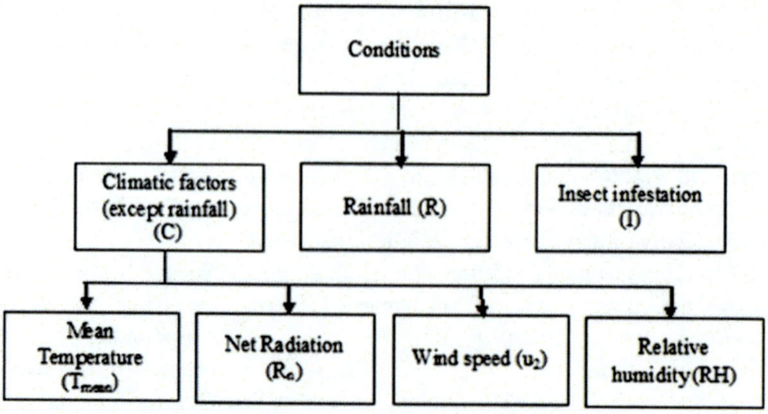

Fig. 3.3 Structure of Multi-Criteria Decision Analysis (MCDA) selection process

matrix for the AHP (also used for FAHP) is selected based on the relative importance of each of the variables on the structures to each other. For the pairwise comparison matrix, the equal importance, moderate importance, strong importance, and very strong importance are given by the numbers 1, 3, 5, 7, and 9, respectively, and 2, 4, 6, and 8 are provided as intermediate values. The final pairwise comparison matrices for the climatic variables namely are stated as follows.

$$\begin{array}{c} \\ T_{mean} \\ R_n \\ u_2 \\ RH \end{array} \begin{array}{c} \begin{matrix} T_{mean} & R_n & u_2 & RH \end{matrix} \\ \begin{bmatrix} 1 & \frac{1}{2} & 7 & 4 \\ 2 & 1 & 8 & 5 \\ \frac{1}{7} & \frac{1}{2} & 1 & \frac{1}{2} \\ \frac{1}{4} & \frac{1}{5} & 2 & 1 \end{bmatrix} \end{array}$$

The pairwise comparison matrix generates the weights for the climatic variable for further analysis of the rankings. After the weight generation of the variables, the final ranking is done based on Weight Aggregated Sum Product Assessment (WASPAS) (Zavadskas et al. 2012, 2013) which is a combination of the weighted sum model (WSM) (Fishburn 1967; Triantaphyllou 2000) and weighted product model (WPM) (Triantaphyllou and Mann 1989; Triantaphyllou 2000). It is given as.

$$Q_i = \lambda S_i + (1 - \lambda)P_i \tag{3.16}$$

where λ = the weight imposed for weighted sum model, and $(1 - \lambda)$ = the weight associated with weighted product model in the Weight Aggregated Sum Product Assessment.

Where S_i and P_i are the weighted sum and weighted product values of the criteria for different structures.

$$S_i = \sum_{j=1}^{m} W_j \cdot X_{ij} \tag{3.17}$$

$$P_i = \prod_{j=1}^{m} (X_{ij})^{W_j} \tag{3.18}$$

W_j is the criteria weight of the criteria.

where j is the number of variables (i.e., the number of columns varies from 1 to m) W_j, X_{ij} is the normalized value of the variables, and i denotes the number of protected structural conditions (varies from 1 to 5). Normalization is done by the Max–Min normalization (Esteves et al. 2009).

For beneficial criteria, it is given as follows.

$$X_{ij} = \frac{x_{ij} - x_j^{\min}}{x_j^{\max} - x_j^{\min}} \tag{3.19}$$

For non-beneficial criteria, it is given as

$$X_{ij} = \frac{x_j^{\max} - x_{ij}}{x_j^{\max} - x_j^{\min}} \quad (3.20)$$

After determination of the effect of the suitability against climatic factors (*C*), the importance of the rainfall (*R*) and insect infestations (*I*) is determined linguistically along with digital numbers as 'worst', 'poor', 'below average', 'intermediate', 'good', and 'excellent' as 0, 1, 2, 3, 4, and 5, respectively. Rainfall is used here as a linguistic term because here, we only consider the rainfall penetrability inside the structures qualitatively, not the amount of rainfall variations. The pairwise comparison matrix for the three factors *C*, *R*, and *I* is for AHP, and FAHP is given as follows.

$$\begin{array}{c|ccc} & C & R & I \\ \hline C & 1 & 2 & 3 \\ R & \frac{1}{2} & 1 & 2 \\ I & \frac{1}{3} & \frac{1}{2} & 1 \end{array}$$

In this present chapter, the final weights of the variables were calculated depending on the individual weights obtained by the three different MCDA techniques mentioned above. Again, the final WASPAS was applied after the calculation of the weighted sums by using WSM and WPM, which was discussed in the results and discussions part of the MCDA approach.

3.3 Results and Discussions

Evapotranspiration (ET_0) for different protected cultivation structures and open fields is presented in Table 3.1 and Fig. 3.4.

From Table 3.1 and Fig. 3.4, it is clear that maximum values of ET_0 are observed during April–May (summer) and minimum during December (winter). The average value of ET_0 is maximum for open conditions and minimum for the polyhouse. The maximum ET_0 is higher for the polytunnel due to the storage of high heat in summer months and no ventilation for air passage. Lesser ET_0 is found in the polyhouse as it is partly covered with a shade net and a ventilation system in it, and also it consists of fan systems for cooling in it.

3.3.1 Sensitivity Study

Sensitivity study checks the effect of a single climatic variable for the change in ET_0, keeping the other variables fixed. The sensitivity analysis of ET_0 results for the different climatic variable is discussed in Fig. 3.5, and the sensitivity study in different protected cultivation structures is discussed in Fig. 3.6. The percentage change in ET_0 due to percentage change in climatic variables under different protected cultivation structures is discussed in Table 3.2.

From Fig. 3.6 and Table 3.2, it is observed that for all the experimental protected cultivation structures as well as in the open field with the increasing and decreasing of mean temperature, solar radiation, and wind speed, the ET_0 is increasing and decreasing, respectively, and the reverse is happening in case of RH. This is because RH has an opposite relation with the ET_0, whereas the other variables are related directly to the ET_0. From the figures and the table, it is also observed that solar radiation $(-24.39 \leq \delta ET_{0-R_n}(\%) \leq 24.39)$ is most sensitive to ET_0, followed by mean temperature $(-20.32 \leq \delta ET_{0-T_{mean}}(\%) \leq 22.46)$, relative humidity $(7.36 \geq \delta ET_{0-RH}(\%) \geq -6.38)$, and wind speed $(-3.06 \leq \delta ET_{0-u_2}(\%) \leq 3.2)$. Again, maximum changes in ET_0 due to changes in R_n, T_{mean}, and RH occurred for the polytunnel as ventilation is absent in the poly tunnel and is fully cladded with polythene film. Therefore, relatively more thermal energy is stored due to the entrapment of solar radiation and RH, as there is no ventilation for airflow. Changes in ET_0 due to wind speed are maximum in the open field because the wind speed is much less in the protected structures. Hence, the other structures with lower wind speeds affect the ET_0 very little.

Table 3.1 Monthly average of daily reference evapotranspiration (ET_0) for open fields and different protected cultivation structures

Month	ET_0 Open	ET_0 Polyhouse	ET_0 Polytunnel	ET_0 Shade net house	ET_0 Shadow hall
Jan	2.87	2.50	2.44	2.61	2.55
Feb	3.66	3.18	3.20	3.35	3.28
Mar	4.71	4.62	4.58	4.50	4.88
Apr	6.02	5.55	5.86	5.87	5.89
May	5.97	5.72	6.35	5.88	6.34
Jun	6.04	5.40	6.00	6.11	5.84
Jul	5.14	5.09	5.53	5.66	5.56
Aug	5.32	4.74	5.17	5.16	5.27
Sep	5.34	5.02	5.29	4.84	4.96
Oct	4.43	3.96	4.08	4.20	3.98
Nov	3.66	2.76	3.04	3.20	2.97
Dec	2.56	2.09	2.22	2.45	2.38
Max	6.04	5.72	6.35	6.11	6.34
Min	2.56	2.09	2.22	2.45	2.38
Mean	4.64	4.22	4.48	4.49	4.49

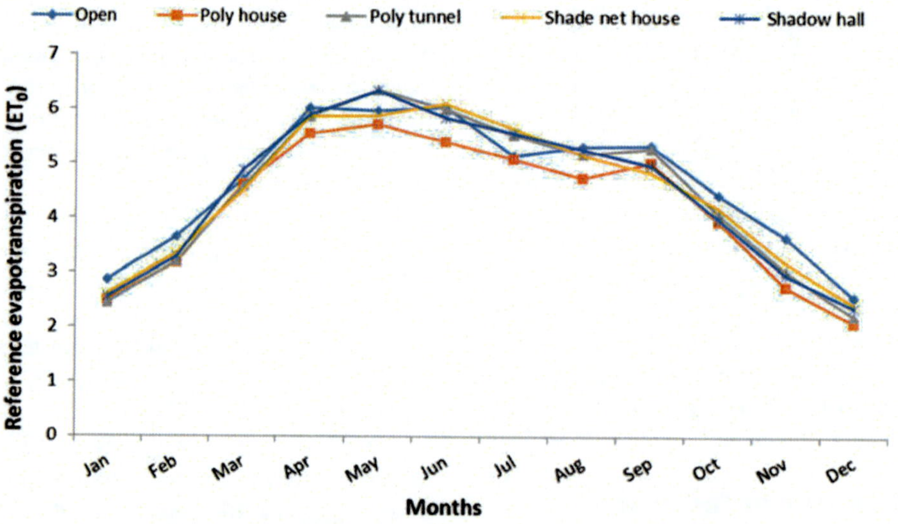

Fig. 3.4 Monthly average ET_0 for different protected cultivation structures

From Fig. 3.7, it is also found that sensitivity coefficients for T_{mean} and R_n are high range, where that for u_2 is in a very small range and negative for RH (Table 3.3). The sensitivity coefficient for T_{mean} fluctuated during the whole year for the protected structures. The above-said fact is a reason for the dynamic differences in the inside temperature of the protected structures, and as the wind speed inside those structures is much less, the sensitivity of ET_0 for u_2 is also much less. The summary of the monthly average sensitivity coefficients is given in the table (Table 3.4).

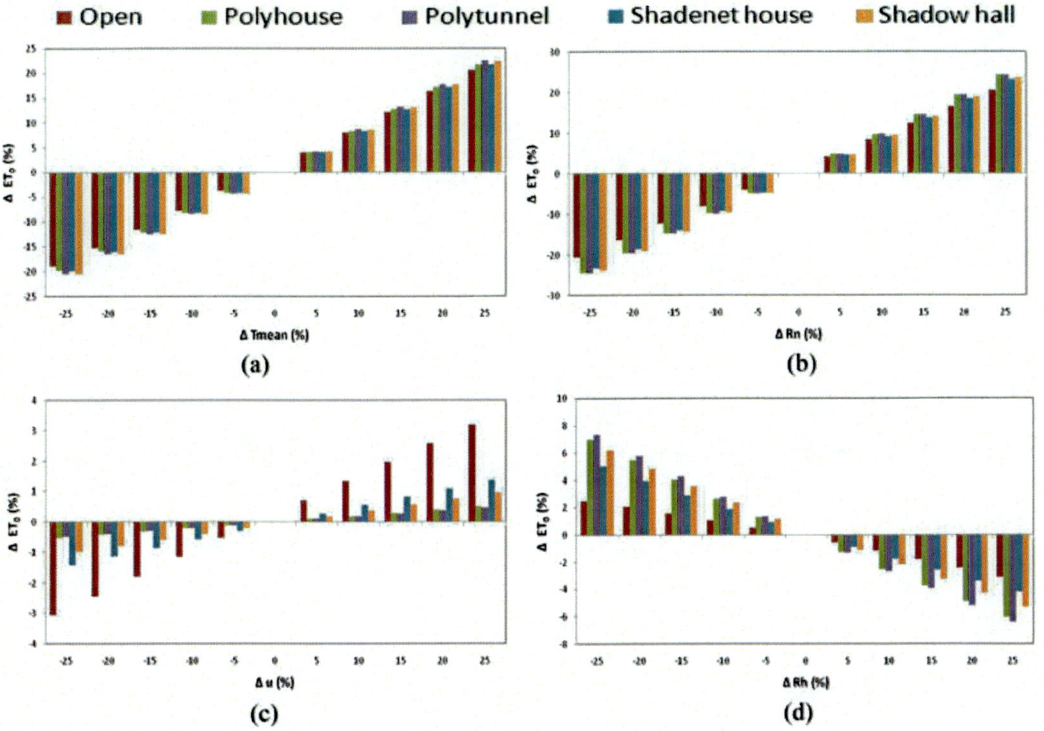

Fig. 3.5 Sensitivity study of ET_0 results with different climatic variables; **a** Δ T_{mean} (%), **b** Δ R_n (%), **c** Δ u (%), and **d** Δ RH (%)

3.3.2 Results of Multi-Criteria Decision Analysis

In this chapter, three different MCDA methods, namely CRITIC, AHP, and FAHP, were applied for the weight determination after determining the ET_0 and its sensitivity analysis with the climatic variables. Weights obtained by the CRITIC methods were 0.24, 0.26, 0.25, and 0.25 for climatic variables, namely T_{mean}, R_n, u_2, and RH. Similarly, weights were obtained as 0.33, 0.51, 0.05, and 0.10 by AHP and 0.34, 0.50, 0.05, and 0.10 by FAHP for the same climatic variables. As AHP and FAHP show nearly equal weights and the CRITIC method gives almost similar weightage to all variables, therefore we gave more importance to the weights generated by AHP (40%) and FAHP (40%) and less importance to CRITIC (20%) method. Therefore, the final values of the weights for climatic variables were obtained as 0.32, 0.46, 0.09, and 0.13 for T_{mean}, R_n, u_2, and RH, respectively. After obtaining the weights, the WASPAS was used with the value of λ taken as 0.75 because the weighted product model provides multiple same rankings. In three conditions, the value of the weighted product comes to zero. Therefore, more emphasis is given to weighted sum values (Q_i). The final rank of the structures is given in Table 3.5.

After determination of the effect of the suitability against climatic factors I, the importance of the rainfall I and insect infestations (I) is determined linguistically along with digital numbers as 'worst', 'poor', 'below average', 'intermediate', 'good', and 'excellent' as 0, 1, 2, 3, 4, and 5, respectively. Rainfall is used here as a linguistic term because here, we consider the rainfall penetrability inside the structures only qualitatively, not the amount of rainfall variations. Hence, the rainfall is also separated from the other climatic variables. Therefore, the final table (Table 3.6) is generated.

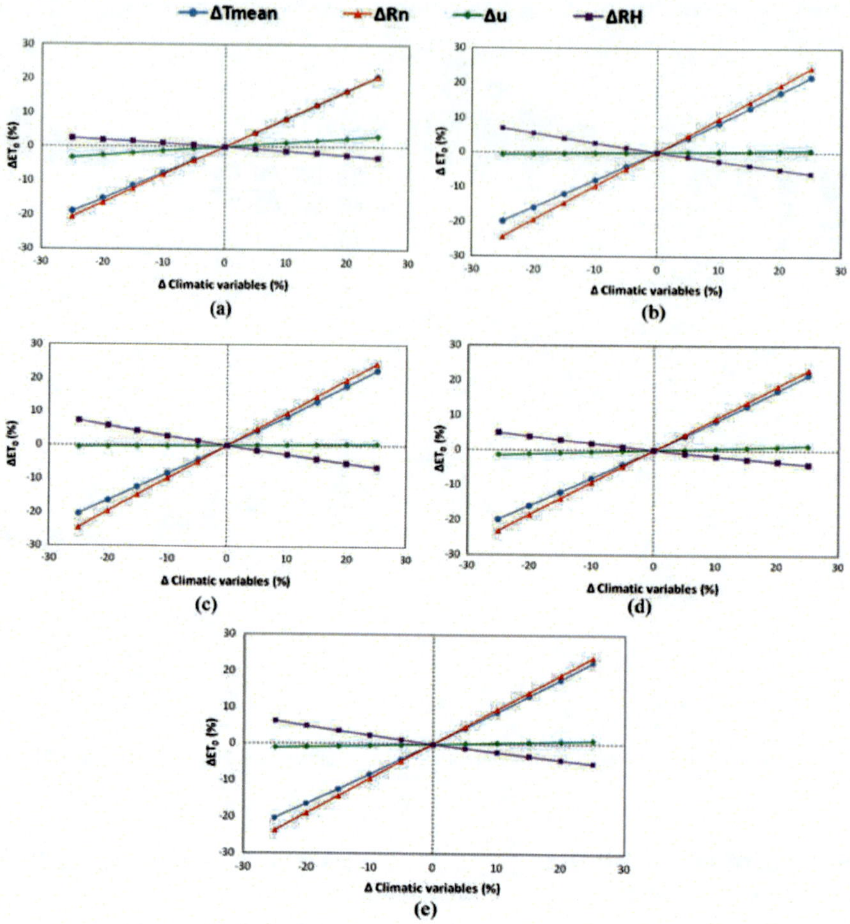

Fig. 3.6 Change in ET_0 with respect to climatic variables in different structures; **a** open field, **b** polyhouse, **c** polytunnel, **d** shade net house, and **e** shadow hall

Table 3.2 Percentage change in ET_0 due to percentage change in climatic variables under different protected cultivation structures

Climatic variables	δ	δET_0 Open	δET_0 Polyhouse	δET_0 Polytunnel	δET_0 Shade net house	δET_0 Shadow hall
T_{mean}	− 25	− 18.95	− 19.75	− 20.32	− 19.82	− 20.33
	− 20	− 15.25	− 15.94	− 16.42	− 16	− 16.42
	− 15	− 11.49	− 12.07	− 12.43	− 12.1	− 12.43
	− 10	− 7.68	− 8.12	− 8.37	− 8.14	− 8.37
	− 5	− 3.81	− 4.1	− 4.23	− 4.11	− 4.23
	0	0	0	0	0	0
	5	4.1	4.17	4.31	4.18	4.31
	10	8.14	8.43	8.71	8.44	8.7
	15	12.23	12.76	13.2	12.77	13.18
	20	16.39	17.18	17.78	17.18	17.75
	25	20.6	21.68	22.46	21.67	22.41

(continued)

Table 3.2 (continued)

Climatic variables	δ	δET_0 Open	δET_0 Polyhouse	δET_0 Polytunnel	δET_0 Shade net house	δET_0 Shadow hall
R_n	− 25	− 20.48	− 24.35	− 24.39	− 23.15	− 23.76
	− 20	− 16.36	− 19.48	− 19.51	− 18.52	− 19.01
	− 15	− 12.24	− 14.61	− 14.63	− 13.89	− 14.25
	− 10	− 8.12	− 9.74	− 9.76	− 9.26	− 9.5
	− 5	− 4	− 4.87	− 4.88	− 4.63	− 4.75
	0	0	0	0	0	0
	5	4.24	4.87	4.88	4.63	4.75
	10	8.36	9.74	9.76	9.26	9.5
	15	12.48	14.61	14.63	13.89	14.25
	20	16.59	19.48	19.51	18.52	19.01
	25	20.71	24.35	24.39	23.15	23.76
u_2	− 25	− 3.06	− 0.51	− 0.48	− 1.39	− 0.97
	− 20	− 2.42	− 0.41	− 0.38	− 1.11	− 0.77
	− 15	− 1.78	− 0.31	− 0.29	− 0.83	− 0.58
	− 10	− 1.14	− 0.2	− 0.19	− 0.55	− 0.39
	− 5	− 0.51	− 0.1	− 0.1	− 0.28	− 0.19
	0	0	0	0	0	0
	5	0.74	0.1	0.1	0.28	0.19
	10	1.36	0.2	0.19	0.55	0.38
	15	1.98	0.3	0.29	0.83	0.58
	20	2.59	0.41	0.38	1.1	0.77
	25	3.2	0.51	0.48	1.38	0.96
RH	− 25	2.51	6.99	7.36	5.04	6.22
	− 20	2.07	5.5	5.79	3.94	4.88
	− 15	1.59	4.06	4.27	2.89	3.59
	− 10	1.09	2.66	2.8	1.89	2.35
	− 5	0.56	1.31	1.38	0.92	1.16
	0	0	0	0	0	0
	5	− 0.58	− 1.27	− 1.34	− 0.89	− 1.12
	10	− 1.18	− 2.51	− 2.65	− 1.75	− 2.2
	15	− 1.8	− 3.72	− 3.92	− 2.57	− 3.26
	20	− 2.44	− 4.9	− 5.17	− 3.37	− 4.28
	25	− 3.09	− 6.05	− 6.38	− 4.14	− 5.27

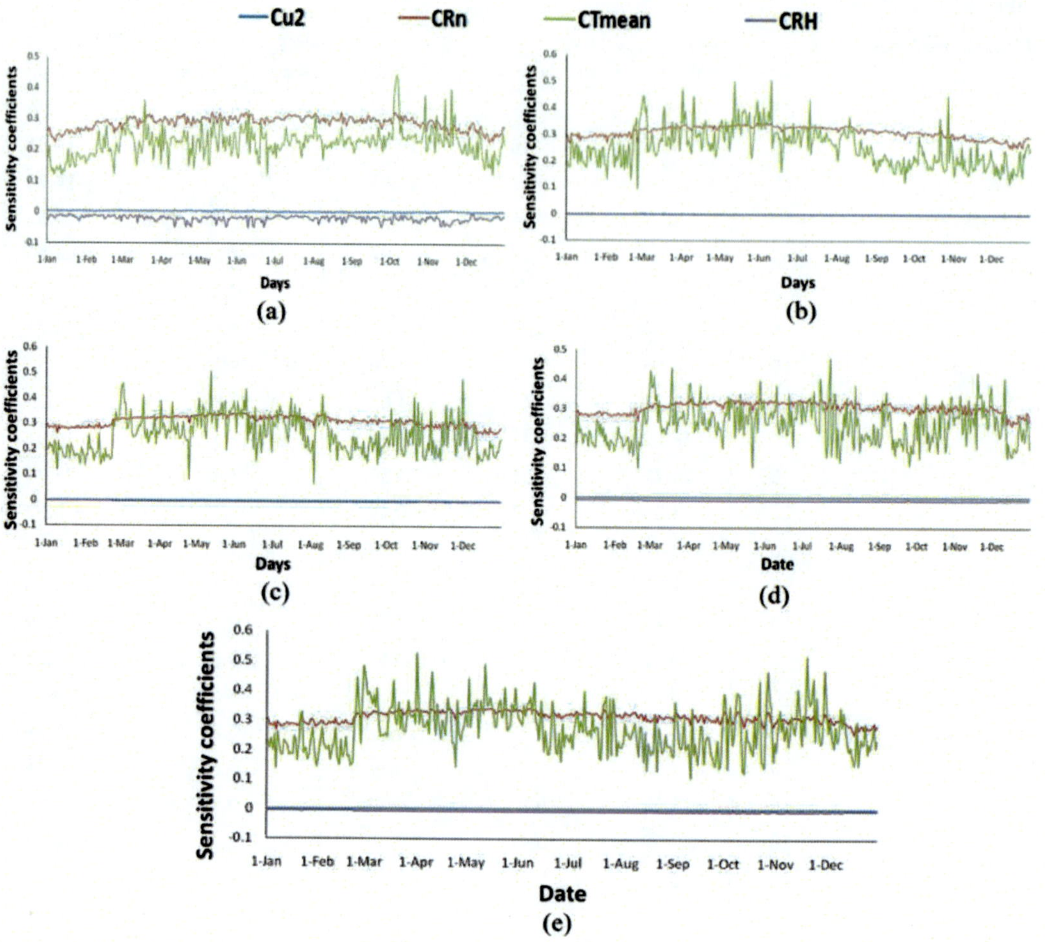

Fig. 3.7 Daily sensitivity coefficients for; **a** open field, **b** polyhouse, **c** polytunnel, **d** shade net house, and **e** shadow hall

Table 3.3 Final sensitivity ranges derived from the sensitivity analysis

Structures	Climatic variables sensitivity range (%)			
	T_{mean}	R_n	u_2	RH
Open	39.55	41.19	6.26	5.60
Polyhouse	41.43	48.70	1.02	13.04
Polytunnel	42.78	48.78	0.96	13.74
Shade net house	41.49	46.3	2.77	9.18
Shadow hall	42.74	47.52	1.93	11.49

Table 3.4 Monthly average value for sensitivity coefficients for T_{mean}, R_n, u_2, and RH for open field and other four structures

Structures	Sensitivity coefficients	Jan	Feb	Mar	Apr	May	Jun	Jul	Aug	Sep	Oct	Nov	Dec
Open	C_{Tmean}	0.1569	0.2017	0.2338	0.2221	0.2316	0.2286	0.2283	0.2334	0.232	0.2727	0.2511	0.1926
	C_{Rn}	0.253	0.2773	0.2957	0.2961	0.3027	0.3008	0.3082	0.3024	0.2982	0.2987	0.2776	0.2614
	C_{u2}	0.0024	0.0024	0.0027	0.0034	0.0031	0.003	0.0017	0.0023	0.0025	0.0028	0.004	0.0025
	C_{RH}	− 0.0155	− 0.0153	− 0.0187	− 0.0253	− 0.0228	− 0.0231	− 0.0136	− 0.0175	− 0.0187	− 0.0195	− 0.0251	- − 0.0153
Polyhouse	C_{Tmean}	0.2179	0.2413	0.2983	0.2991	0.3401	0.3119	0.2933	0.2631	0.1912	0.2233	0.1962	0.1879
	C_{Rn}	0.2919	0.2994	0.3219	0.3291	0.335	0.332	0.3288	0.3222	0.3096	0.3058	0.2867	0.2755
	C_{u2}	0.0004	0.0004	0.0004	0.0004	0.0004	0.0004	0.0004	0.0004	0.0003	0.0004	0.0004	0.0004
	C_{RH}	− 0.0024	− 0.0025	− 0.0028	− 0.0029	− 0.003	− 0.0029	− 0.0029	− 0.0028	− 0.0026	− 0.0026	− 0.0024	− 0.0022
Polytunnel	C_{Tmean}	0.1879	0.2175	0.295	0.2753	0.337	0.2951	0.2975	0.2506	0.2117	0.2511	0.2552	0.2121
	C_{Rn}	0.2829	0.2937	0.3216	0.327	0.3362	0.3313	0.3302	0.3208	0.3151	0.3117	0.3017	0.2835
	C_{u2}	0.0004	0.0004	0.0004	0.0004	0.0004	0.0004	0.0004	0.0004	0.0004	0.0004	0.0004	0.0004
	C_{RH}	− 0.0023	− 0.0024	− 0.0028	− 0.0028	− 0.003	− 0.0029	− 0.0029	− 0.0028	− 0.0027	− 0.0027	− 0.0025	− 0.0023
Shade net house	C_{Tmean}	0.2039	0.2093	0.2885	0.2678	0.2855	0.2703	0.2776	0.2274	0.2166	0.2433	0.2884	0.2224
	C_{Rn}	0.2823	0.2872	0.3149	0.321	0.3237	0.3247	0.3224	0.3121	0.3077	0.3055	0.3044	0.2806
	C_{u2}	0.0011	0.0011	0.0012	0.0011	0.0011	0.0011	0.0011	0.001	0.001	0.0011	0.0012	0.0011
	C_{RH}	− 0.007	− 0.0071	− 0.0081	− 0.0083	− 0.0085	− 0.0085	− 0.0084	− 0.008	− 0.0079	− 0.0078	− 0.0077	− 0.0069
Shadow hall	C_{Tmean}	0.2132	0.2393	0.3242	0.3013	0.3391	0.2807	0.2756	0.2291	0.2195	0.2725	0.3091	0.2413
	C_{Rn}	0.2872	0.2961	0.3242	0.3278	0.3343	0.3264	0.325	0.3159	0.3117	0.3109	0.3095	0.2895
	C_{u2}	0.0007	0.0008	0.0008	0.0008	0.0008	0.0007	0.0007	0.0007	0.0007	0.0008	0.0008	0.0008
	C_{RH}	− 0.0047	− 0.0049	− 0.0056	− 0.0057	− 0.0059	− 0.0057	− 0.0057	− 0.0054	− 0.0053	− 0.0053	− 0.0053	− 0.0048

Table 3.5 Suitability of structures against climatic variables by rank

Structures	Suitability against climatic factors I by rank					
	WSM		WPM		WASPAS	
	S_i	Rank	P_i	Rank	Q_i	Rank
Open	0.31	3	0	3	0.23	3
Polyhouse	0.23	4	0	3	0.17	4
Polytunnel	0.11	5	0	3	0.08	5
Shade net house	0.89	1	0.85	1	0.88	1
Shadow hall	0.64	2	0.62	2	0.64	2

Here, *C* is opposite to *R* and *I*, as for C, a lower value denotes a more preferable, whereas for *R* and *I*, a higher value denotes a more preferable. Similarly, for the final selection of the cultivation structures, *R* and *I* are taken as beneficial as their importance increases from lower to higher values for the cultivation, *C* value is taken as non-beneficial as *C* value is based upon the ranking of structures, lower rank seems best importance, and higher rank seems poor importance for the structures. Weights obtained by CRITIC methods were 0.54, 0.30, and 0.16 for *C*, *R*, and *I*. Similarly, weights were obtained as 0.47, 0.26, and 0.27 by AHP and 0.47, 0.32, and 0.21 by FAHP for the same *C*, *R*, and *I*, respectively. As in all the methods, there is not much difference is found in the variable weights, and for '*R*' and '*I*', CRITIC and FAHP methods give nearly equal weights, whereas for '*C*', AHP and FAHP give equal weights; therefore, in this case the final weights are determined by taking the mean of the weights assigned by using the three methods mentioned above. Therefore, the final weights of the variables *C*, *R*, and *I* were obtained as 0.50, 0.30, and 0.20, respectively. The final ranking of the structures based on WASPAS is described in Table 3.7. The results showed that the shadowed hall is the best suitable structural condition, followed by shade net house, polyhouse, polytunnel, and open condition (least suitable).

Table 3.6 Suitability of structures against climatic factors I, the importance of the rainfall I, and insect infestations (*I*)

Structures	*C*	*R*	*I*
Open	3	0	1
Polyhouse	4	4	5
Polytunnel	5	5	5
Shade net house	1	1	2
Shadow hall	2	4	4

Table 3.7 Final ranking of the cultivation structures based on WASPAS

Structures	Final rank					
	WSM		WPM		WASPAS	
	S_i	Rank	P_i	Rank	Q_i	Rank
Open	0.25	5	0	3	0.19	5
Polyhouse	0.57	3	0.47	2	0.55	3
Polytunnel	0.50	4	0	3	0.38	4
Shade net house	0.61	2	0.47	2	0.58	2
Shadow hall	0.77	1	0.76	1	0.77	1

3.4 Conclusions

The present investigation aimed to compute ET_0 by FAO 56 Penman–Monteith technique and to evaluate the sensitivity of ET_0_FAO-PM to average temperature, solar radiation, wind speed, and relative humidity for the open field as well as in the polyhouse, polytunnel, shade net house, and shadow hall in IIT Kharagpur. A deviation of $\pm$ 25% has been done, and new dataset allowed for ascertaining the coefficient of sensitivity. The minimum and maximum deviations of ET_0 were observed at 2 and 6.5 mm/day for various structures, respectively. In April and May, the maximum values of ET_0 and the minimum values of ET_0 were observed in December. The sensitivity study revealed that a change in ET_0_FAO-PM is very sensitive compared to changes in solar radiation followed by average temperature. The impact of relative humidity and wind speed is the less significant. The sensitivity coefficients for four different types of protected cultivation structures and open field condition vary from 0.25 to 0.34 for solar radiation, from 0.15 to 0.34 for the mean temperature, from − 0.0023 to − 0.025 for the RH, and from 0.0004 to 0.0034 for wind speed, respectively. Therefore, based on the MCDM analysis, we can conclude that the shadowed hall is the best structure, whereas the open condition is the worst for cultivation purposes. This MCDM approach may further be increased by incorporating more inside microclimatic factors, as well as soil parameters and cost economics which will be beneficial for decision-makers to select suitable cultivation structures for crop cultivation and management practices in the future.

References

Agarwal E, Agarwal R, Garg RD, Garg PK (2013) Delineation of groundwater potential zone: an AHP/ANP approach. J Earth Syst Sci 122(3):887–898. https://doi.org/10.1007/s12040-013-0309-8

Allen RG, Pereira LS, Raes D, Smith M (1998) FAO Irrigation and drainage paper No. 56. Food and Agriculture Organization of the United Nations, Rome 56(97):e156. Retrieved from http://www.climasouth.eu/sites/default/files/FAO%2056.pdf

Ambas VT, Baltas E (2012) Sensitivity analysis of different evapotranspiration methods using a new sensitivity coefficient. Glob NEST J 14(3):335–343. https://doi.org/10.30955/gnj.000882

Banihabib ME, Shabestari MH (2017) Fuzzy hybrid MCDM model for ranking the agricultural water demand management strategies in arid areas. Water Resour Manage 31(1):495–513. https://doi.org/10.1007/s11269-016-1544-y

Basak A, Das J, Rahman ATM, Pham QB (2021) An integrated approach for delineating and characterizing groundwater depletion hotspots in a coastal state of India. J Geol Soc India 97(11):1429–1440. https://doi.org/10.1007/s12594-021-1883-z

Djaman K, Tabari H, Balde AB, Diop L, Futakuchi K, Irmak S (2016) Analyses, calibration and validation of evapotranspiration models to predict grass-reference evapotranspiration in the Senegal river delta. J Hydrol Reg Stud 8:82–94. https://doi.org/10.1016/j.ejrh.2016.06.003

Estévez J, Gavilán P, Berengena J (2009) Sensitivity analysis of a Penman-Monteith type equation to estimate reference evapotranspiration in southern Spain. Hydrol Process Int J 23(23):3342–3353. https://doi.org/10.1002/hyp.7439

Fishburn PC (1967) Letter to the editor—additive utilities with incomplete product sets: application to priorities and assignments. Oper Res 15(3):537–542. https://doi.org/10.1287/opre.15.3.537

Geng G, Wardlaw R (2013) Application of multi-criterion decision making analysis to integrated water resources management. Water Resour Manage 27(8):3191–3207. https://doi.org/10.1007/s11269-013-0343-y

Gómez-Limón JA, Berbel J, Arriaza M (2007) MCDM farm system analysis for public management of irrigated agriculture. In: Handbook of operations research in natural resources, Springer, Boston, MA, pp 93–114. https://doi.org/10.1007/978-0-387-71815-6_6

Gong L, Xu CY, Chen D, Halldin S, Chen YD (2006) Sensitivity of the Penman-Monteith reference evapotranspiration to key climatic variables in the Changjiang (Yangtze River) basin. J Hydrol 329(3–4):620–629. https://doi.org/10.1016/j.jhydrol.2006.03.027

Hemming S, Mohammadkhani V, Dueck T (2008) Diffuse greenhouse covering materials-material technology, measurements and evaluation of optical properties. In: International workshop on greenhouse environmental control and crop production in semi-arid regions, vol 797, pp 469–475. https://doi.org/10.17660/ActaHortic.2008.797.68.

Irmak S, Payero JO, Martin DL, Irmak A, Howell TA (2006) Sensitivity analyses and sensitivity coefficients of standardized daily ASCE-Penman-Monteith equation. J Irrig Drain Eng 132(6):564–578. https://doi.org/10.1061/(ASCE)0733-9437(2006)132:6(564)

Kumar T, Gautam AK, Kumar T (2014) Appraising the accuracy of GIS-based multi-criteria decision-making technique for delineation of groundwater potential zones. Water Resour Manage 28(13):4449–4466. https://doi.org/10.1007/s11269-014-0663-6

Liang L, Li L, Zhang L, Li J, Li B (2008) Sensitivity of penman-monteith reference crop evapotranspiration in Tao'er River Basin of northeastern China. Chin Geogra Sci 18(4):340–347. https://doi.org/10.1007/s11769-008-0340-x

Machiwal D, Jha MK, Mal BC (2011) Assessment of groundwater potential in a semi-arid region of India using remote sensing, GIS and MCDM techniques. Water Resour Manage 25(5):1359–1386. https://doi.org/10.1007/s11269-010-9749-y

Mandal T, Saha S, Das J, Sarkar A (2022) Groundwater depletion susceptibility zonation using TOPSIS model in Bhagirathi river basin, India. Model Earth Syst Environ 8(2):1711–1731. https://doi.org/10.1007/s40808-021-01176-7

Mitra R, Das J (2022) A comparative assessment of flood susceptibility modelling of GIS-based TOPSIS, VIKOR, and EDAS techniques in the Sub-Himalayan foothills region of Eastern India. Environ SciPollut Res. https://doi.org/10.1007/s11356-022-23168-5

Mitra R, Saha P, Das J (2022) Assessment of the performance of GIS-based analytical hierarchical process (AHP) approach for flood modelling in Uttar Dinajpur district of West Bengal, India. Geomat Nat Haz Risk 13(1):2183–2226. https://doi.org/10.1080/19475705.2022.2112094

Mosaedi A, Sough MG, Sadeghi SH, Mooshakhian Y, Bannayan M (2017) Sensitivity analysis of monthly reference crop evapotranspiration trends in Iran: a qualitative approach. Theoret Appl Climatol 128(3–4):857–873. https://doi.org/10.1007/s00704-016-1740-y

Ndiaye MP, Bodian A, Diop L, Djaman K (2017) Sensitivity analysis of the Penman-Monteith reference evapotranspiration to climatic variables: case of Burkina Faso. J Water Resour Prot 9:1364–1376. https://doi.org/10.4236/jwarp.2017.912087

Osman SA, Das J (2023) GIS-based flood risk assessment using multi-criteria decision analysis of Shebelle River Basin in Southern Somalia. SN Appl Sci 5:134.https://doi.org/10.1007/s42452-023-05360-5

Pehlivan NY, Paksoy T, Çalik A (2017) Comparison of methods in FAHP with application in supplier selection. In: Fuzzy analytic hierarchy process. Chapman and Hall/CRC, pp 67–98. https://doi.org/10.1201/9781315369884

Saaty TL (2008) Decision making with the analytic hierarchy process. Int J Serv Sci 1(1):83–98. https://doi.org/10.1504/IJSSCI.2008.017590

Singh VK, Tiwari KN, Santosh DT (2016) Estimation of crop coefficient and water requirement of Dutch Roses (Rosa hybrida) under greenhouse and open field conditions. Irrig Drainage Syst Eng 5(169):2. https://doi.org/10.4172/2168-9768.1000169

Tabari H, Talaee PH (2014) Sensitivity of evapotranspiration to climatic change in different climates. Glob Planet Change 115:16–23. https://doi.org/10.1016/j.gloplacha.2014.01.006

Triantaphyllou E, Mann SH (1989) An examination of the effectiveness of multi-dimensional decision-making methods: a decision-making paradox. Decis Support Syst 5(3):303–312. https://doi.org/10.1016/0167-9236(89)90037-7

Triantaphyllou E (2000) Multi-criteria decision making methods. In: Multi-criteria decision making methods: a comparative study. Springer, Boston, MA, pp 5–21. https://doi.org/10.1007/978-1-4757-3157-6

Zavadskas EK, Turskis Z, Antucheviciene J, Zakarevicius A (2012) Optimization of weighted aggregated sum product assessment. Elektronikairelektrotechnika 122(6):3–6. https://doi.org/10.5755/j01.eee.122.6.1810

Zavadskas EK, Antucheviciene J, Saparauskas J, Turskis Z (2013) MCDM methods WASPAS and MULTIMOORA: verification of robustness of methods when assessing alternative solutions. Econom Comput Econom Cybernet Stud Res 47(2):5–20

Zhu F, Zhong PA, Cao Q, Chen J, Sun Y, Fu J (2019) A stochastic multi-criteria decision making framework for robust water resources management under uncertainty. J Hydrol 576:287–298. https://doi.org/10.1016/j.jhydrol.2019.06.049

Check for updates

4 Spatio-Temporal Agricultural Drought Monitoring Using Remote Sensing Indices

Syed Sadath Ali, Koyel Mukherjee, Papia Kundu, and Piu Saha

Abstract

Drought is an intricate weather phenomenon; it directly affects food security and agricultural productivity. Accurate prediction of agricultural drought helps to take mitigation steps for reducing production losses. In the present study, agricultural drought was assessed by using the Normalized Difference Vegetation Index (NDVI), Vegetation Condition Index (VCI), Temperature Condition Index (TCI), and Vegetation Health Index (VHI) based on Landsat 8 and 9 data from 2013–2022. The LULC maps were also prepared using the supervised classification based on the maximum likelihood algorithm by the semi-automatic classification plugin (SCP) in QGIS from Sentinel-2 images. The remote sensing indices were calculated using a raster calculator in ArcGIS software. The results of VCI indicate that 2014 and 2017 years were highly affected by drought, whereas 2016 was the most vulnerable year according to TCI. In 2017, the entire district was badly affected by VCI and TCI. The VHI results showed that 2015, 2016, and 2018 were the most drought-prone years. The spatial agricultural drought result shows that Chattna, Bankura I, Onda, and Ranibudh were extreme drought-affected blocks. Drought greatly impacts agriculture, so satellite-based drought data would benefit the understanding of the drought of Bankura district risk within the entire geographical area.

Keywords

Agricultural drought · Temperature Condition Index · Vegetation Condition Index · Vegetation Health Index · Bankura

S. S. Ali
Civil Engineering Department, Ballari Institute of Technology and Management, Ballari, Karnataka, India

K. Mukherjee
Department of Geography, Rampurhat College, PO. Rampurhat, Dist. Birbhum, Rampurhat, West Bengal 731224, India

P. Kundu · P. Saha (✉)
Department of Geography and Applied Geography, University of North Bengal, PO. North Bengal University, Dist. Darjeeling, Siliguri, West Bengal 734013, India
e-mail: piusaha.nbu@gmail.com

4.1 Introduction

Drought is a recurrent natural hazard that adversely affects the ecosystem, livelihoods, cultivation, and livestock farming (Alam et al. 2023; Ayugi et al. 2022), causing huge economic loss throughout the world (Guo et al. 2021; Zeng et al. 2022). It grows very slowly in the beginning but later it affects a large area with its severity (Liu et al. 2021). Based on physical

J. Das and S. Halder (eds.), *Advancement of GI-Science and Sustainable Agriculture*, GIScience and Geo-environmental Modelling, https://doi.org/10.1007/978-3-031-36825-7_4

aspects, drought is classified as a meteorological, hydrological, and agricultural drought, among which 'agricultural drought' is characterized by insufficient moisture in the soil for cultivation at a particular time (Basak et al. 2022; Das et al. 2020). The intensity of 'agricultural drought' changes with space and time, and it is more challenging than other kinds of drought because it adversely degrades a particular region's agricultural activity. In India, especially in the western part of West Bengal, agricultural drought has a crucial effect on agricultural production and productivity by disturbing the balance between food supply and demand (Gidey et al. 2018). Between 1900 and 2020, a drought event in India had a significant impact on over 1.4 billion people, resulting in a threatening situation for water resources and food security. This event has revealed that agricultural drought affects more than 68% of India's land area (Nath et al. 2017), largely due to the rising trend of mean temperature, geo-environmental conditions, and climate change.

Under this critical situation, appraisal of agricultural drought could be helpful by using different spatio-temporal data from different sources like vegetation, hydrology, meteorology, etc. The spatial and non-spatial datasets are used for drought risk assessment (Apurv and Cai 2021; Hoque et al. 2021a; Kim et al., 2021). Remote sensing techniques are used in spatial analysis to support all the procedures (Hoque et al. 2021b; Zeng et al. 2022). The most popular remote sensing-based vegetation indices, such as Normalized Difference Vegetation Index (NDVI), Temperature Condition Index (TCI), Vegetation Condition Index (VCI), and Vegetation Health Index (VHI) have been used for the drought monitoring system (Hadri et al. 2021; Kogan 1997). The VHI is the most helpful satellite index to monitor agricultural drought (Wang and Yu 2021; Zhang et al. 2013). The Vegetation Health Index correlates with crop yield, the problem of crop health, and crop growth (Alahacoon et al. 2021; Zhao et al. 2022).

NDVI is one of the most popular vegetation health indices that analyze activities like respiration, transpiration productivity, temperature variability, etc. (Pei et al. 2018). For instance, Nejadrekabi et al. (2022) observed the moisture period using NDVI in the Khuzestan province. Vegetation growth in China from 1982 to 2010 was evaluated using NDVI (Peng & Gitelson 2011). Agricultural drought monitoring for three months was calculated using NDVI in Raya of northern Ethiopia (Gidey et al. 2018). This chapter states that VCI and TCI can be used to delineate the seasonal and inter-annual drought, while Sultana et al. (2021) assessed agricultural drought severity in the northwestern part of Bangladesh from 1990 to 2018 by applying TCI, VCI, and VHI. Zambrano et al. (2016) measured the agricultural drought in the cropland of the Biobio region in Chile from 2000 to 2015 by analyzing the temporal and spatial variation of vegetation conditions with stress due to scarcity of rainfall with VCI.

The novelty of this current endeavor is to identify and monitor agricultural drought in the Bankura district of West Bengal using remote sensing data. Evaluation of drought indices, calculation intensity, severity, and duration of drought are the prime concern of this chapter which helps make a proper plan for mitigation and irrigation practice in drought-vulnerable areas through establishing an integrated relationship among NDVI, LST, VCI, TCI, and VHI methods which provide a guideline for future drought.

4.2 Study Area

Bankura is the fourth largest district of West Bengal, located between 22°30′N to 23°30′N latitudes and 87°00′E to 87°30′E longitudes, having a 6,882 km^2 area. The total population of the Bankura district is 3,992,309 persons, and the population density is 523 persons/km^2 (Census 2011). Bankura is a connecting link between the plain of West Bengal and the Plateau of Chotanagpur. The Purulia district in the west surrounds it, Purba Bardhhaman and Paschim Bardhhaman districts in the north, Jhargram and Paschim Medinipur in the south, and the Hugli district in the southeast. Darakeswar, Damodar,

Kangsabati, Silabati, and Gandhewari rivers drain the district. Geomorphologically, the Bankura district is a part of the Chotanagpur plateau (Fig. 4.1). There are three types of topographical terrain; i.e., the western part is a hilly region characterized by large granite rock covered by natural vegetation, the central part is undulating and characterized by red lateritic, and the eastern part is an alluvial plain covered by loamy soil. Following the topographic terrain, land use pattern of this area is also changing from east to west, and the low-lying alluvial plain of the northeast is mainly used for paddy cultivation. The western surface is undulating and gradually rising, so most of the land is covered by jungle. Bankura is part of Rarh, and agriculture is the main economic activity of this concerned study area, but it is challenged by low water availability, climatic change, and reduced annual rainfall. In the last few years, drought incidents and intensity have been increasing (Bhunia et al. 2020; Das et al. 2013). The changing patterns of annual rainfall and 80% rainfall received during four months result in poor moisture in subsoil, which becomes a threat to crops and seriously affects the yields in the study area. The farmers of this area face some socio-economic problems; they lose their job and are forced to migrate. In recent times, Bankura district has become a geographer's attraction due to excessive drought proneness and its relation with the economy, poverty, mitigation, and migration-related scenarios (Raha and Gayen 2020).

4.3 Database and Methods

4.3.1 Database

In the present study, remote sensing data has been utilized from authenticated sources, and therefore, various drought indices are displayed by ESRI ArcGIS (Version 10.4.1) software.

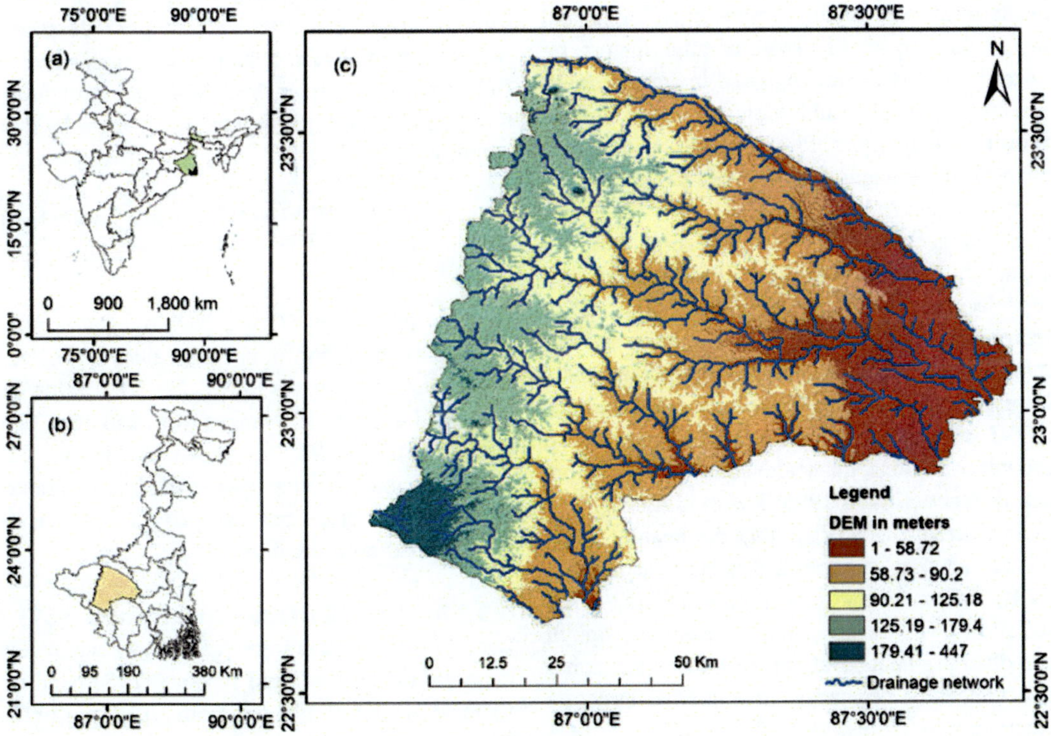

Fig. 4.1 Location map of the study area; **a** India, **b** West Bengal, and **c** Bankura

Landsat 8, Landsat 9 OLI/TIRS collection 2, level 1 images (Path: 139, Row: 044) were used for NDVI, VCI, LST, TCI, and VHI indices, obtained from the USGS Earth explorer website (https://earthexplorer.usgs.gov). The land use/land cover maps were prepared from Sentinel 2 images downloaded from https://scihub.copernicus.eu using a training sample and the maximum likelihood method in a QGIS semi-automatic classification plugin (SCP).

4.3.2 Methods

4.3.2.1 Normalized Difference Vegetation Index (NDVI)

The Normalized Difference Vegetation Index (NDVI) is a widely used remote sensing index to assess vegetation density and health. NDVI measures the difference between the reflectance of near-infrared (NIR) and visible red (VIS) light, which is correlated with the amount of vegetation present in an area (Glenn and Tabb 2019). NDVI can be used to monitor vegetation growth and health over time, detect changes in land use, and assess the impact of environmental factors on vegetation. It is commonly used in agriculture to assess crop health and yield potential and in forestry to monitor forest health and detect changes due to natural or man-made disturbances (Nejadrekabi et al. 2022).

Every geographical space has some carrying capacity (Kogan 1995). For estimating carrying capacity, we used NDVI. The maximum NDVI represents the highest carrying capacity, and the minimum NDVI represents a geographical area's lowest carrying capacity (ecosystem potential). NDVI also helps monitor crop yields, crop growth conditions, the health status of vegetation, and drought (Kogan 1995; Liu et al. 2021). The main concept of NDVI is that the healthy green leaves' internal mesophyll reflects near-infrared (NIR), whereas a large proportion of visible red radiation (VIS) is absorbed by leaf chlorophyll and other pigments. But in the case of water stress and unhealthy vegetation, the internal structure reacts reverse (Moisa et al. 2022).

$$\mathrm{NDVI} = (\mathrm{NIR} - \mathrm{RED})/(\mathrm{NIR} + \mathrm{RED}).$$

NDVI is calculated between the difference between near-infrared (NIR) and visible red bands of the electromagnetic spectrum. The index ranges from -1 to $+1$, with values closer to $+1$ indicating higher levels of healthy vegetation and values closer to -1 indicating little to no vegetation. In tropical and temperate rain forests, the value of NDVI ranges between 0.6 to 0.8, and in barren rock, sand, or snow area, it is below 0.1 (Dutta et al. 2015). There are some noise problems in NDVI. Sensor degradation, satellite change, change of satellite orbital drift, cloud, and aerosol are the sources of error (Kogan 1995). These weather-related NDVI problems must be overcome, and thus why Kogan (1995) suggested the Vegetation Condition Index (VCI).

4.3.2.2 Vegetation Condition Index (VCI)

The Vegetation Condition Index (VCI) is derived from remote sensing data developed for monitoring drought characteristics such as duration, intensity, spatial extent, and severity assessment. In this present chapter, VCI was used to monitor the Bankura district's agricultural drought by this equation.

$$\mathrm{VCI} = (\mathrm{NDVI} - \mathrm{NDVI_{MIN}})/(\mathrm{NDVI_{MAX}} - \mathrm{NDVI_{MIN}}) * 100,$$

whereas NDVI, $\mathrm{NDVI_{MIN}}$, $\mathrm{NDVI_{MAX}}$ are multiyear maximum and minimum values of NDVI. According to Kogan (1995), the VCI value is measured in percentile ranges from 0 to 100. The classification of VCI is shown in Table 4.2. When the value is near 100, it defines favorable condition for crop, but the value 0 or near 0 indicate bad crop condition.

4.3.2.3 Land Surface Temperature (LST)

Land surface temperature (LST) was calculated from the thermal infrared sensor (TIRS) band of Landsat 8 and 9 images from 2013 to 2022. The LST value ranges between 7500 and 65,535

Table 4.2 Detailing the threshold value of VCI, TCI, and VHI

Range	Dryness level
0–10	Extreme drought
10–20	Severe drought
20–30	Moderate drought
30–40	Light drought
> 40	No drought

(Wan 2006), and it was reclosed by 0.02 to convert into Kelvin unit. It represents the radiative skin temperature of the land surface from solar radiation. In this chapter, LST was converted and rescaled into degree Celsius.

LST is gained by these equations.

$$LST = (BT/(1 + (\lambda * BT/\rho) * Ln(\varepsilon))) - 273.15,$$

where LST = Land surface temperature in Celsius (°C).

BT = Sensor brightness temperature in (°C).

λ = Wavelength of thermal band of various Landsat satellite.

ε = Emissivity of the land surface.

$\rho = (h \times (c/\sigma))$, which is equal to 1.438×10^{-2} mK.

In which, σ is the Boltzmann constant (1.380649×10^{-23} J/K), h is Plank's constant ($6.62607015 \times 10^{-34}$ J.s), and c is the velocity of light (3 × 108 m/s).

4.3.2.4 Temperature Condition Index (TCI)

The Temperature Condition Index (TCI) is a remote sensing index that provides an estimation of the vegetation's response to temperature stress. It measures the deviation of land surface temperature (LST) from its long-term average value and is based on the assumption that vegetation is sensitive to temperature anomalies (Swain et al. 2011). TCI is obtained by this equation.

$$TCI = (LST_{MAX} - LST)/(LST - LST_{MIN}) * 100,$$

where LST is the value of the land surface temperature of a particular month, and LST_{MAX} and LST_{MIN} is the temperature of the studying period. LST provides information about the vegetation of the area. If LST increases, then the evapotranspiration of plants also increases, and surface soil moisture also reduces, which is a good indicator of vegetation stress (Kogan 1995; Seiler et al. 1998). The TCI value ranges between 0 and 100. A high value of TCI indicates a favorable condition for a crop, whereas a low value of TCI indicates an adverse effect on vegetation or drought conditions.

4.3.2.5 Vegetation Health Index (VHI)

Vegetation Health Index (VHI) is the outcome of the combination of products extracted from vegetation signals, namely NDVI. It combines VCI and TCI (Orlovsky et al. 2011).

$$VHI = a \times VCI + (1 - a) \times TCI,$$

where VHI represents the vegetation health index, $a = 0.5$ similar contribution of VCI and TCI, VCI is the vegetation condition index, and TCI is the temperature condition index.

4.4 Results and Discussion

4.4.1 Land Use and Land Cover (LULC) Changes

The supervised classification with maximum algorithm method was employed to prepare land use and land cover (LULC) maps using Sentinel 2 images in the QGIS SCP plugin (Patil et al.

2012). The maps were categorized into six major types, including water body, vegetation cover, agricultural land, build-up area, bare ground, and range land, for the years 2018, 2019, 2020, 2021, and 2022, as depicted in Fig. 4.2 and Table 4.3.

The areal coverage or extension and areal changes from 2018 to 2022 have been detected through ArcGIS software, and it has been summarized in (Table 4.3). The LULC classification in 2018 (Fig. 4.2a) depicts that the majority of the area in Bankura district was under agricultural land (64.28%), the rest 18.24, 8.04, 2.72, and 1.10% areas are under vegetation cover, range land, build-up area, water body, and bare ground, respectively. Similarly, in 2019, the greatest share of land was occupied by also agricultural land (64.21%), and the trend of occupancy remained the same, i.e., land under vegetation cover (18.53%), range land (8.04%), build-up area (5.62%), water body (2.72%), and bare ground (1.02%) individually. Considering the trend of extension and rate (in %) of changes of each LULC from 2018 to 2022, vegetation cover and build-up areas increased by 3.90%, partly in Barjora, Sonamukhi, Taldangra, Onda, Vishnupur, Ranibundh blocks of this district and 27.21% in some parts of blocks like Indus, Kotalpur, Patrasayar, Jaypur, Vishnupur, Bankura I sequentially, whereas percentage of land occupancy in bare ground, range of land, and water body decreased by − 32.555 and − 19.75% partly in Mejhia, Gangajalghati, Chhatna, and − 9.84% at Ranibundh and Hirbundh blocks in the same period. Taking into consideration the overall study period, vegetation cover and build-up areas have shown their areal increment. In contrast, water bodies, agricultural land, bare ground, and range land have harshly diminished in the same period due to many unscientific activities like unplanned settlement, massive grazing, and resultant soil degradation. Unlike build-up areas and vegetation cover, the land share of water bodies, agricultural land, bare ground, and range land has been increased. It is also a remarkable point that build-up areas has increased in a far larger percentage than vegetation cover, proving that most of the bare ground, range land, water body, and agricultural land are

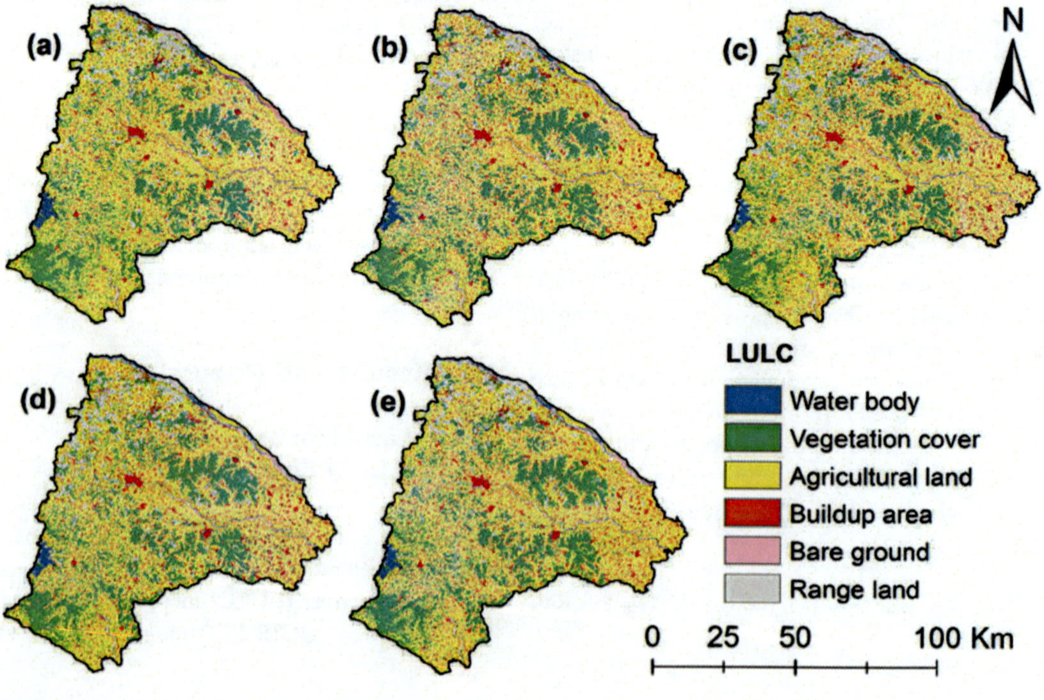

Fig. 4.2 Land use and land cover (LULC); **a** 2018, **b** 2019, **c** 2020, **d** 2021, and **e** 2022

Table 4.3 Land use and land cover (LULC) extent and change detection between 2018 and 2022

Sl No	Land use	2018	2019	2020	2021	2022	% of change
1	Water body	2.72	2.36	1.98	2.72	2.45	− 9.84
2	Vegetation cover	18.24	18.53	17.22	19.64	18.95	3.90
3	Agricultural land	64.28	64.21	63.04	59.82	64.25	− 0.04
4	Build-up area	5.62	6.71	6.86	7.88	7.15	27.21
5	Bare ground	1.10	1.02	0.91	0.79	0.74	− 32.55
6	Range land	8.04	7.17	10.00	9.16	6.46	− 19.75

converted to build-up areas but rationally maintains the vegetation-covered area by not unscientifically destruction of trees and continue the afforestation program. Considering the overall study period, the pattern of land use changes demonstrates that land occupancy of water body was increased from 2018 to 2019, then decreased in 2020, and again increased in 2021, but a little bit decreased in 2022. Land share of vegetation cover increased from 2018 to 2019, then decreased and continue to increase in 2021 which decreased in 2022. Agricultural land area decreased from 2018 to 2019, 2020, and 2021 but increased in 2022. In the build-up area, the land occupancy increased from 2018 to 2021 but decreased in 2022. The area under bare and range land has gradually reduced from 2018 to 2019, and so on. Generally, the result has revealed that a series of LULC changes in the study area for five years (2018–2022) shows the fact that build-up areas are most dominating in this area, indicating the continuous increment of human residence by maintaining green areas, whereas other land use pattern shows declining nature.

4.4.2 Normalized Difference Vegetation Index (NDVI)

It measures the photosynthetic activities of vegetation by indicating favorable vegetation conditions with high value and unfavorable vegetation conditions associated with low NDVI value (Cunha et al 2015). Five principal changes in vegetation (significant vegetation loss, vegetation loss, no vegetation change, vegetation gain, significant vegetation gain) have been detected from 2013 to 2022. Significant vegetation loss, which accounts for 5.42% of the total area, has been partially detected from 2013 to 2022 in Barjora, Sonamukhi, Patrasayar, Indus, Vishnupur, Sarenga, Raipur, Khatra blocks. The most worrying fact is that 48.17% vegetation loss in Mejhia, Gangajalpati, Barjora, Sonamukhi, Patrasayar, and Indus in the northeastern part of Bankura District, Bankura I and II, Onda in the middle part and Ranibundh, Raipur, Serenga in the southern part of this district is registered due to different unscientific construction work. In the meantime, about 12.64% area of the southwestern part is marked as an unchanged vegetation-covered area. Interestingly, 33.45% area has gained vegetation, whereas only 0.32% area has gained significant vegetation (Fig. 4.3 and Table 4.4).

This chapter has monitored the agricultural drought of the Bankura district from the year 2013 to 2022 by using the VCI technique. Figure 4.4 and Table 4.5 depict area-wise extreme, severe, moderate, and no drought conditions for ten years. Meanwhile, 79.65% (most of the area) area of this district was under extreme drought conditions in 2017 due to erratic rainfall and the low water-holding capacity of the soil. Community development blocks like Indus, Kotulpur, Jaypur, and Serenga accounted for extreme drought in 2014, 2015, and 2018. The maximum area (14.37%) under severe drought was in 2014,

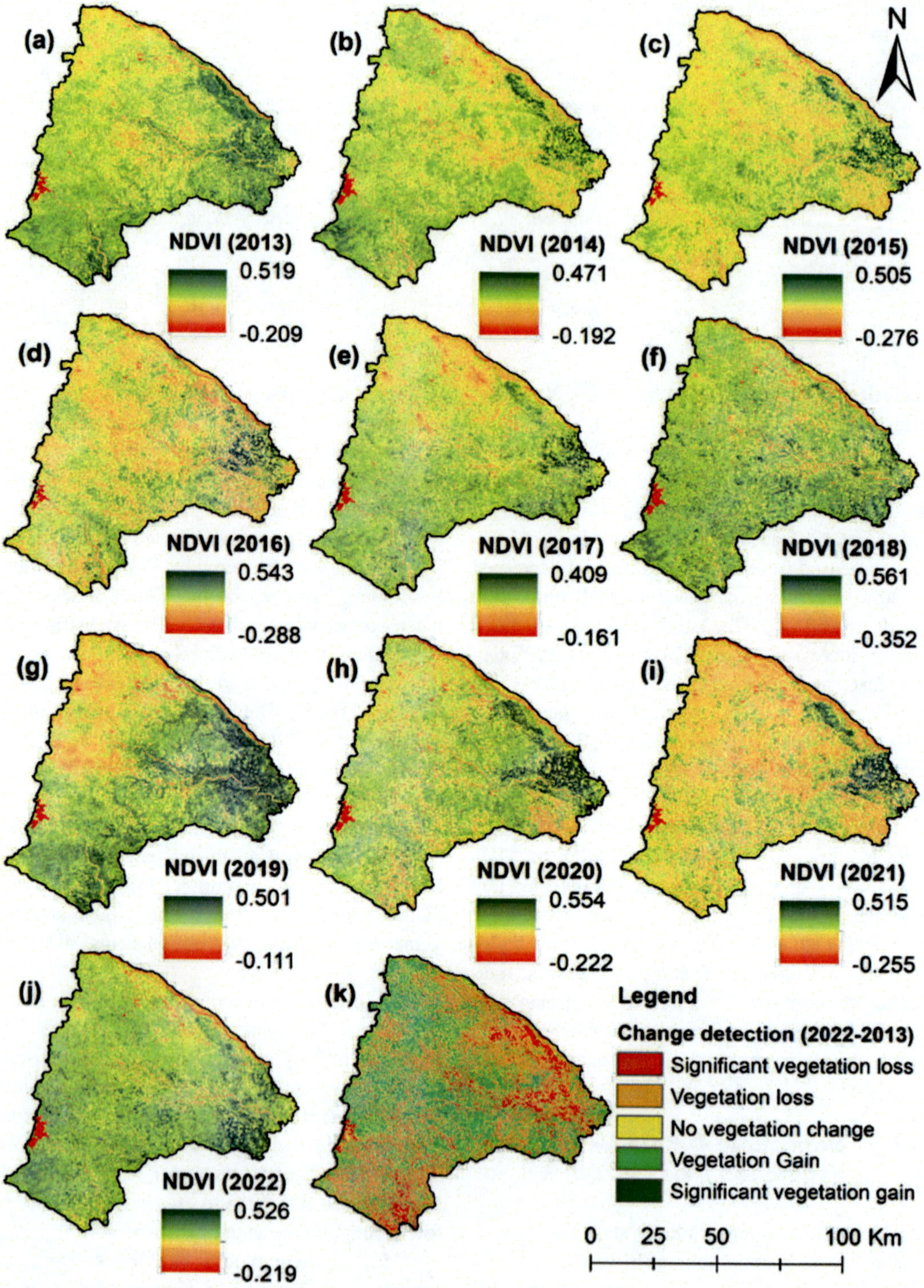

Fig. 4.3 Normalized Difference Vegetation Index (NDVI); **a** 2013, **b** 2014, **c** 2015, **d**, 2016, **e** 2017, **f** 2018, **g** 2019, **h** 2020, **i** 2021, **j** 2022, and **k** Change detection

Table 4.4 Vegetation change detection between 2013 and 2022

Sl. No	Level	Area in sq km	Area in %
1	Significant vegetation loss	373.75	5.42
2	Vegetation loss	3319.68	48.17
3	No vegetation change	871.00	12.64
4	Vegetation gain	2305.29	33.45
5	Significant vegetation gain	21.79	0.32

whereas 18.58 and 18.24% areas of the total were affected by moderate and light drought, respectively, in 2014, and a large area (95.50%) was registered as no drought-affected area in 2019 due to a sufficient amount of rainfall.

Figure 4.5 and Table 4.6 have assessed descriptive statistics (minimum, maximum, and mean temperature in °C) and standard deviation by considering ten years from 2013 to 2022 with the association of the LST technique. It illustrates that the lowest minimum temperature was recorded as 4.95 °C in 2014 and 25.40 °C in 2019. On the other hand, the maximum land surface temperature was 53.83 °C in 2018 and 35.43 °C in 2014. Low land surface temperature indicates dense vegetation cover and a low infiltration rate with minimum soil moisture, whereas high land surface temperature reveals no or thin vegetation cover with a high infiltration rate and maximum soil moisture. The highest mean temperature was recorded as 34.13 in 2016, and the lowest was 22.46 in 2014, respectively. The highest SD was recorded as 3.65 in 2014 and 1.79 in 2019, respectively. Their mean value shows the average fluctuation of temperature throughout ten years by calculating the average value for each year separately, whereas standard deviation shows the consistency among the distribution of mean temperature delicately.

In this chapter, TCI was calculated from Landsat 8, 9, and TIRS band 10 to categorize agricultural drought in five categories, i.e., extreme, severe, moderate, light, and no drought for the above said ten years. In 2016, 7.92% area of Bankura district was under extreme drought, the intensity of which reduced through rest years and remarkably in 2017 to 0.01%. While community development blocks like Chattna, Bankura I, and II, Taldangra, Simlapal, Khatra, Ranibundh, etc., were affected. On the other hand, 32.33% of the area in 2015 was identified as a severely drought-affected area, the areal extent of which was lowered to 0.01% in 2014. 29.65% area, including Chattna, Indpur, Bankura I, and Onda blocks, were under moderate drought-affected areas. 25.98 and 0.05% areas were recognized as the highest and lowest light drought-prone areas, respectively. 99.89% area of the Bankura district was not faced with drought in 2014, but it was reduced to 2.39% in 2017 (Fig. 4.6 and Table 4.7).

VHI, a combined indicator of vegetation health, depicts spatio-temporal drought variation in the Bankura district from 2013 to 2022 and is classified into five types. Figure 4.7 and Table 4.8 demonstrate that 18.96% area, including Chattna, Bankura I, Onda, and Ranibudh blocks, was extremely drought in 2016. This condition improved in 2014 by showing a 0.02% area under this adverse condition. The severe drought area was 27.92% in 2018 and gradually reduced to 0.04% in 2014. This fact clearly shows the improvement of drought conditions by applying canal irrigation, drilling irrigation, submersible

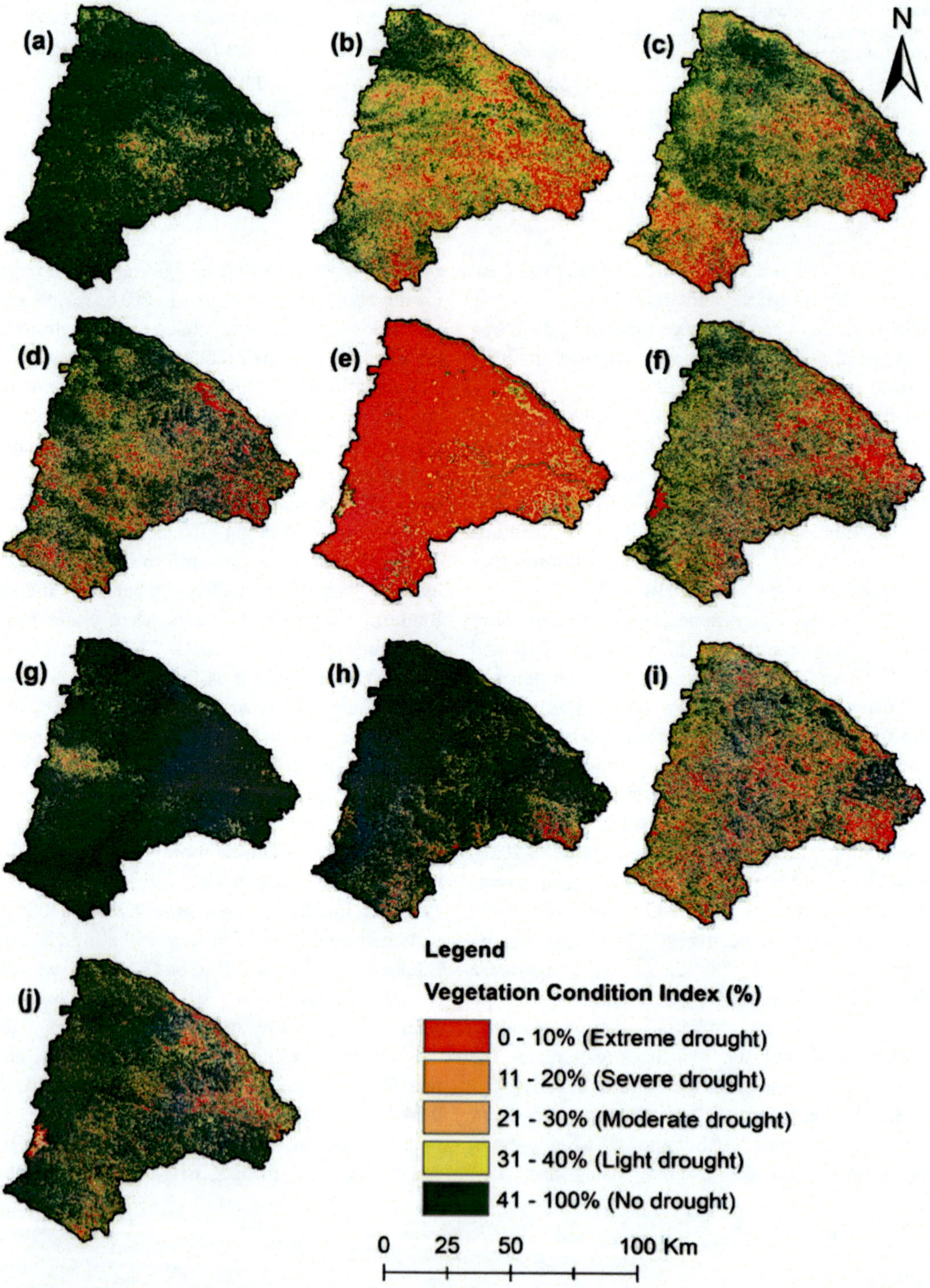

Fig. 4.4 Drought monitoring using Vegetation Condition Index (VCI); **a** 2013, **b** 2014, **c** 2015, **d** 2016, **e** 2017, **f** 2018, **g** 2019, **h** 2020, **i** 2021, and **j** 2022

Table 4.5 Spatio-temporal drought variation using Vegetation Condition Index (VCI)

Year	Extreme drought		Severe drought		Moderate drought		Light drought		No drought	
	Area		Area		Area		Area		Area	
	Sq. km	(%)	Sq. km	(%)	Sq. km	(%)	Sq. km	(%)	Sq. km	(%)
2013	55.51	0.81	90.77	1.32	198.65	2.88	356.79	5.18	6189.79	89.82
2014	1249.24	18.13	990.33	14.37	1280.26	18.58	1257.33	18.24	2114.34	30.68
2015	1075.03	15.60	920.88	13.36	1058.05	15.35	1128.15	16.37	2709.39	39.31
2016	833.95	12.10	631.40	9.16	712.77	10.34	777.86	11.29	3935.53	57.11
2017	5489.40	79.65	628.74	9.12	380.27	5.52	201.14	2.92	191.96	2.79
2018	1028.42	14.92	748.81	10.87	1070.95	15.54	1207.80	17.53	2835.52	41.15
2019	16.42	0.24	36.16	0.52	88.92	1.29	168.66	2.45	6581.36	95.50
2020	146.08	2.12	173.31	2.51	203.32	2.95	246.33	3.57	6122.48	88.84
2021	1086.36	15.76	980.26	14.22	1097.40	15.92	1034.36	15.01	2693.13	39.08
2022	427.40	6.20	344.26	5.00	466.24	6.77	611.26	8.87	5042.35	73.17

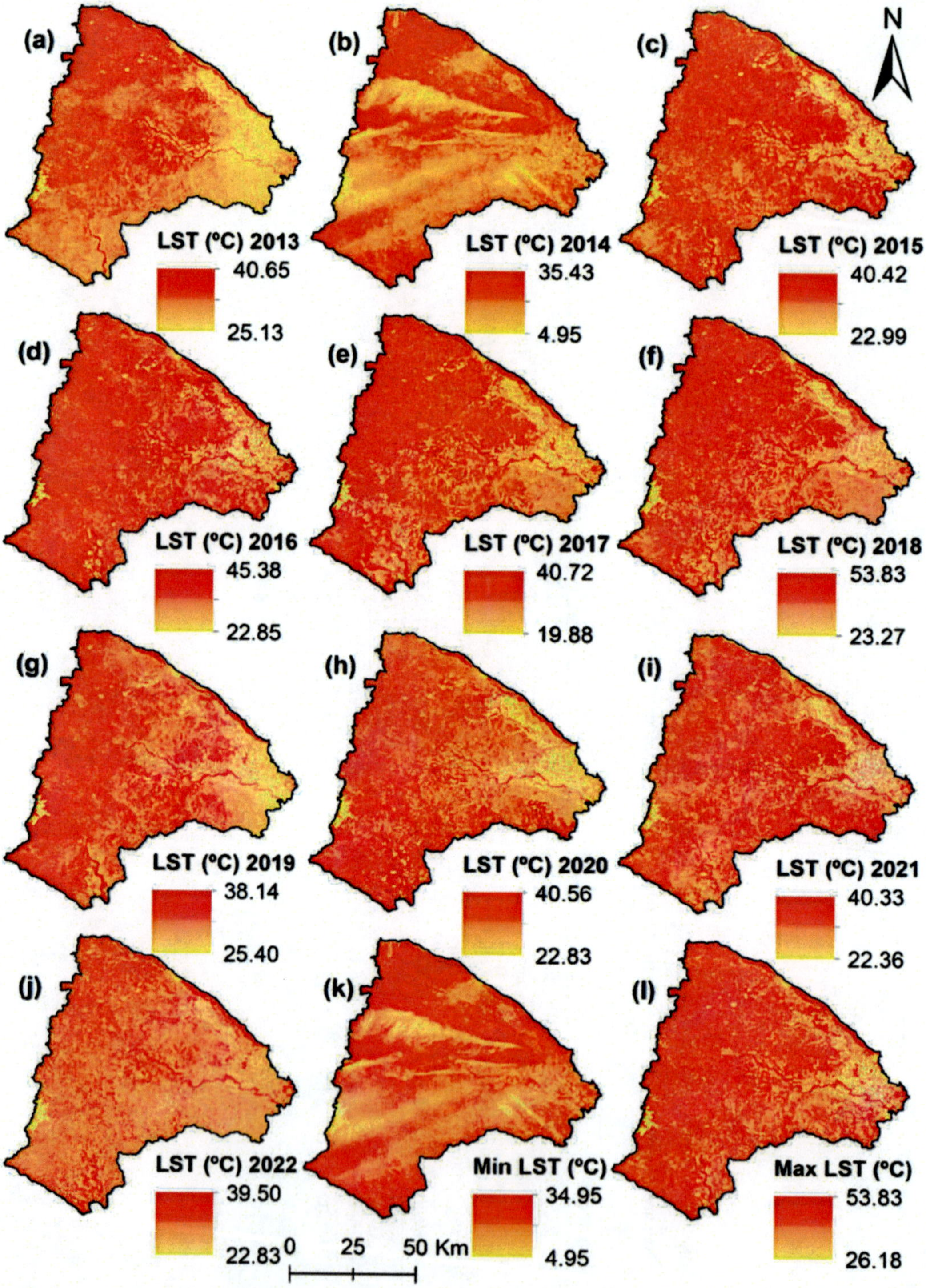

Fig. 4.5 Land surface temperature (LST); **a** 2013, **b** 2014, **c** 2015, **d** 2016, **e** 2017, **f** 2018, **g** 2019, **h** 2020, **i** 2021, **j** 2022, **k** minimum temperature, and **j** maximum temperature

Table 4.6 Descriptive statistics of land surface temperature from 2013 to 2022

Year	Minimum	Maximum	Mean	SD
2013	25.13	40.65	31.56	2.26
2014	4.95	35.43	22.46	3.65
2015	22.99	40.42	32.34	2.37
2016	22.85	45.38	34.13	2.78
2017	19.88	40.72	28.44	2.08
2018	23.27	53.83	33.36	2.88
2019	25.40	38.14	31.44	1.79
2020	22.83	40.56	29.84	2.22
2021	22.36	40.33	31.34	2.22
2022	22.84	39.50	29.19	1.83

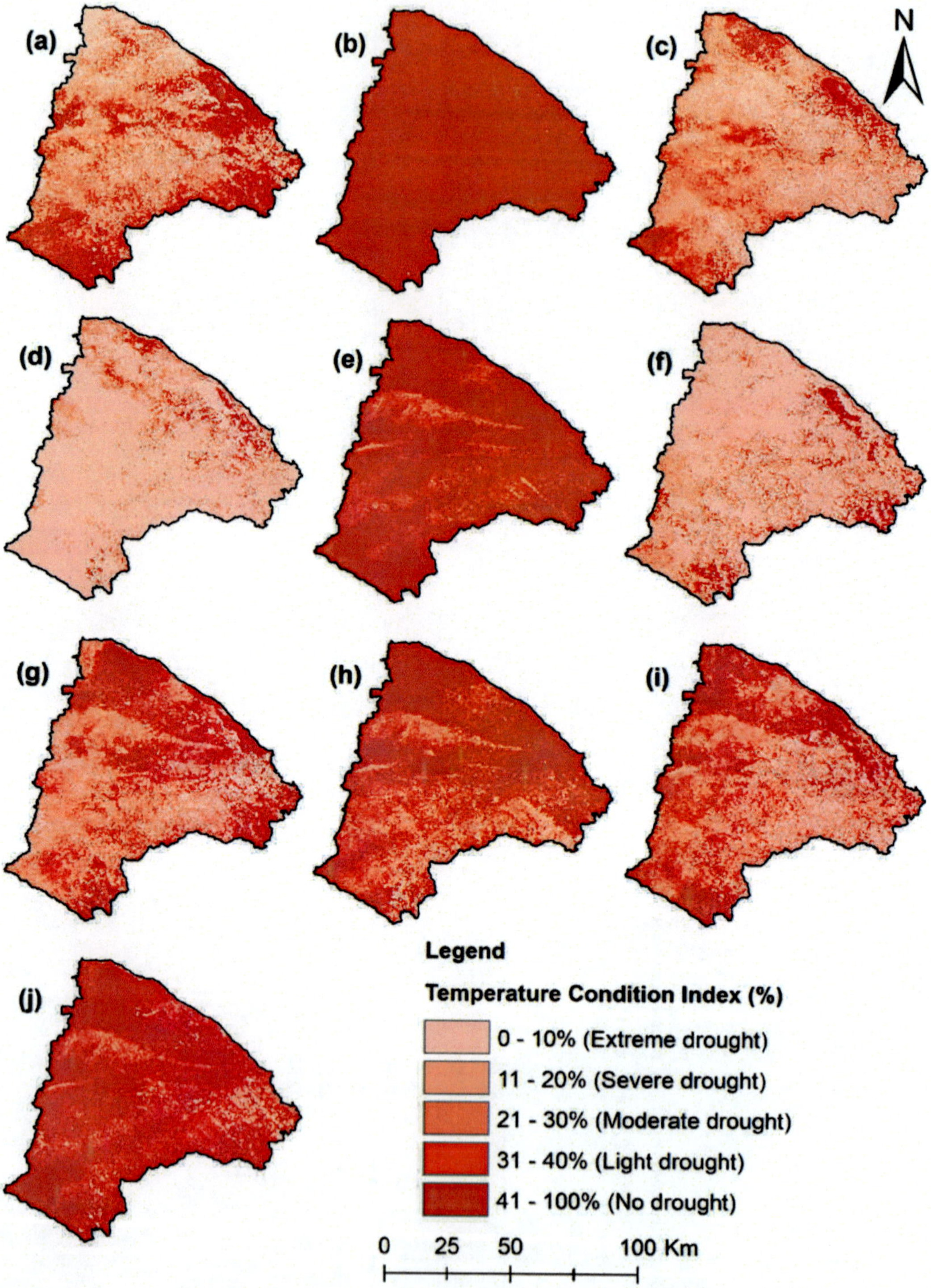

Fig. 4.6 Drought monitoring using Temperature Condition Index (TCI); **a** 2013, **b** 2014, **c** 2015, **d** 2016, **e** 2017, **f** 2018, **g** 2019, **h** 2020, **i** 2021, and **j** 2022

Table 4.7 Spatio-temporal drought variation using temperature condition index (TCI)

Year	Extreme drought		Severe drought		Moderate drought		Light drought		No drought	
	Area		Area		Area		Area		Area	
	Sq. km	(%)	Sq. km	(%)	Sq. km	(%)	Sq. km	(%)	Sq. km	(%)
2013	1070.19	15.53	1629.41	23.64	1749.54	25.39	1137.66	16.51	1304.71	18.93
2014	1.06	0.02	0.91	0.01	1.83	0.03	3.48	0.05	6884.23	99.89
2015	1363.97	19.79	2228.30	32.33	2043.04	29.65	783.74	11.37	472.47	6.86
2016	5232.01	75.92	881.13	12.79	428.21	6.21	185.25	2.69	164.90	2.39
2017	0.86	0.01	4.25	0.06	201.21	2.92	1208.40	17.53	5476.80	79.47
2018	4031.84	58.50	1462.82	21.23	699.42	10.15	317.08	4.60	380.35	5.52
2019	634.80	9.21	1515.30	21.99	1932.91	28.05	1315.07	19.08	1493.43	21.67
2020	97.43	1.41	359.93	5.22	1352.10	19.62	1790.62	25.98	3291.43	47.76
2021	709.47	10.29	1382.04	20.05	1906.45	27.66	1417.24	20.57	1476.30	21.42
2022	64.33	0.93	175.87	2.55	619.79	8.99	1502.11	21.80	4529.42	65.72

irrigation, etc. While, 39.73% area in 2017, including partly Indpur, Onda, Vishnupur, Taldangra, Khatra, Simlapal, the northern part of Ranibundh, Raipur blocks were recognized as the highest percentage of moderate drought-affected area and reduced at 0.16% in 2014. 24.84% area and was accounted for light drought-affected area in 2017. Most areas in this district faced no drought in 2014, which (positively) increased substantial water sources for agriculture.

4.5 Conclusion

In conclusion, the study presents a comprehensive analysis of spatio-temporal agricultural drought monitoring using remote sensing indices in Bankura district of West Bengal. The results show that severe drought conditions have affected various parts of the study area, leading to the abandonment of agricultural lands due to water and soil moisture scarcity. The study identifies specific areas that require urgent attention, including Chattna, Bankura I, Onda, Ranibudh, Indus, Kotulpur, Jaypur, and Serenga blocks, where water shortage is a significant concern for sustainable agricultural practices. The findings suggest that strengthening water resource infrastructure, adopting agricultural water-saving technologies, and promoting seasonal rainwater harvesting could mitigate the impacts of drought in the region. Moreover, the study recommends the implementation of sustainable drought policies, including alteration of sowing and planting times, preservation agriculture, and zero tillage, modification of agricultural practices, to improve resilience toward the effects of drought. The study's results could inform policymakers, farmers, and other stakeholders in addressing local and regional drought issues, and prospective researchers could use them to advance knowledge in the field. Overall, the study provides valuable insights into the spatio-temporal dynamics of agricultural droughts and underscores the need for a comprehensive approach toward managing drought risks.

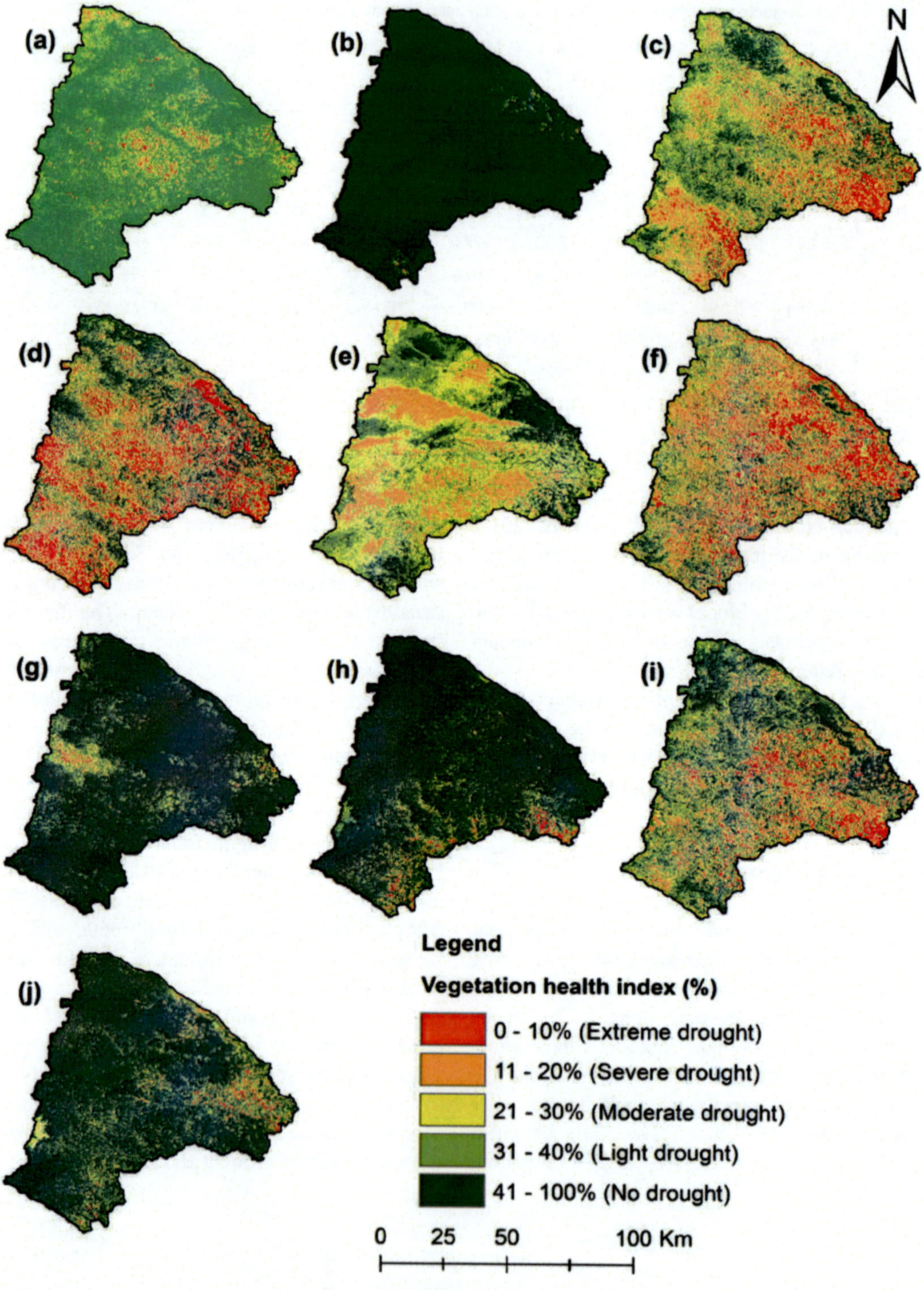

Fig. 4.7 Drought monitoring using Vegetation Health Index (VHI) **a** 2013, **b** 2014, **c** 2015, **d** 2016, **e** 2017, **f** 2018, **g** 2019, **h** 2020, **i** 2021 and **j** 2022

Table 4.8 Spatio-temporal drought variation using Vegetation Health Index (VHI)

Year	Extreme drought		Severe drought		Moderate drought		Light drought		No drought	
	Area		Area		Area		Area		Area	
	Sq. km	(%)	Sq. km	(%)	Sq. km	(%)	Sq. km	(%)	Sq. km	(%)
2013	51.87	0.75	201.21	2.92	556.18	8.07	1104.93	16.03	4977.32	72.22
2014	1.62	0.02	2.96	0.04	10.77	0.16	33.51	0.49	6842.65	99.29
2015	690.57	10.02	1477.71	21.44	1916.88	27.82	1570.98	22.80	1235.37	17.93
2016	1306.43	18.96	1382.12	20.06	1430.71	20.76	1186.07	17.21	1586.17	23.02
2017	3.73	0.05	1167.35	16.94	2737.72	39.73	1711.64	24.84	1271.07	18.44
2018	1051.92	15.26	1923.97	27.92	1922.39	27.90	1078.19	15.65	915.04	13.28
2019	13.90	0.20	72.76	1.06	256.25	3.72	674.78	9.79	5873.82	85.23
2020	68.91	1.00	153.92	2.23	252.58	3.67	454.08	6.59	5962.02	86.51
2021	447.32	6.49	1153.87	16.74	1681.43	24.40	1605.22	23.29	2003.67	29.07
2022	59.99	0.87	202.81	2.94	527.86	7.66	855.45	12.41	5245.41	76.11

References

Alahacoon N, Edirisinghe M, Ranagalage M (2021) Satellite-based meteorological and agricultural drought monitoring for agricultural sustainability in Sri Lanka. Sustainability 13(6):3427. https://doi.org/10.3390/su13063427

Alam J, Saha P, Mitra R, Das J (2023) Investigation of spatio-temporal variability of meteorological drought in the Luni River Basin, Rajasthan, India. Arab J Geosci 16(3):201. https://doi.org/10.1007/s12517-023-11290-8

Apurv T, Cai X (2021) Regional drought risk in the contiguous United States. Geophys Res Lett 48(5): e2020GL092200-1–e2020GL092200-12. https://doi.org/10.1029/2020GL092200

Ayugi B, Eresanya EO, Onyango AO, Ogou FK, Okoro EC, Okoye CO, Ongoma V et al (2022) Review of meteorological drought in Africa: historical trends, impacts, mitigation measures, and prospects. Pure Appl Geophys 179(4):1365–1386

Basak A, Rahman ATMS, Das J, Hosonod T, Kisi O (2022) Drought forecasting using the Prophet Model in semi-arid climate region of western India. Hydrol Sci J 67(9):1397–1417. https://doi.org/10.1080/02626667.2022.2082876

Bhunia P, Das P, Maiti R (2020) Meteorological drought study through SPI in three drought prone districts of West Bengal, India. Earth Syst Environ 4(1):43–55. https://doi.org/10.1007/s41748-019-00137-6

Census of India (2011) District census handbook, Bankura, Government of India

Cunha APM, Alvalá RC, Nobre CA, Carvalho MA (2015) Monitoring vegetative drought dynamics in the Brazilian semiarid region. Agric for Meteorol 214–215:494–505

Das S, Choudhury MR, Nanda S (2013) Geospatial assessment of agricultural drought (a case study of Bankura District, West Bengal). Int J Agric Sci Res (IJASR) 3(2):1–27

Das J, Gayen A, Saha P, Bhattacharya SK (2020) Meteorological drought analysis using Standardized Precipitation Index over Luni River Basin in Rajasthan, India. SN Appl Sci 2(9):1–17. https://doi.org/10.1007/s42452-020-03321-w

Dutta D, Kundu A, Patel NR, Saha SK, Siddiqui AR (2015) Assessment of agricultural drought in Rajasthan (India) using remote sensing derived Vegetation Condition Index (VCI) and Standardized Precipitation Index (SPI). Egypt J Remote Sens Space Sci 18(1):53–63. https://doi.org/10.1016/j.ejrs.2015.03.006

Gidey E, Dikinya O, Sebego R, Segosebe E, Zenebe A (2018) Analysis of the long-term agricultural drought onset, cessation, duration, frequency, severity and spatial extent using Vegetation Health Index (VHI) in Raya and its environs, Northern Ethiopia. Environ Syst Res 7(1):1–18

Glenn DM, Tabb A (2019) Evaluation of five methods to measure normalized difference vegetation index (NDVI) in apple and citrus. Int J Fruit Sci 19 (2):191–210

Guo H, Wang R, Garfin GM, Zhang A, Lin D (2021) Rice drought risk assessment under climate change: based on physical vulnerability a quantitative assessment method. Sci Total Environ 751:141481. https://doi.org/10.1016/j.scitotenv.2020.141481

Hadri A, Saidi MEM, Boudhar A (2021) Multiscale drought monitoring and comparison using remote sensing in a Mediterranean arid region: a case study from west-central Morocco. Arab J Geosci 14(2):1–18. https://doi.org/10.1007/s12517-021-06493-w

Hoque M, Pradhan B, Ahmed N, Alamri A (2021a) Drought vulnerability assessment using geospatial techniques in Southern Queensland, Australia. Sensors 21(20):6896. https://doi.org/10.3390/s21206896

Hoque MAA, Pradhan B, Ahmed N, Sohel MSI (2021b) Agricultural drought risk assessment of Northern New South Wales, Australia using geospatial techniques. Sci Total Environ 756:143600. https://doi.org/10.1016/j.scitotenv.2020.143600

Kim JE, Yu J, Ryu JH, Lee JH, Kim TW (2021) Assessment of regional drought vulnerability and risk using principal component analysis and a Gaussian mixture model. Nat Hazards 109(1):707–724. https://doi.org/10.1007/s11069-021-04854-y

Kogan FN (1995) Application of vegetation index and brightness temperature for drought detection. Adv Space Res 15(11):91–100. https://doi.org/10.1016/0273-1177(95)00079-T

Kogan FN (1997) Global drought watch from space. Bull Am Meteor Soc 78(4):621–636

Liu Q, Zhang J, Zhang H, Yao F, Bai Y, Zhang S, Liu Q (2021) Evaluating the performance of eight drought indices for capturing soil moisture dynamics in various vegetation regions over China. Sci Total Environ 789:147803. https://doi.org/10.1016/j.scitotenv.2021.147803

Moisa MB, Merga BB, Gemeda DO (2022) Multiple indices-based assessment of agricultural drought: a case study in Gilgel Gibe Sub-basin, Southern Ethiopia. Theor Appl Climatol 148(1):455–464. https://doi.org/10.1007/s00704-022-03962-4

Nath R, Nath D, Li Q, Chen W, Cui X (2017) Impact of drought on agriculture in the Indo-Gangetic Plain, India. Adv Atmos Sci 34(3):335–346

Nejadrekabi M, Eslamian S, Zareian MJ (2022) Spatial statistics techniques for SPEI and NDVI drought indices: a case study of Khuzestan Province. Int J Environ Sci Technol 19:6573–6594

Orlovsky L, Kogan F, Eshed E, Dugarjav C (2011) Monitoring droughts and pastures productivity in Mongolia using NOAA-AVHRR data. In: Use of satellite and in-situ data to improve sustainability. Springer, Dordrecht, pp 69–79

Patil MB, Desai CG, Umrikar BN (2012) Image classification tool for land use/land cover analysis: a comparative study of maximum likelihood and minimum distance method. Int J Geol Earth Environ Sci 2(3):189–196

Pei F, Wu C, Liu X, Li X, Yang K, Zhou Y, Wang K, Xu L, Xia G (2018) Monitoring the vegetation activity in China using vegetation health indices. Agric for Meteorol 248:215–227

Peng Y, Gitelson AA (2011) Application of chlorophyll-related vegetation indices for remote estimation of maize productivity. Agric For Meteorol 151(9):1267–1276. https://doi.org/10.1016/j.agrformet.2011.05.005

Raha S, Gayen SK (2020) Simulation of meteorological drought using exponential smoothing models: a study on Bankura District, West Bengal, India. SN Appl Sci 2(5):909

Seiler RA, Kogan F, Sullivan J (1998) AVHRR-based vegetation and temperature condition indices for drought detection in Argentina. Adv Space Res 21(3):481–484

Sultana MS, Gazi MY, Mia MB (2021) Multiple indices based agricultural drought assessment in the north-western part of Bangladesh using geospatial techniques. Environ Challenges 4:100120-1–100120-17

Swain S, Wardlow BD, Narumalani S, Tadesse T, Callahan K (2011) Assessment of vegetation response to drought in Nebraska using Terra-MODIS land surface temperature and normalized difference vegetation index. Gisci Remote Sens 48(3):432–455

Wan Z (2006) MODIS land surface temperature products users'guide. Institute for Computational Earth System Science, University of California, SantaBarbara

Wang JL, Yu YH (2021) Comprehensive drought monitoring in Yunnan Province, China using multisource remote sensing data. J Mt Sci 18(6):1537–1549. https://doi.org/10.1007/s11629-020-6333-7

Zambrano F, Lillo-Saavedra M, Verbist K, Lagos O (2016) Sixteen years of agricultural drought assessment of the BioBío region in Chile using a 250 m resolution Vegetation Condition Index (VCI). Remote Sens 8(6):530

Zeng J, Zhang R, Qu Y, Bento VA, Zhou T, Lin Y, Wang Q (2022) Improving the drought monitoring capability of VHI at the global scale via ensemble indices for various vegetation types from 2001 to 2018. Weather Clim Extremes 35:100412. https://doi.org/10.1016/j.wace.2022.100412

Zhang B, Wu P, Zhao X, Wang Y, Gao X, Cao X (2013) A drought hazard assessment index based on the VIC–PDSI model and its application on the Loess Plateau, China. Theor Appl Climatol 114(1):125–138. https://doi.org/10.1007/s00704-012-0826-4

Zhao X, Xia H, Liu B, Jiao W (2022) Spatiotemporal comparison of drought in Shaanxi–Gansu–Ningxia from 2003 to 2020 using various drought indices in Google Earth Engine. Remote Sens 14(7):1570. https://doi.org/10.3390/rs14071570

Recent Trends of Meteorological Variables and Impacts on Agriculture in Northwest Bangladesh

5

J. M. Adeeb Salman Chowdhury, Md. Abdul Khalek, and Md. Kamruzzaman

Abstract

The study focused on two meteorological variables (rainfall and temperature) and their trend variations from 1960 to 2021. The recorded data was extracted from five regional weather stations in northwest area under Bangladesh Meteorological Department (BMD). By trend analysis, Rajshahi station found the lowest annual average rainfall (1460 mm) and the highest annual average temperature (25.30 °C). The precipitation concentration index (PCI) in the study area described that most of the stations carried varying precipitation concentration (16–20) except Dinajpur station. Linear trend analysis, the nonparametric Mann–Kendall (MK) test, Kendall's tau, Sen's slope estimator, and Spearman's rho (SR) test were used to define whether there were any trend fluctuations and calculate the magnitude of changes at the selected stations. Further, the Sequential Mann–Kendall test was executed to distinguish trend differences and abrupt deviations over time. During the study, Rajshahi station showed the highest significant decreasing trend with the degree of change assessed by Sen's slope estimator which was − 5.50 mm/year. Through cropping seasonal rainfall analysis, it had been observed that only the *Rabi* season (80%) found a declining trend across all stations. According to temperature trend analysis, except for Dinajpur station, all stations showed increasing trends in the annual and seasonal analysis by MK test and SR test. On the other hand, only Bogura and Syedpur stations were found significant at 5% level of significance. The magnitude of change discovered by the Sen's slope estimates varied from − 0.003 to 0.017 °C/year. The *Kharif* season temperature was significantly observed with a positive trend in all stations; however, the *Rabi* and pre-*Kharif* seasons temperature continued to show both increasing and decreasing trends. Furthermore, we checked the time series properties of meteorological data and constructed an appropriate model to forecast next five years based on the previous 60 years' recorded data. The research outcomes will be valuable for the sustainable agronomic growth of the country and reduce agricultural crop vulnerability during drought in the northwest region.

J. M. Adeeb Salman Chowdhury · Md.Kamruzzaman (✉)
Institute of Bangladesh Studies,
University of Rajshahi, Rajshahi, Bangladesh
e-mail: mkzaman@ru.ac.bd

J. M. Adeeb Salman Chowdhury
e-mail: asc_2010@cou.ac.bd

Md.Abdul Khalek
Department of Statistics, University of Rajshahi,
Rajshahi, Bangladesh
e-mail: mak.stat09@gmail.com

J. Das and S. Halder (eds.), *Advancement of GI-Science and Sustainable Agriculture*,
GIScience and Geo-environmental Modelling, https://doi.org/10.1007/978-3-031-36825-7_5

Keywords

Trend analysis · Mann–Kendall test · Spearman's rho test · Sen's slope estimator · Sequential Mann–Kendall test · ARIMA model · Bangladesh

5.1 Introduction

Climate change has been recognized as a worldwide issue in recent years. Now the global environment is possessed by unpredictable weather conditions resulting in the climate variability and change (Alam et al. 2023). Despite the fact that climate change is a global phenomenon, its effects will not be equally distributed (Shahid and Khairulmaini 2009; Shaw et al. 2013; Karim et al. 2020). According to research, underdeveloped countries are more at risk from the effects of climate change than developed ones. This is mostly due to a poor and constrained capacity to adapt to climate change (Basak et al. 2022; Das et al. 2020a, b, c; Rahman and Lateh 2017). As a developing nation, Bangladesh is particularly in danger due to high climate inconsistency, extreme temperature, huge population density, extreme poverty, unorganized infrastructure, insufficient money supply, and weak educational system (Kamruzzaman et al. 2018). The Intergovernmental Panel on Climate Change (IPCC) report found that precipitation and temperature were identified worldwide significantly, but with varying magnitudes. Every year the nation experiences some sort of disaster, such as floods, droughts, river bank erosion, and cyclones threatening country's development efforts and resulting poor livelihood and huge agricultural loss (Adams et al. 1998). According to the IPCC (2014), globally the average temperature is increased by 0.85 °C between the year 1880 and 2012, and it will increase 0.3–4.8 °C by 2100 (Karim et al. 2020; Mullick et al. 2019). It is observed that Bangladesh has experienced with a significant amount of precipitation in the monsoon season and less rainfall in other seasons of the year, which also affects the food security and economic growth of the country (Akter and Rahman 2012; Mondol et al. 2018; Rahman et al. 2017; Rahman et al. 2016b; Salam et al. 2020).

Agriculture is heavily relying on country's economy, but the water constraint (shortage) hinders agricultural production. In our country, agriculture is the most prevalent occupation in rural areas, with approximately 51.88% of people actively involved in agricultural (BBS 2019b). It is a crucial sector contributing 13.82% to the country's GDP (BBS 2019a). The rural population, roughly 70% of the total, is extremely in danger due to natural disasters (Islam et al. 2019a). Bangladesh is the fourth-largest rice producer in the world, with an annual production of over 34.7 million metric tons. Nearly 80% of all cultivated land (11.7 million hectares) is planted with rice (Mainuddin et al. 2022). Rice is used to make 91% of all food grains and 60% of overall agricultural products (Rahman et al. 2016a, b; Yu et al. 2010).

Despite advancements in technology, such as better crop varieties, irrigation practices, and other sustainable adaptation strategies have established food security, climate conditions still contribute unpredictability to agricultural productivity of the country (Mainuddin and Kirby 2015; Kirby et al. 2015). As a result, rainfall and temperature trend analysis and future prediction have become important subjects of research in Bangladesh, as accurate trend detection directly effects on establishment of long-term food security, an agro-based economy, and infrastructural development. Environmental researchers conduct various types of research to understand the rainfall and temperature patterns of Bangladesh. A number of studies exhibited significant rising or falling trends of temperature and rainfall on a large scale using the MK test, Sen's slope estimator, and Kendall's tau (Shahid and Khairulmaini 2009; Zhang et al. 2009; Shahid 2010; Rahman and Lateh 2016; Rahman et al. 2017; Kamruzzaman et al. 2018; Das et al. 2021a, b). Shahid (2010) demonstrated a significant upward trend in annual and pre-season rainfall at 17 rain gauge stations in Bangladesh by using the MK test and Sen's slope estimator between 1958 and 2007. Similarly, Hossain et al. (2014) perceived rising tendencies in seasonal rainfall variability in the southwest coastal area of

Bangladesh during the years 1948–2007. Rahman and Lateh, (2016) constructed a regression model for trend detection from 1971 to 2010 and forecasting ten-year period of rainfall by the ARIMA method from 34 meteorological stations in Bangladesh. Their findings showed that annual rainfall was forecasted to decline by 153 mm between 2011 and 2020. As a result, almost dry conditions would continue during pre- and post-monsoon season in Bangladesh's northwestern, western, and southwestern regions. Rahman et al. (2017) used 60 years of monthly rainfall data to estimate rainfall trends of selected stations by applying the Mann–Kendall test. Additionally, they found unnoticeable trends in precipitation rather than rising trends for Khulna, Satkhira, Cox's Bazar stations, but they discovered falling trends for Srimongal stations by using the ARIMA approach. Similar types of research were found in hydro-climatic research based on global climate problem (Islam et al. 2021; Kamruzzaman et al. 2018, 2022; Yu et al. 2017).

Although the northwest zone produces huge amount of crops, the agriculture mostly relies on rainwater. In addition, inadequate rainfall, extreme temperature, surface water scarcity, and groundwater depletion for irrigation purposes are common problems in this region (Salam et al. 2020; Karim et al. 2020; Uddin et al. 2020a, 2020b). To address this problem, several studies have been done locally and country-wide to determine future trends. Kamruzzaman et al. (2018) identified the changing rainfall and temperature patterns in different cropping seasons in the NW Bangladesh. Almost identical results were found in different studies on the NW region of Bangladesh (Bariet al. 2016; Rahman et al. 2016a, b). It is traditionally accepted that drought is the outcome of decreasing trends of rainfall and increasing trends of extreme temperature. However, many environmental researchers do not agree with the above statement. To address the issue, many previous studies were conducted based on zone-wise or country-wise methods. However, some limitations have been found in the local and regional studies. The present chapter aims to identifying annual and seasonal trend fluctuations of climatic variables of five meteorological stations (*i.e.*, Bogura, Syedpur, Dinajpur, Rajshahi, and Ishwardi) during 1961–2021 and predicts monthly behavioral change to obtain approximate results. Furthermore, the study looks into the development of agro-economic activities and infrastructure that will strengthen the ability of farmers, policymakers, and researchers to deal with meteorological change.

5.2 Materials and Methods

5.2.1 Area of Study

Sixteen administrative districts, consisting 34,359 km^2 and residing 38 million people, from north and northwest Bangladesh are considered as study area of this research (Uddin et al. 2020a, BBS 2019a). According to Banglapedia (2003), the country belongs to a sub-tropical climate region with hot, humid, and erratic rainfall occurring throughout the year, mostly influenced by monsoon weather as well as pre- and post-monsoon exchanges. The country undergoes three climatic seasons such as (1) the month of November to February known as dry winter season, (2) the month of March to May known as pre-monsoon hot summer season, and (3) lastly the month of June to October regarded as rainy monsoon season (Rashid 1991). In winter season, the mean temperature lies between 18.5 and 21.0 °C, and hot summer season mean temperature ranges from 27.8 to 29.0 °C. The overall annual average rainfall from the northwest to the northeast region is 1329–4338 mm, which indicates that northwest region of Bangladesh receives less amount of rainfall (Shahid and Behrawan 2008; Shahid and Khairulmaini 2009).

During the month of April–May, the average temperature was experienced between 27 and 31 °C in the east and south and west-central portion of the country. However, the western regions of Bangladesh become very hot with a maximum of 40 °C temperature (Shahid 2010; Shahid et al. 2012; Khan et al. 2019). It is experienced that the average maximum temperature of northwest zone goes up to 24 °C in January and 34 °C in June. Meanwhile, the average temperature does not go beyond 27 °C from November to February but exceeds 32 °C from April to October (Bhuyan et al. 2018). The district-wise meteorological stations location is presented in (Fig. 5.1). According to Banglapedia (2003), in our country the agricultural calendar is based on three cropping seasons: *Rabi*, pre-*Kharif*, and *Kharif*. The *Rabi* period, which spans from early November to the end of February, is characterized by a dry climate. The pre-*Kharif* period starts in early March and lasts until the end of June, whereas the *Kharif* begins in the month July and ends in the month October when there is enough moisture from rainfall to support rain-fed or non-irrigated crops (Banglapedia 2003; Kamruzzaman et al. 2018; Shahid and Behrawan 2008).

5.2.2 Data

In this study, Bangladesh Meteorological Department (BMD), Dhaka, has provided rainfall and temperature data from 1960 to 2021. Besides, required and relevant records are retrieved from World Meteorological Organization (WMO), Bangladesh Water Development Board, and online resources from different websites.

5.2.3 Estimation of Missing Data

During study, every station has a number of missing values. Due to the global nature of missing daily temperature and rainfall records, many climatological algorithms have encountered difficulties. There are three broad categories of missing data estimation methods, such as: (a) within-stations, (b) between station, and (c) regression-based (Allen and DeGaetano 2001). The within-stations approach forecasts missing temperatures based on the previous and subsequent days' data. For example, to calculate the missing minimum temperature on *October 1*, the minimum temperatures between *September 30* and *October 2* could be averaged (Oliver 1980).The within-station method is utilized to calculate the daily readings that are lost for a maximum of three days temperature. When more than three days of data are unavailable, regression-based methods should be used. This method uses data from nearby stations. Regression-based approaches are also appropriate, when monthly data is lost, because both monthly and daily data are constructed from this method (Eiseheid et al. 1995). According to Kemp et al. (1983), several regression techniques tend to provide more accurate results compared to other missing data estimation procedures. Multiple imputation techniques are utilized to both daily (Recha et al. 2012) and monthly (Ingsrisawang and Potawee 2012) data for missing rainfall records because the standard error is correctly adjusted for missing data estimation (Enders 2010).

5.2.4 Rainfall Climatological Features

5.2.4.1 Coefficient of Rainfall Variability

The coefficient of rainfall variability CV (%) is determined using the following formula:

$$\mathrm{CV}(\%) = \frac{\sigma}{\mu} \times 100 \tag{5.1}$$

where CV is the coefficient of variation in percentage, μ is the annual average precipitation, and σ is the standard deviation.

5.2.4.2 Precipitation Concentration Index (PCI)

The concentration of rainfall (as major form of precipitation, in the selected study area) variability has been computed by applying the following formula of precipitation concentration index (Oliver 1980; Michiels et al. 1992):

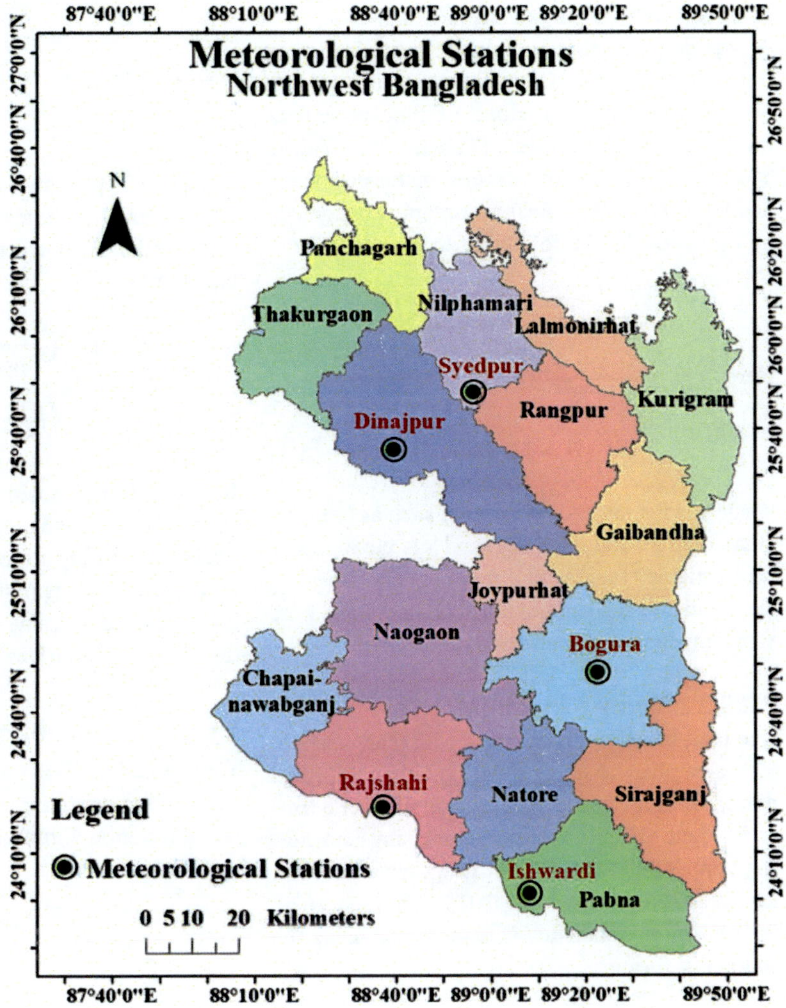

Fig. 5.1 Meteorological stations map in the nothwest region

$$\mathrm{PCI}_{\mathrm{annual}} = \frac{\sum_{i=1}^{12} p_i^2}{\left(\sum_{i=1}^{12} p_i\right)^2} \quad (5.2)$$

PCI represents the monthly precipitation concentration index, and p_i is the monthly precipitation in the month i. It is observed that when PCI values are < 10%, the precipitation concentration should be identical or low precipitation; if PCI value range is 11–15%, it indicates moderate concentration of precipitation; if it ranges 16–20%, then precipitation continues irregularly (Oliver 1980). If a PCI value > 20%, it means that the concentration of precipitation is either high or strongly irregular.

5.2.5 Rainfall and Temperature Trend Analysis

Different statistical techniques and trend detection formula have been implemented for the required study. Mainly descriptive statistical analysis (maximum, minimum, mean/average, standard deviation, and coefficient of variation) has been executed to present the general characteristic of temperature and rainfall time series data.

5.2.5.1 Linear Regression

A popular parametric technique for finding linear trends in time series data is linear regression. It is used to determine the slope of time-dependent meteorological variables (Tabari and Talaee 2011). Positive slope specifies increasing trend and vice versa. The linear monotonic change can be calculated as the following form:

$$y = ax + b \tag{5.3}$$

Here, $a = \frac{\sum (x-\bar{x})(y-\bar{y})}{\sum (x-\bar{x})^2}$ and

$$b = \bar{y} - a\bar{x} \tag{5.4}$$

where y is the dependent variable such as (temperature, rainfall, and humidity) and x is the independent variable (year); a and b are the coefficient of slope and intercept term of the trend line; $\bar{x}$ and $\bar{y}$ are sample means (Rahman and Lateh 2016).

5.2.5.2 Mann–Kendall (MK) Test

The rank-based nonparametric MK (Mann 1945) test is applied as the primary method of trend detection. It is often used to check whether a time series data exhibits a monotonic rising and falling trend. (Das et al. 2021a; Kamruzzaman et al. 2018, 2022; Sa'adi et al. 2019; Rahman et al. 2017). The test statistic *(S)* is identified using the following formula:

$$S = \sum_{K=1}^{n-1} \sum_{j=k+1}^{n} \operatorname{sgn}(x_j - x_k) \tag{5.5}$$

$$\operatorname{sgn}(x_j - x_k) = \begin{cases} +1, & \textit{if} \quad (x_j - x_k) > 0 \\ 0, & \textit{if} \quad (x_j - x_k) = 0 \\ -1, & \textit{if} \quad (x_j - x_k) > 0 \end{cases} \tag{5.6}$$

Here, S means the trend detection. A rising trend is indicated by a positive value of S and vice versa. When sample size is greater than 10, then the test statistics $\mathbf{S}$ is normally distributed, with the following variance (Das et al. 2019):

$$\operatorname{Var}(S) = \frac{[(n(n-1)(2n+5)] - \sum_{i=1}^{m} t_i(t_i - 1)(2t_i + 5)}{18} \tag{5.7}$$

where m represents the total number of tied groups and t_i denotes the size of the *ith* group. The $\mathbf{Z}$ statistic is calculated using the following equation:

$$Z = \begin{cases} \frac{S-1}{\sqrt{\operatorname{Var}(S)}}, & \text{for } S > 0 \\ 0, & \text{for } S = 0 \\ \frac{S+1}{\sqrt{\operatorname{Var}(S)}}, & \text{for } S < 0 \end{cases} \tag{5.8}$$

In general, the positive and negative Z values represent the rising and falling trends, respectively. According to the null hypothesis, there is no trend in records (either accepted or rejected) if the estimated Z value is smaller or greater than the critical value (Das et al. 2021b).

5.2.5.3 Kendall's Tau

Tau (Kendall 1948) assesses the intensity of x and y's monotonic relationship. The Kendall's tau correlation formula is given by,

$$\tau = \frac{S}{n(n-1)/2} \tag{5.9}$$

5.2.5.4 Sen's Slope Estimator

Sen's slope estimator (Sen 1968), which provides the robust estimation of time series, is used to evaluate the degree of change (slope Q) (Das and Bhattacharya 2018; Das et al. 2020b). The estimation slope can be attained from N pairs of data as:

$$Q_i = \frac{x_k - x_j}{k - j}, \qquad i = 1, 2, 3 \ldots N, \quad k > j \tag{5.10}$$

where x_k and x_j represent values of data at k and j times and Q_i is the median slope, respectively.

5.2.5.5 Spearman's Rho (SR) Test

The Spearman's rho (SR) test (Spearman 2010) is another technique that uniformly detects tends and clearly indicates the lack of trends (Das et al. 2020b; Shadmani et al. 2012; Tonkaz et al. 2007; Yue et al. 2002a). According to Yue et al. (2002b), maximum trend test has same strength for long-term data when it comes to detecting monotonic hydrological time series trends. The test statistic D and standardized test statistic Z_{SR} are defined by:

$$D = 1 - \frac{6\sum_{1}^{n}[R(X_i - 1]^2}{n(n^2 - 1)} \tag{5.11}$$

$$Z_{SR} = D\sqrt{\frac{(n-2)}{(1-D^2)}} \tag{5.12}$$

where R is the ith observation rank order in X_i time series and n is the sample size. The positive values Z_{SR} suggest rising trends and vice versa.

5.2.5.6 Sequential Mann–Kendall (SMK) Test

At the end of any period, it is well known that trend identification using the MK approach does not produce an accurate trend image for entire series. By using the sequential test for each distinct time series, it is possible to identify any fluctuations in the trend during the course of the inquiry (Sneyers 1990). This test establishes 2 series: [i] a progressive series $u(t_i)$ and [ii] a backward series $u'(t_i)$. . There is a statistically noteworthy change point that exists if they cross each other and diverge over predetermined definite threshold value (95% confidence limit). Their intersection point provides an approximate idea of the trend beginning year (Mosmann et al. 2004).

The SMK test is calibrated using rank values. The actual values in the series $(x_1, x_2, x_3 \ldots x_n)$ with the rank value y_i and the magnitudes of y_i $(i = 1, 2, 3\ldots\ldots n)$ are compared with y_j $(j = 1, 2, 3\ldots i-1)$. The cases $y_i > y_j$ are enumerated and indicated for each comparison by $n_{i.}$ Here, t_i is test statistic of SMK test by applying the following formula:

$$t_i = \sum_{j}^{i} n_i$$

The distribution of test statistic t_i has a mean $E(t_i) = \frac{i(i-1)}{4}$ and variance $\mathrm{Var}(t_i) = \frac{i(i-1)(2i+5)}{72}$; then the following formula is employed to obtain the forward sequential results: (Sneyers 1990).

$$u(t_i) = \frac{[t_i - E(t_i)]}{\sqrt{\mathrm{Var}(t_i)}} \tag{5.13}$$

Similar approaches are used to find the backward sequential statistics $u'(t_i)$, which starts at the end of the sequence. Occasionally, the positive and negative trends are canceled by each other in SMK analysis but do not provide a substantial trend turning point (Nalley et al. 2013). This technique was widely used by many researchers to identify trends from the beginning point (Esteban-Parra et al. 1995; Makokha and Shisanya 2010) rather than whole trend detection. Nalley et al. (2013) applied this technique in order to determine the periodicities in the temperature time series.

5.2.5.7 Spatial Distribution Analysis

The Inverse Distance Weighing (IDW) method was employed to evaluate the spatial pattern of trend magnitude (MK test and Sen's slope estimates). This procedure is popular and simple for spatial interpolation (Shahid and Behrawan 2008; Rahman et al. 2017; Kamruzzaman et al. 2018; Islam et al. 2019a, b), and it is very reliable for displaying the spatial analysis of hydrometeorological records (Praveen et al. 2020). Compared to other interpolation techniques, the IDW method is more effective (Islam et al. 2021). We have used the ArcGIS 10.0 software for performing the IDW method in this study.

5.2.5.8 ARIMA Modeling and Prediction

The ARIMA approach is recognized as a sophisticated and broadly used data analysis technique for forecasting time series (Box and Jenkins 1976). In hydrological research, the ARIMA models are utilized to discover complexity in data series and forecast upcoming scenarios (Dimri et al. 2020; Rahman et al. 2017). Typically, ARIMA model consists of three components: AR is the autoregressive component, MA is the moving average component, and the differencing order (D) integrated with the model both AR and MA. The ARIMA model ($p.d.q$) includes p = autoregressive term, d = trend term, and q = moving average term. After attaining data stationary, ARIMA method usually proceeds via four major steps: model identification, parameter estimates, model adequacy diagnostics, and forecasting. For the seasonal ARIMA model, notation is like ARIMA $(p, d, q) \times (P, D, Q)S$ where p = non-seasonal autoregressive (AR) order, d = non-seasonal differencing, q = non-seasonal moving average (MA) order, P = seasonal AR order, D = seasonal differencing, Q = seasonal MA order, and S = time span of repeating seasonal pattern. The fitted model performance is based on model selection criteria like log-likelihood, Akaike Information Criteria (AIC), and Bayesian Information Criteria (BIC). For this study, the Box Jenkins technique is implemented to forecast monthly rainfall and temperature data for the following five years based on 1960–2021.

5.3 Results and Discussion

5.3.1 Annual Rainfall Features

Rainfall is not constant across the country. Historically, in our country the northwest region received the least amount of rainfall as compared to the northeast. Around 75% of rainfall happens during the monsoon season (Alamgir et al. 2015). The results in Table 5.1 indicated that the mean annual rainfall ranges from 2171 to 1460 mm with mean value of 1767.80 mm. In 2010, Rajshahi station received the lowest amount of annual rainfall, 792 mm. Meanwhile, Syedpur station received the highest amount of annual precipitation 3748 mm in 1984. Moreover, Syedpur station received a maximum average annual rainfall amount of 2171 mm, whereas the other stations received less than 2000 mm amount of annual rainfall. During study period, the rainfall data was found positively skewed. The coefficient of variation is low in Rajshahi station (19.93%) and high in Ishwardi station (26.12%) which means maximum rainfall variability happened in Ishwardi station. The standard deviation of annual rainfall is high in Syedpur station, with 463.87 mm, and low in Rajshahi station, with 291 mm (Fig. 5.2 a). From the statistical analysis, about 80% stations in the research area were covered by irregular precipitation concentration index (PCI = 16–20), except Dinajpur station that carried a high precipitation concentration, which was more than 20%. Finally, all the stations were found to be positively skewed. The study area's annual rainfall statistics are provided in the following tables (Table 5.1).

5.3.2 Trend Analysis of Annual Rainfall

The MK test result revealed that 40% stations exhibit a positive trend, while the rest of the stations showed negative trend. Moreover, in annual rainfall trend analysis, a positive slope specified increasing amount of rainfall, while a negative slope designated decreasing trends (Table 5.2). According to MK test result *(Z)*, Rajshahi station showed the highest amount of significant decreasing trend (− 2.654) at 5% level of significance, followed by Bogura (− 0.176) and Ishwardi (− 0.808), but they were found statistically insignificant. The magnitude of change in annual average rainfall was estimated at the rates of − 5.50, − 0.59, and − 1.71 mm/year at Rajshahi, Bogura, and Ishwardi stations, respectively, but only Rajshahi station showed statistically significant. In contrast, Syedpur and Dinajpur stations found insignificant positive trend, and the degree of

Table 5.1 Descriptive statistics of annual rainfall in the study area (1960–2021)

Stations	Annual rainfall (mm)							
	Min	Max	Mean	CV (%)	SD	Skewness	Kurtosis	PCI_{Annual}
Bogura	1081	2601	1719	21.92	376.77	0.307	− 0.635	19.47
Dinajpur	920	3179	1906	23.77	453.08	0.395	0.525	20.58
Ishwardi	893	3021	1583	26.12	413.42	1.30	2.530	17.80
Rajshahi	792	2241	1460	19.93	291.00	0.069	0.114	18.16
Syedpur	1231	3748	2171	21.37	463.87	0.905	1.510	19.63

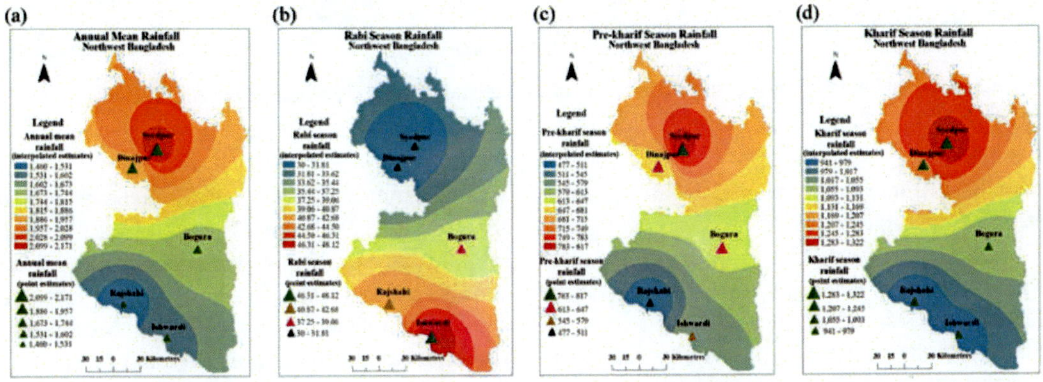

Fig. 5.2 Distribution of **a** Annual average rainfall **b** *Rabi* rainfall **c** Pre-*Kharif* rainfall **d** *Kharif* rainfall

change of these two stations was 3.68 and 2.59 mm/year, respectively. Besides, Kendall's tau revealed a monotonic trend of weak relationships in all stations. The linear regression trends of annual precipitation graph are shown in (Fig. 5.5). The MK test results (Z statistic) and the Sen's estimator of different stations are presented by the spatial distribution analysis (Fig. 5.3a). It is observed that annual average rainfall decreases at the Rajshahi division (Rajshahi, Bogura, and Ishwardi) but increases at the Rangpur division (Dinajpur and Syedpur).

5.3.3 Trend Analysis of Seasonal Rainfall

5.3.3.1 Rabi Season

During the *Rabi* season, the northwest zone received 37.8 mm of annual rainfall, which was significantly less than other parts of Bangladesh. In the study domain, annual average rainfall ranged from 30 mm to less than 50 mm in all stations. According to the regional distribution (Fig. 5.2b) of the *Rabi* season analysis, Dinajpur station found the lowest quantity of annual

Table 5.2 Long-term annual rainfall trend test results

Stations	MK test	Sen's slope	Kendall's tau	SR test
Bogura	− 0.176	− 0.59	− 0.016	− 0.195
Dinajpur	0.692	2.59	0.061	0.501
Ishwardi	− 0.808	− 1.71	− 0.071	− 0.830
Rajshahi	− 2.654*	− 5.50*	− 0.232*	− 2.813*
Syedpur	1.178	3.68	0.103	1.063

* Significant trends at 5% significance level, Sen's slope unit is in mm per year

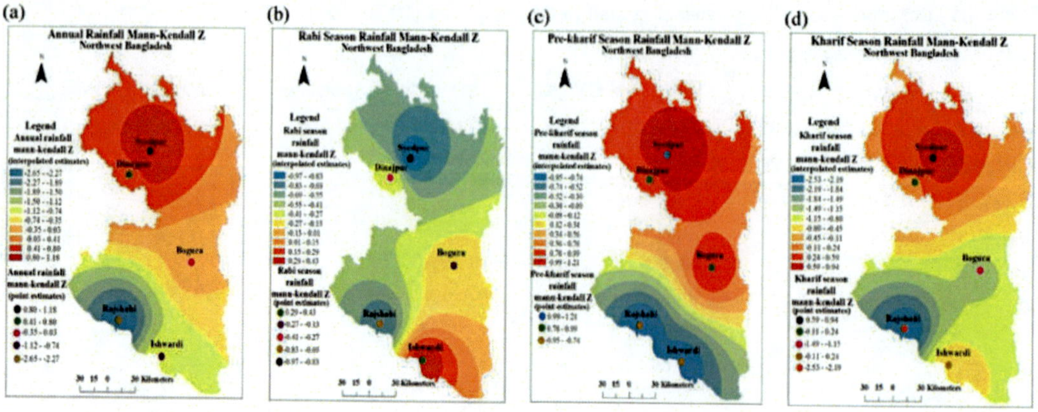

Fig. 5.3 Distribution of *Z* statistics estimates **a** Annual average rainfall **b** *Rabi* rainfall **c** Pre-*Kharif* rainfall **d** *Kharif* rainfall

rainfall (30 mm), whereas Ishwardi station received the highest quantity of rainfall, 48 mm. The standard deviation was high in Rajshahi station (38 mm) and low in Dinajpur station (25.85 mm). Moreover, the coefficient of variation revealed that 95.22% variation occurred in Bogura station, whereas Ishwardi station found low amount of rainfall variation (78.37%). The results of the MK test (Z statistics) exhibited that almost all stations shown an insignificant downward trends ranged from − 0.176 to − 0.966 at 95% confidence interval. Analysis of Sen's slope estimations revealed declining trends of 0.029, − 0.034, − 0.167, and − 0.136 mm/year in Bogura, Dinajpur, Rajshahi, and Syedpur, respectively, except Ishwardi station observed at a rate of change 0.107 mm/year (Table 5.3). Spearman rho's test also showed similar trend in all the station that ranged from 0.537 to − 1008. The distribution map of Z statistics and Sen's slope estimates of the *Rabi* season rainfall is depicted in (Fig. 5.3b and 5.4b). The findings of the study exhibited that the northwest region experienced a declining amount of rainfall in the Rabi season, which is particularly harmful for Boro rice production.

5.3.3.2 Pre-Kharif Season

The annual (mean or average) rainfall at different stations, in Bangladesh, in the pre-*Kharif* period was 623 mm, and the standard deviation varied from 153 to 240 mm. The pre-Kharif season rainfall analyses of spatial distribution (Fig. 5.2c) clearly showed that Syedpur station received significant rainfall with 1667 mm. Meanwhile, Rajshahi station found only 96 mm of rainfall in the study period (1960–2021). The northwest region had experienced both rising and falling trends in pre-*Kharif* season. The result derived from different trend analysis test (MK test, Sen's slope, and SR test) points out that 60% of stations showed increasing trend and 40% found insignificant decreasing trend. The MK test outcomes *(Z)* revealed that rainfall varied from − 0.905 (Ishwardi) to 1.208 (Syedpur). The Sen's slope estimates of Bogura, Dinajpur, and Syedpur stations identified insignificant rising trends at a rate of changes 1.73, 1.59, and 1.54 mm/year, respectively, while stations like Ishwardi and Rajshahi achieved insignificant falling trends at a magnitude of change − 0.905 and − 0.954 mm/year, respectively (Table 5.3). Spearman's rank correlation test (SR test) observed the same increasing and decreasing trends for the all stations. The spatial analysis of Z statistics and Sen's slope estimates is displayed in Figs. 5.3c and 5.4c. From above discussion, it is clear that the pre-*Kharif* season rainfall in Bangladesh shows downward trend in the northwest but strongly predicts upward trend in the north.

Table 5.3 Long-term seasonal rainfall trend test results

Stations	Rabi season			Pre-Kharif season			Kharif season		
	MK test	Sen's slope	SR test	MK test	Sen's slope	SR test	MK test	Sen's slope	SR test
Bogura	− 0.176	− 0.029	− 0.147	0.966	1.73	1.008	− 1.227	− 2.18	− 1.432
Dinajpur	− 0.255	− 0.034	− 0.031	0.947	1.59	0.995	0.20	0.50	− 0.007
Ishwardi	0.431	0.107	0.537	− 0.905	− 1.105	− 0.838	− 0.486	− 1.24	− 0.421
Rajshahi	− 0.838	− 0.167	− 0.886	− 0.954	− 1.00	− 0.903	− 2.53*	− 3.80*	− 2.711*
Syedpur	− 0.966	− 0.136	− 1.008	1.208	1.54	1.330	0.94	2.53	0.717

* Significant trends at 5% level, Sen's slope unit is in mm/year

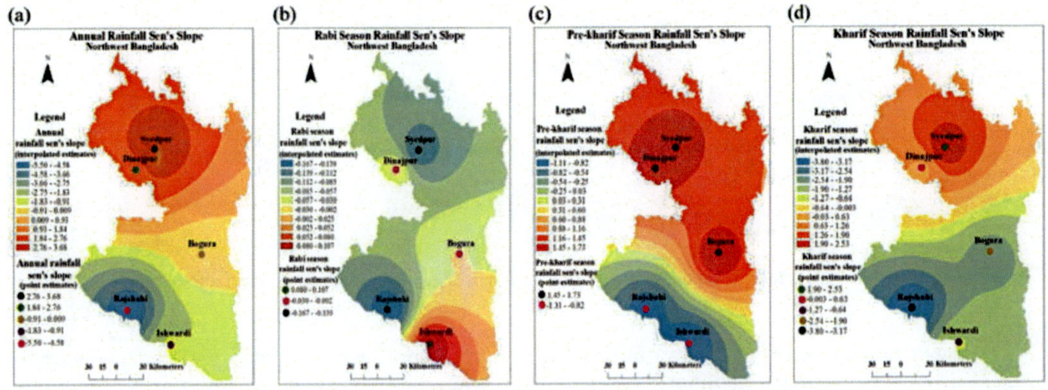

Fig. 5.4 Distribution of Sen's slope estimates **a** Annual average rainfall **b** *Rabi* rainfall **c** Pre-*Kharif* rainfall **d** *Kharif* rainfall

5.3.3.3 Kharif Season

The *Kharif* rainfall is very important for Aman rice production because rice is the main staple food in our country. The average annual rainfall during this season was 1104.8 mm. In this season, Rajshahi station received the least amount of maximum rainfall (1583 mm), and Syedpur station found maximum rainfall (2440 mm). The coefficient of variation is high (29.87%) at Dinajpur station and low (24.98%) at Rajshahi station (Fig. 5.2d). During the month of July to October is considered as the *Kharif* season in our country. In the present study, long-term *Kharif* season was dominated by a mixed rainfall trends in all stations. The MK test and SR test results were detected with same trend directions in the northwest region. From trend analysis, Rajshahi station showed the most significant decreasing trend with − 2.53 at 5% level of significance; on the contrary, Bogura and Ishwardi stations also observed decreasing trend in all trend analysis test, but they were not statistically significant. Moreover, the degree of change obtained by Sen's slope results was − 3.80, − 2.18, and − 1.24 mm/year at Rajshahi, Bogura, and Ishwardi stations, respectively. Dinajpur and Syedpur stations exhibited insignificant positive trends of 0.20 and 0.935 in accordance with the MK test statistic (*Z*), and the magnitude of change assessed by Sen's slope estimates of these two stations was 0.50 and 2.53 mm/year, respectively. According to the SR test results, all stations found a downward trend varied from − 2.711 to 0.717 (Table 5.3). The spatial distribution of MK *(Z)* statistic and Sen's slope estimates of the *Kharif* rainfall trends is shown in (Figs. 5.3d and 5.4d).

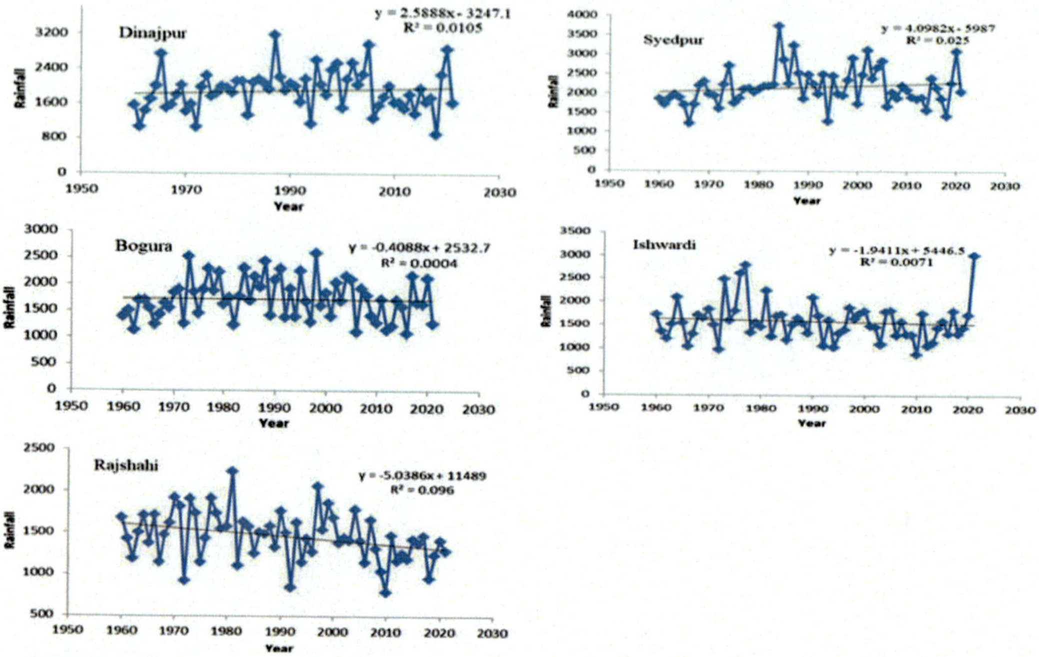

Fig. 5.5 Linear regression trends of annual average rainfall in northwest Bangladesh at five stations during 1960–2021

The linear trend line of yearly rainfall in the northwest region during 1960–2021 is displayed in (Fig. 5.5). The graph indicates that the trend line is decreasing at Bogura, Ishwardi, and Rajshahi stations, while an increasing annual average rainfall at Syedpur and Dinajpur stations.

5.3.4 Annual Temperature Features

The long-term annual mean temperature in the northwest recorded stations was 24.96 °C, and the temperature ranged from 24.30 °C (Syedpur) to 25.30 °C (Rajshahi). Based on historical temperature data, the warmer average extreme temperature varied from 26.57 °C to 25.35 °C, and the cooler average minimum temperature lied between 20.64 °C and 24.67 °C. It was evident that (Table 5.4) along with other descriptive analysis of annual temperature, the standard deviation was high (0.75 °C) at Syedpur station and low (0.31 °C) at Bogura station. The coefficient of variation of temperature data varied from 1.24% to 3.07% at different stations. It means Syedpur station observed maximum temperature variation. The long-term annual mean temperature is shown in (Fig. 5.6a) by spatial distribution analysis. The temperature dataset

Table 5.4 Descriptive statistics of the annual mean temperature of Bangladesh (1960–2021)

Stations	Annual temperature (°C)						
	Min	Max	Mean	CV (%)	SD	Skewness	Kurtosis
Bogura	24.67	26.08	25.24	1.24	0.31	0.613	0.204
Dinajpur	24.10	26.03	24.90	1.75	0.43	0.835	0.521
Ishwardi	24.12	26.26	25.08	1.52	0.38	0.242	1.369
Rajshahi	24.43	26.57	25.30	1.84	0.47	0.856	0.369
Syedpur	20.64	25.35	24.30	3.07	0.75	− 2.504	9.380

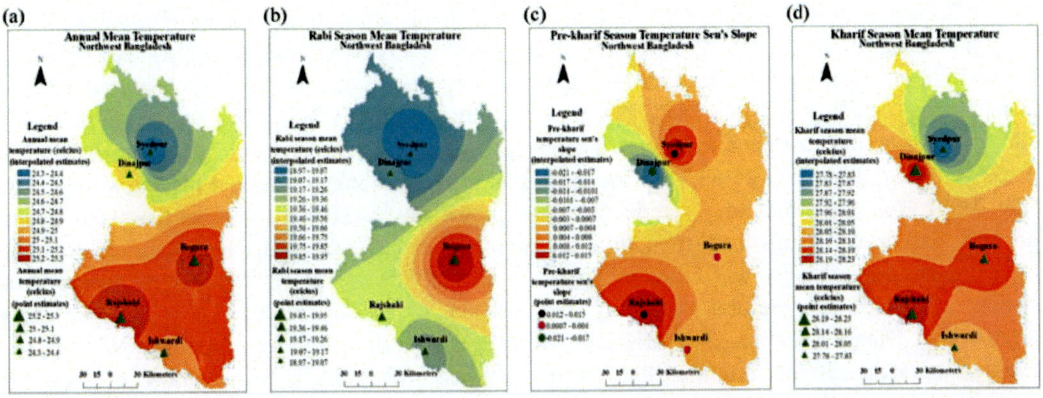

Fig. 5.6 Distributions of **a** Annual average temperature **b** *Rabi* temperature **c** Pre-*Kharif* temperature **d** *Kharif* temperature

was positively skewed in four stations, except Syedpur stations which found negative skewness. Similar results were identified for kurtosis. From the analysis, it is clearly said that the datasets are leptokurtic not mesokurtic.

5.3.5 Trend Analysis of Annual Temperature

The annual mean temperature observed that 80% of the stations displayed an increasing trend, except Dinajpur station which found decreasing trend (Table 5.5). During 60 year period, the highest significant rising trend was found at Bogura and Syedpur stations by the MK test statistic (Z) and SR test, and the magnitude of changes discovered by the Sen's slope estimates were 0.0095 °C/year and 0.017 °C/year, respectively. Ishwardi and Rajshahi stations also presented rising tendencies in annual temperature analysis, but not statistically significant. Apart from that, only Dinajpur station was shown falling trend with the MK test (− 1.014) and SR test (− 1.655), and the rate of change assessed by the Sen's estimates was − 0.003 °C/year. Meanwhile, Kendall's tau received positive relationships over time. The spatial distribution of average annual temperature is displayed in (Figs. 5.7a and 5.8a). Over the years, upward trend of temperature is dominating in this part of the country which will create more dry years and extreme temperature in the future. Linear trend analyses of the long-term annual temperature are shown in (Fig. 5.9).

5.3.6 Trend Analysis of Seasonal Temperature

5.3.6.1 Rabi Season

This season is characterized by the coldest temperature in Bangladesh, generally referred to as winter. The average temperature during *Rabi*

Table 5.5 Long-term annual temperature trend test results

Stations	MK test	Sen's slope	Kendall's tau	SR test
Bogura	4.514*	0.0095*	0.395*	5.040
Dinajpur	− 1.014	− 0.003	− 0.089	− 1.655
Ishwardi	1.451	0.004	0.130	1.390
Rajshahi	1.233	0.004	0.109	1.044
Syedpur	5.674*	0.017*	0.495*	7.259*

* Significant trends at 5% level, Sen's slope unit is in mm per year

season fluctuates from 18.97 to 19.95 °C with an average value 19.32 °C. In this season, the minimum temperature at different stations varied from 16.91 °C (Syedpur) to 18.78 °C (Bogura), and the maximum temperature ranged from 20.60 °C (Syedpur) to 21.52 °C (Rajshahi) (Fig. 5.6b). The coefficient of variation of temperature data varied from 2.51 to 4.16%; i.e., low-temperature variability was found at Bogura station, and high-temperature variability was described at Dinajpur station. The MK and SR test findings indicated that the trend fluctuations were identified by both positive and negative directions. According to the 60 years of historical data, 60% stations (Dinajpur, Ishwardi, and Rajshahi) showed a decreasing trend and 40% stations (Syedpur and Dinajpur) experienced a rising trend with the MK test results or SR test analysis (Table 5.6). The values of MK test statistic (Z) ranged from − 0.83 to 2.81. The degree of change evaluated by Sen's slope estimates was − 0.09, − 0.004, and − 0.01 °C/year at Dinajpur, Ishwardi, and Rajshahi stations, respectively, but statistically insignificant (Table 5.6). From the study, only Syedpur station revealed a significant rate of change with 0.014 °C/year. The spatial distribution of MK test statistic (Z) and Sen's slope estimates is displayed in (Figs. 5.7b and 5.8b).

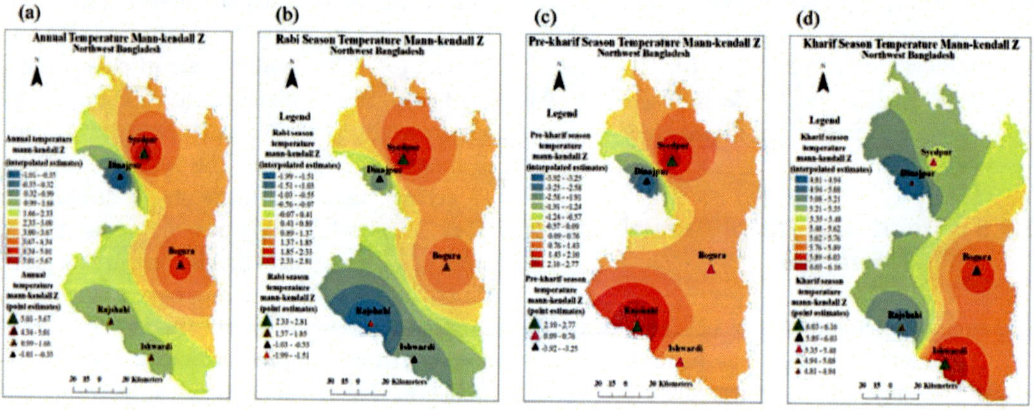

Fig. 5.7 Distributions of Z statistics estimates **a** Annual average temperature **b** *Rabi* temperature **c** Pre-*Kharif* temperature **d** *Kharif* temperature

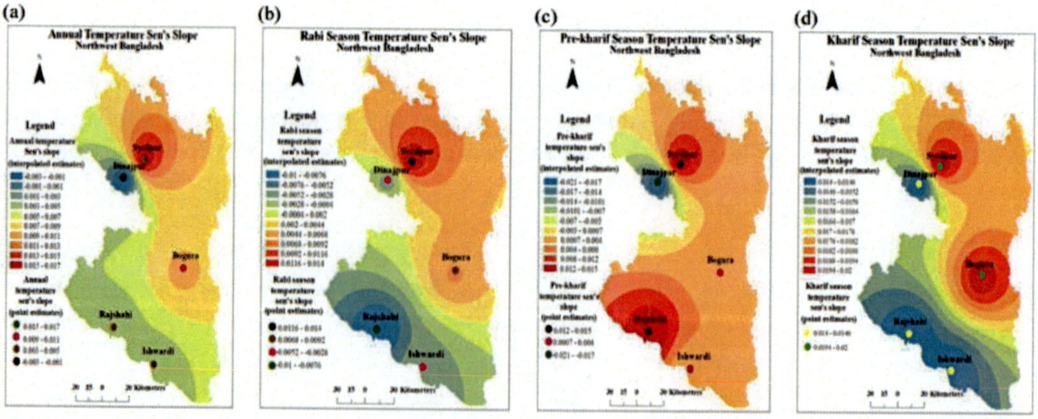

Fig. 5.8 Distribution of Sen's slope estimates **a** Annual average temperature **b** *Rabi* temperature **c** Pre-*Kharif* temperature **d** *Kharif* temperature

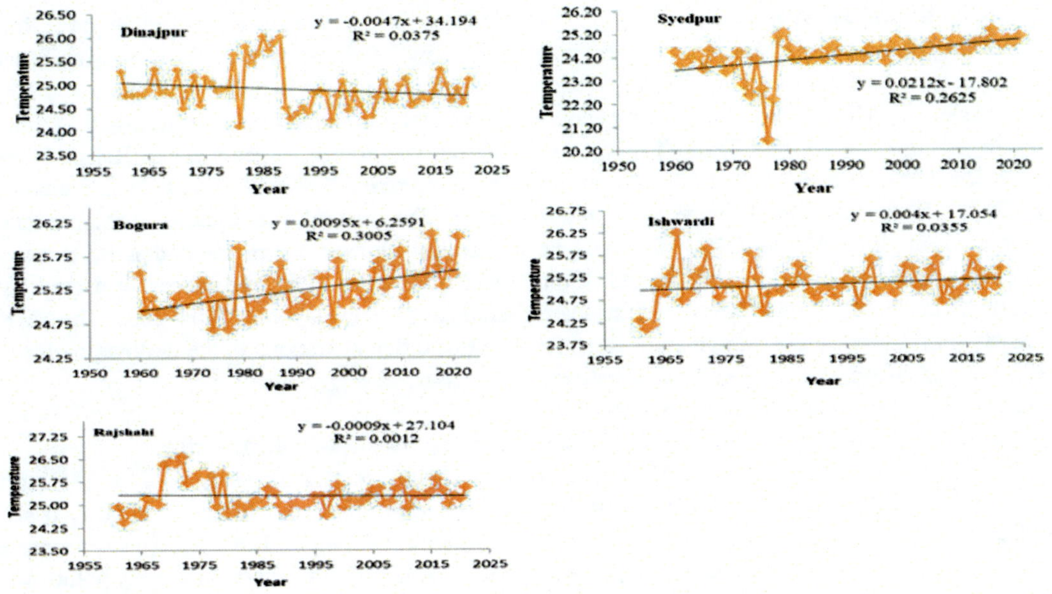

Fig. 5.9 Linear regression trends of annual average temperature during 1960–2021

5.3.6.2 Pre-Kharif Season

During the pre-*Kharif* season temperature analysis, Rajshahi station found average maximum temperature of 30.66 °C, while Syedpur station received 20.68 °C minimum average temperature. The overall average temperature ranged from 26.14 °C to 28.25 °C with a mean value of 27.47 °C. The standard deviation of pre-*Kharif* season temperature varied from 0.63 °C to 1.27 °C, and the CV showed 2.27% at Bogura station and high 4.85% at Syedpur station (Fig. 5.6c). In the study area, 80% of stations observed a strongly positive trend, while 20% station showed negative trend. From trend analysis, Dinajpur, Rajshahi, and Syedpur were statistically significant, but other two stations (Bagura, Ishwardi) were insignificant with increasing trends. Both the MK test and SR test revealed identical result, but only the Rajshahi station showed decreasing trend of − 0.744 in the SR test (Table 5.6). The proportion of change assessed by the Sen's slope estimates ranged from − 0.02 to 0.02 °C/year. However, a station like Dinajpur showed only a significant downward trend at 95% confidence interval, and the magnitude of change was –0.02 °C/year (Table 5.6). The areal dissemination of different stations indicates that in the future temperature trend is more increasing in the northwest region (Figs. 5.7c and 5.8c).

Table 5.6 Long-term seasonal temperature trend test results

Stations	Rabi season			Pre-Kharif season			Kharif season		
	MK test	Sen's slope	SR test	MK test	Sen's slope	SR test	MK test	Sen's slope	SR test
Bogura	1.85	0.007	1.875	0.48	0.003	0.512	5.95*	0.02*	7.542*
Dinajpur	− 0.890	− 0.09	− 1.130	− 3.92*	− 0.02*	− 4.9*	4.81*	0.01*	5.529*
Ishwardi	− 0.834	− 0.004	− 0.799	0.24	0.001	0.227	6.16*	0.01*	8.016*
Rajshahi	− 1.99	− 0.01	− 1.671	2.77*	0.015*	− 0.744	4.96*	0.01*	6.118*
Syedpur	2.81*	0.014*	2.956*	2.77*	0.02*	2.753*	5.41*	0.02*	6.216*

* Significant trends at 5% level, Sen's slope unit is in mm per yer

5.3.6.3 *Kharif* Season

The *Kharif* season, which runs from July through October, is primarily referred to as a rainy season of Bangladesh. The annual average temperature during this time varied from 27.28 °C (Syedpur) to 28.23 °C (Dinajpur), with a mean value of 28.08 °C. The maximum average temperature ranged from 29.33 °C to 29.03 °C, and the minimum average temperature varied from 21.78 °C to 26.70 °C (Fig. 5.6d). The standard deviation of *Kharif* season temperature was low (0.35 °C) at Ishwardi station and comparatively high (1.13 °C) at Syedpur station. Among all the stations, the coefficient of variation was maximum at Syedpur station with 4.08%. Trend analysis of different stations revealed that significant rising trends were found at all five stations, which means the temperature was increased rapidly in the *Kharif* season. The MK test result showed that Ishwardi station experienced the highest positive trend (6.16), while Dinajpur station found the least positive trend (4.81) at 95% confidence interval. Almost similar result was predicted by the SR test analysis. Sen's slope analysis points out that the degree of change varied from 0.01 to 0.02 °C/year with an average 0.014 °C/year (Table 5.6). The spatial map of MK test and Sen's slope estimates is displayed in (Figs. 5.7d and 5.8d) which indicates that significant rising trend continues in the northwest region.

The linear trend analysis during 1960–2021 is displayed in (Fig. 5.9). The graphical analysis indicates that annual average temperature trend lines are decreasing at Dinajpur and Rajshahi stations and increasing at Syedpur, Bogura, and Iswardi stations.

5.3.7 Sequential Mann–Kendall (SMK) Analysis for Annual Rainfall and Temperature

5.3.7.1 Annual Rainfall

The SMK test determines the approximate year where the trend begins at 95% confidence limit. Results from the SMK test revealed that considering $u(t)$ statistics had almost similar to the MK test with decreasing trend at Bogura, Rajshahi, and Ishwardi stations. Here, no,-significant change of points is depicted in Bogura and Ishwardi station, but some abrupt change in annual rainfall was found 15 years ago at Rajshahi station approximately with a significant change point. The rest of the stations (Dinajpur and Syedpur) showed rising trend that starts early year of 1960 to 1965 but noticeable decreasing changes also display after 2004–2005. The SMK plot of different stations in the northwest zone is displayed in (Fig. 5.10).

5.3.7.2 Annual Temperature

The annual temperature analysis of the SMK test exhibited an upward tendency in almost all stations except Dinajpur. From trend analysis, more than one non-significant trend turning points was described at Dinajpur, Ishwardi, and Rajshahi stations, but only Dinajpur station discovered decreasing trend which starts from 1988 to 1990 and it continued till 2020. On the other hand, Rajshahi and Ishwardi stations also found periodic fluctuations between the late 1990s and 2020. A periodic significant upward trend was detected at Bogura and Syedpur stations defining only a trend turning point in the year 2003–2004, and the increment continued. It means that the temperature trend analysis method is dominant after 2003–2004 for Bogura and Syedpur stations. The following Fig. 5.11 represents SMK test analysis trend of annual temperature of different station.

5.3.8 Forecast from ARIMA Models

According to different trend analysis methods, the northwest zone experiences high temperatures and less rainfall throughout the year. The country's economy mostly depends on agricultural production, and about 80% of the population, directly and indirectly, depends on primary economic activities (Kamruzzaman et al. 2018). In our country, the summer monsoon receives the most precipitation, which directly affects our agricultural productivity. Adequate information on total rainfall and temperature characteristics helps our

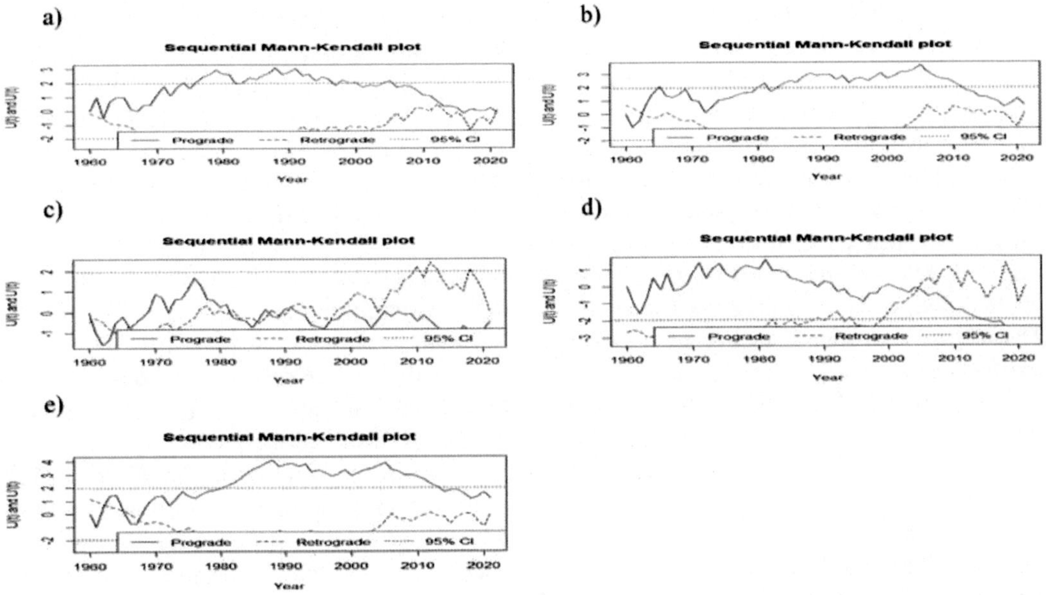

Fig. 5.10 Detection of annual rainfall trends **a** Bogura **b** Dinajpur **c** Ishwardi **d** Rajshahi **e** Syedpur by sequential MK analysis

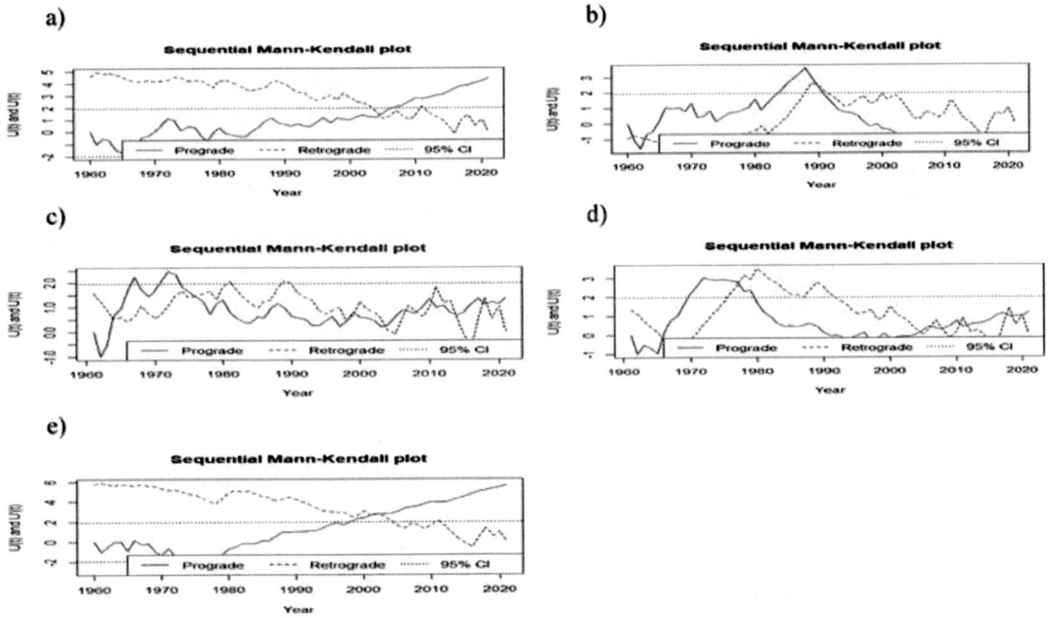

Fig. 5.11 Detection of annual temperature trends **a** Bogura **b** Dinajpur **c** Ishwardi **d** Rajshahi **e** Syedpur by sequential MK analysis

farmers and policymakers to draw a comprehensive picture of the monsoon and dry seasons. Early climate prediction lessens significant crop damage from drought, flood, and other climatic calamities. Moreover, if the total annual rainfall is normal, the intra-seasonal variability, such as early monsoon onset or usual monsoon throughout the year, may cause unavoidable disruption of agricultural activities, hydro-electric power supply, groundwater depletion, or even drinking water supply. The ARIMA technique is applied here to forecast monthly rainfall and temperature data for future five years (60 months) based on sixty years data. Previous years' data is cast-off to articulate the seasonal ARIMA model in determining model parameters.

5.3.8.1 Monthly Rainfall Forecast

Sixty years of monthly rainfall data is analyzed and used to predict future rainfall scenarios. The following table provides the estimated parameters for the ARIMA model of monthly rainfall (Table 5.7).

The station-wise forecasting values of the next five years monthly rainfall of different stations are displayed in (Fig. 5.12). In our study, monthly rainfall varied one station to another due to the diverse geographical locations.

5.3.8.2 Monthly Temperature Forecast

The monthly temperature predictions of all stations are detailed below (Table 5.8). The expected parameters of the ARIMA model for the next five years are based on the previous sixty years data.

It has been detected that the monthly predicting temperature not only fluctuates in one station with a small margin (Fig. 5.13) but also fluctuates from one station to another. Moreover, the temperature has been rising over the years in the northwest zone, and gradually it will increase in the near future also.

5.4 Conclusion

The present chapter elaborates on the latest trends in rainfall variability and temperature during 1960–2021 at five distinguished weather stations in the northwest Bangladesh. The MK test and SR test are equally effective in assessing rainfall and temperature trends. Sen's estimation also gives excellent results, and it is almost identical to the MK test statistic and SR test results. From the recent research work, it is established that the average annual rainfall is 1768 mm and temperature is 24.96 °C. Whereas Syedpur station shows the highest amount of annual average rainfall, Rajshahi station receives the maximum average temperature in the northwest zone. The study findings discover that only Rajshahi station shows maximum significant downward trend of annual rainfall (− 5.50 mm/year), as well as the *Kharif* season (− 3.80 mm/year) also by Sen's estimation. However, other stations represent increasing/decreasing trend by the MK test, SR test, and Sen's slope estimator, but they are not statistically significant. The total rainfall statistic exhibits a substantial change during the research period. These kinds of changes in most of the

Table 5.7 ARIMA models of monthly rainfall data

Station	Month January 1960–December 2021				
	Model	Sigma^2 estimated	Log-likelihood	AIC	BIC
Bogura	ARIMA(0,0,1)(1,1,0)[12]	14,253	− 4540.04	9086.09	9099.87
Dinajpur	ARIMA(0,0,3)(2,1,0)[12]	17,660	− 4618.73	9249.45	9277.0
Ishwardi	ARIMA(1,0,0)(1,1,0)[12]	13,235	− 4513.08	9032.16	9045.95
Rajshahi	ARIMA(1,0,0)(1,1,0)[12] with drift	9365	− 4386.2	8780.4	8798.78
Syedpur	ARIMA(1,0,0)(1,1,0)[12] with drift	21,813	− 4695.41	9398.83	9417.21

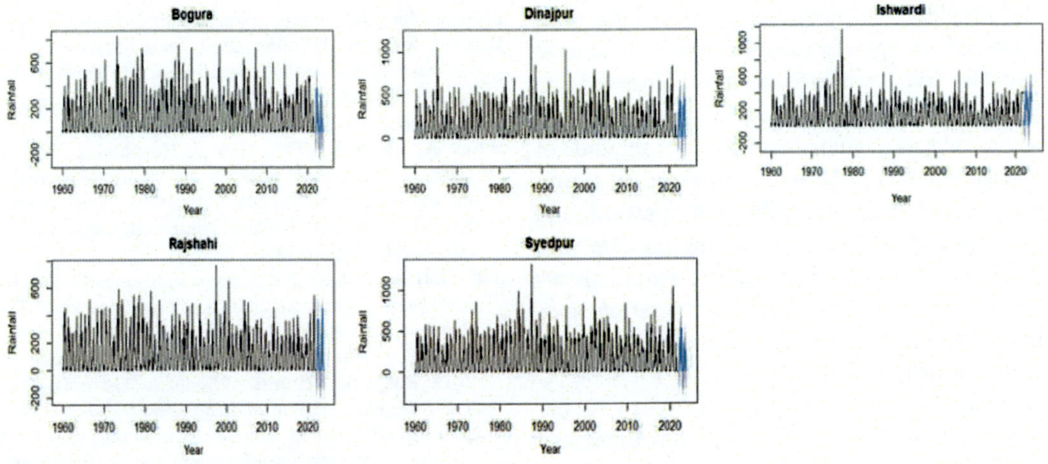

Fig. 5.12 Forecasting monthly rainfall of different stations

Table 5.8 ARIMA models of monthly temperature data

Station	Month January 1960–December 2021				
	Model	Sigma^2 estimated	Log-likelihood	AIC	BIC
Bogura	ARIMA(2,0,2)(2,1,0)[12] with drift	1.103	− 1073.5	2163	2199.76
Dinajpur	ARIMA(0,0,3)(2,1,2)[12]	0.9458	− 1023.63	2063.27	2100.03
Ishwardi	ARIMA(2,0,2)(2,1,2)[12]	0.6953	− 897.94	1813.89	1855.1
Rajshahi	ARIMA(0,0,2)(0,1,1)[12] with drift	0.8151	− 954.94	1919.88	1942.78
Syedpur	ARIMA(0,0,1)(0,1,1)[12] with drift	1.317	− 1146.91	2301.82	2320.2

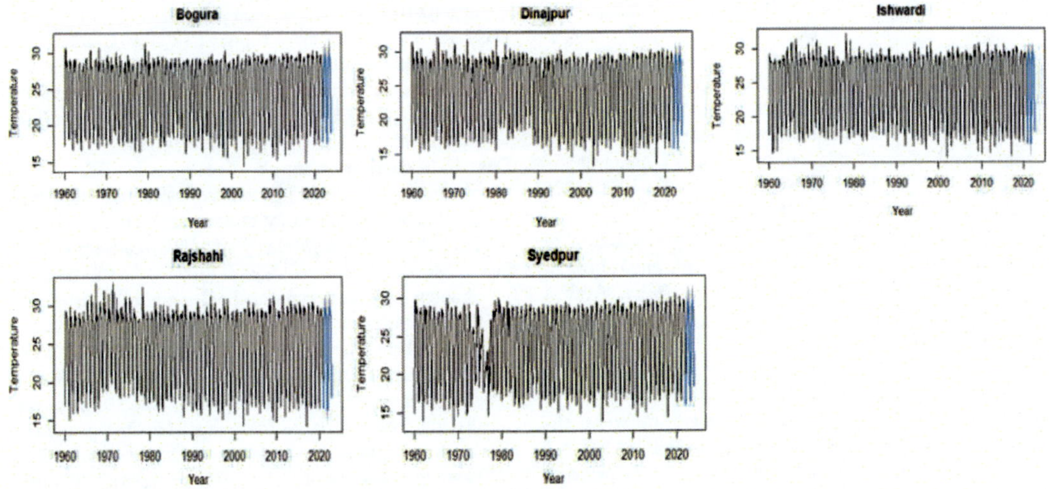

Fig. 5.13 Forecasting monthly temperature of different stations

stations placed the northwest region in danger and increased drought frequency. However, climate variable like temperature also observed an increasing trend 80% of stations, but Dinajpur station reveals a falling trend with a magnitude of change − 0.003 °C/year. From cropping season temperature trend analysis, both upward and downward trends are found during the *Rabi* and pre-*Kharif* seasons. But *Kharif* season shows noteworthy trend of increment at 5% significance level. The degree of change is assessed by Sen's slope estimates which differ from 0.01 to 0.02 °C per year with an average value of 0.014 °C per year. Additionally, SMK analysis has illustrated the starting and turning points of the annual trends. In this chapter, the ARIMA models are used for long-term monthly rainfall and temperature data to forecast future trend in the northwest region. Based on the findings, the information is gathered consistently with the previous studies and will be supportive for farmers, policymakers and researchers on local-scale scheduling on climate change scenarios of the country. Therefore, sustainable adaptation practices are mandatory for agricultural development in the northwest area.

References

Adams RM, Hurd BH, Lenhart S et al (1998) Effects of global climate change on agriculture: an interpretative review. Climate Res 11(1):19–30

Akter KS, Rahman M (2012) Spatio-temporal quantification and characterization of drought patterns in Bangladesh. J Water Environ Technol 10(3):277–288. https://doi.org/10.2965/jwet.2012.277

Alam J, Saha P, Mitra R, Das J (2023) Investigation of spatio-temporal variability of meteorological drought in the Luni River Basin, Rajasthan, India. Arab J Geosci 16(3):201. https://doi.org/10.1007/s12517-023-11290-8

Alamgir M, Shahid S, Hazarika MK, Nashrrullah S, Harun SB, Shamsudin S et al (2015) Analysis of meteorological drought pattern during different climatic and cropping seasons in Bangladesh. JAWRA J Am Water Resour Assoc 51(3):794–806. https://doi.org/10.1111/jawr.12276

Allen RJ, DeGaetano AT (2001) Estimating missing daily temperature extremes using an optimized regression approach. Int J Climatol 21(11):1305–1319. https://doi.org/10.1002/joc.679

Banglapedia (2003) Bangladesh Asiatic Society, Dhaka

Bari SH, Rahman MTU, Hoque MA et al (2016) Analysis of seasonal and annual rainfall trends in the northern region of Bangladesh. Atmos Res 176–177. https://doi.org/10.1016/j.atmosres.2016.02.008

Basak A, Rahman ATMS, Das J, Hosonod T, Kisi O (2022) Drought forecasting using the Prophet Model in semi-arid climate region of western India. Hydrol Sci J 67(9):1397–1417. https://doi.org/10.1080/02626667.2022.2082876

BBS (2019a) Statistical Yearbook of Bangladesh 2018. Bangladesh Bureau of Statistics. Government of The People's Republic of Bangladesh, Dhaka

BBS (2019b) Report on Agriculture and Rural Statistics 2018. Bangladesh Bureau of Statistics. Government of The People's Republic of Bangladesh, Dhaka

Bhuyan Md DI, Islam M, Bhuiyan K et al (2018) A trend analysis of temperature and rainfall to predict climate change for Northwestern region of Bangladesh. Am J Clim Change 07(02):115–134. https://doi.org/10.4236/ajcc.2018.72009

Box EP, Jenkins GM (1976) Time series analysis: forecasting and control (Revised edition). San Francisco, Holden Day

Das J, Bhattacharya SK (2018) Trend analysis of long-term climatic parameters in Dinhata of Koch Bihar district, West Bengal. Spatial Inf Res 26(3):271–280. https://doi.org/10.1007/s41324-018-0173-3

Das J, Mandal T, Saha P (2019) Spatio-temporal trend and change point detection of winter temperature of North Bengal, India. Spatial Inf Res 27(4):411–424. https://doi.org/10.1007/s41324-019-00241-9

Das J, Gayen A, Saha P, Bhattacharya SK (2020a) Meteorological drought analysis using Standardized Precipitation Index over Luni River Basin in Rajasthan, India. SN Appl Sci 2(9):1–17. https://doi.org/10.1007/s42452-020-03321-w

Das J, Rahman AS, Mandal T, Saha P (2020b) Challenges of sustainable groundwater management for large scale irrigation under changing climate in Lower Ganga River basin in India. Groundw Sustain Dev 11(1–11):100449. https://doi.org/10.1016/j.gsd.2020.100449

Das J, Mandal T, Saha P, Bhattacharya SK (2020b) Variability and trends of rainfall using non-parametric approaches: a case study of semi-arid area. MAUSAM 71(1):33–44. https://doi.org/10.54302/mausam.v71i1.4

Das J, Mandal T, Rahman S et al (2021a) Spatio-temporal characterization of rainfall in Bangladesh: an innovative trend and discrete wavelet transformation approaches. Theoret Appl Climatol 143(3–4):1557–1579. https://doi.org/10.1007/s00704-020-03508-6

Das J, Rahman AS, Mandal T, Saha P (2021b) Exploring driving forces of large-scale unsustainable groundwater development for irrigation in lower Ganga River basin in India. Environ Dev Sustain 23(5):7289–7309. https://doi.org/10.1007/s10668-020-00917-5

Dimri T, Ahmad S, Sharif M et al (2020) Time series analysis of climate variables using seasonal ARIMA approach. J Earth Syst Sci 129(1):149. https://doi.org/10.1007/s12040-020-01408-x

Eischeid JK, Baker CB, Karl TR et al (1995) The quality control of long-term climatological data using objective data analysis. J Appl Meteorol 34:2787–2795

Enders CK (2010) Applied missing data analysis. The Guilford Press, New York. http://books.google.co.ke/. ISBN: 978-1 60623-639-0 Assessed 20 Nov 2022

Esteban-Parra MJ, Rodrigo FS, Castro-Díez Y et al (1995) Temperature trends and change points in the northern Spanish Plateau during the last 100 years. Int J Climatol 15(9):1031–1042. https://doi.org/10.1002/joc.3370150909

Hossain M, Roy K, Datta D et al (2014) Spatial and temporal variability of rainfall over the South-West Coast of Bangladesh. Climate 2(2):28–46. https://doi.org/10.3390/cli2020028

Ingsrisawang L, Potawee D (2012) Multiple imputations for missing data in repeated measurements using MCMC and copulas. In: Proceedings of the international multi conference of engineers and computer scientist 2012. Assessed 20 July 2013

IPCC (2014) Climate change 2014: synthesis report. Contribution of Working Groups I, II and III to the fifth assessment report of the intergovernmental panel on climate change [Core Writing Team, Pachauri RK, Meyer LA (eds)]. IPCC, Geneva, Switzerland, p 151

Islam M, Hossain M, Sikder M et al (2019a) Farmers' adaptation strategies to drought and their determinants in Barind tract, Bangladesh. SAARC J Agric 17 (1):161–174. https://doi.org/10.3329/sja.v17i1.42769

Islam MS, Hossain MZ, Sikder MB et al (2019b) Drought adaptation measures and their effectiveness at Barind Tract in northwest Bangladesh: a perception study. Nat Hazards 97(3):1253–1276. https://doi.org/10.1007/s11069-019-03704-2

Islam T, Karim MR, Mondol AH et al (2021) Appraising trends and forecasting of hydroclimatic variables in the north and northeast regions of Bangladesh. Theor Appl Climatol 143(1–2):33–50. https://doi.org/10.1007/s00704-020-03411-0

Kamruzzaman M, Rahman MS, Ahmed MS et al (2018) Spatio-temporal analysis of climatic variables in the western part of Bangladesh. Environ Dev Sustain 20 (1):89–108. https://doi.org/10.1007/s10668-016-9872-x

Kamruzzaman M, Rahman MS, Basak A et al (2022) Assessment and adaptation strategies of climate change through the prism of farmers' perception: a case study. Int J Environ Sci Technol. https://doi.org/10.1007/s13762-022-04254-0

Karim F, Mainuddin M, Hasan M et al (2020) Assessing the potential impacts of climate changes on rainfall and evapotranspiration in the Northwest Region of Bangladesh. Climate 8(8):94. https://doi.org/10.3390/cli8080094

Kemp WP, Burnell DG, Everson DO, Thomson AJ et al (1983) Estimating missing daily maximum and minimum temperatures. J App Meteorol Climatol 22 (9):1587–1593

Kendall MG (1948) Rank correlation methods. 4th edition. Charles Griffin, London

Khan HR, Rahman A, Luo C, Kumar S et al (2019) Detection of changes and trends in climatic variables in Bangladesh during 1988–2017. Heliyon 5(3):e01268. https://doi.org/10.1016/j.heliyon.2019.e01268

Kirby JM, Ahmad MD, Mainuddin M et al (2015) The impact of irrigation development on regional groundwater resources in Bangladesh. Agric Water Manag 159: 264–276. https://doi.org/10.1016/j.agwat.2015.05.026

Mainuddin M, Kirby M (2015) National food security in Bangladesh to 2050. Food Secur 7(3):633–646. https://doi.org/10.1007/s12571-015-0465-6

Mainuddin M, Peña-Arancibia JL, Karim F et al (2022) Long-term spatio-temporal variability and trends in rainfall and temperature extremes and their potential risk to rice production in Bangladesh. PLOS Climate 1 (3):e0000009. https://doi.org/10.1371/journal.pclm.0000009

Makokha GL, Shisanya CA (2010) Trends in mean annual minimum and maximum near surface temperature in Nairobi City, Kenya. Adv Meteorol 2010:1–6. https://doi.org/10.1155/2010/676041

Mann HB (1945) Nonparametric tests against trend. Econometrica 13(3):245. https://doi.org/10.2307/1907187

Michiels P, Gabriels D, Hartmann R et al (1992) Using the seasonal and temporal Precipitation concentration index for characterizing the monthly rainfall distribution in Spain. CATENA19(1):43–58. https://doi.org/10.1016/0341-8162(92)90016-5

Mondol AH, Al-Mamun, Iqbal Metal (2018) Precipitation concentration in Bangladesh over different temporal periods. Adv Meteorol 2018:1–18. https://doi.org/10.1155/2018/1849050

Mosmann V, Castro A, Fraile R (2004) Detection of statistically significant trends in the summer precipitation of mainland Spain. Atmos Res 70(1):43–53. https://doi.org/10.1016/j.atmosres.2003.11.002

Mullick MRA, Nur RM, Alam MJ et al (2019) Observed trends in temperature and rainfall in Bangladesh using pre-whitening approach. Global Planet Change 172:104–113. https://doi.org/10.1016/j.gloplacha.2018.10.001

Nalley D, Adamowski J, Khalil B et al (2013) Trend detection in surface air temperature in Ontario and Quebec, Canada during 1967–2006 using the discrete wavelet transform. Atmos Res 375–398. https://doi.org/10.1016/j.atmosres.2013.06.011

Oliver JE (1980) Monthly precipitation distribution: a comparative index. Prof Geogr 32(3):300–309. https://doi.org/10.1111/j.0033-0124.1980.00300.x

Praveen B, Talukdar S, Shahfahad et al (2020) Analyzing trend and forecasting of rainfall changes in India using non-parametrical and machine learning approaches. Sci Rep 10(1):10342. https://doi.org/10.1038/s41598-020-67228-7

Rahman MR, Lateh H (2016) Spatio-temporal analysis of warming in Bangladesh using recent observed temperature data and GIS. Clim Dyn 46(9–10):2943–2960. https://doi.org/10.1007/s00382-015-2742-7

Rahman MR, Lateh H (2017) Climate change in Bangladesh: a spatio-temporal analysis and simulation of recent temperature and rainfall data using GIS and time series analysis model. Theoret Appl Climatol 128 (1–2):27–41. https://doi.org/10.1007/s00704-015-1688-3

Rahman AS, Kamruzzama Md, Jahan CS et al (2016a) Long-term trend analysis of water table using 'MAKESENS' model and sustainability of groundwater resources in drought prone Barind area, NW Bangladesh. J Geol Soc India 87(2):179–193. https://doi.org/10.1007/s12594-016-0386-9

Rahman MS, Kamruzzaman M, Jahan CS et al (2016b) Evaluation of spatio-temporal dynamics of water table in NW Bangladesh: an integrated approach of GIS and Statistics. Sustain Water Resour Manag 2(3):297–312. https://doi.org/10.1007/s40899-016-0057-4

Rahman MA, Yunsheng L, Sultana N et al (2017) Analysis and prediction of rainfall trends over Bangladesh using Mann-Kendall, Spearman's rho tests and ARIMA model. Meteorol Atmos Phys 129(4):409–424. https://doi.org/10.1007/s00703-016-0479-4

Rashid HE (1991) Geography of Bangladesh. University Press Ltd., Dhaka (Bangladesh)

Recha CW, Makokha GL, Traore PS et al (2012) Determination of seasonal rainfall variability, onset and cessation in semi-arid Tharaka district, Kenya. Theor Appl Climatol 108(3–4):479–494. https://doi.org/10.1007/s00704-011-0544-3

Sa'adi Z, Shahid S, Ismail T, Chung E-S et al (2019) Trends analysis of rainfall and rainfall extremes in Sarawak, Malaysia using modified Mann-Kendall test. Meteorol Atmos Phys 131(3):263–277. https://doi.org/10.1007/s00703-017-0564-3

Salam R, Towfiqul AR, Islam S et al (2020) Spatiotemporal distribution and prediction of groundwater level linked to ENSO teleconnection indices in the northwestern region of Bangladesh. Environ Dev Sustain 22(5):4509–4535. https://doi.org/10.1007/s10668-019-00395-4

Sen PK (1968) Estimates of the regression coefficient based on Kendall's Tau. J Am Stat Assoc 63 (324):1379–1389. https://doi.org/10.1080/01621459.1968.10480934

Shadmani M, Marofi S, Roknian M et al (2012) Trend analysis in reference evapotranspiration using Mann-Kendall and Spearman's Rho tests in Arid Regions of Iran. Water Resour Manage 26(1):211–224. https://doi.org/10.1007/s11269-011-9913-z

Shahid S (2010) Rainfall variability and the trends of wet and dry periods in Bangladesh. Int J Climatol 30 (15):2299–2313. https://doi.org/10.1002/joc.2053

Shahid S, Behrawan H (2008) Drought risk assessment in the western part of Bangladesh. Nat Hazards 46(3):391–413. https://doi.org/10.1007/s11069-007-9191-5

Shahid S, Khairulmaini OS (2009) Spatio-Temporal variability of rainfall over Bangladesh during the time period 1969–2003. Asia-Pac J Atmos Sci 45(3):375–389

Shahid S, Harun SB, Katimon A et al (2012) Changes in diurnal temperature range in Bangladesh during the time period 1961–2008. Atmos Res 118:260–270. https://doi.org/10.1016/j.atmosres.2012.07.008

Shaw R, Mallick F, Islam, A (eds) (2013) Climate change adaptation actions in Bangladesh. Springer, Japan. https://doi.org/10.1007/978-4-431-54249-0

Sneyers R (1990) On the statistical analysis of series of observations. Secr. of the World Meteorological Organization. World Meteorological Organization, ISBN 978-92-63-10415-1

Spearman C (2010) The proof and measurement of association between two things. Int J Epidemiol 39 (5):1137–1150. https://doi.org/10.1093/ije/dyq191

Tabari H, Talaee PH (2011) Temporal variability of precipitation over Iran: 1966–2005. J Hydrol 396(3–4):313–320. https://doi.org/10.1016/j.jhydrol.2010.11.034

Tonkaz T, Çetin M, Tülücü K et al (2007) The impact of water resources development projects on water vapor pressure trends in a semi-arid region, Turkey. Clim Change 82(1–2):195–209. https://doi.org/10.1007/s10584-006-9160-0

Uddin MA, Kamal AM, Shahid S et al (2020a) Volatility in rainfall and predictability of droughts in Northwest Bangladesh. Sustainability 12(23):9810. https://doi.org/10.3390/su12239810

Uddin MJ, Hu J, Islam MT et al (2020b) A comprehensive statistical assessment of drought indices to monitor drought status in Bangladesh. Arab J Geosci 13(9):323. https://doi.org/10.1007/s12517-020-05302-0

Yu WH, Alam M, Hassan A et al (2010) Climate change risks and food security in Bangladesh. Earthscan London, Washington, DC

Yu Z, Lei G, Jiang Z, Liu F et al (2017) ARIMA modeling and forecasting of water level in the middle reach of the Yangtze River. In: 2017 4th International Conference on Transportation Information and Safety (ICTIS), pp 172–177. https://doi.org/10.1109/ICTIS.2017.8047762

Yue S, Pilon P, Cavadias G (2002a) Power of the Mann-Kendall and Spearman's rho tests for detecting monotonic trends in hydrological series. J Hydrol 259(1–4):254–271. https://doi.org/10.1016/S0022-1694(01)00594-7

Yue S, Pilon P, Phinney B et al (2002b) The influence of autocorrelation on the ability to detect trend in hydrological series. Hydrol Process 16(9):1807–1829. https://doi.org/10.1002/hyp.1095

Zhang Q, Xu C-Y, Zhang Z (2009) Observed changes of drought/wetness episodes in the Pearl River basin, China, using the standardized precipitation index and aridity index. Theoret Appl Climatol 98(1–2):89–99. https://doi.org/10.1007/s00704-008-0095-4

6 Application of RS-GIS-Based Multi-Criteria Decision-Making Model (MCDM) on Site Suitability Analysis for Potato Cultivation in Jalpaiguri District, West Bengal, India

Indrajit Poddar, Amiya Basak, Jiarul Alam, Jayanta Das, and Asraful Alam

Abstract

Contemporarily, among the crucial techniques for generating precise land suitability map site suitability analysis (SSA) is the most acceptable one. This chapter intended to determine the fittest areas for producing potato crops in the Jalpaiguri district of West Bengal, India. Twenty-four criteria were selected for the study, and their weights were determined using the GIS-based fuzzy analytical hierarchy process (F-AHP) approach and expert judgment. Based on the results, soil texture (0.1349), soil organic carbon (0.0614), geomorphology (0.1523), slope (0.1043), soil moisture index (0.0845), temperature (0.0769), nitrogen (0.0529), and soil pH (0.0432) and rainfall (0.0635) are among the most crucial criteria for producing SSA for potato production. It has been observed that regions for potato production are extremely favorable (15.30%), highly suitable (25.11%), moderately suitable (28.55%), marginally suitable (22.52%), and not suitable (8.52%), and their geographical extend covers 508.83 km^2, 835.05 km^2, 949.35 km^2, 749.04 km^2, and 283.43 km^2 area, respectively. The area confirms the F-AHP approach's reliability in SSA under curve (AUC), and the value of AUC is 0.83. The results also have been verified with the help of GPS and an interview survey from the potato cultivators. The studies indicate expected scenarios for agricultural activity. Thus, an appropriate sustainable plan should be prepared to increase the regional agrarian output, and the local farmers should use targeted adaptable farming practice(s). Farmers, regional planners, and government representatives may utilize the proposed map, containing information about the agriculture suitability, to make essential choices for the area, including identifying viable agricultural areas, promoting agricultural development, and supporting local independent companies for potato farming.

I. Poddar (✉) · A. Basak · J. Alam
Department of Geography and Applied Geography, University of North Bengal, NBU, Darjeeling 734013, India
e-mail: indrajit.poddar.cob@gmail.com

A. Basak
e-mail: amiyabasak4@gmail.com

J. Alam
e-mail: geojia94@gmail.com

J. Das · A. Alam
Department of Geography, Rampurhat College, Rampurhat, Birbhum 731224, India
e-mail: jayanta.daas@gmail.com

A. Alam
e-mail: geoasraful@gmail.com

J. Das and S. Halder (eds.), *Advancement of GI-Science and Sustainable Agriculture*, GIScience and Geo-environmental Modelling, https://doi.org/10.1007/978-3-031-36825-7_6

Keywords

Agricultural suitability · Potato cultivation · Remote sensing · GIS · F-AHP · MCDM · Jalpaiguri

6.1 Introduction

The population of the world is increasing, particularly in developing countries. As a result, the rising population with rapid urbanization imposes extravagant pressure on agricultural and natural resources to meet their higher economic returns (Feizizadeh and Blaschke 2013; Ahmed et al. 2016). Excessive consumption or exploitation of natural resources due to population increase is causing land degradation (Maddahi et al. 2014), a severe threat to the agricultural system. In the Indian scenario, approximately 44% of total farm land has been degraded in recent times. The increasing trend of such farmland degradation threatens food security and the agricultural system (Roy et al. 2022). In response to these risks, contemporary agricultural landscapes must be redesigned to sustain agricultural production sustainably in the face of limited resources and climate change. Contextually, a comprehensive land suitability analysis (LSA) is critical for the agricultural sector in formulating land-use policies that promote a self-sufficient agricultural system on a sustainable basis.

In general, LSA for agricultural crop production refers to a land's ability or fitness to cultivate a specific crop efficiently and sustainably (Halder 2013; Jamil et al. 2018; Karimi et al. 2018). It is the function of proper land management (Zolekar 2018) by which we made (broader) characterization and quality assessment of the land and, accordingly, the suitable crop also be judged (Amara et al. 2016). LSA enables planners, decision-makers, and agricultural managers to build sustainable agricultural systems to enhance land productivity. It also allows for identifying the major limiting parameters of the land for certain crop production (El Baroudy 2016).

LSA in a region is often accomplished by efficiently connecting land features with land utilization (Kahsay et al. 2018). To conduct a systematic LSA for a certain crop, we must first understand the pedological, climatological, water resource, topographical, agricultural process, and technological aspects involved in land classification (Ahmed et al. 2016; Karimi et al. 2018), which can be gained through expert opinion or existing literature. Hence, we chose 14 parameters (geomorphology, slope, elevation, rainfall, temperature, soil texture, soil pH, organic carbon, boron, carbon, potassium, nitrogen, phosphorus, and zinc) from a varied collection of domains to conduct LSA for potato cultivation. These parameters are chosen based on expert judgment, the uniqueness of the subject area, and the available literature. The influence of these parameters vary depending on the quality of the land. Moreover, each parameter has a different effect based on the type of cultivable crop.

Geographic information systems and remote sensing (GIS and RS) are recognized as powerful technologies that can improve the availability, analysis, validation, and production of geospatial information (data) (Sarkar et al. 2021; Mitra et al. 2022). When GIS and RS are used with MCDM methodologies, it provides a useful tool for determining the suitable location for specific land use. Therefore, integrated approaches (i.e., MCDM approach in GIS environment) can aid land-use managers, administrators, and planners in making appropriate decisions. Influencing parameters can be calculated using GIS and RS. MCDM, on the other hand, efficiently combines a diverse set of influencing parameters into a suitability map (Mustafa et al. 2011). Consequently, MCDM approaches in a GIS environment have been widely used to identify suitable sites for producing various crops. For example, the suitability map for paddy (Everest et al. 2022; Kihoro et al. 2013; Özkan et al. 2019; Roy and Saha 2018; Sarkar et al. 2021), wheat (Amiri Kia and Naji Domirani 2018; Bagherzadeh and Gholizadeh 2016; Dadhich et al. 2017; Mokarram et al. n.d.; Sarkar et al. 2014), soybean (Radočaj et al. 2020), maize (Habibie et al. 2021; Kenzong et al. 2022; Ramamurthy et al. 2020), sorghum (Al-Mashreki et al. 2011; Elaalem 2012; Kahsay et al. 2018; Tadesse and Negese

2020), sugarcane (Jamil et al. 2018), barley (Fekadu and Negese 2020), etc., has been developed by a considerable number of researchers across the world.

Jalpaiguri is the northern district of West Bengal, and its economy is predominantly agrarian, while small and medium-scale enterprises are gradually expanding in the district (Datta 2020). After rice, wheat, and maize, potato is the fourth vital food crop in this concerned district. Farmers need much money to grow this crop; likewise, farmers make enough money by selling this crop. However, the price of potatoes sometimes depends on productivity; if there is a large production in one year, there is a chance that the price of potatoes will be low, resulting in a fall in income or a huge loss for farmers (Datta 2020). Sometimes, they often face huge losses for growing potatoes without testing the soil in the same way, judging the quality of the land. Hence, before cultivating this crop, the most important thing is to determine whether the land is suited for potato production. A systemic LSA is very much essential for this district. Consequently, the present work was pursued on a prospective site for the potato crop in the Jalpaiguri district. The current chapter's goal was to use the F-AHP method to identify suitable potato production sites.

6.2 Study Area

The most important district, Jalpaiguri, has an area of roughly 3386.18 km^2 and is situated on the southern side of the Foothills of the Himalayas. The district is shaped like a circle, with moderately uneven older and active flood plains, tea cultivation fields, dispersed forests in the north, and a flat stretch of ground with excellent agricultural prospects in the south (Roy et al. 2022). Jalpaiguri and Alipurduar districts were formed in 2014 when the Jalpaiguri district was divided into two districts (Das 2022). The ALOS PALSAR DEM for the terrain of the sub-Himalayan Jalpaiguri district produced by NASA indicates that the area's height ranges from 44 to 576 m. Several significant rivers drain the entire region, including the Tista, Mahananda, Jaldhaka, and others. As a result, a network of well drainage systems is seen in the research region. Jalpaiguri district has seven C.D. Blocks. The selected C.D. Blocks are Rajganj, Jalpaiguri, Mal, Maynaguri, Dhupguri, Nagrakata, and Matiali. The statistics of total population of the study area is 2,381,596 persons, while a rapid population growth in urban as well as rural areas is observed in recent times. Because of this, there is a tremendous strain on both agricultural areas and natural resources (Chandramouli and General 2011). A hot, humid climate with monsoonal rainfall prevails in the Jalpaiguri district. The mean annual rainfall ranges from 2561 to 3817 mm, and the average number of rainy days are 133, annually. The distribution of mean annual rainfall decreases from the northwest to the southwest. The yearly average temperature varies from 13.60 to 24.10 °C. Approximately 89.51% of the precipitation falls between June and September. The environment is quite muggy in the summertime, with a moisture content of 71–73%.

6.3 Materials and Methods

6.3.1 Data Sources

The present chapter of land suitability analysis exclusively focusing potato production is carried out by utilizing the available physicochemical qualities of soil and climatic data from the selected region obtained from various sources. For land suitability assessment (LSA), we have selected 14 factors (Table 6.1). A suitable place with a favorable temperature and soil compatibility for crop cultivation is essential for increasing crop productivity, and these criteria must be taken into account for field cultivation. The elevation and slope maps were produced using the SRTM DEM. The average yearly rainfall and temperature were acquired from the Indian Meteorological Department (IMD). The soil moisture dataset (2022) was collected from MOSDAC; the links are below (Table 6.1). Sixty soil samples were taken from seven community development blocks (CD blocks) in the Jalpaiguri district (Fig. 6.1), using a stratified random sampling technique, in order to collect information about its soil chemical properties, viz. soil pH, nitrogen (N), phosphorus (P), potassium (K), zinc (Zn), and boron (Br) (Table 6.1). The collected samples were examined in the Faculty of Tea Science, University of North Bengal's Soil Testing Laboratory using the standard methods as follows. The pipette method of mechanical investigation of soils based on improved dispersion was employed to evaluate the soil texture (Olmstead et al. 1930). An HM Digital pH meter (Model pH-80) was used to examine the soil pH, followed by Richard's 1: 2.5 soil: water ratio approach (Richards 1944), and the Walkley–Black wet digestion strategy was employed (Walkley and Black 1934) to determine organic carbon (OC). The nitrogen levels were determined using the alkaline potassium permanganate ($KMnO_4$) procedure (Emmert 1934). The Olsen method calculated the available phosphorous in a 0.5 M $NaHCO_3$ extract (Pierzynski 2000). The obtainable K was determined using a flame photometer and the neutral ammonium acetate technique (Jackson et al. 1973). The available micronutrient cations (Zn) were mined using the DTPA-CaC12 extractant at pH 7.3 (Lindsay and Norvell 1978) and measured using an atomic absorption spectrophotometer (AAS).

6.3.2 Parameters

Generally, the major crop like potatoes is ideally produced in sandy, well-drained soil. A poorly

Table 6.1 Parameters used for the generation of potato cultivation suitability map, their descriptions, and web sources

Thematic layers	Source of the parameters	Description of the data types	Weblink	Year of acquisition
Geomorphology	Geological Survey of India (GSI)	Shapefile, 1:1,182,847	https://www.gsi.gov.in	2022
Rainfall	Meteorological Department of India (IMD)	High-resolution binary rainfall data: resolution (0.25° × 0.25°)	https://www.imdpune.gov.in	2022
Temperature	India Meteorological Department (IMD)	High-resolution binary temperature data: resolution (0.5° × 0.5°)	https://www.imdpune.gov.in	2022
Soil moisture index	MOSDAC	L-band radiometric data	https://www.mosdac.gov.in/opendata/soil_moisture/smi/	2022
Elevation	ALOS PALSAR DEM	Raster layer, 12.5 m × 12.5 m	https://search.earthdata.nasa.gov	2022
Slope	ALOS PALSAR DEM	Raster layer, 12.5 m × 12.5 m	https://search.earthdata.nasa.gov	2022
Soil Texture	ESRI shape file	Digital Soil Map of the World (DSMW), FAO	www.fao.org	2022
Nitrogen (N), phosphorus (K), potassium (P), zinc (Zn), boron (Br)	Attribute data	Interpolation through IDW in ArcGIS vs.10.4.0	Soil Testing Laboratory, Department of Tea Science, University of North Bengal, Siliguri, West Bengal	2022
Location of the soil sample point	Interview survey from the tea cultivators	Geographic position system (GPS) location	Field survey by researchers	2022

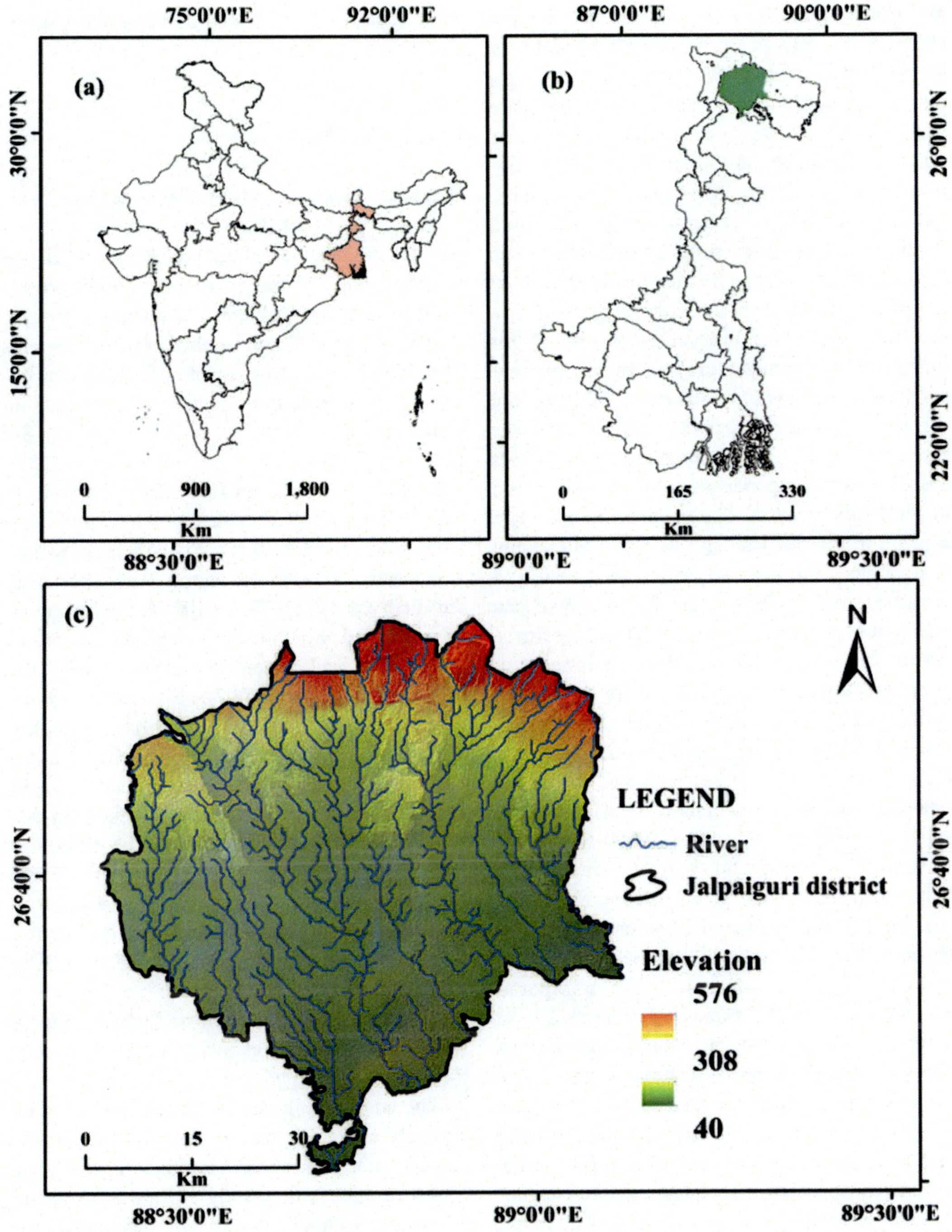

Fig. 6.1 Study area **a** India, **b** West Bengal, and **c** Jalpaiguri district

drained soil is more prone to produce sick tubers. The ideal soil pH range is from 6 to 6.5 for potato production, but they may withstand pH levels as low as 5 (Yusianto et al. 2020). The potato crop is susceptible to dryness due to its shallow root system. Maintaining optimum quantities of available soil moisture throughout the growing season is the only way to generate high yields of good-quality potatoes. Without consistent rainfall, irrigation is required regularly. Soil moisture should be above 70% at all phases; stress occurs when the available soil water falls below 65% (Dubois et al. 2021). However, soil fertility, nutrition, and natural ecosystems, such as food production, depending on the amount of organic substances (the carbon in organic compounds) are present in the soil. As a result, soil care is essential. Moreover, restoration is critical for long-term development; thus, this is an important parameter for potato cultivation. Among the various important chemical parameters, the researchers have selected some important factors based on expert experiences, which are discussed in the below section. Boron helps to maintain calcium in the cell wall and works with calcium to promote plant resilience to disease, pests, and environmental challenges. In this regard, it can aid in the reduction of sprout apical necrosis. Boron deficiency inhibits the occurrence of enzymatic browning in potato tubers (Mahmoud et al. 2019; Shirur et al. 2021). Nitrogen is essential for promoting growth and producing high harvests, and it is essential for leaf development and tuber growth and yield because it supports adequate glucose production in the leaves. Nitrogen utilization at later growth stages helps keep the canopy green and increases output. Phosphorus is essential for early root and shoots growth because it provides energy for plant functions like ion uptake and transportation. An appropriate supply of phosphorus at tuber commencement ensures the formation of an optimal number of tubers (Shirur et al. 2021). Potassium is essential for water interactions and overall plant health. Throughout the growing season, potato plants consume a lot of potassium. Potassium impacts nutrient transport and carbohydrate transfer from the leaf to the tuber. Potassium is essential for good-yielding potatoes (Das et al. 2017). Zinc plays an important role in nitrogen metabolism, and zinc-deficient crops will have lower protein levels. Zinc affects starch content as well. Zinc is also required for auxin production, as well as cell division and elongation (Yusianto et al. 2020).

6.3.3 Methodology

6.3.3.1 Fuzzy Analytical Hierarchy Process (F-AHP)

The fuzzy analytic hierarchy process (F-AHP*)* is a multifaceted algorithm to solve multi-criteria decision-making problems. This method creates its identity in different applied research spheres like flood risk assessment, farming sector, industrial production, reducing waterlogging problems, etc. (Chang et al. 2007; van der Sar et al. 2015; van Laarhoven and Pedrycz 1983). After Saaty's work, van Laarhoven and Pedrycz (1983) remodeled previous research and used fuzzy triangular numbers for fuzzification and set a pair-wise comparison matrix (van Laarhoven and Pedrycz 1983). This AHP Method is generally directed by human perception, involving subjectivity and uncertainty (Das et al. 2017; Chan and Kumar 2007). Sometimes multi-criteria decision-making method faces complexity in the comparison process and fuzzy utility ranking and can give unreliable results. Therefore, several complexities can be generated during dealing with this method. The theory of 'fuzzy set' was introduced to eradicate the lacuna of AHP (Jakhar and Barua 2013). Many studies have got their significance with valid outcomes using this method. Therefore, we use the F-AHP method to conduct our study with a trustworthy result. Several steps are followed to calculate the weightage of the factors and sub-factors used for F-AHP.

The analytical hierarchy process (AHP) is the simplest and most recommendable method for solving multi-criteria decision-making problems. This technique is frequently utilized across many industries, including land suitability classes for agriculture, flash floods, vulnerability assessments, and flood susceptibility (Li et al. 2009; Lin et al. 2020; Poddar et al. 2023; Sarkar et al. 2021; Veisi et al. 2016). The AHP approach often relies on subjective and imprecise human perception (Chan and Kumar 2007). Hence, several issues have emerged in the AHP method. According to Jakhar and Barua (2013), to eradicate the shortcomings of AHP, *fuzzy set theory* was proposed. The fuzzy number set theory can successfully handle the challenges that emerged from the AHP technique (Chang et al. 2007). Simultaneously, a good number of studies have proven the reliability of the F-AHP method. Consequently, we used the F-AHP approach to locate sites that would be ideal for potato production. (Jamil et al. 2018; Sridhar and Ganapuram 2021). When employing F-AHP, there are multiple phases involved in determining the weight of factors and sub-factors. They are as follows:

Step I: In this step, we clearly defined the land suitability map for potato cultivation in the study area.

Step II: The fuzzy set theory is often helpful in providing valuable information for defining issues that arise in vague and confusing surroundings. In the theory of fuzzy set, if *X* denotes a group of objects and '*x*'s with values (×1, 2, × 3,..., *xn*) indicates the generic element of '*X*,' then the fuzzy set '*M*' for this object set is designated by {(*E*, *M*(*x*))|*x* ∈ *X*} (Dubois and Prade 1979). Moreover, *M* (*x*) defines its function, which usually occurs on the scale of real numbers and ranges from the interval [0,1]. In addition, the triangular fuzzy number, known as the triangle fuzzy number (TFN), is a specific type of fuzzy number, and this is the most common choice for practical uses. The membership function of any TFN (*a*, *b*, *c*) is expressed mathematically (*M*(*x*)) as Eq. (6.1).

$$\mu_M(x) = \begin{Bmatrix} 0, & x \le a, \\ \frac{x-a}{b-a}, & x \in [a, b] \\ \frac{x-c}{b-c}, & x \in [b, c], \\ 0, & x > c, \end{Bmatrix}, \quad (6.1)$$

where *a*, *b*, and *c* replicate the lower boundary, mean, and upper boundary of TFN, respectively.

Step III: The hierarchical structure should be derived in this step. The pair-wise calculation matrix is created by taking into account expert linguistic judgments.

Step IV: We developed a fuzzy positive matrix in this phase. To convert the language phrases into definite values, the pair-wise assessment matrices are relieved by correspondent positive triangular fuzzy numbers, written as *K* = [*zij*] *n* × *m nm*. Furthermore, in fuzzy positive matrices, the fuzzy elements are represented by *zij* = (*aij*, *bij*, *cij*), and in this connection, positive fuzzy numbers fulfill the following property as Eq. (6.2):

$$a_{ij} = \frac{1}{a_j}, b_{ij} = \frac{1}{b_j}, c_{ij} = \frac{1}{c_j}, \text{ where, } \quad i \,\text{and}\, j = 1, 2, \ldots k. \quad (6.2)$$

Step V: In this stage, the preferred weight for each component is determined. For experts to calibrate preference weights for each component and its corresponding sub-classes and determine their degree of importance, they must transform fuzzy set theory into membership values. We calculated the weight of components and sub-factors using the geometric mean approach. In this chapter, the maximum Eigen Value (γ_{max}) = 18.23636137, RI = 1.537, CI = 0.054657254, CR = 0.031516718, and *n* = 14. However, the CR value should be less than 0.10; in our study, we obtained a CR value of = 0.040627609. Consequently, these above-mentioned parameters are regarded to be indispensable in the F-AHP technique (Figs. 6.2 and 6.3).

6.4 Results and Discussion

6.4.1 Evaluation of Certain LSA Development Criteria

The highest Eigen Value in this investigation was 18.23636137, the RI was 1.537, the CI was 0.054657254, the CR was 0.031516718, and the n was 14. The calculated value must typically be under 0.10, and this investigation found that it was = 0.031506518. All the thematic layers are thus seen as being crucial for this applied method (F-AHP). According to experts' findings, every component was assigned a certain value of weightage. Soil texture (0.1349), organic carbon

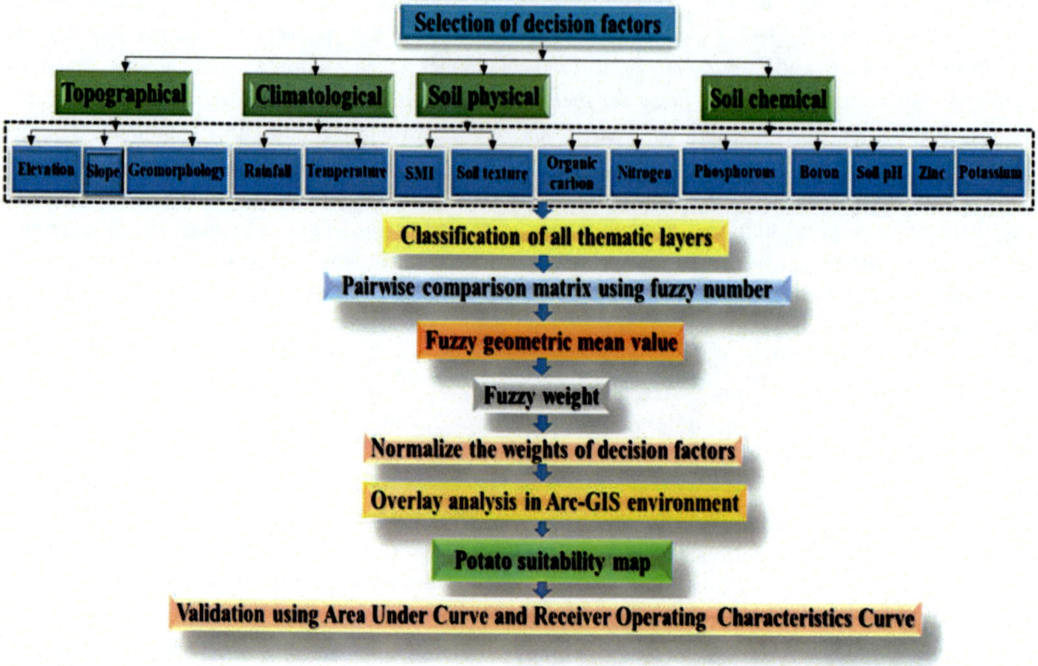

Fig. 6.2 Methodological flowchart for generation of land suitability classes for potato cultivation

(0.0614), geomorphology (0.1523), slope (0.1043), soil moisture index (0.0845), temperature (0.0769), nitrogen (0.0529), and pH (0.0432), rainfall (0.0635), were the most crucial variables for establishing LSA for potato cultivation (Table 6.2). Contrarily, the least significant components were zinc (0.0391) and elevation (0.2742) boron (0.0245). Fine and coarse loamy soil textures were found in the southwestern and middle portion of the entire research area, which is more suitable for potato cultivation. Soil moisture index is medium in the middle part of the study region, and nitrogen in the soil was all greater in the western part of the Rajganj block, and Jalpaiguri Sadar block, the northern part of the Matiali block is very much suitable for potato cultivation. It can be seen in the geomorphology map (*n*), older flood plain and younger alluvial plain are very suitable for potato cultivation, and this region is situated in the southern part of the Mal and Matiali block and the southern part of the Dhupguri block. A similar pattern can be seen in altitude and degree of slope, which were greater in the western and northwestern regions and progressively dropped from north to south. The Rajganj block, Maynaguri block, Dhupguri block, Southern part of the Mal and Matiali block, and Jalpaiguri Sadar block have pH levels between 5 and 7, which are appropriate for the production of potatoes, according to the geographical map of soil pH (FAO 1976). The western part of the Rajganj block, the northern zone of the Jalpaiguri Sadar block, Dhupguri, and Maynaguri have greater phosphorous contents (Fig. 6.4k). The northern part of the Mal and Matiali blocks southern part of the Dhupguri block has greater potassium contents (Fig. 6.5n).

6.4.1.1 Geographical Distribution Land Suitability Zones for Potato Cultivation

The output of fourteen raster layers has been combined to create a land suitability map for potato cultivation, viz. elevation, geomorphology, rainfall, temperature, soil texture, soil moisture index, soil pH, soil organic carbon, boron, nitrogen, potassium, phosphorus, and

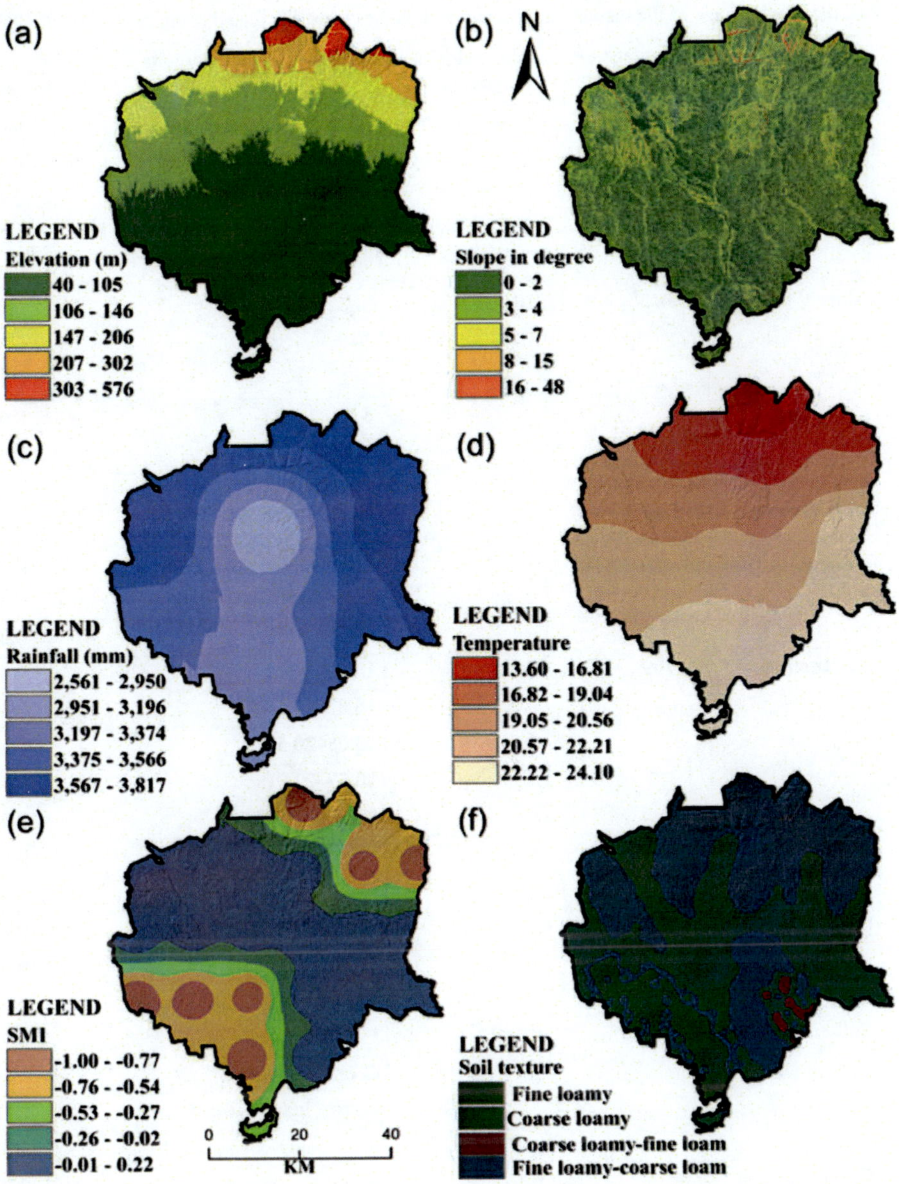

Fig. 6.3 Thematic layers of **a** elevation, **b** slope, **c** rainfall, **d** temperature, **e** soil moisture index (SMI), and **f** soil texture

zinc. Finally, the five land suitability classes for potato cultivation have been identified in the Jalpaiguri district. Identifying adequate and appropriate components is a crucial LSA component for a given crop. The parameters' relative importance in the LSA for potato cultivation determined the weights they should be given. The land suitability map for potato cultivation was built to show which areas are suitable for potato farming. In accordance with FAO (1976), we divided the 'potato cultivation land suitability region' into five categories: very highly suitable (S1), highly suitable (S2), suitable (S3), marginally suitable (S4), and not appropriate (N). The findings show regions for potato farming that are extremely favorable (15.30%), highly

Table 6.2 Criteria weightage of the thematic layers and sub-classes were calculated using the F-AHP MCDM method

SL. No	Parameters	Weightage of the criteria	Percentage of weightage	Sub-classes of the criteria	Weightage of sub-classes
1	Elevation (m)	0.2742	2.74	40–105	0.361
				106–146	0.294
				147–206	0.162
				207–302	0.131
				303–576	0.052
2	Slope (°)	0.1043	10.43	0–2	0.432
				3–4	0.32
				5–7	0.13
				8–15	0.062
				16–48	0.056
3	Rainfall (mm)	0.0635	6.35	2561–2950	0.47
				2951–3196	0.35
				3197–3374	0.08
				3375–3566	0.07
				3567–3817	0.03
4	Temperature (° C)	0.0769	7.69	13.60–16.81	0.411
				16.82–19.04	0.369
				19.05–20.56	0.105
				20.57–22.21	0.102
				22.22–24.10	0.013
5	Soil moisture index	0.0845	8.45	− 1.00 to − 0.77	0.421
				− 0.76 to − 0.54	0.306
				− 0.53 to − 0.27	0.135
				− 0.26 to − 0.02	0.101
				− 0.01 to 0.22	0.037
6	Soil texture	0.1349	13.49	Fine loamy	0.512
				Coarse loamy	0.302
				Coarse loamy—fine loam	0.131
				Fine loamy-coarse loam	0.055
7	Organic carbon (kg/ha)	0.0614	6.14	0	0.015
				0.01–220.00	0.131
				220.01–274.00	0.154
				274.01–364.00	0.236
				364.01–564.00	0.464
8	Nitrogen (kg/ha)	0.0529	5.29	< 310.00	0.034
				310.01–342.67	0.085
				342.68–376.67	0.164
				376.68–412.00	0.352
				412.01 <	0.365

(continued)

Table 6.2 (continued)

SL. No	Parameters	Weightage of the criteria	Percentage of weightage	Sub-classes of the criteria	Weightage of sub-classes
9	Phosphorous (kg/ha)	0.0627	6.27	< 66.18	0.039
				66.19–82.94	0.085
				82.95–96.76	0.167
				96.77–109.41	0.305
				109.42 <	0.404
10	Boron (kg/ha)	0.0245	2.45	< 0.20	0.071
				0.21–0.26	0.098
				0.27–0.26	0.167
				0.27–0.31	0.299
				0.38 <	0.365
11	Soil pH	0.0432	4.32	< 5.04	0.082
				5.05–5.59	0.273
				5.60–6.17	0.404
				6.18–6.78	0.205
				6.79 <	0.036
12	Zinc (kg/ha)	0.0391	3.91	< 1.29	0.039
				1.30–2.12	0.127
				2.13–2.99	0.195
				3.00–3.96	0.264
				3.97 <	0.375
13	Potassium (kg/ha)	0.0724	7.24	< 240	0.034
				241–280	0.085
				281–320	0.164
				321–360	0.352
				361 <	0.365
14	Geomorphology	0.1523	15.23	Active flood plain	0.204
				Dam and reservoir	0.012
				Embankment	0.011
				Highly dissected structural hills and valleys	0.013
				Mass wasting products	0.014
				Moderately dissected Structural hills and valleys	0.021
				Older flood plain	0.277
				Piedmont alluvial plain	0.217
				Pond	0.01
				River	0.01
				Waterbodies—unclassified	0.01
				Younger alluvial plain	0.201

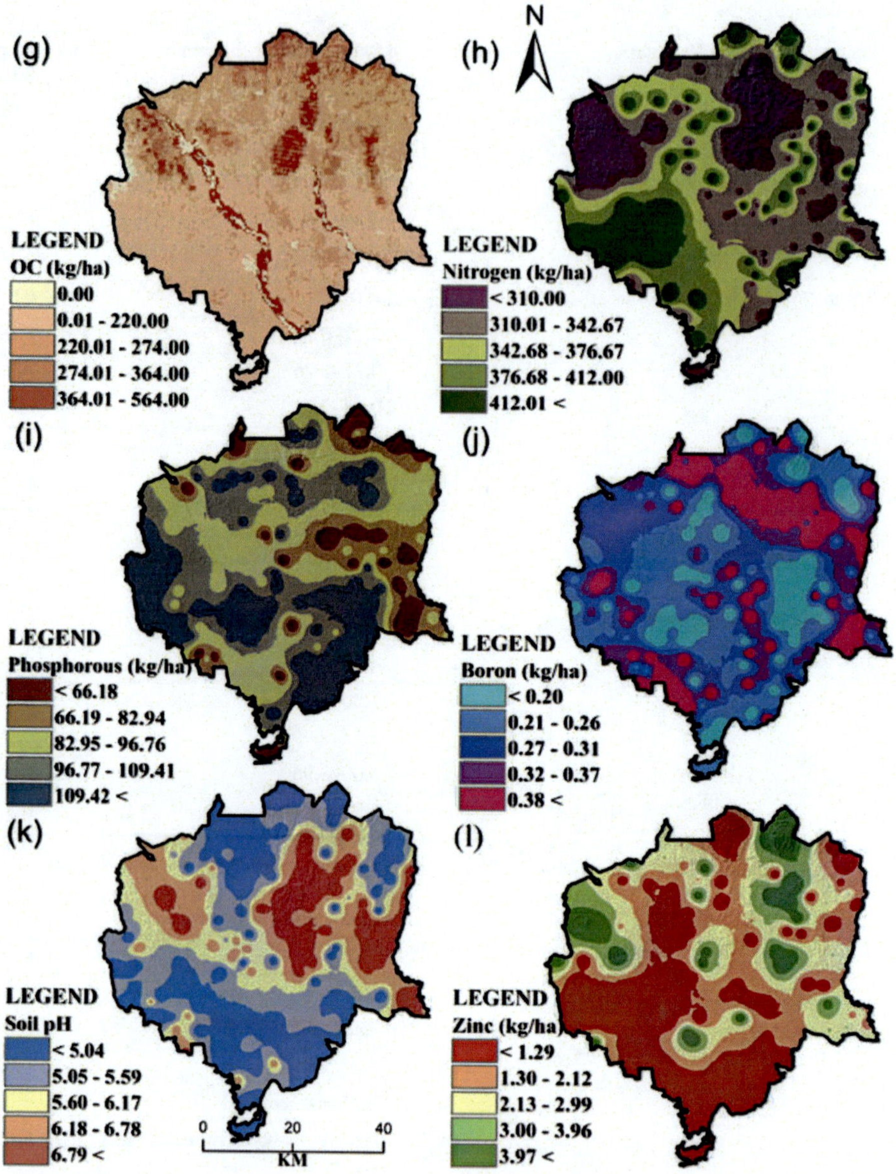

Fig. 6.4 Thematic layers of **g** soil organic carbon (OC), **h** nitrogen, **i** phosphorus, **j** boron, **k** soil pH, and **l** zinc

suitable (25.11%), suitable (28.55%), marginally suitable (22.52%), and not suitable (8.52%), and their spatial extend covers 508.83 km^2, 835.05 km^2, 949.35 km^2, 749.04 km^2, and 283.43 km^2 area, respectively. Though, the composed map underlined that a very high suitability land for potato cultivation zone covers 15.30% area, which includes the southern portion of the Matiali and Nagrakata block, the northern portion of the Nagrakata and Maynaguri block, the northeastern portion of the Jalpaiguri block, the southern portion of the Dhupguri western portion of the Rajganj block, and the northeastern portion of the Nagrakata block. Gentle slopes influence significant land suitability zones for potato cultivation with deep soils, fine loamy soil texture, very high nitrogen, potassium, and phosphorus, soil organic carbon contents, not

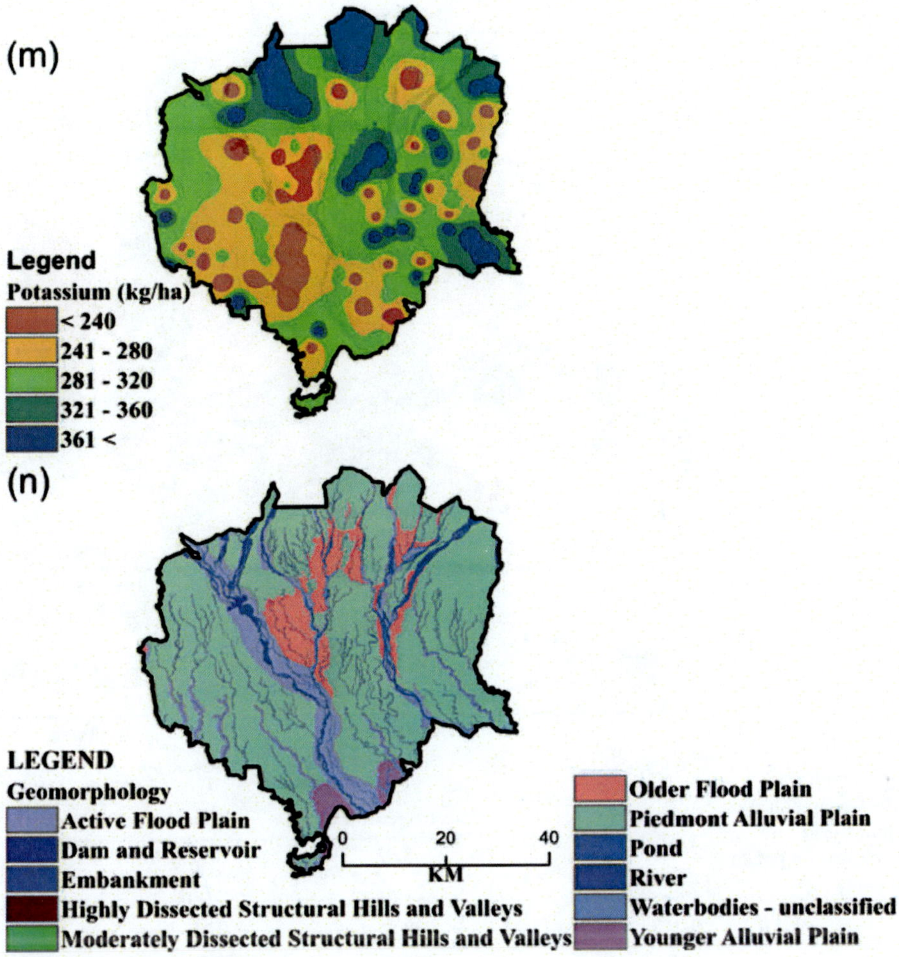

Fig. 6.5 Thematic layers of **m** potassium and **n** geomorphology

excessive rainfall, older and active flood plains, moderate soil moisture index, lower elevation, and moderate temperature.

The high suitability land for potato cultivation zone covers an area of 25.11%, with the western and central portion of the Dhupguri block, the southern portion of the Matiali and Nagrakata block, the northern portion of the Nagrakata and Maynaguri block, the northeastern portion of the Jalpaiguri block, the southern portion of the Dhupguri block western portion of the Rajganj block, and the northeastern portion of the Nagrakata block. Major features of high land suitability classes for potato cultivation contain gentle to moderate slopes with deep soils, fine loamy soil texture, high nitrogen, potassium, and phosphorus contents, moderate rainfall, piedmont alluvial plains, moderate soil moisture index, moderate temperature and high soil organic carbon contents. The moderately suitable areas of potato cultivation have 28.55% area coverage and are located scattered throughout the Rajganj block, the western part of the Dhupguri block, the central part of the Nagrakata block, the central part of the Matiali block, and the maximum part of the Jalpaiguri Sadar block. The medium land suitability classes for the potato cultivation zone seem to have a moderate to gentle slope, course loamy soil texture, piedmont alluvial plains, high rainfall intensity, moderate soil

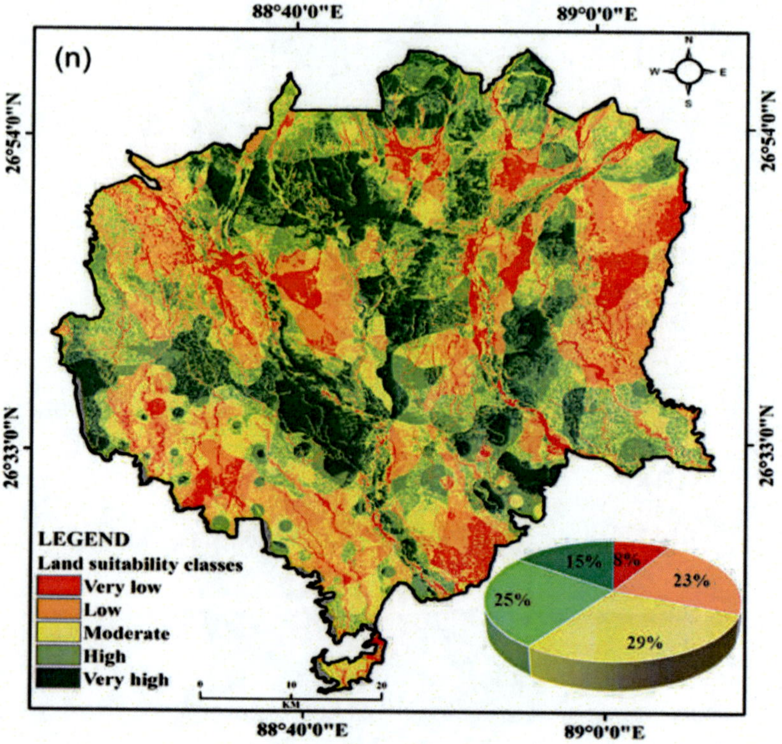

Fig. 6.6 Land suitability map for potato cultivation classes and their geographical distribution in percentage

Table 6.3 Geographical distribution of land suitability classes for potato cultivation area

Area in km^2	Percentage of area	Land suitability classes
283.43	8.52	Very low
749.04	22.52	Low
949.35	28.55	Moderate
835.05	25.11	High
508.83	15.30	Very high

moisture index, moderate nitrogen, phosphorous, potassium, zinc, boron, soil pH, and soil organic carbon contents.

The low land suitability classes for potato cultivation occupy 22.52% of the total geographical region under the Jalpaiguri district. The region is scattered throughout the Rajganj and Jalpaiguri blocks, the northwestern portion of the Mal block, the northeastern portion of the Rajganj block, the middle portion of the Dhupguri block, the middle portion of the Nagrakata block, the southern portion of the Maynaguri block, and middle portion of the Matiali block. The low land suitability classes for potato cultivation zone seem to have a steep slope, course loamy soil texture, piedmont alluvial plains, high rainfall intensity, high soil moisture index, low nitrogen, phosphorous, potassium, zinc, boron, soil pH, and low soil organic carbon contents.

The not suitable potato cultivation zone occupies 8.52% of the entire geographical region (Jalpaiguri district). The region is scattered throughout the Rajganj and Jalpaiguri blocks, the northwestern portion of the Mal block, the northeastern portion of the Rajganj block, the middle portion of the Dhupguri block, the middle portion of the Nagrakata block, the southern portion of the Maynaguri block, and middle

Table 6.4 Geographical distribution of potato cultivation areas in different community development blocks of Jalpaiguri district

Land suitability classes for potato cultivation	Area in km^2	Area in %	Geographical distribution of potato cultivation areas in different community development blocks
Very high	508.83	15.30	Southern part of the Matiali and Nagrakata block, northern part of the Nagrakata and Maynaguri C.D. block, northeastern part of the Jalpaiguri block, southern part of the Dhupguri C.D. block western part of the Rajganj C.D. block, northeastern part of the Nagrakata C.D. block
High	835.05	25.11	Western central part of the Dhupguri block, southern part of the Matiali and Nagrakata block, northern part of the Nagrakata and Maynaguri block, northeastern part of the Jalpaiguri block, southern part of the Dhupguri C.D. block western part of the Rajganj block, northeastern part of the Nagrakata C.D. block
Medium	949.35	28.55	Located scattered throughout the Rajganj block, western part of the Dhupguri block, central part of the Nagrakata C.D. block, central part of the Matiali C.D. block and maximum part of the Jalpaiguri C.D. block
Low	749.04	22.52	Located scattered throughout the Rajganj and Jalpaiguri block, northwestern part of the Mal block, northeastern part of the Rajganj C.D. block, middle part of the Dhupguri C.D. block, middle part of the Nagrakata C.D. block, southern part of the Maynaguri C.D. block and middle part of the Matiali C.D. block
Very low	283.43	8.52	Located scattered throughout the Rajganj and Jalpaiguri block, northwestern part of the Mal C.D. block, northeastern part of the Rajganj C.D. block, middle part of the Dhupguri C.D. block, middle part of the Nagrakata C.D. block, southern part of the Maynaguri C.D. block and middle part of the Matiali C.D. block
Total	3325.699	100	

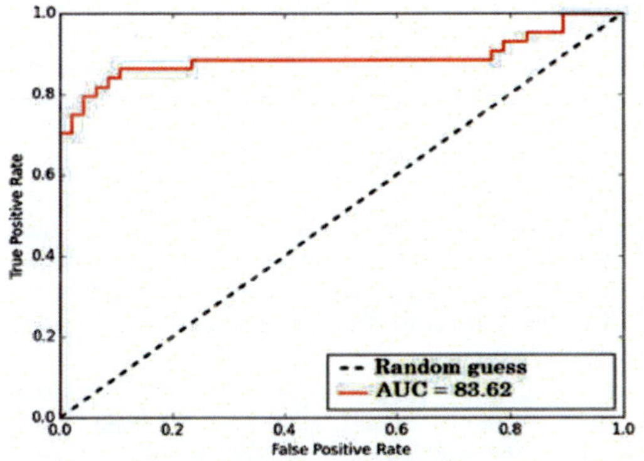

Fig. 6.7 Model validation using the AUC and ROC curve

portion of the Matiali block. The not suitable for potato cultivation zone classes for potato cultivation zone seem to have a steep slope, course loamy soil texture, pond, river, very high rainfall intensity, very high soil moisture index, very low nitrogen, phosphorous, potassium, zinc, boron, soil pH, and very low soil organic carbon contents (Fig. 6.6; Tables 6.3 and 6.4).

6.4.1.2 Validation

Any model's validity is assessed by comparing its anticipated outcomes to a collection of facts from the actual world, looking at how well it achieves its objectives, and assuring the model is appropriate (Poddar et al. 2023; Mitra and Das 2022). The fuzzy analytic hierarchy process (F-AHP) model has been validated using the receiver operating characteristics (ROC) curve. The area defines the accuracy of prediction or classification under the ROC curve (AUC) (Zhou et al. 2020). In this study, we have run and validated this model based on field survey means and locating the points on Google earth images with the help of a GPS survey. The AUC value ranges between 0 and 1. If the value is less than 0.5, it suggests that the model's output is unsuitable and must be classified again, while a value near 1 shows that the output is well defined (Mallick et al. 2021). The ROC curve was generated for this study to validate this research's results (Fig. 6.7). The result depicts that the obtained AUC value was 83.62. Therefore, this validation method is highly recommended for the future research of site suitability analysis for crop cultivation.

6.5 Conclusion

Although the area has a sufficient amount of potential farmland, irresponsible farming practices and overpopulation have recently led to environmental degradation and a decline in soil productivity. Additionally, the region's tiny, dispersed, and sparse landholdings need to produce more grain to meet the region's rising demand. The critically endangered area needs immediate action to enhance farmland quality and create livelihood opportunities through regional resource planning. To preserve soil quality, advance habitat, and provide a sizable source of income for the area's producers, various methods like organic agriculture and smart farming should be taken into account in marginal and currently appropriate places. Additionally, climate pattern estimates would have been a great way to identify which places are perfectly adapted for resilient and environmentally friendly agricultural methods since such assessment might help create a successful likelihood strategic framework. Additionally, it is encouraged to embrace and encourage recreational and tourism activities in regions that need more productive sites. In conclusion, the present chapter proclaims an essential toolkit for deciding how farmland use will develop in the sub-Himalayan area and for closing the food security gap. To determine agriculturally suitable places for long-term land management, the results of this chapter may be utilized as a reference for legislation. Lastly, state and municipal authorities must offer locals financial, infrastructural, and technical support to improve land utilization planning.

References

Ahmed GB, Shariff ARM, Balasundram SK, bin Abdullah AF (2016) Agriculture land suitability analysis evaluation based multi criteria and GIS approach. IOP Conf Ser Earth Environ Sci 37(1):012044

Al-Mashreki MH, Akhir JBM, Rahim SA, Lihan K, Haider AR (2011) GIS-based sensitivity analysis of multi-criteria weights for land suitability evaluation of sorghum crop in the Ibb Governorate Republic of Yemen. J Basic Appl Sci Res 1(9):1102–1111

Amara DMK, Kamanda PJ, Patil PL, Kamara AM (2016) Land suitability assessment for maize and paddy production in Bogur microwatershed using remote sensing and GIS techniques. Int J Environ Agric Biotechnol 1(3):238561

Amiri Kia F, Naji Domirani S (2018) Land suitability evaluation for dryland wheat production in Fars province based on climatic and physiographic factors and integrated TOPSIS-AHP model in GIS environment. Appl Field Crops Res 30(4):74–92

Bagherzadeh A, Gholizadeh A (2016) Modeling land suitability evaluation for wheat production by parametric and TOPSIS approaches using GIS, northeast of Iran. Model Earth Syst Environ 2(3):1–11

Chan FTS, Kumar N (2007) Global supplier development considering risk factors using fuzzy extended AHP-based approach. Omega 35(4):417–431. https://doi.org/10.1016/j.omega.2005.08.004

Chandramouli C, General R (2011) Census of India. Rural urban distribution of population, Provisional population total. Office of the Registrar General and Census Commissioner, New Delhi, India. hl=en&as_sdt=0%2C5&q=census+of+india+2011&btnG=

Chang CW, Wu CR, Lin CT, Chen HC (2007) An application of AHP and sensitivity analysis for selecting the best slicing machine. Comput Ind Eng 52(2):296–307. https://doi.org/10.1016/j.cie.2006.11.006

Dadhich G, Patel PR, Kalubarme MH (2017) Agriculture land suitability evaluation for wheat cultivation using geomatics for Patna District, India. Int J Agric Resour Gov Ecol 13(1):91. https://doi.org/10.1504/IJARGE.2017.084038

Das M (2022) An Appraisal of spatio-temporal occurrence of fluvial hazard: case study of flood in the districts of Jalpaiguri and Alipurduar (West Bengal, India). In: Islam A, Das P, Ghosh S, Mukhopadhyay A, Das Gupta A, Kumar Singh A (eds) Fluvial systems in the anthropocene (pp 265–275). Springer International Publishing. https://doi.org/10.1007/978-3-031-11181-5_14

Das J, Gayen A, Saha S, Bhattacharya SK (2017) Modelling of alternative crops suitability to tobacco based on Analytical hierarchy process in Dinhata subdivision of Koch Bihar district, West Bengal. Model Earth Syst Environ 3(4):1571–1587. https://doi.org/10.1007/s40808-017-0392-y

Datta K (2020) Economics behind potato production-with special reference to Jalpaiguri district of West Bengal. Asian J Econ Bus 1(1)

Dubois A, Teytaud F, Verel S (2021) Short term soil moisture forecasts for potato crop farming: a machine learning approach. Comput Electron Agric 180:105902. https://doi.org/10.1016/j.compag.2020.105902

El Baroudy AA (2016) Mapping and evaluating land suitability using a GIS-based model. CATENA 140:96–104

Elaalem M (2012) Land suitability evaluation for Sorghum based on Boolean and Fuzzy-multi-criteria decision analysis methods. Int J Environ Sci Dev 3 (4):357

Emmert EM (1934) A rapid method for the determination of total nitrogen in soil. Soil Sci 38(2):139–142. https://doi.org/10.1097/00010694-193408000-00005

Everest T, Sungur A, Özcan H (2022) Applying the best-worst method for land evaluation: a case study for paddy cultivation in northwest Turkey. Int J Environ Sci Technol 19(4):3233–3246

FAO (1976) World Health Organization (Technical Series No. 593; Methodology of Nutritional Suveillancee (pp 41–59). Joint FAO/UNICEF/WHO Expert committee. http://www.fao.org/3/x5310e/x5310e00.htm

Feizizadeh B, Blaschke T (2013) Land suitability analysis for Tabriz County, Iran: a multi-criteria evaluation approach using GIS. J Environ Planning Manage 56 (1):1–23

Fekadu E, Negese A (2020) GIS assisted suitability analysis for wheat and barley crops through AHP approach at Yikalo sub-watershed, Ethiopia. Cogent Food Agric 6(1):1743623

Habibie MI, Noguchi R, Shusuke M, Ahamed T (2021) Land suitability analysis for maize production in Indonesia using satellite remote sensing and GIS-based multicriteria decision support system. GeoJournal 86(2):777–807. https://doi.org/10.1007/s10708-019-10091-5

Halder JC (2013) Land suitability assessment for crop cultivation by using remote sensing and GIS. J Geogr Geol 5(3):p65. https://doi.org/10.5539/jgg.v5n3p65

Jackson WA, Flesher D, Hageman RH (1973) Nitrate uptake by dark-grown corn seedlings: some characteristics of apparent induction. Plant Physiol 51 (1):120–127. https://doi.org/10.1104/pp.51.1.120

Jakhar SK, Barua MK (2013) Supply chain agility for firm's performance: a study of textile-apparel-retail supply chain network. Int J Agile Syst Manag 6 (3):215. https://doi.org/10.1504/IJASM.2013.054975

Jamil M, Ahmed R, Sajjad H (2018) Land suitability assessment for sugarcane cultivation in Bijnor district, India using geographic information system and fuzzy analytical hierarchy process. GeoJournal 83(3):595–611. https://doi.org/10.1007/s10708-017-9788-5

Kahsay A, Haile M, Gebresamuel G, Mohammed M (2018) Land suitability analysis for sorghum crop production in northern semi-arid Ethiopia: application of GIS-based fuzzy AHP approach. Cogent Food Agric 4(1):1507184. https://doi.org/10.1080/23311932.2018.1507184

Karimi F, Sultana S, Shirzadi Babakan A, Royall D (2018) Land suitability evaluation for organic agriculture of wheat using GIS and multi-criteria analysis. Papers Appl Geogr 4(3):326–342. https://doi.org/10.1080/23754931.2018.1448715

Kenzong B, Bitondo D, Tamfuh PA, Kameni GSK, Vounang JG, Enang RK, Temgoua E, Bitom D (2022) Assessing and mapping land suitability units for maize production using integrated DEMATEL-ANP model and GIS in the Foumbot agricultural basin (Cameroon Western Highlands). J Geosci Environ Prot 10 (06):57–85. https://doi.org/10.4236/gep.2022.106005

Kihoro J, Bosco NJ, Murage H (2013) Suitability analysis for rice growing sites using a multicriteria evaluation and GIS approach in great Mwea region, Kenya. Springerplus 2(1):265. https://doi.org/10.1186/2193-1801-2-265

Li L, Shi ZH, Yin W, Zhu D, Ng SL, Cai CF, Lei AL (2009) A fuzzy analytic hierarchy process (FAHP) approach to eco-environmental vulnerability assessment for the danjiangkou reservoir area, China. Ecol Model 220(23):3439–3447. https://doi.org/10.1016/j.ecolmodel.2009.09.005

Lin K, Chen H, Xu CY, Yan P, Lan T, Liu Z, Dong C (2020) Assessment of flash flood risk based on improved analytic hierarchy process method and integrated maximum likelihood clustering algorithm. J Hydrol 584:124696. https://doi.org/10.1016/j.jhydrol.2020.124696

Lindsay WL, Norvell WA (1978) Development of a DTPA soil test for zinc, iron, manganese, and copper. Soil Sci Soc Am J 42(3):421–428. https://doi.org/10.2136/sssaj1978.03615995004200030009x

Mahmoud AWM, Abdeldaym EA, Abdelaziz SM, El-Sawy MBI, Mottaleb SA (2019) Synergetic effects of zinc, boron, silicon, and zeolite nanoparticles on confer tolerance in potato plants subjected to salinity. Agronomy 10(1):19. https://doi.org/10.3390/agronomy10010019

Mitra R, Das J (2022) A comparative assessment of flood susceptibility modelling of GIS-based TOPSIS, VIKOR, and EDAS techniques in the Sub-Himalayan foothills region of Eastern India. Environ SciPollut Res. https://doi.org/10.1007/s11356-022-23168-5

Mitra R, Saha P, Das J (2022) Assessment of the performance of GIS-based analytical hierarchical process (AHP) approach for flood modelling in Uttar Dinajpur district of West Bengal, India. Geomat Nat Haz Risk 13(1):2183–2226. https://doi.org/10.1080/19475705.2022.2112094

Mokarram M, Rangzan K, Moezzi A, Baninemeh J (n.d.) Land suitability evaluation for wheat cultivation by fuzzy theory approach as compared with parametric method 38

Olmstead LB, Alexander LT, Middleton HE (1930) A pipette method of mechanical analysis of soils based on improved dispersion procedure. United States

Department of Agriculture, 1488-2016-123973:1–23. hl=en&as_sdt=0%2C5&q=Olmstead+et+al.+1930&btnG

Özkan B, Dengiz O, Demirağ Tİ (2019) Site suitability assessment and mapping for rice cultivation using multi-criteria decision analysis based on fuzzy-AHP and TOPSIS approaches under semi-humid ecological condition in delta plain. Paddy Water Environ 17(4): 665–676. https://doi.org/10.1007/s10333-019-00692-8

Pierzynski GM (2000) Methods of phosphorus analysis for soils, sediments, residuals, and waters. North Carolina State University

Poddar I, Alam J, Basak A, Mitra R, Das J (2023) Application of a geospatial-based subjective MCDM method for flood susceptibility modeling in Teesta River Basin, West Bengal, India. In: Das J, Bhattacharya SK (eds) Monitoring and managing multi-hazards (pp 135–152). Springer International Publishing. https://doi.org/10.1007/978-3-031-15377-8_10

Radočaj D, Jurišić M, Gašparović M, Plaščak I (2020) Optimal soybean (*Glycine max* L.) Land suitability using GIS-based multicriteria analysis and Sentinel-2 multitemporal images. Remote Sens 12(9):1463. https://doi.org/10.3390/rs12091463

Ramamurthy V, Reddy GPO, Kumar N (2020) Assessment of land suitability for maize (*Zea mays* L) in semi-arid ecosystem of southern India using integrated AHP and GIS approach. Comput Electron Agric 179:105806. https://doi.org/10.1016/j.compag.2020.105806

Richards LA (1944) Moisture retention by some irrigated soils as related to soil-moisture tension (69). J Agric Res. hl=en&lr=&id=BZVUHdFQLbIC&oi=fnd&pg=PA215&dq=Richard%27s+1:+2.5+soil:+water+ratio+approach&ots=jxSLLnqE0D&sig=xaeoeV6JeDzxFfdzZrRAMzJr1jI&redir_esc=y#v=onepage&q&f=false

Roy J, Saha S (2018) Assessment of land suitability for the paddy cultivation using analytical hierarchical process (AHP): a study on Hinglo river basin, Eastern India. Model Earth Syst Environ 4(2):601–618. https://doi.org/10.1007/s40808-018-0467-4

Roy S, Singha N, Bose A, Basak D, Chowdhury IR (2022) Multi-influencing factor (MIF) and RS–GIS-based determination of agriculture site suitability for achieving sustainable development of Sub-Himalayan region, India. Environ Dev Sustain https://doi.org/10.1007/s10668-022-02360-0

Sarkar B, Das P, Islam N, Basak A, Debnath M, Roy R (2021) Land suitability analysis for paddy crop using GIS-based Fuzzy-AHP (F-AHP) method in Koch Bihar district, West Bengal. Geocarto Int 1–27. https://doi.org/10.1080/10106049.2021.2007299

Sarkar A, Ghosh A, Banik P (2014) Multi-criteria land evaluation for suitability analysis of wheat: a case study of a watershed in eastern plateau region, India. Geo-Spatial Inf Sci 17(2):119–128. https://doi.org/10.1080/10095020.2013.774106

Shirur V, Channakeshava S, Bhairappanavar ST (2021) Effect of different levels of gypsum and boron on yield, nutrient content and uptake by potato (*Solanum tuberosum* L.). J Pharmacognosy Phytochem 10(1): 1163–1169. https://www.phytojournal.com/archives/2021/vol10issue1/PartQ/9-6-360-750.pdf

Sridhar P, Ganapuram S (2021) Morphometric analysis using fuzzy analytical hierarchy process (FAHP) and geographic information systems (GIS) for the prioritization of watersheds. Arab J Geosci 14(4):236. https://doi.org/10.1007/s12517-021-06539-z

Tadesse M, Negese A (2020) Land suitability evaluation for sorghum crop by using GIS and AHP techniques in Agamsa sub-watershed, Ethiopia. Cogent Food Agric 6(1):1743624. https://doi.org/10.1080/23311932.2020.1743624

van der Sar T, Casola F, Walsworth R, Yacoby A (2015) Nanometre-scale probing of spin waves using single electron spins. Nat Commun 6(1):7886. https://doi.org/10.1038/ncomms8886

van Laarhoven PJM, Pedrycz W (1983) A fuzzy extension of Saaty's priority theory. Fuzzy Sets Syst 11(1–3): 229–241. https://doi.org/10.1016/S0165-0114(83)80082-7

Veisi H, Liaghati H, Alipour A (2016) Developing an ethics-based approach to indicators of sustainable agriculture using analytic hierarchy process (AHP). Ecol Ind 60:644–654. https://doi.org/10.1016/j.ecolind.2015.08.012

Walkley A, Black IA (1934) An examination of the Degtjareff method for determining soil organic matter, and a proposed modification of the chromic acid titration method. Soil Sci 37(1):29–38. https://doi.org/10.1097/00010694-193401000-00003

Yusianto R, Marimin S, Hardjomidjojo H (2020) Spatial analysis for crop land suitability evaluation: a case study of potatoes cultivation in Wonosobo, Indonesia. Int Semin Appl Technol Inf Commun (ISemantic) 2020:313–319. https://doi.org/10.1109/iSemantic50169.2020.9234284

Zolekar RB (2018) Integrative approach of RS and GIS in characterization of land suitability for agriculture: a case study of Darna catchment. Arab J Geosci 11(24):780. https://doi.org/10.1007/s12517-018-4148-4

Check for updates

Comparative Assessment of Projected Suitability of Finger Millet Crops in Tamil Nadu and Parambikulam Aliyar Basin Using ECOCROP Model: A Geospatial Approach

7

P. Dhanya, T. Sankar, and V. Geethalakshmi

Abstract

Climate change is anticipated to pose extensive adverse effects on the range and suitability of various crop-growing areas across the tropics. The agro-climatic suitability of the finger millet (Ragi) crop in Tamil Nadu and Parambikulam Aliyar Basin (PAB) is simulated using an updated version of the FAO's ECOCROP model integrated with DIVA-GIS software. The WorldClim dataset is used worldwide in climate change-related suitability studies. For this purpose, CCSM4 datasets, one of the global circulation models (GCMs) whose climate projections were employed in the IPCC's Fifth Assessment Report, are utilized based on RCP 2.6. The RCP 2.6 scenario projects an average global warming increase of around 1 °C by the end century. The results indicate that DIVA-GIS embedded in the ECOCROP model is an efficient tool in identifying the agro-climatic suitability of finger millet crops for the state as a whole and at the PAB basin scale. Changes in the current suitability (as a mean value for the period between 1970 and 2000) and future simulated suitability conditions for the year 2050 show that there may be a positive change of about 30–40% more areas under excellent future suitability class for the finger millet crop in the state and PAB. The findings reveal that Tirupur, the northern part of Krishnagiri, a small patch in Erode, parts of Kanyakumari, Thoothukudi, etc., have shown unsuitable or very marginally suitable for finger millet crops in the future under the purview of climate change. The places that are likely to be excellent for cultivation in the future are Nilgiris, southwest of Coimbatore, Theni, Dharmapuri, Salem, and north part of Namakkal. Vellore, Nagapattinam, Thiruvarur, Cudullore, Kancheepuram, Thiruvallur, and others come under excellent to very suitable categories. The outcomes from the secondary data analysis based on efficient cropping zones and ECOCROP model-based outcomes have been further evaluated for better comparative analysis. As per the current situation, most of the state districts fall under highly inefficient zones except 5–6. It is alarming to note that even though there is current climatic suitability as per ECOCROP, finger millet cultivation in coastal areas of Kancheepuram,

P. Dhanya (✉)
Directorate of Crop Management, Tamil Nadu Agriculture University, Coimbatore, Tamil Nadu, India
e-mail: dhanyaeptri@gmail.com

T. Sankar
Agro Climate Research Centre, Directorate of Crop Management, Tamil Nadu Agriculture University, Coimbatore, Tamil Nadu, India

V. Geethalakshmi
Director Crop Management, Tamil Nadu Agriculture University, Coimbatore, Tamil Nadu, India

J. Das and S. Halder (eds.), *Advancement of GI-Science and Sustainable Agriculture*,
GIScience and Geo-environmental Modelling, https://doi.org/10.1007/978-3-031-36825-7_7

Thiruvannamalai, Villupuram, Cuddalore, Ariyalur, Thiruvarur, and Nagapattinam is completely neglected. It is significant to note that currently, there are no most efficient cropping zone (MECZ) in the state. Hence, ample scope for wider scaling up of its cultivation. India, with its National Mission on Sustainable Agriculture and Food Security, is working toward food and nutritional security. Hence enhancing more areas under finger millets in suitable zones would be highly appreciated.

Keywords

Crop suitability · Agro-climatic zones · ECOCROP model · Geospatial modeling · Finger millet · Parambikulam Aliyar Basin

7.1 Introduction

Geospatial technologies provide huge potential for climate change impact, vulnerability assessments, and land use planning (IPCC 2012). Worldwide, geospatial tools, particularly maps, are extensively utilized to convey spatial disparities in climate change and land degradation (Fish 2020; Quach and Jenny 2020). Geospatial models support crucial spatial data analysis and modeling to evaluate climate change's current and anticipated implications and variability on major crops (Das et al. 2017). Several scholars have proposed conducting an elaborate investigation at the local level to measure the effects and enhance our understanding of crop susceptibility and societal reactions to climate change. (Knight and Messer 2012; Ramachandran et al. 2017).

Rain-fed farming in India faces multiple problems due to the vagaries of monsoons, rising temperatures due to climate change, water shortages, and so forth (Sivakumar et al. 2005). In the semi-arid tropics of South Asia, subsistence farmers cultivate finger millet (*Eleusine coracana* [L.]) as a principal food crop. It constitutes an essential element of farming systems in arid and semi-arid regions. In India, the promotion of rice and wheat cultivation was prevalent during the Green Revolution in the 1960s and 70s. However, the production and consumption of wide traditional millet varieties started diminishing across the nation (Rurinda et al. 2014). Nonetheless, as almost 60% of the cultivated land in the country is rain-fed and not irrigated, farmers growing rice and wheat are excessively dependent on climatic conditions, which are increasingly becoming unfavorable for farming due to the impacts of climate change. Hence more climate-resilient crops have to be promoted in the future. The cultivation of major cereals is surpassed by millets in terms of tolerance to environmental stresses due to several morpho-physiological, molecular, and biochemical characteristics possessed by the latter (Bandyopadhyay et al. 2017). Millets, a climate-resilient crop, can increase farming communities' income and food security in arid and semi-arid regions. Their extensive root systems enable them to endure various ecological conditions even in the face of water scarcity. (Satyavathi et al. 2021).

Karnataka, Andhra Pradesh, Odisha, and Tamil Nadu are the main states where finger millet is predominantly cultivated in India. Side by side, the finger millets can be grown as a cereal crop in arid and semi-arid regions. This crop is known to have the capacity to withstand drought conditions (Gupta et al. 2017). It is a wonder crop because it is rich in calcium, about ten times that of rice or wheat (Shukla et al. 2015). Millets are cultivated over an area of 15.48 million hectares in India, resulting in a production of 17.2 million tons and a yield of 1111 kg/ha (Directorate of Economics and Statistics 2015). The chapter endeavored to understand the potentiality of the geospatial model eco crop to assess the suitability of drought hardy crops for the state as a whole and the PAB in particular based on the suitability of finger millet presently and in the future.

Tamil Nadu's geographic location renders it susceptible to natural calamities, including cyclones, droughts, floods, and tsunamis caused by earthquakes (Varadan and Kumar 2015). Increased frequency of unexpected dry spells and rising temperatures are likely to negatively impact soil moisture availability, thereby increasing the risk of crop failure in semi-arid south India. Research on the impact of climate change on the phenology and grain yield of cereal crops has been focused mainly for the eastern delta regions of Tamil Nadu (Yadav et al. 2015). In Tamil Nadu, finger millet (Ragi) is cultivated as a Rabi season crop from October to March. In the state, finger millet is known as Ragi or 'Kelvaragu.' It is farmed as a dry land crop during the northeast monsoon seasons in Tamil Nadu. Hence the current necessity is to discover crops and cropping systems that are more climate-resilient and better suited agro ecologically. It must also provide better performance in terms of economic yield and livelihood security for the farmers of Parambikulam Aliyar Basin (PAB) and the state.

7.2 Materials and Methods

7.2.1 Study Area

Tamil Nadu, which falls under the semi-arid tracts of south India, is being considered for the study. The state of interest is located in southern India, bordered on the north and northwest by the states of Andhra Pradesh, Karnataka, and Kerala and the Bay of Bengal and the Indian Ocean in the eastern and southern parts, respectively. In the present chapter, crop suitability and efficient cropping zone identification for finger millet crops for 38 districts is carried out in general. In order to understand the changing in crop suitability at the basin level Parambikulam Aliyar Basin (PAB), in Coimbatore district, Tamil Nadu, is taken into consideration (Fig. 7.1). This research was conducted as part of the climate change and crop simulation study proposed under DST project entitled "Enhancing climate change adaptive capacity and agricultural productivity in Parambikulam Aliyar Basin areas through ICTs and other technological interventions (2019–2022). Understanding the impact of climate change on climate-resilient crops is one of the study's main components. The study area is located around the foothills of Valparai, in the Anamalai Hills of the Western Ghats. During the southwest monsoon season, this area receives good amount of rain, unlike other parts of state; however, a large portion of the land falls within regions that experience a rain shadow effect. Even though there are two dams, such as Aliyar and Thirumurthy dams, on the rivers Aliyar and Palar, respectively, to serve the irrigation needs of the basin, many parts in the northeastern parts of the basin are under rain-fed cultivation. These areas' major seasonal crop cultivation is groundnuts, maize, sorghum, and finger millets.

7.2.2 Agro-climatic Requirement of Finger Millet Crop

For ideal growth, it is necessary to have daytime temperatures ranging from 30 to 34 °C and nighttime temperatures ranging from 22 to 25 °C, in addition to ample sunlight. The optimal conditions for its growth are typically found in regions with an annual precipitation of approximately 100 cm. It is possible to cultivate this crop in various types of soil, including nutrient-rich loam and shallow upland soils with sufficient organic matter. Nonetheless, the cultivation of this crop is best suited to porous, light red loam, or sandy loam soils with efficient internal drainage. The cultivation of this crop can also be considered in black soils with adequate drainage, as it can tolerate some degree of water stagnation. Soils with a pH ranging from 4.5 to 8.0 are most suitable for the optimal growth of finger millet crop, while heavy clay soils with inadequate drainage capacities should be avoided.

7.2.3 ECOCROP Modeling Method

Crop suitability analysis was conducted for the state of Tamil Nadu and Parambikulam Aliyar Basin using the global meteorological data provided by WorldClim. The Eco Crop Model is a mechanistic model that utilizes temperature and rainfall ranges from the FAO-EcoCrop database, which are based on expert knowledge (http://ecocrop.fao.org). The model takes climatic factors

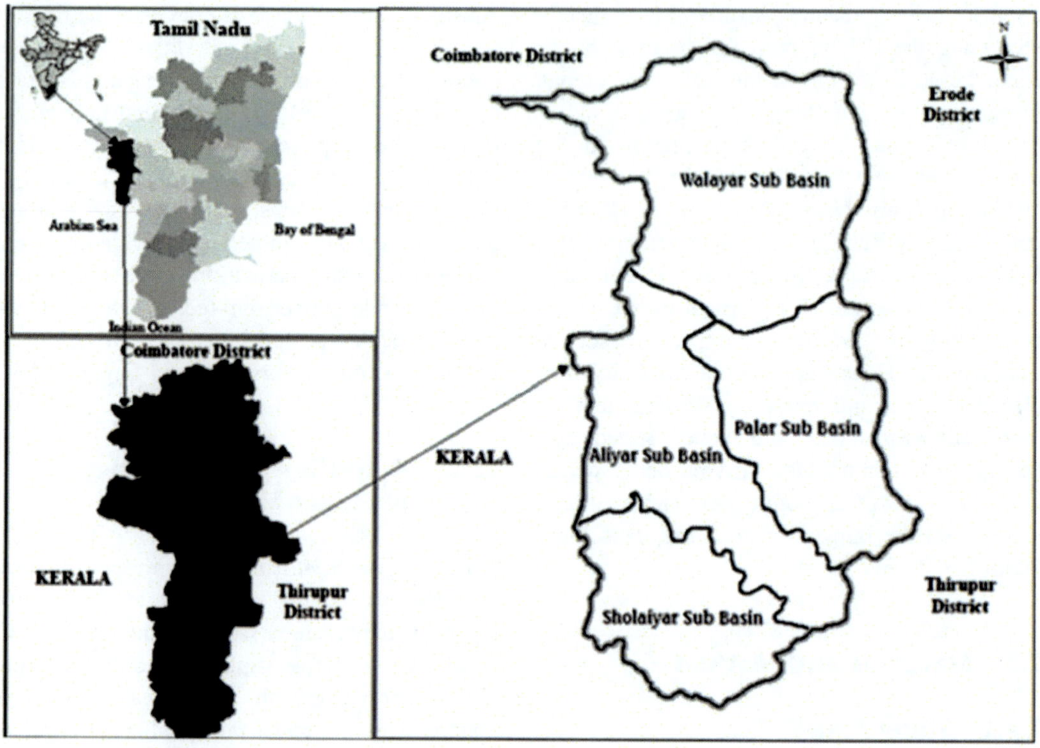

Fig. 7.1 The geographical area of interest in the research

as inputs to identify the appropriate environmental conditions for a crop, and generates a score indicating its suitability as output (Ramirez-Villegas et al. 2013). The range of agroecological suitability is fixed mainly based on a pair of variables, i.e., temperature and rainfall. The ranges of temperature and rainfall are characterized by both the absolute range, which includes the minimum and maximum temperatures and rainfall necessary for crop growth, as well as the optimum range, which includes the minimum and maximum temperatures and rainfall that are ideal for crop growth. By employing gridded temperature and rainfall statistics, the algorithm of the model calculates the environmental conditions throughout the growing season at a specific location. When the conditions are beyond the absolute thresholds, the suitability index is zero (not suitable); when they are between absolute and optimum thresholds, the suitability score ranges from 1 to 99, and when they are within the optimum current climate (WorldClim ver. 1.4) and future climate (CCSM4 RCP 2.6—2050). The RCP 2.6 pathway represents a low emissions scenario in which global greenhouse gas emissions peak around 2020 and then rapidly decline, leading to a stabilization of atmospheric concentrations at around 450 parts per million (ppm) CO_2 equivalent by 2100. The RCP 2.6 scenario is often used as a benchmark for assessing the impacts of climate policies and mitigation efforts on future climate change. After generating the suitability scores, they are classified into different categories based on their level of suitability, which includes very marginal (1–20%), marginal (20–40%), suitable (40–60%), very suitable (60–80%), and excellent (80–100%). Climate datasets for the current climate, WorldClim version 1.4, available at http://www.worldclim.org, was used. The global climate layers (climate grids) with a 30 arc-second spatial resolution (approximately 1 km × 1 km) are utilized to depict the monthly climatology (maximum, minimum, and mean temperatures and total monthly rainfall) representative of the years 1970–

2000. These datasets, were downloaded from http://worldclim.org/cmip5_30s. The downscaled versions of the CCSM4 GCM projections were calibrated, or bias-corrected, against the WorldClim 1.4 baseline for present-day climate.

7.2.4 Identification of Efficient Cropping Zones

Time series data on Crop production and productivity were gathered for a period of 30 years (1985–2015) from the Crop Production Statistics Information System, as well as season and crop reports (SCR) specific to the area. Additionally, data on the total cultivable area of Tamil Nadu were collected for a period of 15 years (2000–2015). Efficient Cropping Zones were identified based on a temporal database of area and production for various districts in Tamil Nadu; RSI (Relative Spread Index) and RYI (Relative Yield Index) were calculated as decadal average periods. The statistical formulas utilized in this study to compute the Relative Yield Index (RYI) and Relative Spread Index (RSI), as well as the criteria employed to categorize the Efficient Cropping Zone (EFC), adapted from researches of Pradip et al. (2018), Sankar et al. (2019), Kowshika and Sankar (2020), are given as follows (Table 7.1).

$$\text{RSI} = \frac{\text{Area of particular crop expressed as \% of the total cultivable area in the district}}{\text{Area of crop expressed as a percentage to the total cultivable area in the State}} * 100$$

$$\text{RYI} = \frac{\text{Mean yield of a particular crop in the district}}{\text{mean yield of a particular crop in the state}} * 100$$

7.2.5 Mapping of Efficient Cropping Zones

District-level Efficient Cropping Zones of finger millet (Ragi) crop over Tamil Nadu state for three decadal maps of 1985–1995, 1996–2005, and 2006–2015 were mapped and compared with the average period (1985–2015). ArcGIS v10.5 software was employed to generate all of the maps used in this study.

7.3 Results and Discussion

7.3.1 Changes of Area and Yield

Data shows that the area under finger millet showed a general declining trend in the state until 2012–2013. After that, it showed a slight increasing trend in the state (Fig. 7.2). Cultivated areas under finger millet showed a drastic decline in the Coimbatore district (Fig. 7.3).

The average productivity of finger millet in the state is 3338 kg/Ha, and in the Coimbatore

Table 7.1 Criteria for classification of efficient cropping zones

Sl. No	RSI	RYE	Cropping zone
1	> 100 (High)	> 100 (High)	Most Efficient Cropping Zone (MECZ)
2	> 100 (High)	< 100 (Low)	Efficient cropping zone (ECZ)
3	< 100 (Low)	> 100 (High)	Not efficient cropping zone (NECZ)
4	< 100 (Low)	< 100 (Low)	Highly inefficient cropping zone (HICZ)

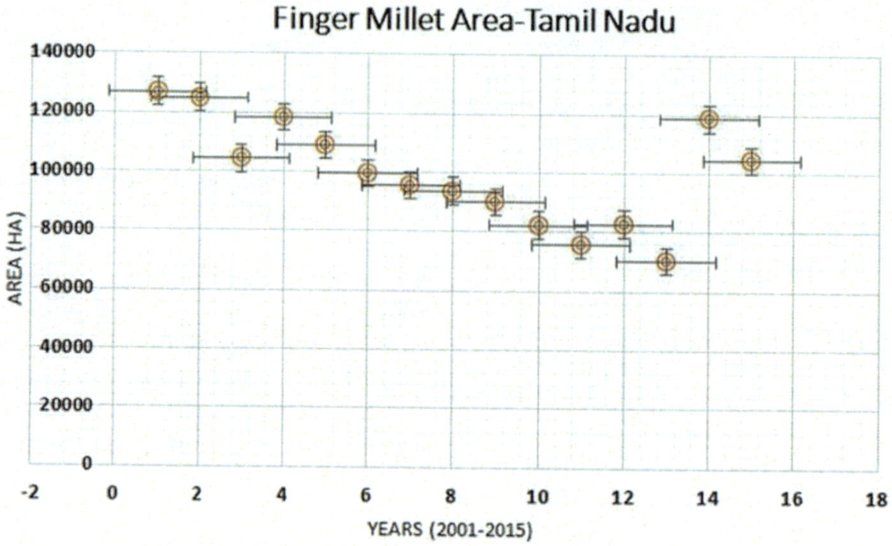

Fig. 7.2 Trend of finger millet growing areas in the state from 2001 to 2015

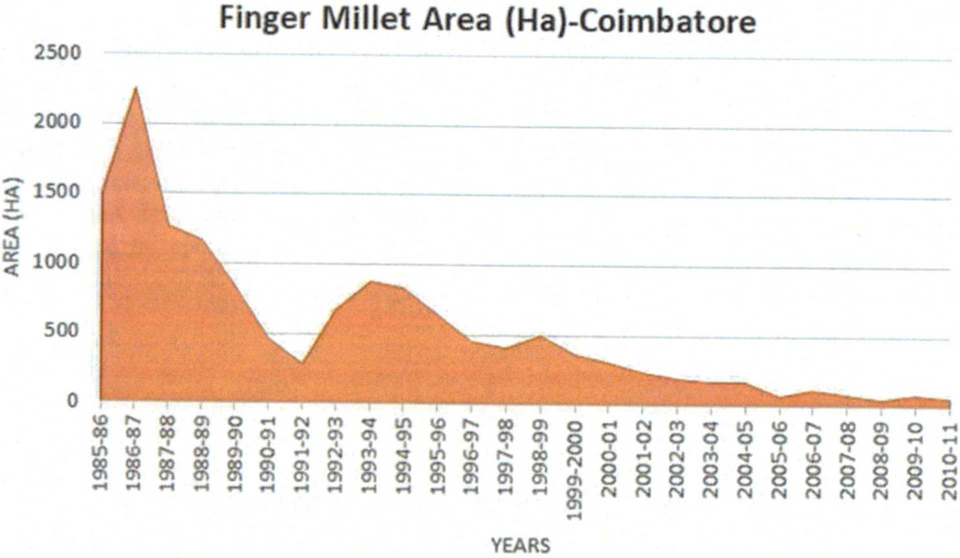

Fig. 7.3 Trend of finger millet growing areas in the Coimbatore from 2001 to 2015

district, it is 2401 kg/Ha from 2011–2012. There is a huge spatial disparity in the yield of Ragi crops in the state (Figs. 7.4 and 7.5). The data clearly explains how climate variability impacts crop area and yield. 2002–2003 show a striking reduction in crop yield and area due to drought.

7.3.2 Finger Millet Current and Future Suitability

This study involved simulating the current and potential suitability of finger millet crops in the future based on the ECOCROP model.

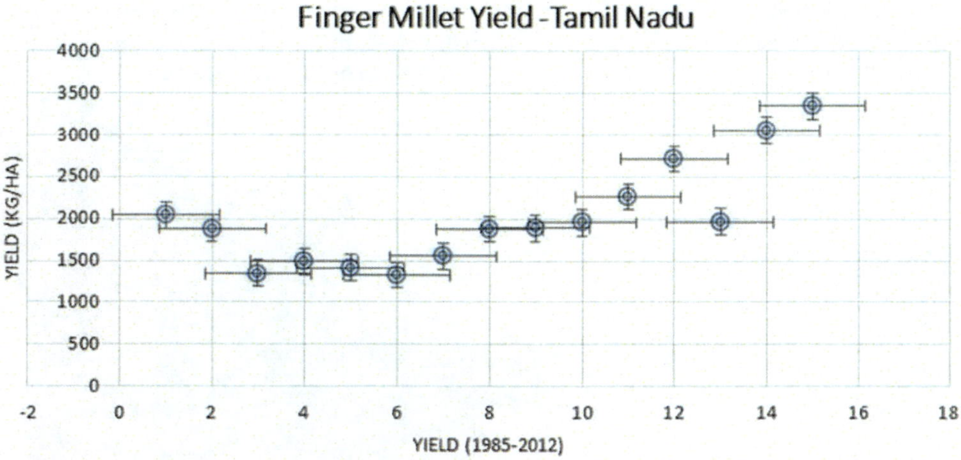

Fig. 7.4 Trend of finger millet yield in the state from 2001 to 2015

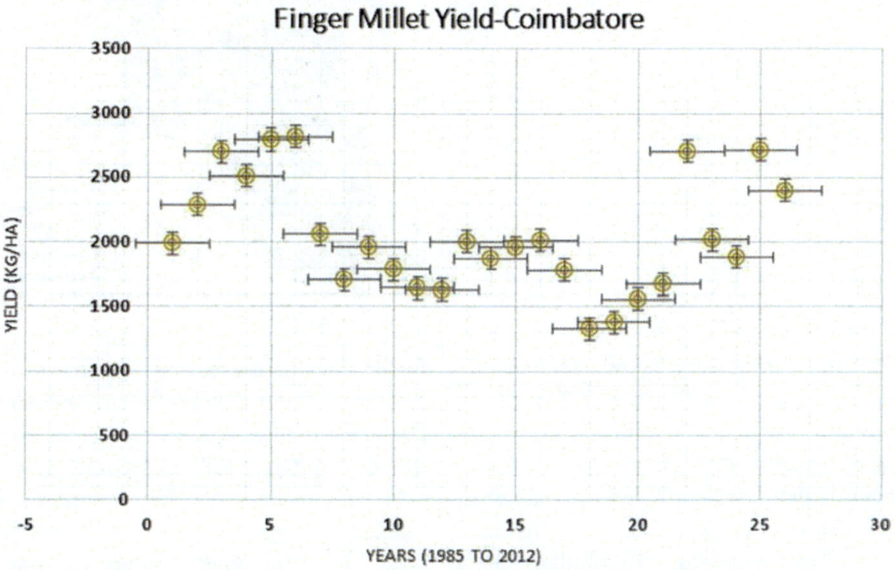

Fig. 7.5 The trend of finger millet yield in Coimbatore district from 2001 to 2015

Regarding current agro-climatic suitability, northeast agro-climatic zones are found to be more suitable for finger millet cultivation. However, Thoothukudi, Virudhanagar, Madhurai, Thiruppur, Karur, Erode, central and north Coimbatore, Salem, Namakkal, and Perambalur comes under not suitable categories based on the current climatic conditions, agro-climatic suitability. The districts, Sivagangai, east of Pudukottai, Thiruvannamalai, Villupuram, and so on come under marginally suitable classes (Fig. 7.6).

Regarding future suitability, Tirupur, the northern part of Krishnagiri, a small patch in Erode, the eastern part of Krishnagiri, parts of Kanyakumari, Thoothukudi, and so on have

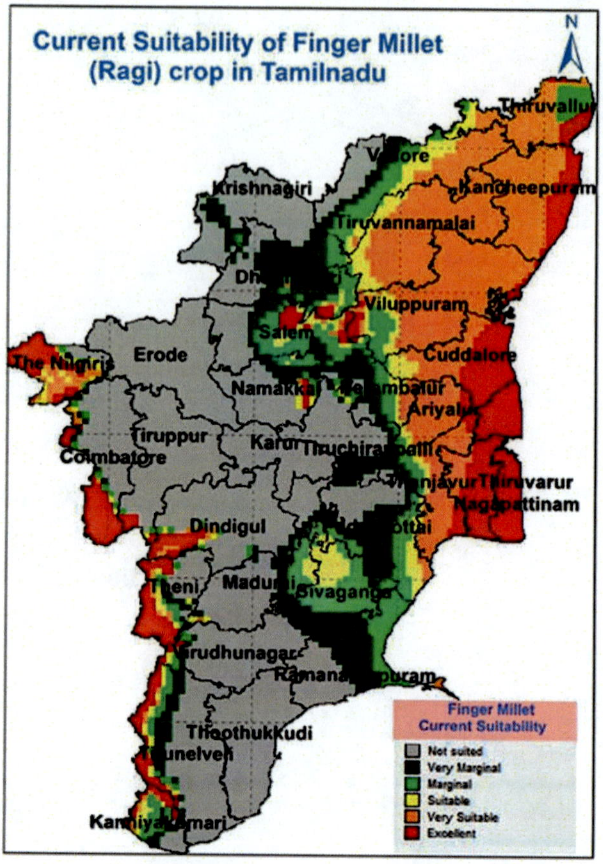

Fig. 7.6 Current suitability of finger millet (Ragi) crop over Tamil Nadu

shown not suitable or very marginally suitable for finger millet crops in the future. The places which are excellent for cultivation in the future are Nilgiris, southwest of Coimbatore, Theni, Dharmapuri, Salem, and north part of Namakkal, whereas Vellore, Nagapattinam, Thiruvarur, Cudullore, Kancheepuram, Thiruvallur, etc., come under excellent to very suitable categories (Fig. 7.7).

7.3.3 Current and Future Suitability Finger Millet: Parambikulam Aliyar Basin

As far as Parambikulam Aliyar Basin (PAB) is concerned, more areas may come under excellent, very suitable, and suitable categories in the future than in the current situations. As far as current climatic conditions are concerned, only 50–60% of the grids are under excellent and very suitable for finger millet cropping; however, 70–80% of the areas can be anticipated to be an excellent category in the PAB in the future (Fig. 7.8). The areas around the northeastern parts of the PAB come under marginally to very marginally and not suitable for the cultivation of finger millets in the current conditions and under projected future climatic conditions based on IPCC's RCP 2.6.

7.3.4 Efficient Cropping Zones

Multitudes of factors affect an area to be an efficient cropping zone (ECZ) for a particular crop. Among all factors, climate, weather, soil

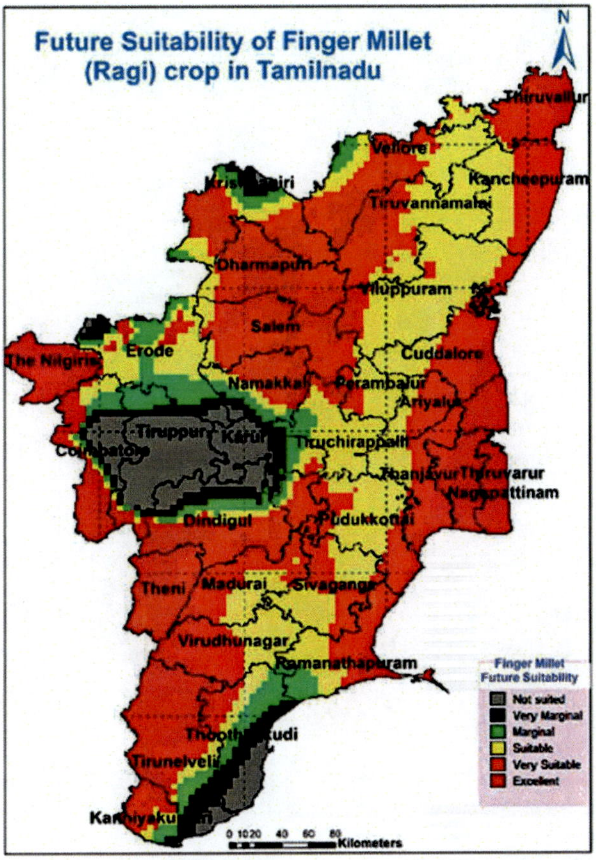

Fig. 7.7 Future suitability of finger millet (Ragi) crop over Tamil Nadu

quality, and land management, higher yield plays a critical role in deciding the efficacy of an area to be suitable for the further spread of a particular crop. ECZs can be classified based on their high Relative Spread Index (RSI) and Relative Yield Index (RYI). RSI is influenced by factors such as the availability of seeds, fertilizers, water, pesticides, labor, technology, and government policies, whereas RYI is determined mainly by the climate and weather conditions specific to the region (Sankar & Kowshika, 2020). Based on the secondary data, districts that come under efficient cropping zones are Erode, Salem, Dharmapuri, and Vellore from 1985 to 1995. However, during 1996–2005, Vellore and Thiruvallur districts also came into the category. Data showed that Krishnagiri also fell into the efficient cropping zone category from 2006 to 2015. (Figs. 7.9 and 7.10). If we consider the overall suitability of finger millet crop for 30 year period, all five districts, namely Erode, Salem, Dharmapuri and Vellore, and Krishnagiri, come under efficient cropping zones (Figs. 7.11 and 7.12).

The outcomes from the secondary data analysis and ECOCROP model-based analysis have been evaluated for comparative analysis. As per the current situation, all other districts in the state fall under highly inefficient zones. It is alarming to note that even though the area has been identified as climatically suitable as per ECOCROP model-based analysis (Fig. 7.6), finger millet cultivation in coastal areas of Kancheepuram, Thiruvannamalai, Villupuram, Cuddalore, Ariyalur, Thiruvarur, and Nagapattinam is completely neglected. It is significant to note that the state has no most efficient cropping zone (MECZ).

The WorldClim dataset is used worldwide in climate change-related suitability studies

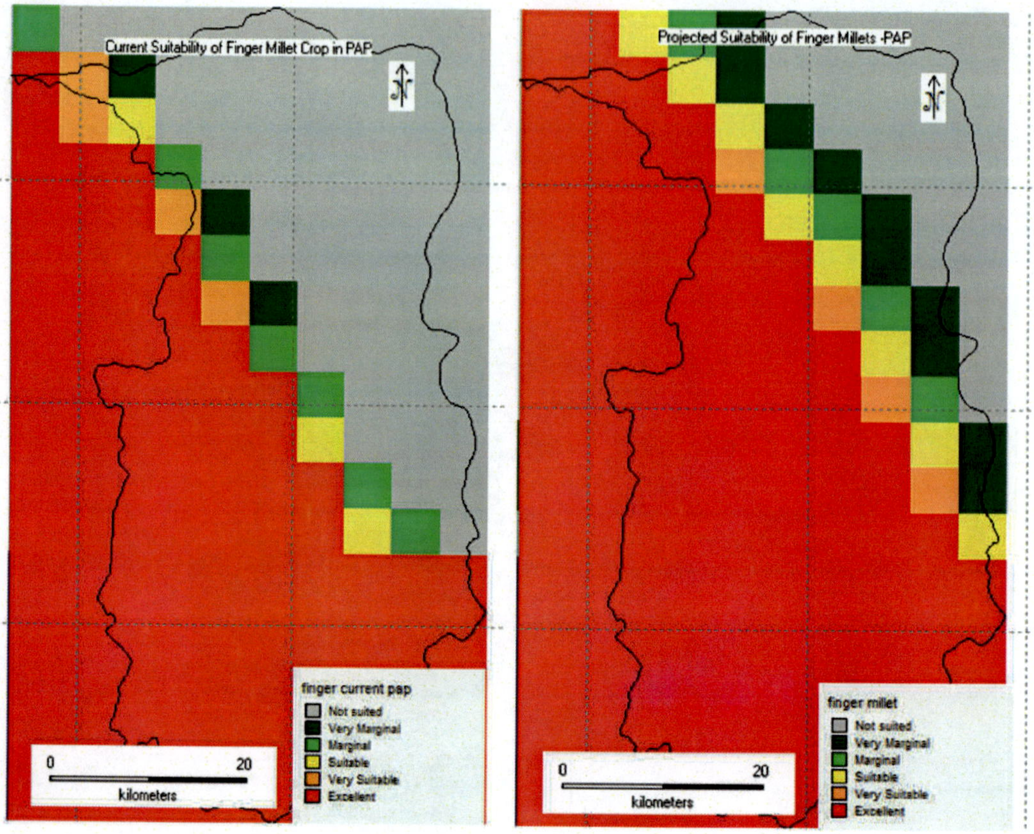

Fig. 7.8 Current and future suitability of finger millet crop to climate change in PAB basin

(Hijmans et al. 2001), such as those by Lane and Jarvis (2007), Ramirez-Villegas et al. (2013). The CCSM4 is one of the GCMs whose climate projections were used in the Fifth Assessment IPCC report. The RCP 2.6 scenario projects an average global warming increase of 1 °C in the end century period. Two sets of crop suitability maps were generated for the state of Tamil Nadu and the PAB using the FAO crop ecological database. These maps represented the suitability of crops under both present and future climate conditions. Agro-meteorological settings cause wide instabilities in finger millet crop growth, development, and harvest. It shows a conducive environment for the future spread of finger millet crops growing in 70–80% of areas of the state of Tamil Nadu.

Finger millet is a highly nutritious cereal. Finger millet is rich in vitamins, minerals, and fiber, which can offer several health benefits. The high potassium content in finger millet promotes healthy kidney and heart function, while also aiding in the transmission of nerve signals. In addition, finger millet is an excellent source of B vitamins that support brain function and healthy cell division. Knowing its significance, to ensure sustainable growth in millet production, FAO (2012) recommends the immediate adoption of innovative practices. These practices are crucial for providing access to nutritious food for those suffering from malnutrition and hunger, improving the global supply chain, and reducing food wastages. Following India's request, FAO has agreed to celebrate 2023 as the 'International Year of Millets.' India's Millet Mission focuses on developing farm gate processing and empowering farmers through collectives, enhancing value-addition, and ensuring that farmers have

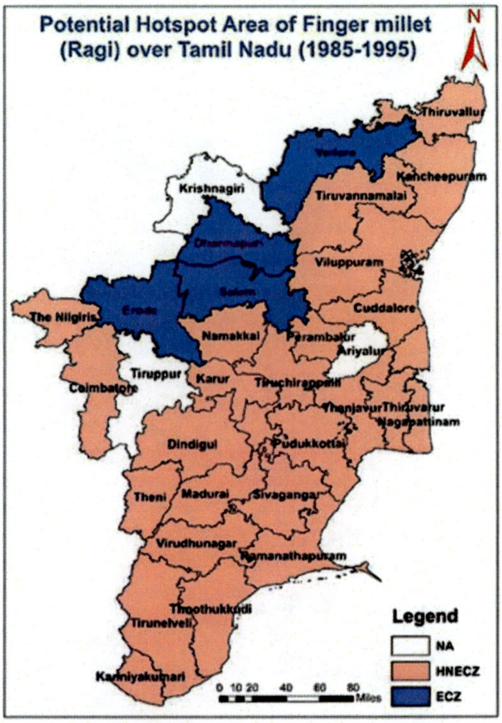

Fig. 7.9 Efficient cropping zone (ECZ) of finger millet (Ragi) crop during 1985–1995 over Tamil Nadu

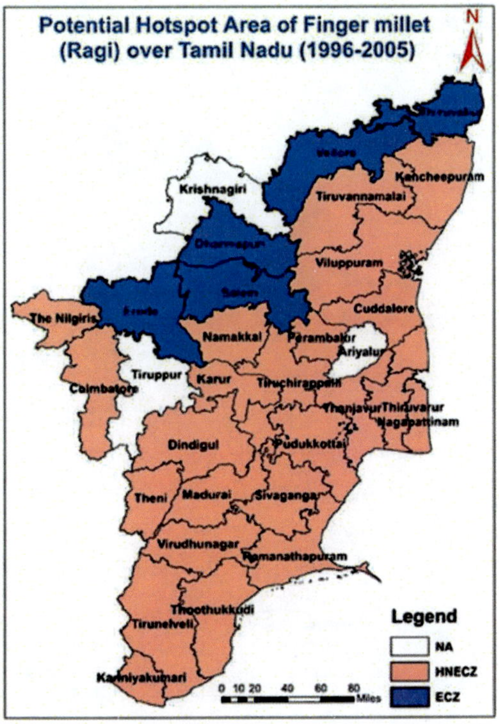

Fig. 7.10 Efficient cropping zone (ECZ) of finger millet (Ragi) crop during 1996–2005 over Tamil Nadu

access to better quality millet seeds. Finger millet yield in central Nepal has been increasing at 7.39 and 36.9 kg/ha yearly, even in lower tropical and upper tropical to subtropical climates, respectively (Luitel et al. 2019). In this purview, geospatial modeling using ECOCROP would assist in selecting the most suitable location for cultivating millet crops (Parthasarathy et al. 2016).

7.4 Conclusion

The suitability of finger millet crops in Tamil Nadu state and the PAB, in particular, has been analyzed using the ECOCROP agro-climatic spatial modeling tool. The analysis outcomes would definitely benefit the farmers, agricultural researchers, policymakers, and other stakeholders in identifying the potential finger millet growing areas. Improvements in inputs, weather early warnings, cultivation practices, post-harvest technologies, and value-added services are critical steps toward improving livelihood and nutritional security. Modeling shows much better suitability for finger millet cultivation in the entire state, especially in northeast coastal areas and Nilgiris, southwest of Coimbatore, Theni, Dharmapuri, Salem, and the north of Namakkal, Vellore districts, etc. With its National Mission on Sustainable Agriculture and Food Security, India is working toward food and nutritional security. It is working together to scale up nutrition in SUN countries. Different organizations are working internationally, nationally, and regionally to achieve and enhance nutritional security by promoting millet varieties. These geospatial modeling analyses enabled us to understand the state and basin-level potential zones.

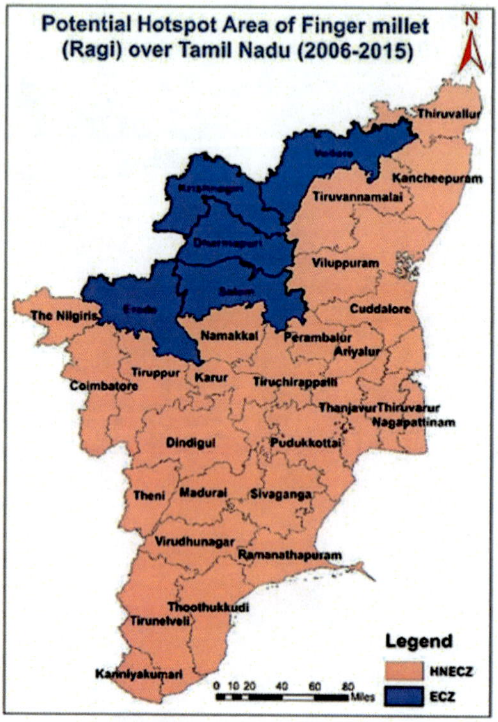

Fig. 7.11 Efficient cropping zone (ECZ) of finger millet (Ragi) crop during 2006–2015 over Tamil Nadu

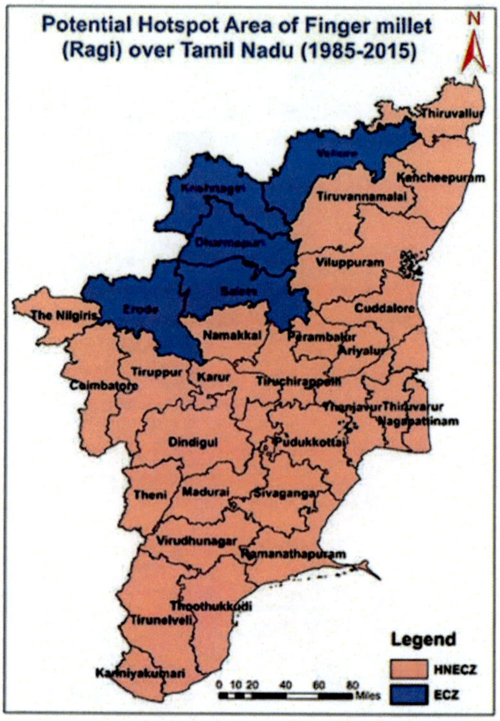

Fig. 7.12 Efficient cropping zone (ECZ) of finger millet (Ragi) crop during 1985–2015 over Tamil Nadu

Acknowledgements The authors thank the Department of Science and Technology, KIRAN/WISE division, Government of India, for their financial support for running the project.

Conflict of Interest The authors state that they do not have any conflicts of interest.

References

Bandyopadhyay T, Muthamilarasan M, Prasad M (2017) Millets for next generation climate-smart agriculture. Front Plant Sci 8:1266. https://doi.org/10.3389/fpls.2017.01266

Das J, Gayen A, Saha S, Bhattacharya SK (2017) Modelling of alternative crops suitability to tobacco based on analytical hierarchy process in Dinhata subdivision of Koch Bihar district, West Bengal. Model Earth Syst Environ 3(4):1571–1587. https://doi.org/10.1007/s40808-017-0392-y

FAO (2012) Food security and nutrition and sustainable agriculture, accessed from Food security and nutrition and sustainable agriculture. Department of Economic and Social Affairs (un.org).

Fish CS (2020) Cartographic content analysis of compelling climate change communication. Cartogr Geogr Inf Sci 47(6):492–507. https://doi.org/10.1080/15230406.2020.1774421

Gupta SM, Arora S, Mirza N, Pande A, Lata C, Puranik S, Kumar J, Kumar A (2017) Finger millet: a "certain" crop for an "uncertain" future and a solution to food insecurity and hidden hunger under stressful environments. Front Plant Sci 8:643. https://doi.org/10.3389/fpls.2017.00643

Hijmans RJ, Guarino L, Cruz M, Rojas E (2001) Computer tools for spatial analysis of plant genetic resources data: 1 DIVA-GIS. Plant Genet Resour Newsl 127:15–19

IPCC (2012) IPCC: Managing the risks of extreme events and disasters to advance climate change adaptation. A special report of working groups I and II of the intergovernmental panel on climate change. In: Field CB, Barros V, Stocker TF, Qin D, Dokken DJ, Ebi KL, Mastrandrea MD, Mach KJ, Plattner G-K, Allen SK, Tignor M, Midgley PM (eds). Cambridge University Press, Cambridge, UK and New York, NY, USA, 582 pp

Knight KW, Messer BL (2012) Environmental concern in cross-national perspective: the effects of affluence, environmental degradation, and world society. Soc Sci Q 93(2):521–537

Koushik N, Sankar T (2020) Potential zones of turmeric and coriander cultivation in Tamilnadu. Int J Environ Clim Change 10(12):20–30. https://doi.org/10.9734/ijecc/2020/v10i1230281

Luitel DR, Siwakoti M, Jha PK (2019) Climate change and finger millet: perception, trend and impact on yield in different ecological regions in Central Nepal. J Mt Sci 16:821–835. https://doi.org/10.1007/s11629-018-5165-1

Parthasarathy U, Nandakishore OP, Jayarajan K, Saji KV, Babu KN (2016) Prediction of crop suitability of certain Indian spices—A GIS approach. In: Raju N (ed) Geostatistical and geospatial approaches for the characterization of natural resources in the environment. Springer, Cham. https://doi.org/10.1007/978-3-319-18663-4_124

Pradip CS, Panneerselvam RD, Bharathy Dheebakaran GA, Geethalakshmi V, Ragunath KP, Kowshika N (2018) Status of Bengal gram over Tamilnadu. Agric Sci Dig 38(3):193–196

Quach Q, Jenny B (2020) Immersive visualization with bar graphics. Cartogr Geogr Inf Sci 47(6):471–480. https://doi.org/10.1080/15230406.2020.1771771

Ramachandran A, Dhanya P, Jaganathan R, Rajalakshmi D, Palanivelu K (2017) Spatiotemporal analysis of projected impacts of climate change on the major C3 and C4 crop yield under representative concentration pathway 4.5: Insight from the coasts of Tamil Nadu, South India. PLoS ONE 12(7):e0180706. https://doi.org/ https://doi.org/10.1371/journal.pone.0180706.

Ramirez-Villegas J, Jarvis A, Läderach P (2013) Empirical approaches for assessing impacts of climate change on agriculture: the EcoCrop model and a case study with grain sorghum. Agric for Meteorol 170:67–78

Sankar T, Arul Prasad S, Dheebakaran GA (2019) Identification of efficient cropping area for groundnut over north-western zone of Tamil Nadu. In: Proceedings of the national seminar on current trends and challenges in sustainable agriculture. ISBN: 978-93-5351-321-4

Sankar T, Kowshika N (2020) Delineating efficient cropping zones of potato and chilli in Tamilnadu. Int J Environ Clim Change 10(11):143–154. https://doi.org/10.9734/ijecc/2020/v10i1130275

Satyavathi CT, Ambawat S, Khandelwal V, Srivastava RK (2021) Pearl millet: A climate-resilient nutricereal for mitigating hidden hunger and provide nutritional security. Front Plant Sci 12:659938. https://doi.org/10.3389/fpls.2021.659938

Shukla A, Lalit A, Sharma V, Vats S, Alam A (2015) Pearl and finger millets: the hope of food security. Appl Res J 1:59–66

Sivakumar MVK, Das HP, Brunini O (2005) Impacts of present and future climate variability and change on agriculture and forestry in the arid and semi-arid tropics. Clim Change 70:31–72. https://doi.org/10.1007/s10584-005-5937-9

Varadan RJ, Kumar P. (2015) Mapping agricultural vulnerability of Tamil Nadu, India to climate change: a dynamic approach to take forward the vulnerability assessment methodology

Yadav MK, Singh RS, Singh KK, Mall RK, Patel CB, Yadav SK, Singh MK (2015) Assessment of climate change impact on productivity of different cereal crops in Varanasi India. J Agrometeorol 17(2):179–184

8 Vegetation Indices-Based Rice and Potato Yield Estimation Through Sentinel 2B Satellite Imagery

Chiranjit Singha and Kishore C. Swain

Abstract

High-resolution optical remote sensing imagery has sound potential for future crop yield estimation. In precision agriculture adoption, these systems can provide valuable information on various factors affecting a farm's production. The current study used different crop vegetation indices, such as NDVI, NDWI, BI, DVI, SAVI, GEMI, PVI, RVI, and LAI, for estimating rice and potato crop yield on a microscale. Optical Sentinel-2B images were used for rice and potato crop estimation during 2018 and 2019 in ArcGIS 10.7 software environment for the rural neighborhood of Tarakeswar region, Hooghly (West Bengal, India). The geostatistical semivariogram analysis with the best fitting of the exponential and spherical models determined the degree of spatial variability of rice and potato yield. In statistical Getis-Ord *Gi** analysis, the clusters of VIs indicated high yield with NDVI, RVI, and SAVI surfaces, while low vegetation indices showed low yield. Furthermore, the support vector machine, random forest, and logistic regression models were positively used in the spatial assessment of rice and potato crop yield estimation, with AUROC values of 80–90%. However, the Naïve Bayes model was categorized as a moderate to good predictor with an AUC value between 60 and 80%. This chapter introduced a novel approach for crop yield prediction and validation with optical satellite imagery for microscale precision and agriculture adoption, which further helps using this method for other crops.

Keywords

Vegetation indices · NDVI · Sentinel-2B · Crop yield · AUROC

8.1 Introduction

In developing countries, the population explosion is the major issue for food security achievement. The future food grain yield is estimated using standard crop and environmental parameters. With the predicted crop yield, the government gets enough time to prepare itself for alternative food sources to face the food security threat (if arises). However, several optical reflectance-based vegetation indices (VIs) are very good evaluators of the local food security scenario well in advance through the earth observation-based remote sensing (RS) techniques (Lambert et al. 2018). Singha and Swain

C. Singha (✉) · K. C. Swain
Department of Agricultural Engineering, Institute of Agriculture, Visva-Bharati, Sriniketan, West Bengal 731236, India
e-mail: singha.chiranjit@gmail.com

J. Das and S. Halder (eds.), *Advancement of GI-Science and Sustainable Agriculture*,
GIScience and Geo-environmental Modelling, https://doi.org/10.1007/978-3-031-36825-7_8

(2016) described the advantages of RS techniques based on multicriteria decision-making analysis, which has greater potential for sustainable agricultural planning. Integrating spatial analysis and sensor technology in the precision agriculture study can provide valuable crop information for field-specific management (ESA 2019). This also assists in monitoring variable fertilizer application, irrigation scheduling, biomass estimation, and harvesting at actual crop maturity by observing various agronomic parameters (Vallentin et al. 2021). Quick and fast crop yield estimation at the micro- to macro-level is very promising through remote sensing, further assisting in sustainable agricultural planning (Yuan et al. 2015). Assessing crop yield is one of the major issues to enhance the farmer's socioeconomic development and optimize the industrial processing demand. Earth observation-based satellite systems could better monitor the changes in crop growth due to management practices, climate change impact, the emergence of pests and diseases, or water stress (ESA 2019; Yuan et al., 2015). Remote sensing resulting in several spectral VIs has helped monitor crop productivity at the farm level (Shammi and Meng 2020). Machine learning (ML)-based regression models were useful for crop yield prediction through the in-situ filed information and RS-derived VIs in US Corn Belt (Ji et al. 2021). The above-said model could provide a reference for both physical and data organization. Several studies have shown that using ML techniques to analyze the data collected from the RS technique improves crop yield prediction (Shammi and Meng 2020; Guo et al. 2021). Locally field-based multitemporal satellite remote sensing of crop VIs has good statistical significance with wheat grain yield up to 2 t ha^{-1} (Ali et al. 2019; Panek et al. 2020). European Space Agency (ESA) provides high-resolution Sentinel 2 (S2) data through open access for the environmental monitoring and assessment of agriculture, wetland vegetation, forestry development, etc. S2 data offers a high spatial, spectral, and temporal resolution which can estimate individual farm-level crop health status, soil moisture, crop stress, and crop yield information through RGB and near-infrared-based spectral VIs. Multiple Linear Regression (MLR), support vector machine (SVM), extreme gradient boosting (XGB), stochastic gradient descent (SGD), and random forest (RF) were employed for field scale prediction of soya yield using cloud-free S2 multispectral images derived from different VIs of specific growing periods (2018, 2019, and 2020) in Upper Austria (Pejak et al. 2022).

The simplified S2-MSI imagery can scientifically create different vegetation biophysical variables from the surface reflectance values such as Normalized Differential Vegetation Index (NDVI) (Rouse et al. 1973); Normalized Differential Water Index (NDWI) (Chandel et al. 2019); Brightness Index (BI) (Cavalaris et al. 2021), Difference Vegetation Index (DVI) (Kussul et al. 2020), Soil-Adjusted Vegetation Index (SAVI) (Nagy et al. 2021); Global Environmental Monitoring Index (GEMI) (Kobayashi et al. 2020); Perpendicular Vegetation Index (PVI) (Lu et al. 2016); Ratio Vegetation Index (RVI) (Quan et al. 2011); and Leaf Area Index (LAI) (Gaso et al. 2019). Using satellite images and VIs allows the farmers to identify different management zones on a commercial farm (Campillo et al. 2018). These indices were used to estimate vegetation characteristics as they mostly serve as indicators for crop dynamics and overall changes in biomass quantity and properties. Moreover, some indices are used to monitor changes in the water content of leaves, while others can suppress the soil's influence or eliminate the influence of the atmosphere.

Several kinds of literature reported that VIs-based crop yield estimation is very popular in RS domains for wheat (Nagy et al. 2018), corn (Panda et al. 2010), maize (Fang et al. 2011), rice (Son et al. 2013), sunflower (Ali et al. 2019). In recent years, vegetation reflectance values have been further analyzed through geostatistical and machine learning techniques as an alternative approach for better crop yield prediction (van Klompenburg et al. 2020). Koutsos et al. (2021) applied the statistical hotspot autocorrelation to identify low-yield areas for special attention to the better management of the agronomic inputs as a sustainable approach. LAI and NDVI are good predictors of crop yield through the logistic regression model (Kunapuli et al. 2015). Marino and Alvino (2021) stated that statistical hotspot analysis is a sound strategy for defining the spatial differentiation of crop yield variation on a small scale with the help of vegetation indices. The US and German organic agriculture used the clustering hotspot algorithm to better support their regional economic development (Marasteanu and Jaenicke 2016). The GIS-based interpolations techniques such as IDW, EBK, and Kriging are the most preferred technique for multi-crop yield mapping (McKinion et al. 2010). RF and SVM allowed predictive models of crop yield estimation using multi-temporal data for site-specific management in different seasons (Shah et al. 2018; Filippi et al. 2019). Even the logistic regression (LR) and Naïve Bayes (NB) models correctly predicted the corn nitrogen stress class with the help of hyperspectral sensors derived VIs (Laacouri et al. 2018; Mupangwa et al. 2020). MODIS-derived LAI estimated rice crop yield on a near real-time basis using gradient boosted regression in India (Arumugam et al. 2021). ML models are very helpful in predicting the number of crop yields for supporting food security in Africa (Chepngetich 2020; Cedric et al. 2022). Different vegetation indices, namely NDVI, OSAVI, RSI, MTCI, and BOP, incorporated with BP neural network, significantly impact the cotton yield prediction mapping in China. Ramos et al. (2020) used multispectral UAV-based 33 VIs to predict maize yield through the ranking-based RF model, where the NDVI, NDRE, and GNDVI VIs are the most precedence indices. On the contrary, Pham et al. (2022) showed that the ML approach is useful for enhancing the rice crop yield through the spatial variability of VCI/TCI data in Vietnam. Five MLS, namely elastic net (EN), linear regression (LR), support vector regression (SVR), and k-nearest neighbor (k-NN) are tested for the potato tuber quality mapping with the NDVI value in Atlantic Canada (Abbas et al. 2020). Sharifi (2020) proposed that integrating satellite remote sensing and ML techniques provides a powerful potential approach for a barley yield prediction map based on seasonal phenological behavior in eastern Iran.

S2 MSI data with 10 m resolution has a greater potential to build a decision-making tool for crop yield estimation through numerous vegetation spectral indices. With the accessibility of open-source Python and SNAP tools along with Geostatistics, ML, and area under the curve (AUC) models, we tried to integrate several biophysical spectral indices for yield estimation of rice and potato crops in Hooghly District (West Bengal, India). Nine major crop biophysical parameters were identified, namely NDVI, NDWI, BI, DVI, SAVI, GEMI, PVI, RVI, and LAI, for accurate crop yield estimation during two crop seasons (summer, 2018, and winter, 2019).

8.2 Materials and Methods

8.2.1 Study Area

The study was carried out in the Hooghly Region (Tarakeswar ~ 2,528,500 to 2,530,600 N and 604,500 to 606,500 E in UTM-WGS84 Zone 45 N India) during the 2018 and 2019 crop seasons (Fig. 8.1). The study region experienced rich fertile alluvial soil with high productivity of rice and potato compared to other districts of West Bengal (Singha et al. 2020). The densely populated region is supported by high cropping intensity (184%). The total study area was around 300 ha with an average MSL of 18 m.

This area is categorized by tropical monsoon climate with an average annual rainfall of 1200–1700 mm) and temperature of 15–35 °C (Singha et al. 2019). Rice and potato are the main crops, along with jute and vegetables. The growing season for *Kharif* rice starts with the arrival of the monsoon in July, and crops are harvested from October to December. In the *Rabi* season, the potato crop is sown in November–December and harvested in March next year. The potato crop has many advantages over other local crops in terms of high crop yield, easy market access, short crop growth, etc.

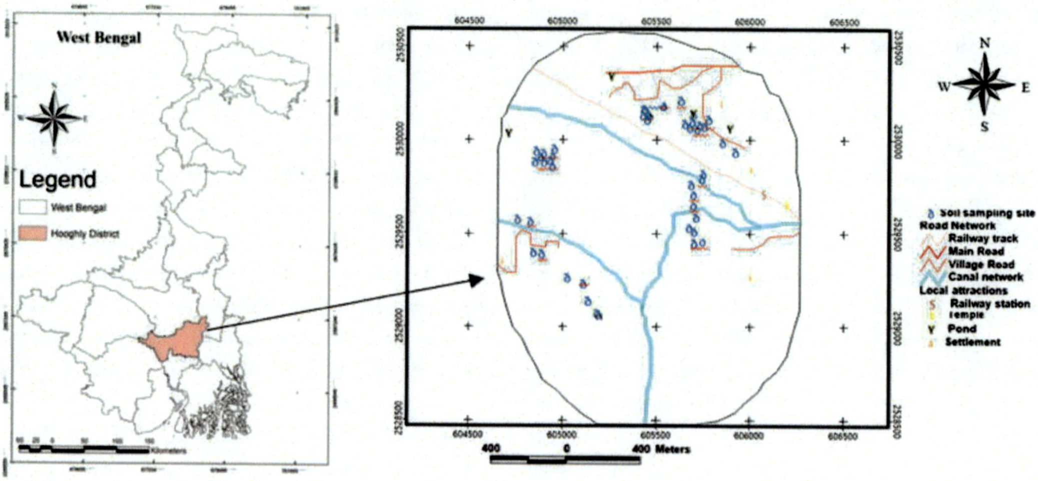

Fig. 8.1 Study area map

8.2.2 Satellite Image Processing

Cloud-free Sentinel-2B MSI imagery was selected for the *Kharif* rice (October 17, 2018) and *Rabi* potato (March 6, 2019) season, with a temporal resolution of 10 days and 290 km swath width. The images were processed for crop biophysical status mapping and yield estimation of two crops. The descending node of S2B MSI offers a 5-day temporal frequency with an orbital overpass time of approximately 10:30 a.m. (Drusch et al. 2012).

The Copernicus Open Access Hub data is easily retrieved by being atmospherically and radiometrically corrected S2B bottom of atmosphere (BOA) reflectance with 10 m pixel size imagery (https://sci-hub.copernicus.eu/). The Sentinel Application Platform v.8.0.0 (SNAP-ESA) was used for image pre-processing. The orthorectified images were geo-referenced in WGS84 UTM zone 45 N with Survey of India (SOI) topographical maps (No. 79B/1, 1:50,000 scale). The nine ground control points (GCPs) developed the final study area location map with a handheld GPS receiver e-Trex 20 Garmin). The detailed workflow of this research methodology is presented in Fig. 8.2.

8.2.3 Vegetation Indices (VIs)

This chapter selected the spatial resolution of 10 m for different VIs as input parameters for estimating the spatial variability of rice and potato yield, namely slope-based- (NDVI and RVI), distance-based—PVI, soil noise-based—SAVI, and NDWI, BI, GEMI, LAI (Hatfield et al. 2008). All the VIs maps were developed through ArcGIS 10.7 software for individual farm plots extracted from S2 images for two peak growing seasons. VIs are developed as the combination of numerous wavebands (red, NIR, SWIR, green, blue) and is also related to various canopy estimated parameters (Table 8.1).

8.2.4 Yield Estimation

During the post-harvest processing, a set of randomly selected 70 farms plot-wise agronomic practice details, along with crop yield information for two crops, were collected. The crop yield data were collected along with a handheld e-Trex GPS receiver (Garmin Ltd., Olathe, Kansas, USA) in October 2018 for *Kharif* rice and March 2019 for *Rabi* potato. Rice yield was estimated

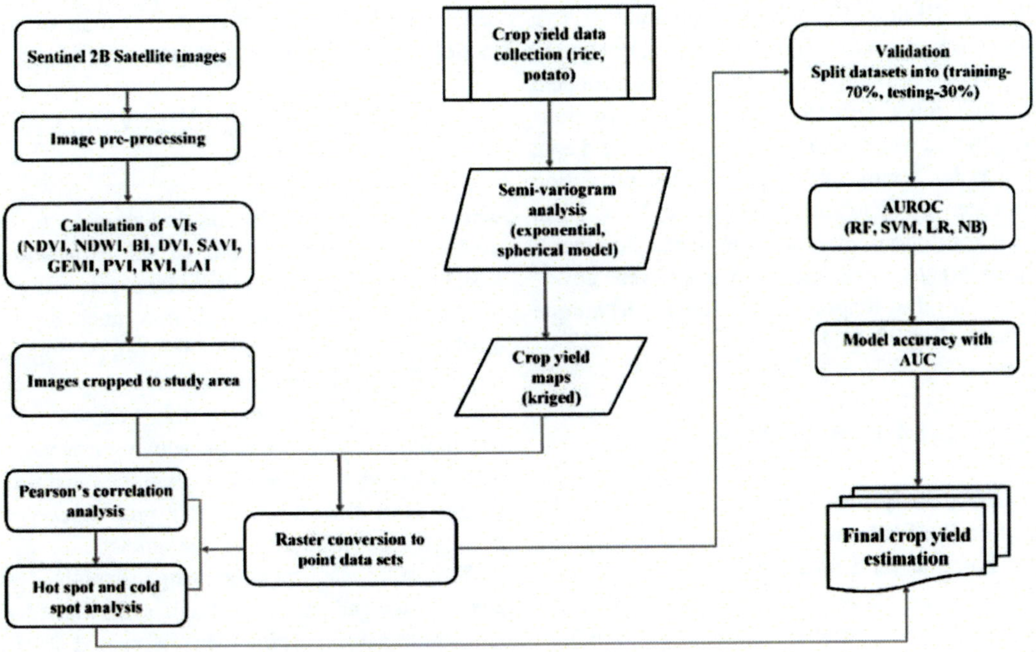

Fig. 8.2 Detailed workflow of the research work

Table 8.1 Details of selected vegetation indices (VIs)

Indices	Explanation	Formula	Source
NDVI	Normalized difference vegetation index	$\rho_{NIR} - \rho_{red}/\rho_{NIR} + \rho_{red}$	Vannoppen et al. (2018)
NDWI	Normalized difference water index	$\rho_{NIR} - \rho_{SWIR}/\rho_{NIR} + \rho_{SWIR}$	Chandel et al. (2019)
BI	Brightness index	$(\rho_{red} - \rho_{SWIR}) - (\rho_{NIR} + \rho_{blue})/(\rho_{red} + \rho_{SWIR}) + (\rho_{NIR} - \rho_{blue})$	Cavalaris et al. (2021)
DVI	Difference vegetation index	$\rho_{NIR} - \rho_{red}$	Kussul et al. (2020)
SAVI	Soil-adjusted vegetation index	$1.5 \times (\rho_{NIR} - \rho_{red})/(\rho_{NIR} + \rho_{red} + 0.5)$	Nagy et al. (2021)
GEMI	Global environmental monitoring index	$\eta(l - 0.25\eta) - (\text{red} - 0.125)/(l - \text{red})$ $\eta = [2(\text{NIR2} - \text{red2}) + 1.5\text{NIR} + 0.5\text{red}]/(\text{NIR} + \text{red} + 0.5)$ 9	Kobayashi et al. (2020)
PVI	Perpendicular vegetation index	$\sqrt{(\rho_{soil} - \rho_{veg})^2_{red} - (\rho_{soil} - \rho_{veg})^2_{NIR}}$	Xue and Su (2017)
RVI	Ratio vegetation index	ρ_{red}/ρ_{NIR}	Quan et al. (2011)
LAI	Leaf area index	*3:618 * EVI - 0:118*	Gaso et al. (2019)

using a 10 m × 10 m square quadrant at five places for each plot. The yield data is converted into kg/ha. Similarly, potato yield was estimated for five rows, and the crop row's dimensions (length and width) were measured. The yield data in a point vector format at a spatial resolution were converted to raster GeoTiff at 10 × 10 m pixel resolution corresponding to the Sentinel-2 image pixels. Then, the yield maps were developed by the Kriging interpolation technique through ArcGIS 10.7.

8.2.5 Statistical Analysis

Statistical analysis was carried out to validate the relationship between the VIs, generic crop yield, and estimate yield data for two crops. The normality of the dataset was organized by descriptive statistics integrating minimum, maximum, range, mean, standard deviation (SD), skewness, kurtosis, and coefficient of variation. All the statistical calculations were performed in the 'R' software v.4.0.5 (University of Auckland, New Zealand) environment.

8.2.5.1 Geostatistical Analysis

ArcGIS Geostatistical Analyst tool was also utilized to study the degree of spatial dependence between Sentinel MSI-based VIs and yield data in the ArcGIS environment. The gradation of consistency for rice and potato yielding was determined using the standard equation using the geostatistical semivariogram model (Eq. 8.1) (McKinion et al. 2010).

$$\gamma(h) = \frac{1}{2N(h)} + \sum_{i=1}^{N(h)} [Z(x_i) - Z(x_i + h)]^2 \quad (8.1)$$

where $\gamma(h)$ represents semivariance, $Z(x_i)$ is the distance between the measured sample at point Xi, and the point $Z(x_i + h)$ is the sum of the pairs separated by the lag h.

The semivariogram was fitted using exponential, spherical, Gaussian, and linear models. The residual sum of squares (RSS) was used to determine the form and goodness of fit of the model semivariogram (Eq. 8.2).

$$\text{RSS} = \sum_{i=1}^{n} (y_i - f(x_i))^2 \quad (8.2)$$

where y_i is the ith value of the variable to be predicted, x_i is the ith value of the explanatory variable, and $f(x_i)$ is the predicted value of y_i

The statistical cross-validation method was used for the performance of the semivariogram model. The prediction accuracy of crop yield and VIs was assessed through the coefficient of determination and root mean square error value. The best model accuracy indicated the maximum R^2 and least RMSE value for a final agreement between crop yield and VIs parameters.

The sill, easily-defined range explains the exponential and spherical models for the plant and soil variability fitting (McKinion et al. 2010). Our study structured the three parameters when fitting the best semivariogram model, namely (i) Nugget C_0 represents short-scale randomness, (ii) Sill ($C_0 + C_1$) of the semivariogram is equal to the variance of the random variable when growing beyond a certain distance and becomes less or steadier around an edge value, (iii) range is defined the correlation between two properties inclined to be equivalent to zero when the distance h becomes too large. In semivariogram analysis, the trends of spatial dependence (SD) are measured by the nugget/sill ratio $C_0/(C_0 + C_1)$, %. It is described in terms of three types of spatial dependency levels; (i) strong SD < 25%; (ii) moderate SD—25–75%; (iii) weak SD > 75% (Singha et al. 2020). The geostatistical kriging analysis is incorporated with the semivariogram that estimates the known values to unknown values for spatial judgment of crop yield variation at a similar level of resolution (Li et al. 2016). Yield maps converted into raster form correspond to the same spatial resolution of S2 images for extracting multi-point values. Then Pearson correlation matrix showed the relationship between the VIs and yield for the two crops.

8.2.5.2 Hotspot Analysis

Hotspot analysis is a statistical procedure that identifies statistically significant cold and hot spots through the Getis-Ord *Gi** statistic. It utilizes a set of features that are weighted to identify these cluster areas. These statistics have three components: *Gi Z* Score, *Gi p*-value, and *Gi*-Bin values of the selected criteria. The resultant *Gi Z* Score and *Gi p*-Value define the high or low values of the cluster in a spatial process. A high *Z* score and a small p-value indicate a significant hot spot positively correlated with a low negative *Z* score. The more intense the hotspot or coldspot clustering, the more it is distributed using kriging interpolation within the region. This approach identifies high and low clusters of yield data correspondence to VIs with similar areas. The ArcGIS spatial statistics tools did spatial autocorrelation of the VIs in hotspots analysis in the mapping cluster approach. The Getis-Ord *Gi** statistic was explained in (Eqs. 8.3, 8.4, and 8.5) (Abdulhafedh 2017).

$$G_i^* = \frac{\sum_{j=1}^{n} w_{i,j}x_j - \overline{X}\sum_{j=1}^{n} w_{i,j}}{s\sqrt{\frac{\left[n\sum_{j=1}^{n} w_{i,j}^2 - \left(\sum_{j=1}^{n} w_{i,j}\right)^2\right]}{n-1}}} \quad (8.3)$$

$$\vec{X} = \frac{\sum_{j=1}^{n} x_j}{n} \quad (8.4)$$

$$S = \sqrt{\frac{\sum_{j=1}^{n} x_j^2}{n} - \left(\overline{X}\right)^2} \quad (8.5)$$

where G^* indicates the G statistics of *i*, x_j is the VIs and yield of *j*, $w_{i,j}$ describes the spatial weight between VIs and yield of *i* and *j*, *X*, and *S* represent mean, variance, and n indicate the sum of the VIs and yield parameters.

When evaluating crop yield and VIs agricultural data with their surrounding neighbors, the hotspots represent positive autocorrelation with high production, and coldspots represent negative autocorrelation with low production in the study area. The attempts were made to carry out hotspot analysis for point data on a microscale.

8.3 Validation

Validation is carried out using the survey training dataset of rice and potato yield information collected from seventy farm plots in the study area. For validation purposes, we use the area under the receiver operating characteristic (AUROC) curve incorporated into the four ML techniques, namely (i) logistic regression (LR), (ii) support vector machine (SVM), (iii) random forest (RF), and (iv) Naïve Bayes (NB) which measured the consistent relationship between yield data and VIs.

LR is a classification algorithm that calculates a predicted probability for a dichotomous dependent variable based on one or more independent variables (Ahmed and Sajjad 2018). Authentication of the LR model is regulated by the binary composition that indicates the probability of incidence with pair sample space, *P* (Eq. 8.6)

$$P(y = \pm 1|x, w) = \frac{1}{1 + \exp(-y(w^T x + b))} \quad (8.6)$$

where *b* signifies the intercept, *T* signifies the transfer matrix, and the *k*-dimensional coefficient, $w = (w_1,\ w_2,\ \ldots\ w_k)^T$, comprises the essential parameters to be assessed. Similarly, SVM, RF, and NB models can be estimated using their respective standard equations.

The model's prediction accuracy was estimated at 70% for training and 30% for testing data through the area under the AUROC curve (Eq. 8.7) (Allen 2015).

$$\text{AUC} = \frac{\sum \text{TP} + \sum \text{TN}}{P + N} \quad (8.7)$$

where TP—true positive and TN—true negative represents the number of farm plots correctly predicted crop yield, *P* is the total number of plots with heavy specs, and *N* is the total number of plots without substantial specs.

All-model performances were optimized for choosing the best tuning hyperparameter to determine the high accuracy (AUC) results (Table 8.2). The AUROC curve was produced by the Anaconda python Jupyter notebook v. 6.0.1.

8.4 Results and Discussion

8.4.1 Descriptive Statistical Analysis

Descriptive statistical analysis was carried out for all the selected VIs and crop yields (Table 8.3). The mean rice and potato yield rate of the study area was 7.11 and 28.04 (t/ha), respectively. The spatial variability of crop yield and VIs data were good estimators of the ecological vegetation process that resulted in a standard deviation (SD), verified a CV value of 23.25% for rice yield and 29.88% for potato crop yield. The VIs of NDVI, NDWI, BI, DVI, GEMI, SAVI, PVI, RVI, and LAI had low CVs in rice crop; similarly, the NDVI, DVI, SAVI, PVI, and RVI had a moderate level of CVs for the potato crop. The NDVI and SAVI ranged between 0.45 to 0.58 and 0.27 to 0.34 for rice crops, and similarly for potato crops between 0.16 to 0.55 and 0.11 to 0.35. The high mean values for RVI were around 3.33, followed by LAI (2.06), GEMI (0.64), and NDWI (0.62) in the rice crop, whereas the potato crop is high mean values of RVI (2.25), followed by LAI (0.80), GEMI (0.56), and NDVI (0.38), respectively. The positive (1.29) and negative (− 2.08) skewness were found with BI and NDWI for the rice crop. However, the negative (− 0.32) skewness found in NDVI and positive (0.44) in the LAI was found in potato crop, creating the symmetrical type distribution. The kurtosis interrelated to DVI, SAVI, and PVI showed platykurtic behavior associated with normal distribution for rice crops. In the case of NDWI and BI, high kurtosis (− 0.89 and 0.27) was observed, which may represent the leptokurtic behavior of distribution for the potato crop.

8.4.2 Relationship Between VIs and Crop Yields

Pearson correlations matrix described the interrelation between the selected VIs and crop yield for two specific crop seasons (Tables 8.4 and 8.5). Higher r^2 values reported the most significance in crop growth and yield patterns around the experimental farm plots. Rice yield was significant/positively correlated with NDVI, and RVI, while a moderate correlation with SAVI, DVI, and PVI. VIs of RVI, NDVI, SAVI, DVI, and PVI were best positively correlated with the actual rice yield with high r values that ranged from 0.80 and 0.92 (Table 8.4).

Similarly, potato yield was strongly correlated with NDVI and moderately correlated with RVI,

Table 8.2 Optimization of parameters for AUROC validation

Model	Tuning parameter	Feature selection
RF	n_estimators	10
	max_features	Auto
	Max depth	5
	Criterion	'gini'
SVM	Kernel	'rbf'
	Gamma	2
	C	1
	Degree	3
NB	var_smoothing	1e–09
LR	C	1
	Solver	'lbfgs'
	Intercept	24.05

Table 8.3 Descriptive statistical analysis of selected VIs and crop yields (tons/hector)

Crops	Parameters	NDVI	NDWI	BI	DVI	GEMI	SAVI	PVI	RVI	LAI	Yield
Rice	Min	0.45	0.42	0.08	0.14	0.55	0.27	0.10	2.64	1.26	3.40
	Max	0.58	0.68	0.11	0.19	0.68	0.34	0.13	3.74	2.68	12.00
	Mean	0.54	0.62	0.09	0.17	0.64	0.31	0.12	3.33	2.06	7.12
	SD	0.03	0.04	0.00	0.01	0.03	0.02	0.01	0.23	0.26	2.01
	Skewness	− 1.19	− 2.08	1.29	− 0.33	− 1.62	− 0.55	− 0.34	− 0.77	− 0.40	0.53
	Kurtosis	2.34	8.21	3.21	− 0.33	3.96	− 0.03	− 0.34	1.18	0.86	0.25
	CV	4.66	7.07	4.44	6.47	3.90	5.75	6.67	6.76	12.83	28.25
Potato	Min	0.16	− 0.02	0.10	0.06	0.43	0.11	0.05	1.46	0.36	9.90
	Max	0.55	0.55	0.14	0.21	0.69	0.35	0.15	3.37	1.43	39.00
	Mean	0.38	0.27	0.12	0.13	0.56	0.22	0.09	2.25	0.80	28.04
	SD	0.09	0.16	0.01	0.04	0.06	0.06	0.03	0.49	0.29	8.38
	Skewness	− 0.32	− 0.25	1.14	0.19	0.08	0.03	0.20	0.31	0.44	− 0.47
	Kurtosis	− 0.37	− 0.90	0.26	− 0.74	− 0.34	− 0.81	− 0.73	− 0.65	− 0.39	− 0.54
	CV	24.23	59.26	8.42	28.50	10.26	27.18	28.50	21.73	36.17	29.88

N.B. *SD* standard deviation; *CV* coefficient of variation; *n* No. of sample

Table 8.4 Pearson correlations matrix between VIs and rice yield (n = 70)

	NDVI	NDWI	BI	DVI	GEMI	SAVI	PVI	RVI	LAI	Rice yield
NDVI	1.000									
NDWI	0.704**	1.000								
BI	− 0.701**	− 0.770**	1.000							
DVI	0.821**	0.353*	− 0.178	1.000						
GEMI	0.542**	0.248	− 0.116	0.661**	1.000					
SAVI	0.908**	0.470**	− 0.332*	0.984**	0.645**	1.000				
PVI	0.819**	0.354*	− 0.175	0.999**	0.665**	0.983**	1.000			
RVI	0.996**	0.688**	− 0.681**	0.836**	0.565**	0.918**	0.833**	1.000		
LAI	0.804**	0.850**	− 0.696**	0.555**	0.615**	0.648**	0.555**	0.813**	1.000	
Rice yield	0.895**	0.551**	− 0.563**	0.806**	0.582**	0.862**	0.800**	0.920**	0.746**	1.000

** Correlation is significant, 0.01 level; * Correlation is significant, 0.05 level

SAVI, DVI, PVI, and LAI of r values from 0.761 to 0.908 (Table 8.5). BI was insignificant/negatively correlated with both the crop yield (rice and potato) because of the variability of canopy reflectance and light use efficiency (LUE) with site-specific field management.

8.4.3 Geostatistical Analysis

Plot-based yield information and extracted VIs datasets were analyzed using the geostatistical semivariogram method (Table 8.6). The lag size of both crops was between 14.90 and 64.46,

Table 8.5 Pearson correlations matrix between VIs and potato yield (n = 70)

	NDVI	NDWI	BI	DVI	GEMI	SAVI	PVI	RVI	LAI	Potato yield t/ha
NDVI	1									
NDWI	0.899**	1								
BI	− 0.854**	− 0.866**	1							
DVI	0.938**	0.917**	− 0.830**	1						
GEMI	0.880**	0.934**	− 0.738**	0.946**	1					
SAVI	0.944**	0.928**	− 0.862**	0.998**	0.940**	1				
PVI	0.938**	0.917**	− 0.830**	1.000**	0.945**	0.998**	1			
RVI	0.934**	0.928**	− 0.857**	0.993**	0.936**	0.994**	0.994**	1		
LAI	0.909**	0.971**	− 0.835**	0.943**	0.950**	0.947**	0.943**	0.961**	1	
Potato yield	0.908**	0.783**	− 0.781**	0.808**	0.741**	0.821**	0.807**	0.801**	0.761**	1

** Correlation is significant, 0.01 level; * Correlation is significant, 0.05 level

that's were depending on the spatial variability of the crop, VI, and soil nutrients. The high lag size was 57.76 for rice (BI) and 64.46 for potato (NDWI) with a spherical semivariogram. Associating range values of rice VIs, the lowest value was measured for GEMI (95.59 m), and the highest value was measured at NDWI (255.01 m). Similarly, the VIs of potato crops varied amply between 246.46 and 349.83 m. Nuggets were highest for PVI (5.29) and lowest for BI, GEMI (0.0001) of rice crops, where the potato crop has the lowest nugget for GEMI variables. The nine VIs of the rice crop revealed a sill size between 0.002 (BI, GEMI) and 6.42 (PVI), which stated a relatively parallel total variance. The VIs of potato crop sill varied from 0.008 (GEMI) to 0.093 (NDVI). The ratio of percentage between the two parts, a nugget to sill variance, ranged noticeably between RVI (0.039%) and PVI (82.37%), where C_0 represented 0.039% and 82.37% of $C_0 + C$ with strong spatial dependency (0.41%) for rice yield, and the other potato crops where C_0 denoted 9.87% to 60.36% of $C_0 + C$ with moderate spatial dependency (34.27%) for potato yield. The exponential and spherical models are very good of a fit in semivariogram analysis with the lowest RMSE value, which means they are highly significant. Exponential models were suitable alternatives to the experimental semivariograms for NDVI, NDWI, DVI, SAVI, PVI, RVI, and LAI, whereas BI and GEMI values were best fitted with a spherical model for rice crops (Fig. 8.3). NDVI and BI are very well-explained by the exponential models; whereas NDWI, DVI, GEMI, SAVI, PVI, RVI, and LAI, are the best suited to the spherical model for potato crops (Fig. 8.3). R^2 was always > 0.75 estimated for the crop yield, with an error of RMSE very low, particularly concerning average VI. The best R^2 values of 0.846, 0.801, and 0.743 were found for RVI, NDVI, and SAVI, respectively, while a weak R^2 was found with NDWI and BI for rice crops (Figs. 8.3 and 8.4). On the other hand, the NDVI, SAVI, DVI, PVI, and RVI were associated with the highest corresponding R^2 values of 0.824, 0.674, 0.653, and 0.652 for the potato crop. The RMSE of VIs ranged between 0.004 and 0.41, concerning rice yield of 1.77 t/ha. Similarly, the VIs of potatoes range between 0.006 and 0.29, with a potato yield of 8.4 t/ha (Table 8.6). The maps indicated that the northeastern field had a lower crop yield than the southwestern part of the field. The spatial variability of the crop was also exhibited in the generated rice and potato yields.

Table 8.6 Semivariograms model performance with the exponential and spherical models for both the crop VIs and yields analysis

		Model	C_0	C_1	$(C_0 + C_1)$	$C_0/(C_0 + C_1)$ (%)	SD	a (m)	Lag size	R^2	RMSE
Rice	NDVI	Exponential	0.001	0.002	0.003	39.604	M	250.126	20.843	0.80	0.024
	NDWI	Exponential	0.001	0.01	0.011	4.757	S	255.01	21.25	0.30	0.035
	BI	Spherical	0.0001	0.001	0.002	18.072	S	171.72	57.76	− 0.32	0.004
	DVI	Exponential	0.003	0.001	0.005	75.764	W	111.36	17.17	0.65	0.012
	GEMI	Spherical	0.0001	0.002	0.002	13.634	S	95.592	14.9	0.74	0.021
	SAVI	Exponential	0.002	0.001	0.004	64.12	M	266.8	22.23	0.34	0.018
	PVI	Exponential	5.286	1.131	6.417	82.375	W	126.35	17.175	0.64	0.008
	RVI	Exponential	0.0022	0.0034	0.0057	0.0392	S	240.629	20.052	0.85	0.0207
	LAI	Exponential	0.001	0.019	0.021	6.978	S	242.186	20.18	0.56	0.206
	Rice	Exponential	0.037	0.053	0.09	0.414	S	203.468	16.995		1.777
Potato	NDVI	Exponential	0.0092	0.084	0.0932	9.871	S	349.834	29.152	0.82	0.074
	NDWI	Spherical	0.0027	0.019	0.0217	12.442	S	279.85	64.464	0.61	0.0815
	BI	Exponential	0.005	0.008	0.013	38.462	M	361.3	30.108	− 0.61	0.0067
	DVI	Spherical	0.01	0.07	0.08	12.5	S	247.15	31.703	0.65	0.0232
	GEMI	Spherical	0.0017	0.007	0.0087	19.54	S	260.74	27.686	0.67	0.0347
	SAVI	Spherical	0.0104	0.0749	0.0853	12.192	S	276.062	28.779	0.55	0.0395
	PVI	Spherical	0.0097	0.0711	0.0808	12.005	S	246.46	31.908	0.65	0.0164
	RVI	Spherical	0.0067	0.033	0.0397	16.877	S	265.344	34.255	0.64	0.2968
	LAI	Spherical	0.0166	0.0109	0.0275	60.364	M	289.486	30.697	0.58	0.1529
	Potato	Exponential	0.0629	0.1206	0.1835	34.278	M	368.364	30.697		8.4

N.B. C_0 Nugget; C_1 Partial sill; $(C_0 + C_1)$ Sill; $C_0/(C_0 + C_1)$ (%) Nugget/Sill; *a* Range (m), *RMSE* root mean square error

Remote sensing-based S2-derived VIs maps and the responses crop yield maps were established for similar spatial resolution (Figs. 8.5 and 8.6). A total of nine VIs are selected through their associations with crop productivity. The study region also observed a spatial pattern consistent with crop yield distribution. High values found in the south–southwest part of the farm plot were associated with high crop yield compared to the northern part of the region.

8.4.4 Hot Spots Analysis

The concept of the *Z* score indicates a significant hotspot and a negative *Z* score indicates a significant coldspot in Getis-Ord *Gi** statistic. The analysis data revealed that the high-yielding areas were mainly concentrated in the RVI, NDVI, and SAVI surfaces. This observation explained that high-yield clusters exist in the RVI, NDVI, and SAVI surfaces with significant cluster distribution.

This analysis suggested that RVI, NDVI, and SAVI indices were superior in guiding maps for ideal demonstration of the spatial variability for rice and potato yield patterns at small farm levels (Figs. 8.7 and 8.8). BI values are highly significant (hotspot) and range between 90 and 95% confidence level in the northeastern region due to soil humidity and pH variability with low crop yield for both crops (Fig. 8.7). Similarly, the other VIs (DVI, GEMI, PVI, and LAI) showed a highly significant (90–95%) cluster with high yield due to better crop practice management in the southwest part of the farm areas for both the

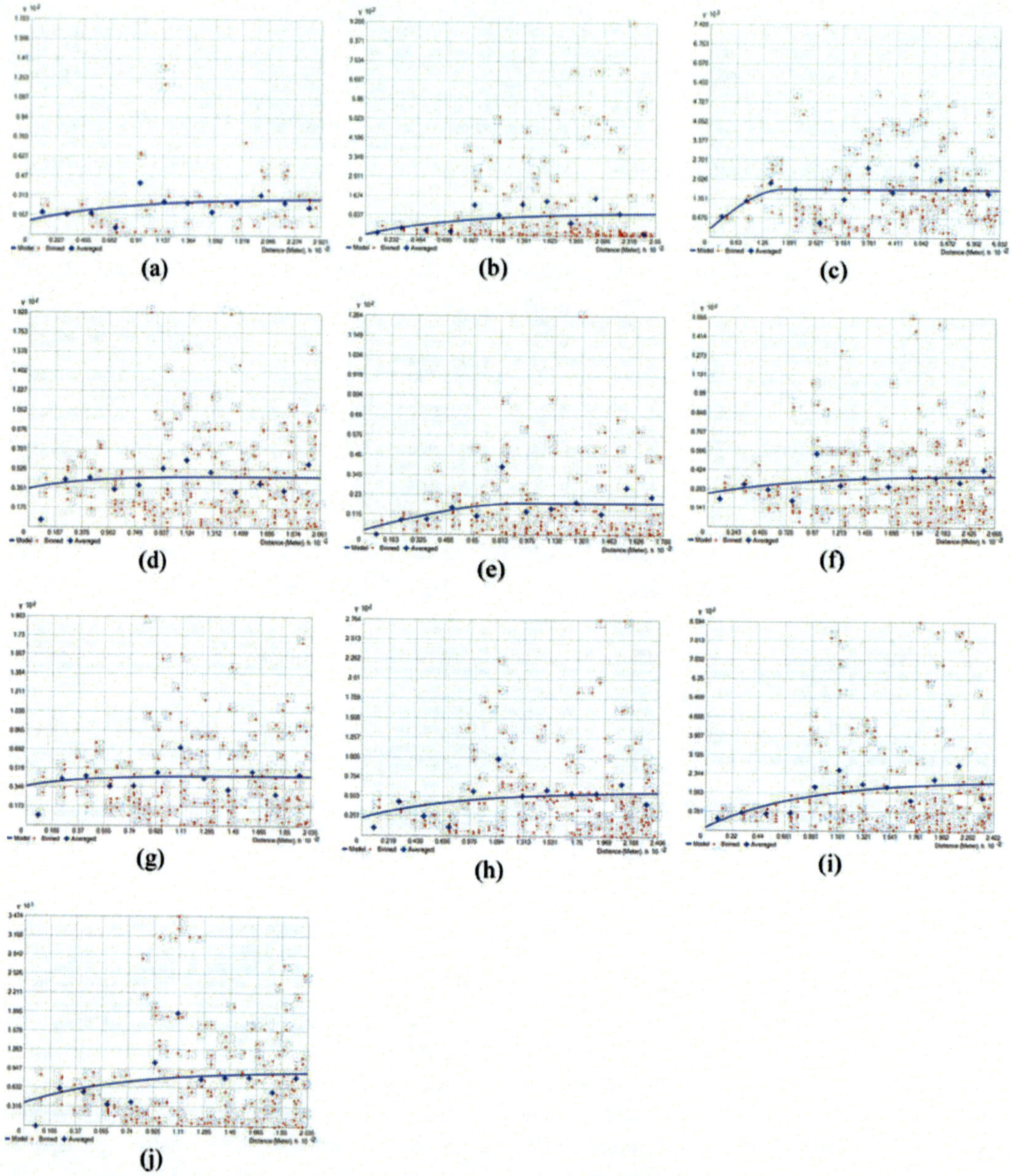

Fig. 8.3 Semivariograms graph for rice crop; **a** NDVI, **b** NDWI, **c** BI, **d** DVI, **e** GEMI, **f** SAVI, **g** PVI, **h** RVI, **i** LAI, and **j** rice yield

crop (Figs. 8.7 and 8.8). In this chapter, the hotspot analysis described the spatial pattern of both crop yield distribution with VIs: coldspot significance level found in the northern part of the farm plot, and in contrary hotspot was in significance level in the south–southwestern side of the farming area. The significance levels of hotspot maps presented a high number of low-distant VIs.

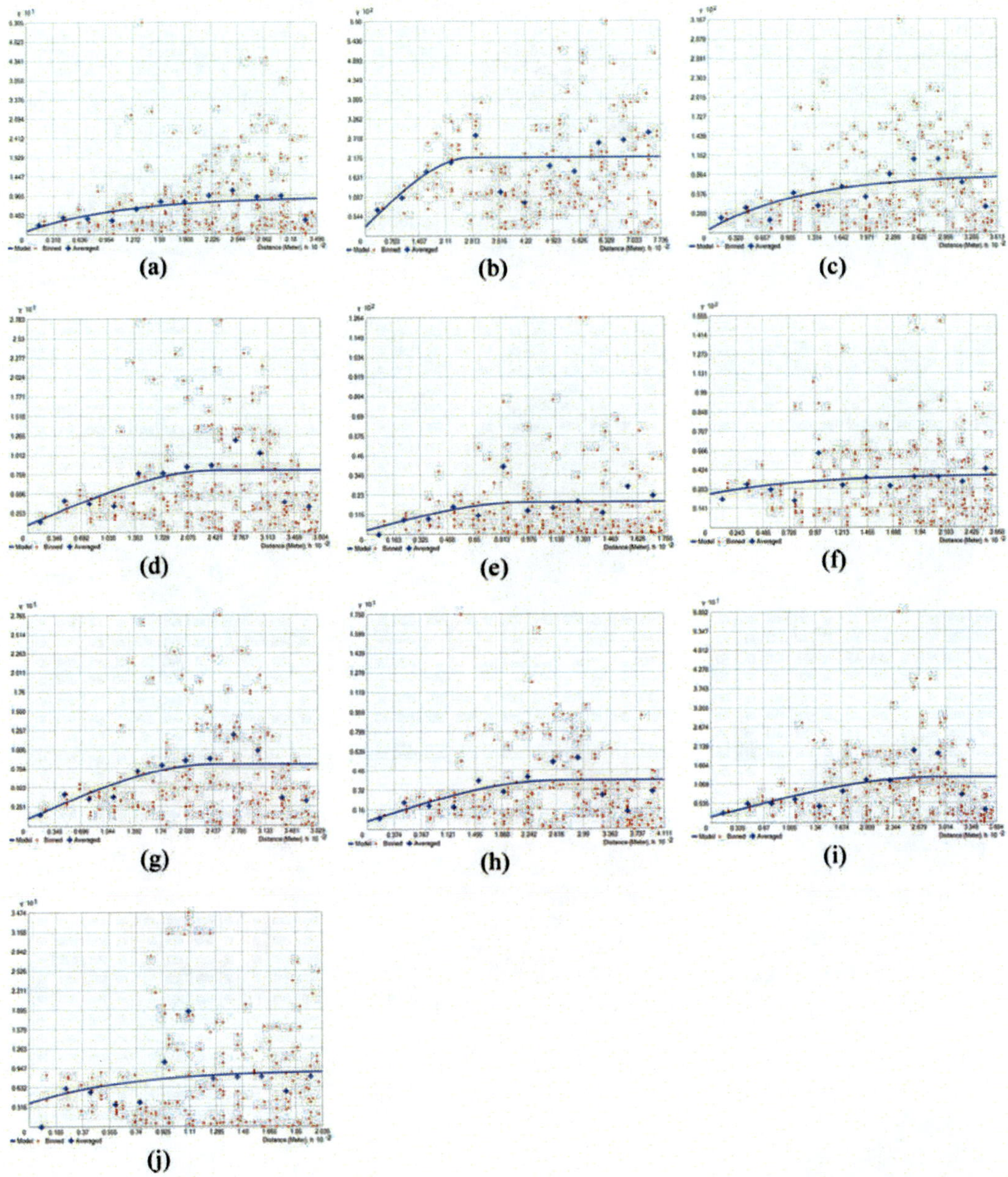

Fig. 8.4 Semivariograms graph for potato crop; **a** NDVI, **b** NDWI, **c** BI, **d** DVI, **e** GEMI, **f** SAVI, **g** PVI, **h** RVI, **i** LAI, and **j** potato yield

8.4.5 AUROC Validation

The outcomes of the validation model of AUROC with four MLAs techniques between the yield and VIs datasets were assessed for both crops (Fig. 8.9). SVM, LR, and RF algorithms led to the spatial assessment of rice and potato crop yield estimation with AUC accuracy values between 80 and 90%, categorizing them as very good predictors, while NB had an accuracy value between 60 and 80%, categorized as a moderate to the good predictor for both the crops. More

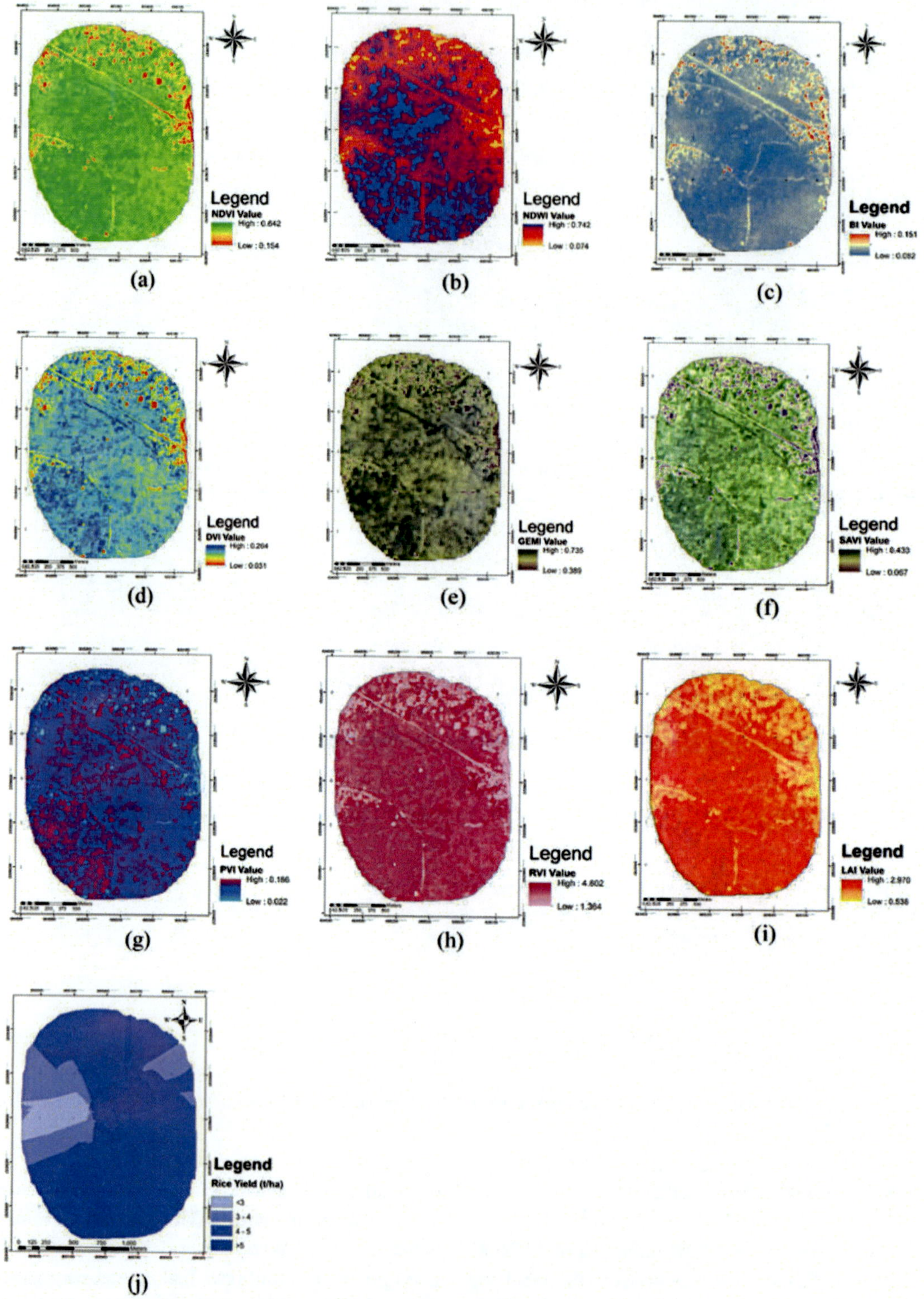

Fig. 8.5 Vegetation indices and yield distribution map for rice crop; **a** NDVI, **b** NDWI, **c** BI, **d** DVI, **e** GEMI, **f** SAVI, **g** PVI, **h** RVI, **i** LAI, and **j** rice yield

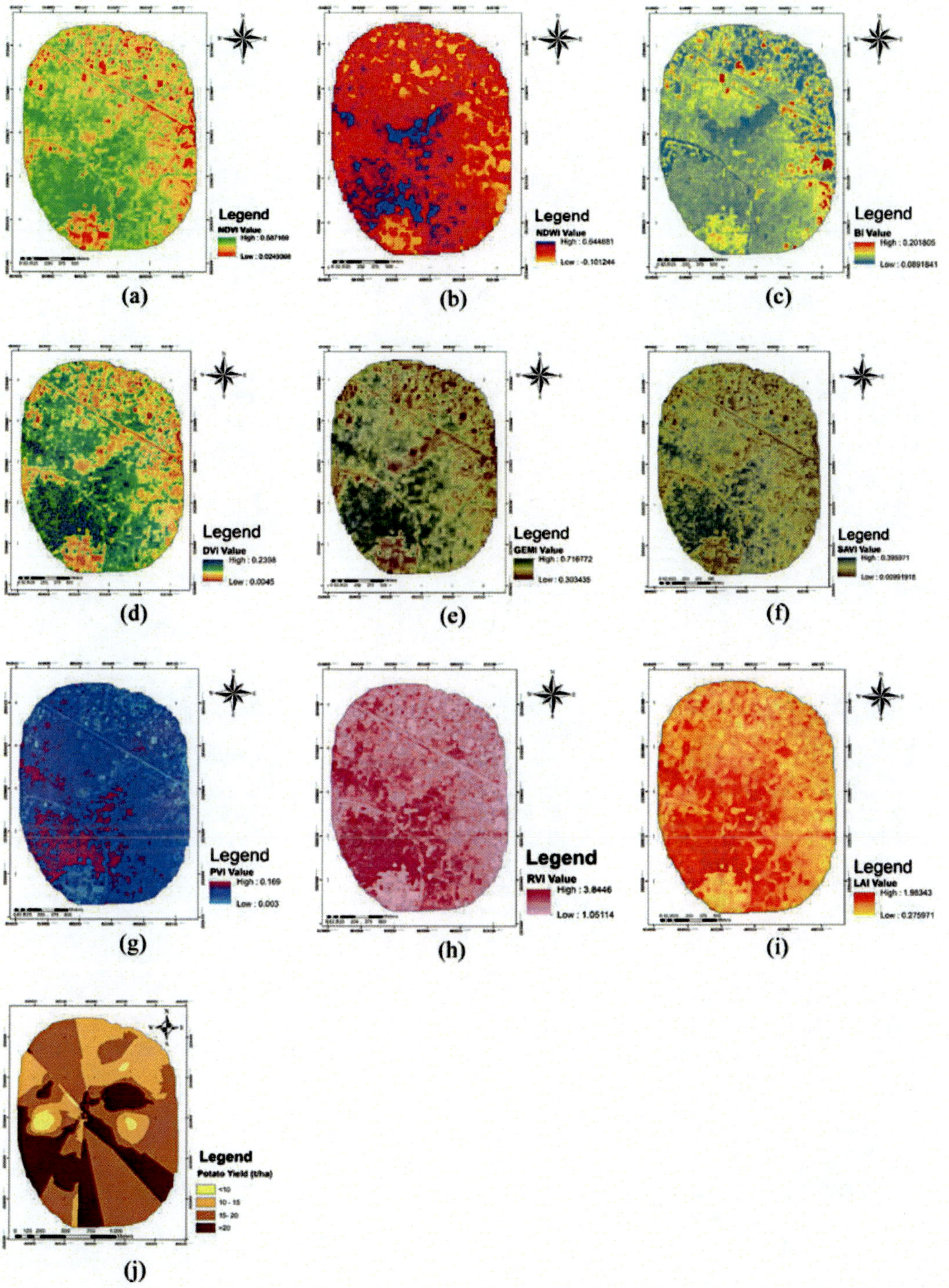

Fig. 8.6 Vegetation indices and yield distribution map for potato crop; **a** NDVI, **b** NDWI, **c** BI, **d** DVI, **e** GEMI, **f** SAVI, **g** PVI, **h** RVI, **i** LAI, and **j** potato yield

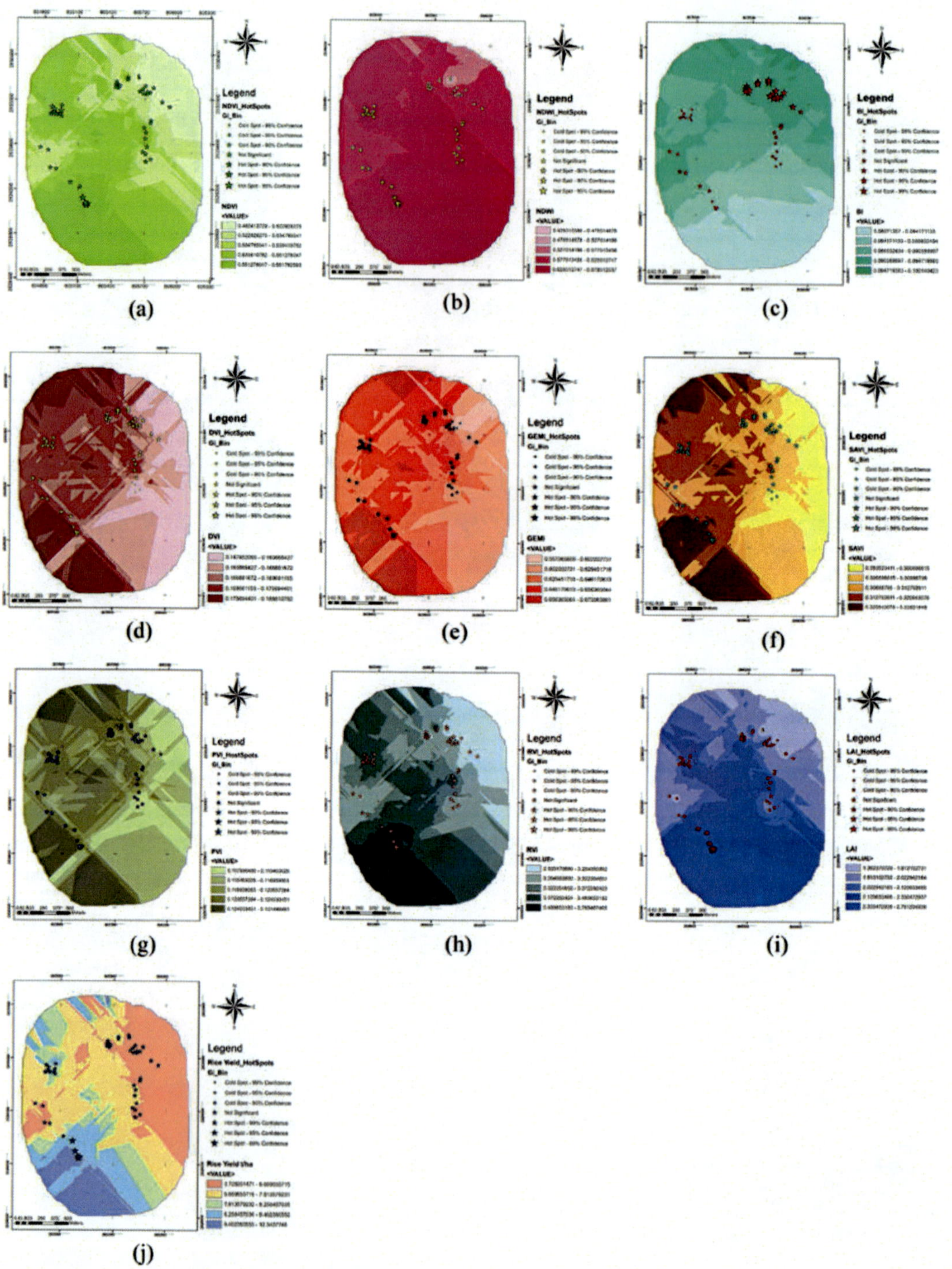

Fig. 8.7 Hotspot analysis for rice crop; **a** NDVI, **b** NDWI, **c** BI, **d** DVI, **e** GEMI, **f** SAVI, **g** PVI, **h** RVI, **i** LAI, and **j** rice yield

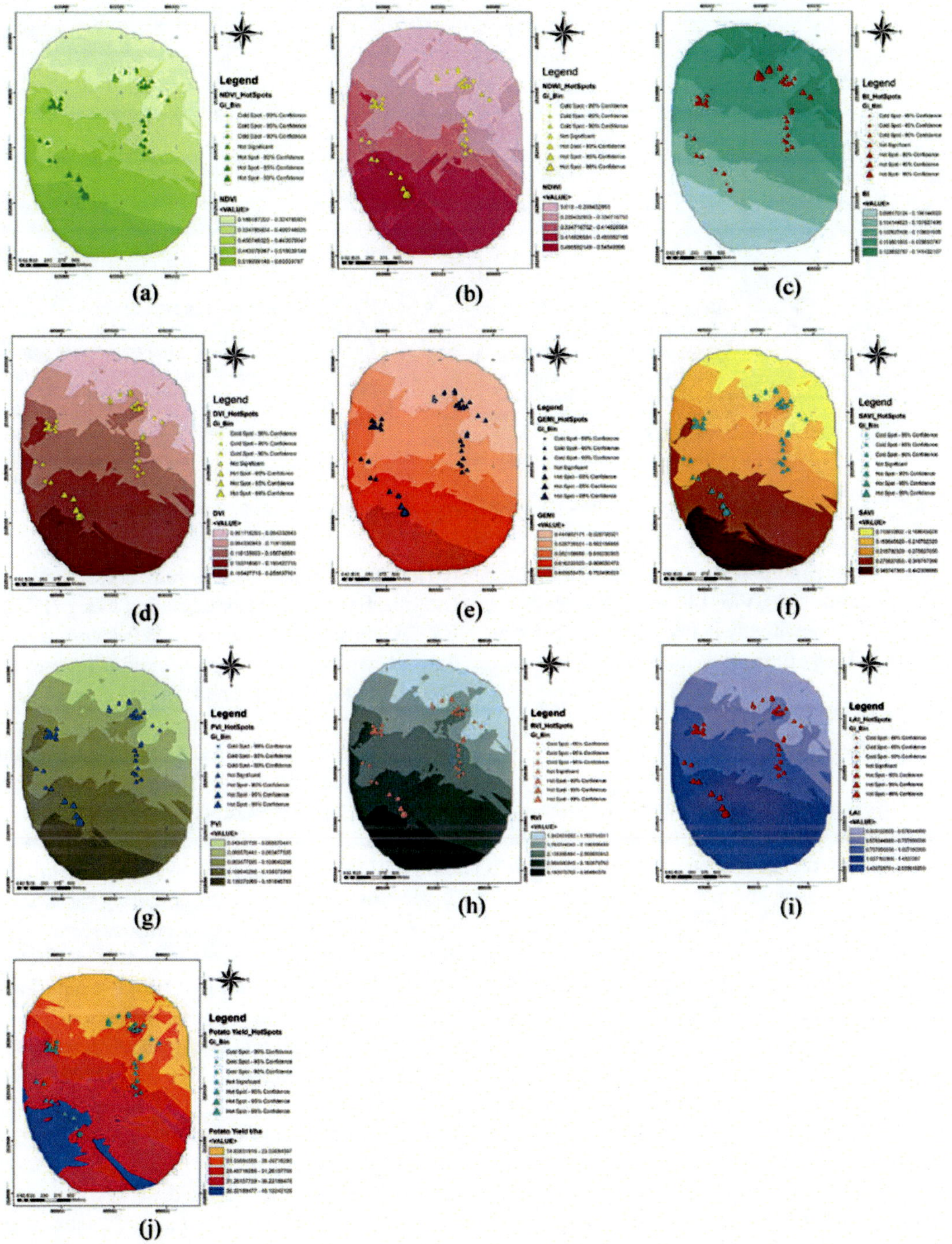

Fig. 8.8 Hotspot analysis for potato crop **a** NDVI, **b** NDWI, **c** BI, **d** DVI, **e** GEMI, **f** SAVI, **g** PVI, **h** RVI, **i** LAI, and **j** potato yield

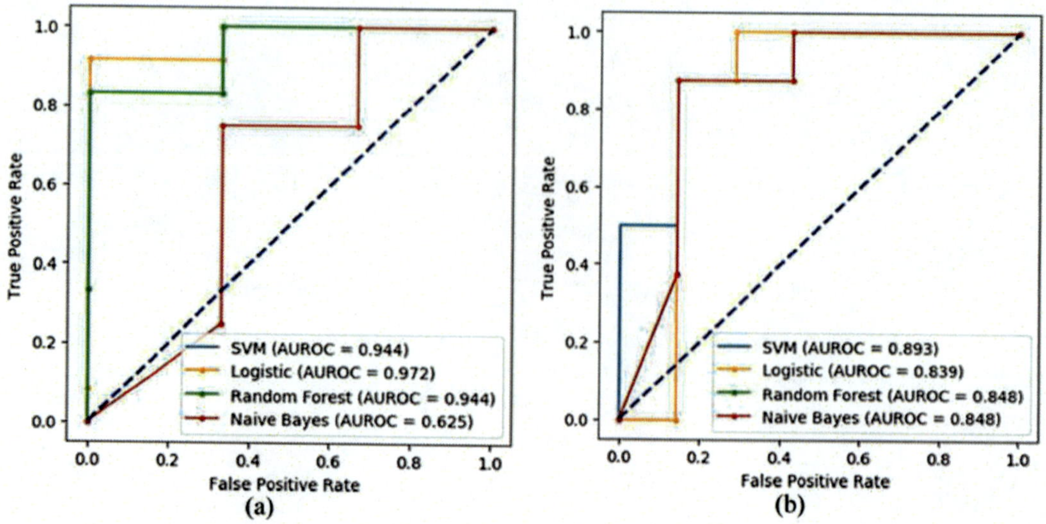

Fig. 8.9 AUROC validation between VIs and crop yield (rice and potato); **a** rice crop, **a** potato crop

precisely, the outputs revealed that the three MLAs techniques, SVM, LR, and RF algorithms, produced competitively better performances than the NB ML algorithm for both the crop yield data.

8.5 Discussion

We have used different spectral vegetation indices, estimated from the Sentinel-2B satellite data, to carry out the geostatistical-based crop yield estimation of rice and potato crops in two different seasons in Tarakeswar, Hooghly region (West Bengal). The spatial variability of crop yielding is delineated by the climate, topography, soil factor, water availability, and farm mechanization parameters (Singha and Swain 2022). A non-destructive yield prediction map was developed using a multiyear dataset. The map was developed using a geostatistical tool and an artificial intelligence model (Panek et al. 2020). A correlation coefficient analysis shows the spatial pattern of the VIs and yield. The VIs of RVI, NDVI, SAVI, DVI, and PVI were found to be significantly positive and correlated to the rice yield in the *Kharif* season.

The most commonly used VIs utilizes the information in the red and near-infrared (NIR) canopy reflectance or radiances. They are combined in ratios: ratio vegetation index (RVI). RVI is the most frequently used VI with a significant correlation with grain yield (Pinter et al. 2003; Ali et al. 2019). The RVI, which measures the relative importance of rice yield to the farm plot, showed the best correlation with the actual rice yield, followed by the NDVI > SAVI > DVI > PVI > LAI > GEMI > NDWI > BI (Fig. 8.4). The plant signals attained from SAVI and NDVI had high correlation with the Sentinel-2-measured canopy reflectance during the *Kharif* rice and *Rabi* potato growing seasons. Similarly, during the *Rabi* seasons, the R^2 values of NDVI, SAVI, DVI, PVI, RVI, NDWI, and LAI decreased toward potato yield distribution. These relationships showed a sound correlation coefficient ranging between, 0.761 and 0.908, with potato crop yield.

A weak correlation was observed between the various parameters, namely NDWI, BI of crop concentration, and actual yield. It was attributed to the fact that the crop's maturity caused an increase in visible reflectance (Kumhálová and Matˇejková 2017). RVI, NDVI, and SAVI indices were the better-controlling factors to variate crop yield. On the other hand, for potato crops, the NDVI, SAVI, DVI, PVI, and RVI indices were the highest R^2 values of 0.824, 0.674, 0.653, and 0.652, respectively (Table 8.6). Seasonal NDVI values indicate the best yield estimation factor for rice and potato crop (yield) predictions. The maximum NDVI was retained as the most significant variable for predicting field-level yield. Borowik et al. (2013) reported that the relationship between vegetation aboveground biomass and NDVI reflects each habitat type in seasonal variation. The selected NDVI and RVI indices have comparatively higher correlations with grain yield than SAVI (Ali et al. 2019).

Geostatistical semivariogram analysis was performed with best-fitting exponential and spherical models to develop the RS-based VIs and crop yield maps (Singha et al. 2020). VIs maps combined with RGB and NIR bands have a greater potentiality for plant sensitivity analysis in different crop growing seasons (Pinter et al. 2003). The ML methods such as support vector regression (SVR) and ensemble multilayer perceptron neural networks are successfully predicting the rice yield in the Cauvery Delta Zone (CDZ) and Gujarat India (Bhojani and Bhatt 2020; Yu et al. 2021). The chapter's contribution is simple and efficient in providing effective structures within appropriate site-specific agronomic management. The sill variance depends on canopy cover reflectance with the plant growth condition assumed by the VIs semivariograms analysis (Pradhan et al. 2014). Data of all VIs and crop yield were signified according to the Getis-Ord *Gi** statistic for identifying the degree of significance in the study area. In statistical hotspot analysis, high clusters of yield were detected in RVI, NDVI, and SAVI surfaces and low yield while the low significant clusters observed in RVI, NDVI, and SAVI surfaces were superior to guide map and best alternatives for rice and potato yield prediction in a small scale of village level. The hotspot analysis was used to define a cluster's higher or lower values in a spatial process, specifying the number of clusters to be detected (Marino and Alvino 2021). VIs map correspondence to yield map have similar trends in hotspot analysis. Generally, the northern part of the study area has some limitations with coldspots due to high soil pH (acidic soil), least organic carbon, high electrical conductivity, and low farm mechanization level. Similarly, high yield and VIs values were found in the northwestern region with good correlations because there has been remarkable crop growth represented by higher canopy biomass in agreement with the final yield.

The validation process was made through the AUROC machine learning model between VIs and crop yield (rice and potato) for two different seasons. SVM, LR, and RF models have good validation results with > 83% accuracy for both crops. Crop yield estimation can be improved using modern advanced techniques such as hybrid ML and deep learning, satellite data fusion, LiDAR, UAV, IoT, SAR, and GEE cloud with a higher number of influential factors (soil factor, climate, topography, hydrology, AGB, farm mechanization level, and socioeconomic background).

8.6 Conclusion

The Sentinel-2 mission has opened up new scenarios for monitoring the performance of smallholder farming systems. The various temporal and spatial resolutions of satellite images were possible to estimate a farm's crop yield using crop observation VIs. Evaluation of farm decisions can then be made based on the data collected by observing different crop practice parameters. High-resolution images can now accurately classify crop biophysical data for individual farmlands. The data collected by the NDVI and SAVI revealed a high correlation between the observed canopy reflectance and the plant signals. The hotspot analysis further identified the areas with high crop yields and problem areas with low crop yields. These signals indicated that the various interventions designed to improve smallholder farms' productivity may not address these communities' diverse socioeconomic conditions. The results of the crop models may be provided to the farmer to take necessary steps to improve crop productivity, particularly in the northern part of the concerned study area.

Acknowledgements The authors acknowledge the contribution of Visva-Bharati (A Central University), West Bengal, India, for facilitating this research work.

References

Abbas F, Afzaal H, Farooque A, Tang S (2020) Crop yield prediction through proximal sensing and machine learning algorithms. Agron 10:1046

Abdulhafedh A (2017) A novel hybrid method for measuring the spatial autocorrelation of vehicular crashes: combining Moran's Index and Getis-Ord G_i^* statistic. Open J Civil Eng 7:208–221. https://doi.org/10.4236/ojce.2017.72013

Ahmed R, Sajjad H (2018) Analyzing factors of groundwater potential and its relation with population in the Lower Barpani Watershed, Assam, India. Nat Resour Res 27:503–515

Allen LM (2015) Influence of corn seeding rate, soil attributes, and topographic characteristics on grain yield, yield components, and grain composition. Graduate Theses and Dissertations. 14949. https://lib.dr.iastate.edu/etd/14949

Ali A, Martelli R, Lupia F, Barbanti L (2019) Assessing multiple years' spatial variability of crop yields using satellite vegetation indices. Remote Sens 11:2384

Arumugam P, Chemura A, Schauberger B, Gornott C (2021) Remote sensing based yield estimation of rice (*Oryza Sativa* L.) using gradient boosted regression in India. Remote Sens 13:2379. https://doi.org/10.3390/rs13122379

Bhojani SH, Bhatt N (2020) Wheat crop yield prediction using new activation functions in neural network. Neu Comput Appl 32:13941–13951

Borowik T, Pettorelli N, Sonnichsen L, Jedrzejewska B (2013) Normalized difference vegetation index (NDVI) as a predictor of forage availability for ungulates in forest and field habitats. Eur J Wild Res 59:675–682

Campillo C, Carrasco J, Gordillo J, Cordoba A, Macua J (2018) Use of satellite images to differentiate productivity zones in commercial processing tomato farms. In: Proceedings of the XV International Symposium on Processing Tomato 1233, Athens, Greece, pp 97–104

Cavalaris C, Megoudi S, Maxouri M, Anatolitis K, Sifakis M, Levizou E, Kyparissis A (2021) Modeling of durum wheat yield based on sentinel-2 imagery. Agron 11:1486. https://doi.org/10.3390/agronomy11081486

Cedric LS, Adoni WYH, Aworka R, Zoueu JT, Mutombo FK, Kimpolo CLM (2022) Crops yield prediction based on machine learning models: case of West African countries. Smart Agric Technol 2:100049. https://doi.org/10.2139/ssrn.4003105

Chandel NS, Tiwari PS, Singh KP, Jat D, Gaikwad BB, Tripathi H, Golhani K (2019) Yield prediction in wheat (*Triticum aestivum* L.) using spectral reflectance indices. Current Sci 116(2):272–278. https://doi.org/10.18520/cs/v116/i2/272-278

Chepngetich J (2020) Vegetation index based crop yield prediction model using convolution neural network: a case study of Kenya. Dissertation paper Strathmore University, pp 1–57

Drusch M, Del Bello U, Carlier S, Colin O, Fernandez V, Gascon F, Hoersch B, Isola C, Laberinti P, Martimort P, Meygret A, Spoto F, Sy O, Marchese F, Bargellini P (2012) Sentinel-2: ESA's optical high-resolution mission for GMES. Oper Serv 120:25–36. https://doi.org/10.1016/j.rse.2011.11.026

ESA (2019) European Space Agency, EO4SD-Earth Observation for Sustainable Development. Agriculture and Rural Development/Service Portfolio (2019). https://www.eo4idi.eu/sites/default/files/eo4sd_agri_portfolio_170529_singlepag.pdf. Accessed 01 July 2020

Fang H, Liang S, Hoogenboom G (2011) Integration of MODIS LAI and vegetation index products with the CSM-CERES-Maize model for corn yield estimation. Int J Remote Sens 32:1039–1065

Filippi P, Jones EJ, Wimalathunge NS, Somarathna PDSN, Pozza LE, Ugbaje SU, Bishop TFA (2019) An approach to forecast grain crop yield using multi-layered, multi-farm data sets and machine learning. Preci Agric. https://doi.org/10.1007/s11119-018-09628-4

Gaso DV, Berger AG, Ciganda VS (2019) Predicting wheat grain yield and spatial variability at field scale using a simple regression or a crop model in conjunction with Landsat images. Comput Electron Agric 159:75–83. https://doi.org/10.1016/j.compag.2019.02.026

Guo Y, Fu Y, Hao F, Zhang X, Wu W, Jin X, Bryant CR, Senthilnath J (2021) Integrated phenology and climate in rice yields prediction using machine learning methods. Ecol Ind 120:106935

Hatfield JL, Gitelson AA, Schepers JS, Walthall C (2008) Application of spectral remote sensing for agronomic decisions. Agron J 100, S-117–S-131

Ji Z, Pan Y, Zhu X, Wang J, Li Q (2021) Prediction of crop yield using phenological information extracted from remote sensing vegetation index. Sensor 21:1406. https://doi.org/10.3390/s21041406

Kobayashi N, Tani H, Wang X, Sonobe R (2020) Crop classification using spectral indices derived from Sentinel-2A imagery. Info Telecom 4(1):67–90. https://doi.org/10.1080/24751839.2019.1694765

Koutsos TM, Menexes GC, Mamolos AP (2021) The use of crop yield autocorrelation data as a sustainable approach to adjust agronomic inputs. Sustain 13:2362. https://doi.org/10.3390/su13042362

Kumhálová J, Matˇejková Š (2017) Yield variability prediction by remote sensing sensors with different spatial resolution. Int Agroph 31:195–202

Kunapuli SS, Rueda-Ayala V, Benavidez-Gutierrez G, Cordova-Cruzatty A, Cabrera A, Fernandez C, Maiguashca J (2015) Yield prediction for precision territorial management in maize using spectral data. In: Precision Agriculture 2015—Papers presented at the 10th European conference on precision agriculture, ECPA 2015, pp 199–206

Kussul N, Lavreniuk M, Kolotii A, Skakun S, Rakoid O, Shumilo L (2020) A workflow for sustainable development goals indicators assessment based on high-resolution satellite data. Int J Digi Earth 13(2):309–321. https://doi.org/10.1080/17538947.2019.1610807

Laacouri A, Nigon T, Mulla D, Yang C (2018) A case study comparing machine learning and vegetation indices for assessing corn nitrogen status in an agricultural field in Minnesota. In: Proceedings of the 14th international conference on precision agriculture. International Society of Precision Agriculture, Monticello, IL

Lambert MJ, Pierre C, Traoréb S, Blaesa X, Bareta P, Defournya P (2018) Estimating smallholder crops production at village level from Sentinel-2 time series in Mali's cotton belt. Remote Sens Environ 216:647–657

Li Z, Liu S, Zhang X, West TO, Ogle SM, Zhou N (2016) Evaluating land cover influences on model uncertainties—A case study of cropland carbon dynamics in the Mid-Continent Intensive Campaign region. Ecol Model 337:176–187. https://doi.org/10.1016/j.ecolmodel.2016.07.002

Lu G, Li C, Yang G, Yu H, Zhao X, Zhang X (2016) Retrieving soybean leaf area index based on high imaging spectrometer. Soyb Sci 35:599–608

Marasteanu I, Jaenicke E (2016) Hot spots and spatial autocorrelation in certified organic operations in the United States. Agric Res Econ Rev 45(3):485–521. https://doi.org/10.1017/age.2016.5

Marino S, Alvino A (2021) Vegetation indices data clustering for dynamic monitoring and classification of wheat yield crop traits. Remote Sens 13:541. https://doi.org/10.3390/rs13040541

McKinion JM, Willers JL, Jenkins JN (2010) Spatial analyses to evaluate multi-crop yield stability for a field. Comput Electron Agric 70:187–198

Mupangwa W, Chipindu L, Nyagumbo I, Mkuhlani S, Sisito G (2020) Evaluating machine learning algorithms for predicting maize yield under conservation agriculture in Eastern and Southern Africa. SN Appl Sci 2:952. https://doi.org/10.1007/s42452-020-2711-6

Nagy A, Fehér J, Tamás J (2018) Wheat and maize yield forecasting for the Tisza river catchment using MODIS NDVI time series and reported crop statistics. Compu Electron Agric 151:41–49

Nagy A, Szabó A, Adeniyi OD, Tamás J (2021) Wheat yield forecasting for the Tisza River catchment using Landsat 8 NDVI and SAVI time series and reported crop statistics. Agron 11:652. https://doi.org/10.3390/agronomy11040652

Panda SS, Ames DP, Panigrahi S (2010) Application of vegetation indices for agricultural crop yield prediction using neural network techniques. Remote Sens 2:673–696

Panek E, Gozdowski D, Stepien M, Samborski S, Rucinski D, Buszke B (2020) Within-field relationships between satellite-derived vegetation indices, grain yield and spike number of winter wheat and triticale. Agron 10:1842

Pejak B, Lugonja P, Antić A, Panić M, Pandžić M, Alexakis E, Mavrepis P, Zhou N, Marko O, Crnojević V (2022) Soya yield prediction on a within-field scale using machine learning models trained on Sentinel-2 and soil data. Remote Sens 14:2256. https://doi.org/10.3390/rs14092256

Pinter PJ, Hatfield JL, Schepers JS, Barnes EM, Moran MS, Daughtry CST, Upchurch DR (2003) Remote sensing for crop management. Photogram Eng Remote Sens 69(6):647–664

Pham HT, Awange J, Kuhn M, Nguyen BV, Bui LK (2022) Enhancing crop yield prediction utilizing machine learning on satellite-based vegetation health indices. Sensors 22(3):719. https://doi.org/10.3390/s22030719

Pradhan S, Bandyopadhyay KK, Sahoo RN, Sehgal VK, Singh R, Gupta VK, Joshi DK (2014) Predicting wheat grain and biomass yield using canopy reflectance of booting stage. J Indian Soc Remote Sens 42(4):711–718. https://doi.org/10.1007/s12524-014-0372-x

Quan Z, Xianfeng Z, Miao J (2011) Eco-environment variable estimation from remote sensed data and eco-environment assessment: models and system. Acta Scien Nat Univ Pekine 47:1073–1080

Ramos APM, Oscob LP, Furuyaa DEG, Gonçalves WN et al (2020) A random forest ranking approach to predict yield in maize with UAV-based vegetation spectral indices. Compu Electron Agric 178:10579. https://doi.org/10.1016/j.compag.2020.105791

Rouse JW, Haas RH, Schell JA, Deering DW (1973) Monitoring vegetation systems in the Great Plains with ERTS (Earth Resources Technology Satellite). In: Proceedings of 3rd earth resources technology satellite symposium, Greenbelt, 10–14 December, SP-351, pp 309–317

Shah A, Dubey A, Hemnani V, Gala D, Kalbande DR (2018) Smart farming system: crop yield prediction using regression techniques. Springer, Singapore, pp 49–56. https://doi.org/10.1007/978-981-10-8339-6_6

Shammi SA, Meng Q (2020) Use time series NDVI and EVI to develop dynamic crop growth metrics for yield modeling. Ecol Ind 121:107124. https://doi.org/10.1016/j.ecolind.2020.107124

Sharifi A (2020) Yield prediction with machine learning algorithms and satellite images. J Sci Food Agric 10696. https://doi.org/10.1002/jsfa.10696

Singha C, Swain KC, Swain SK (2020) Best crop rotation selection with GIS-AHP technique using soil nutrient variability. Agric 10:213

Singha C, Swain KC, Saren B (2019) Land suitability assessment for potato crop using analytic hierarchy process technique and geographic information system. J Agric Eng 56(3):78–87

Singha C, Swain KC (2016) Land suitability evaluation criteria for agricultural crop selection: a review. Agric Rev 37:125–132

Singha C, Swain KC (2022) Rice and potato yield prediction using artificial intelligence techniques. In: Pattnaik PK, Kumar R, Pal S (eds) Internet of things and analytics for agriculture, Volume 3. Studies in Big Data, vol 99. Springer, Singapore. https://doi.org/10.1007/978-981-16-6210-2_9

Son NT, Chen CF, Chen CR, Chang LY, Duc HN, Nguyen LD (2013) Prediction of rice crop yield using MODIS EVI−LAI data in the Mekong delta, Vietnam. Int J Remote Sens 34:7275–7292

Vallentin C, Harfenmeister K, Itzerott S, Kleinschmit B, Conrad C, Spengler D (2021) Suitability of satellite remote sensing data for yield estimation in northeast Germany. Preci Agric 23:52–82. https://doi.org/10.1007/s11119-021-09827-6

Vannoppen Sui J, Qin Q, Ren H, Sun Y, Zhang T, Wang J, Gong S (2018) Winter wheat production estimation based on environmental stress factors from satellite observations. Remote Sens 10(6):962–972. https://doi.org/10.3390/rs10060962

van Klompenburg T, Kassahun A, Catal C (2020) Crop yield prediction using machine learning: a systematic literature review. Comp Electron Agric 177:105709. https://doi.org/10.1016/j.compag.2020.105709

Xue J, Su B (2017) Significant remote sensing vegetation indices: a review of developments and applications. J Sens 1–17. https://doi.org/10.1155/2017/1353691

Yuan W, Cai W, Nguy-Robertson AL, Fang H, Suyker AE, Chen Y, Dong W, Liu S, Zhang H (2015) Uncertainty in simulating gross primary production of cropland ecosystem from satellite-based models. Agric Forest Meteo 207:48–57

Yu J, Wang J, Leblon B (2021) Evaluation of soil properties, topographic metrics, plant height, and unmanned aerial vehicle multispectral imagery using machine learning methods to estimate canopy nitrogen weight in corn. Remote Sens 13:3105. https://doi.org/10.3390/rs13163105

9

A Multi-Criteria Decision-Making Approach for Land Suitability Assessment for Tea Cultivation in Hilly Aizawl District in Mizoram, India

Jonmenjoy Barman and Partha Das

Abstract

Geographic Information System (GIS) can be a strong tool for determining the environmental limitations of sustainable tea cultivation. Aizawl, a hilly district, is mostly covered by steep slopes and considered unsuitable for intensive agricultural practices. The present chapter attempts to find the potential land suitability for tea cultivation in Aizawl using geospatial technology. In this regard, raster information of topographic, climatic, hydrological, and soil properties have been intergraded into the GIS environment. A subjective ranking method has been implemented for weightage measurements of those different properties. Then, a relatively new multi-criteria decision-making approach, Combined Compromise Solution (CoCoSo), is used for the desired solution. After that, the suitability classes are divided into five potential classes using the natural breaking method. The result reveals that around a quarter of the total area is very highly suitable (23.74%) for tea cultivation, followed by high (34.89%), moderate (26.34%), low (12.05%), and very low (2.98%) suitability areas. While the very highly suitable areas are found on the steep slopes, the very low suitability areas are located on the area's central and eastern sides. Due to insufficient ground information, it has yet to be possible to implement proper validation methods. Farmers' perception and location of existing Jhum land are used to validate the Kappa coefficient, which is 87.27%. The present chapter will be helpful for policymakers and the agricultural department for proper economic development in the district.

Keywords

CoCoSo · Land suitability · Aizawl · Kappa index

9.1 Introduction

Land evaluation assessment is a fundamental technique for different types of planning like agriculture, housing, industrialization, etc. Land use denotes the sociocultural function of humans on landmass, whereas land cover defines the biophysical attributes of land in a specific land unit (Layomi Jayasinghe et al. 2019). Agriculture is the backbone of the Indian economy. However, because of their climatic and topographic conditions, the northeastern states of India are limited in

J. Barman (✉)
Department of Geography and RM, Mizoram University, Aizawl 796004, India
e-mail: jonmenjoybarman07@gmail.com

P. Das
Department of Geography, A.B.N. Seal College, Cooch Behar 736157, India

J. Das and S. Halder (eds.), *Advancement of GI-Science and Sustainable Agriculture*, GIScience and Geo-environmental Modelling, https://doi.org/10.1007/978-3-031-36825-7_9

agriculture practices. In the last few decades, multi-criteria decision-making (MCDM) methods have been widely used by researchers in different fields like groundwater potentiality mapping (Barman and Biswas 2022), groundwater delete (Basak et al. 2021; Mandal et al. 2022), flood susceptibility mapping (Mitra and Das 2022; Mitra et al. 2022; Khosravi et al. 2019), landslide susceptibility mapping (Aslam et al. 2022) etc. The MCDM has better accuracy in decision making than single criteria decision making as it intergrades different criteria in a decision (Velasquez and Hester 2013). When choosing a suitable MCDM technique from the many MCDM problems available, there is always some degree of uncertainty (Abrishamchi et al. 2005). As a result, no single method can solve the MCDM issue. Guitouni and Martel (1998) have provided some criteria for choosing an effective MCDM approach. However, the MCDM model, such CoCoSo, has been widely used by several scientists (Peng et al. 2020; Stanujkic et al. 2020; Luo et al. 2021). Evaluating the soil's compatibility with tea is essential to maximize productivity and provide corrective advice for improved crop management. Decision-makers can create appropriate management actions such as new planting, infilling, replanting, adopting climate-friendly best practices, and diversification, to boost the productivity of the land using the suitability analysis to identify limiting variables (Das et al. 2017). The aforesaid is accomplished by identifying marginally suitable locations, which enable decision-makers to recognize such elements and take appropriate action. A thorough investigation must be done to determine the optimum use of the land in unproductive estates while considering other factors. Additionally, this enables stakeholders to pinpoint new sites where tea can be grown per the crop's needs while highlighting the land unit's advantages.

9.2 Background of Tea Cultivation in Mizoram

The Naga, Manipuri, and Lushai hills along the Indo-Burmese border are considered the tea plant's natural habitat in India (Panabokke et al. 2008). Tea was first cultivated in Mizoram in the Biate village headed by Thangchuanga. Later, the cultivated area was increased to one *bigha* plot in 1935 by the villagers on Halzawl hill, followed by those on Tuilut hill in 1943 (The Telegraph online, June 21, 2022). A new land use policy in Mizoram in 1995 brought tea cultivation under the Soil and Water Conservation Department focusing on Biate village, where 325 families directly depended on it. In 2004, tea cultivation was increased to about 343 small gardens existing at the time. It was boosted when the Governor of Mizoram, A. K. Kohli, inaugurated the first tea factory at Ngopa, which had an annual production capacity of 400 tonnes (The Telegraph online, June 21, 2022). However, the Mizoram government took further initiatives for tea cultivation in 2011, where 115 households of Biate village were provided assistance for tea cultivation. As per present statistics, about 877.15 hectares of areas in Mizoram are being cultivated for tea (Eastern Mirror, Feb 24 2022).

9.3 Materials and Methods

9.3.1 Study Area

The district Aizawl is centrally located between 23° 19′N, 92° 38′E and 24° 24′N, 93° 13′E in the northern portion of the northeastern state of Mizoram in India (Fig. 9.1). The dissected and undulated topography is a major barrier for extensive agriculture practices. The temperature in the district is very comfortable. While the summer temperature ranges between 20 and 30 °C, winter temperature varies between 11 and 12 °C (District Disaster Management Plan, Aizawl 2019). The rainfall here is mostly caused by the southwest monsoon and amounts to 2500 mm annually (Barman et al. 2023). Two perennial rivers, namely Tlang and Tuirial, bound the district along its western and eastern sides, respectively. According to the Census of India, 2011, the district has a population of 400,309, and the population density is 112 persons per square kilometer. The local people practice *Jhum* cultivation as their traditional way of life.

According to statistics by the Department of Agriculture (2020–2021), the 19,080 ha area is under *Jhum* land, and their productive power is 1331.71 kg/Ha.

9.3.2 Database

The present chapter includes topographic factors, namely elevation, slope, and aspect; climatic factors like annual average rainfall and temperature; and hydrological factors, namely distance to the river and lastly lithological factors, such as soil organic nitrogen content and (organic) carbon content used to evaluate land suitability for tea cultivation in Aizawl district of Mizoram. An intensive literature survey was pursued to select land suitability conditioning factors (Table 9.1). Firstly, topographic factors are prepared from ASTER 30 m Digital Elevation Model [Earthdata Search|Earthdata Search (nasa.gov)]. Further, elevation, slope, and aspect thematic layer were prepared in ArcGIS 10.4 environment. For annual average rainfall and temperature, high-resolution gridded data sets of the climatic research group were used (Harris et al. 2020).

The growth of a tree is sensitive to various topographic and climatic factors such as slope, elevation, aspect, distance to a river, amount of precipitation, temperature, and nature of the soil, and so forth. The slope is the first controlling factor for tree growth which controls the runoff intensity, soil erosion, and nature of drainage. It also affects the percentage of soil nutrients such as organic carbon, and nitrogen. Generally, the slope between 5° and 25° is considered suitable for tea cultivation, while a slope gradient beyond 25° is unsuitable for tea cultivation (Das et al.

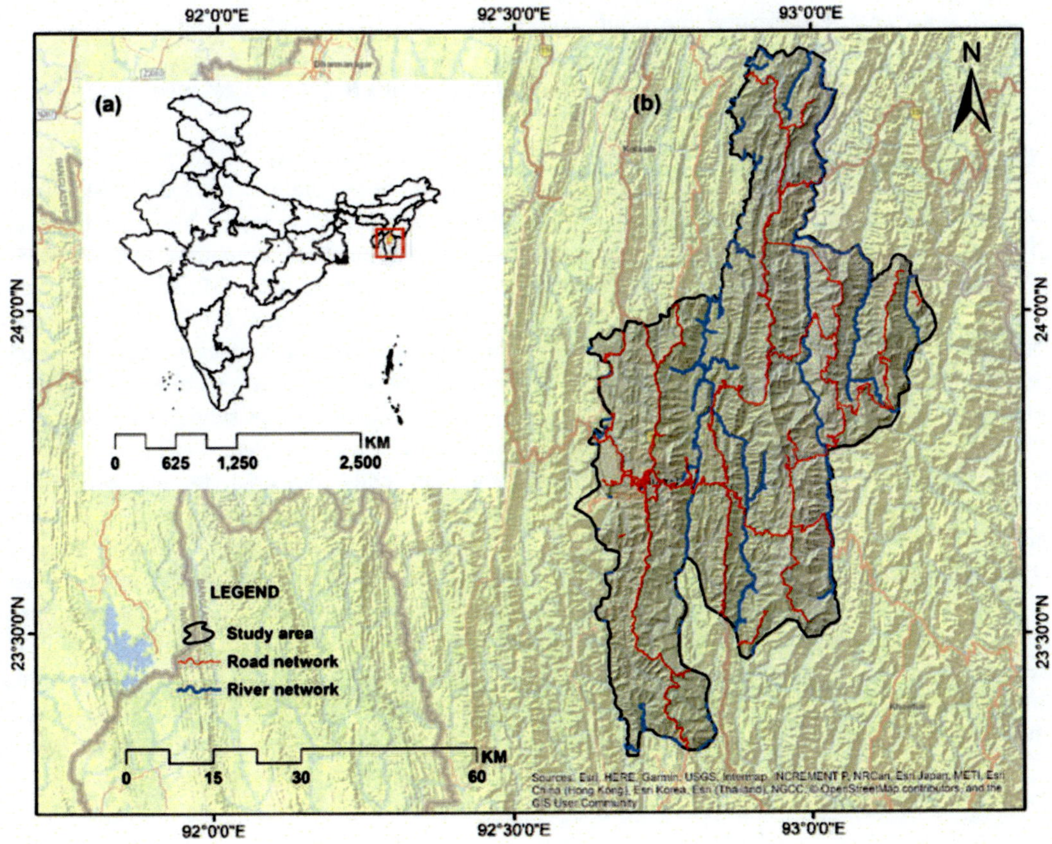

Fig. 9.1 Location of the study area: **a** India and **b** Aizawl district with river and road networks

Table 9.1 Literature survey for selection of conditioning factors

Factors	Slope	Elevation	Aspect	Rainfall	Temperature	Dist. to river	Soil
Layomi Jayasinghe et al. (2019)	✓	✓	✓	✓	✓		✓
Das et al. (2020)	✓	✓		✓	✓	✓	✓
Rahaman and Aruchamy (2022)	✓	✓		✓	✓		✓
Bo et al. (2012)	✓	✓	✓	✓	✓		✓
Panabokke et al. (2008)	✓						✓
Ashraf et al. (2020)	✓	✓	✓	✓		✓	✓
Chen et al. (2022)	✓	✓	✓	✓	✓	✓	✓
Gahlod et al. (2017)							✓
Xing et al. (2022)	✓	✓	✓	✓	✓	✓	✓

2020). Due to the water logging problem, flat areas are unsuitable for cultivation. Elevation influences the soil depth. The tea plant prefers a damp environment, not a waterlogged area because high water can destroy the plant root. Aspect can be defined as the direction a slope faces. Physical factors like the amount of solar radiation received, rainfall received, soil moisture, etc., directly depend on the aspect. In case of climatic factors, annual rainfall and temperature play a significant role in the growth of tea plants. Generally, a minimum of 1000 mm of annual rainfall is required during the growing season, while the optimum amount ranges between 1800 and 2000 mm (Han and Li 2018). It is reported that temperature lower than 13 °C and higher than 30 °C affect the growth of plant branches, and temperature between 19 and 23 °C is suitable for tea. The tea plants cannot tolerate water logging; it is very sensitive and cannot survive in the waterlogged area (Parthasarathy et al. 2006; Mukherjee et al. 2017).

Regarding soil nutrients, the percentage of nitrogen content in the soil is considered one of the land suitability factors, as nitrogen is required for plant growth, specifically for the development and production of leaves and stems (Zolekar and Bhagat 2015). As well as soil, organic carbon content is a significant determinant of tea quality as it is an economic crop generally known for leaves. The soil's carbon content has been reported to benefit branches and leaves growth (Chen et al. 2022).

9.3.3 Detailing on CoCoSo

Combined Compromise Solution (CoCoSo) is a newly developed multi-criteria decision-making model (MCDM) developed by Yazdani et al. (2019). The method combines two MCDMs, namely Weighted Sum Model (WSM) and Weighted Product Model (WPM). Three procedures are used in this strategy to rank the possibilities. The arithmetic means of each alternative score serves as the initial strategy. The scores of each choice are calculated using the second technique and compared to the best ones. Between the first and second strategies, the third one represents a compromise. Utilizing the geometric and mathematical means of the three procedures, the ultimate rank of each possibility is determined (Banihashemi et al. 2021).

The following steps are used to integrate and select the best alternatives.

Step 1: A decision matrix of alternatives (i) and criteria (j) is prepared as an equation

$$X_{ij} = \begin{bmatrix} x_{11} & x_{12} & \cdots & x_{1e} \\ x_{21} & x_{22} & \cdots & x_{2e} \\ \vdots & \vdots & \ddots & \vdots \\ x_{f1} & x_{f2} & \cdots & x_{fe} \end{bmatrix}, \quad (9.1)$$
$$i = 1,2,3,\ldots e,\ j = 1,2,3,\ldots,f$$

Step 2: Raw decision matrix is normalized using equation

$$\text{For beneficial criteria} \quad r_{ij} = \frac{x_{ij} - \min_i x_{ij}}{\max_i x_{ij} - \min_i x_{ij}}, \quad (9.2)$$

$$\text{For cost criteria} \quad r_{ij} = \frac{\min_i x_{ij} - x_{ij}}{\max_i x_{ij} - \min_i x_{ij}}, \quad (9.3)$$

Step 3: Selection of criteria weight is performed using the subjective expert judgment method.
Step 4: Equations are used for performing comparability sequences (Pi) and comparability sequence (Si), respectively.

$$P_i = \prod_{j=1}^{n} (r_{ij})^{w_j} \quad (9.4)$$

$$S_i = \sum_{j=1}^{n} (w_j r_{ij}) \quad (9.5)$$

Step 5: Three aggregation approaches are used for measuring the relative weightage.

$$k_{ia} = \frac{S_i + P_i}{\sum_{i=1}^{m} (P_i + S_i)} \quad (9.6)$$

$$k_{ib} = \frac{S_i}{\min_i S_i} + \frac{P_i}{\min_i P_i} \quad (9.7)$$

$$k_{ic} = \frac{\lambda(S_i) + (1-\lambda)(P_i)}{\left(\lambda \max_i S_i + (1-\lambda) \max_i P_i\right)} \quad (9.8)$$

where K_{ia} is the arithmetic mean of sums of scores, K_{ib} is the relative sum of a relative score, K_{ic} is the balanced compromise, and the value λ ranges between 0 and 1, which is free to choose as decision-maker knowledge. In this study, λ value of 0.5 is taken for the decision solution.
Step 6: Ranking index value of alternatives is calculated as an equation.

$$k_i = (k_{ia}k_{ib}k_{ic})^{\frac{1}{3}} + \frac{1}{3}(k_{ia} + k_{ib} + k_{ic}) \quad (9.9)$$

where K_i represents the conditioning weightage of alternatives, a higher value denotes the best alternatives, and a lower value denotes the worst alternatives.

9.4 Result and Discussion

In the first step, eight conditioning factors have been selected through a literature survey, and a thematic map has been prepared in ArcGIS 10.4. Then, 19,915 random points are selected to extract the raster value (Table 9.2). A decision matrix has been prepared consisting of 19,915 alternatives and eight criteria to make a problem (Table 9.3). The decision matrix has been normalized to get dimensionless values ranging between 0 and 1 (Table 9.4). Furthermore, based on the relationship between conditioning factors and land suitability for tea cultivation, all the criteria are classified into two groups: beneficial and non-beneficial. The criteria having a proportional positive relationship with land suitability are considered beneficial criteria and vice versa. For example, the higher the elevation, the better the suitability for tea cultivation, and it is considered as a beneficial criterion. At the same time, a subjective weighting by three members from the Department of Geography and RM, Mizoram University, assigned weight for conditioning factors as per role as suitable for tea cultivation. The factor of topsoil nitrogen content is credited with the highest weightage, followed by topsoil organic carbon content, distance to a river, elevation, slope, annual average temperature, annual average rainfall, and aspect (Fig. 9.3). In the final stage, Inverse Distance Weightage (IDW) interpolation techniques were implemented to develop spatial land suitability

Table 9.2 Statistical table of investigation factors in tea plantations

Factors	Range	Minimum	Maximum	Mean	SD	Variance	Skewness
Slope	65.400	0.000	65.400	22.565	10.168	103.384	0.186
Elevation	1818.000	25.000	1843.000	652.015	316.145	99,947.440	0.634
Aspect	360.779	− 1.000	359.779	180.943	104.207	10,859.139	− 0.041
Annual rainfall	351.710	2269.900	2621.610	2502.257	50.720	2572.496	− 0.255
Temperature	2.561	20.999	23.559	22.492	0.471	0.222	− 0.533
Distance to river	9085.460	0.000	9085.460	2632.040	1805.176	3,258,660.684	0.512
N_contain	0.240	0.230	0.470	0.360	0.120	0.014	− 0.165
OC_contain	0.580	3.280	3.860	3.594	0.289	0.084	− 0.165

Table 9.3 Decision matrix

Alternatives	Criteria							
	Slope	Elevation	Aspect	Annual rainfall	Temperature	Dist. to river	N_contain	OC_contain
	10.50	25.00	159	2609.270	22.989	0.000	0.230	3.280
	11.43	30.00	78	2593.900	22.872	239.008	0.230	3.280
	24.37	35.00	57	2606.060	22.914	0.000	0.230	3.280
	0.00	36.00	− 1	2593.700	22.871	0.000	0.230	3.280
	0.00	37.00	− 1	2591.490	22.869	0.000	0.230	3.280
	…							…
	38.179	1821.000	54.392	2502.830	22.295	3250.860	0.470	3.860
	26.562	1823.000	52.374	2509.250	22.369	6980.520	0.470	3.860
	5.957	1825.000	198.435	2505.450	22.359	5392.270	0.230	3.280
	13.232	1835.000	201.615	2466.130	22.015	5640.800	0.470	3.860
	22.169	1843.000	68.629	2466.230	22.019	5241.860	0.470	3.860

Table 9.4 Normalized decision matrix

Slope	Elevation	Aspect	Annual rainfall	Temperature	Dist. to river	N_contain	OC_contain
0.16	0.00	0.44	0.96	0.78	1.00	0.00	0.00
0.17	0.00	0.22	0.92	0.73	0.97	0.00	0.00
0.37	0.01	0.16	0.96	0.75	1.00	0.00	0.00
0.00	0.01	0.00	0.92	0.73	1.00	0.00	0.00
0.00	0.01	0.00	0.91	0.73	1.00	0.00	0.00
…							….
0.58	0.99	0.15	0.66	0.51	0.64	1.00	1.00
0.41	0.99	0.15	0.68	0.54	0.23	1.00	1.00
0.09	0.99	0.55	0.67	0.53	0.41	0.00	0.00
0.20	1.00	0.56	0.56	0.40	0.38	1.00	1.00
0.34	1.00	0.19	0.56	0.40	0.42	1.00	1.00

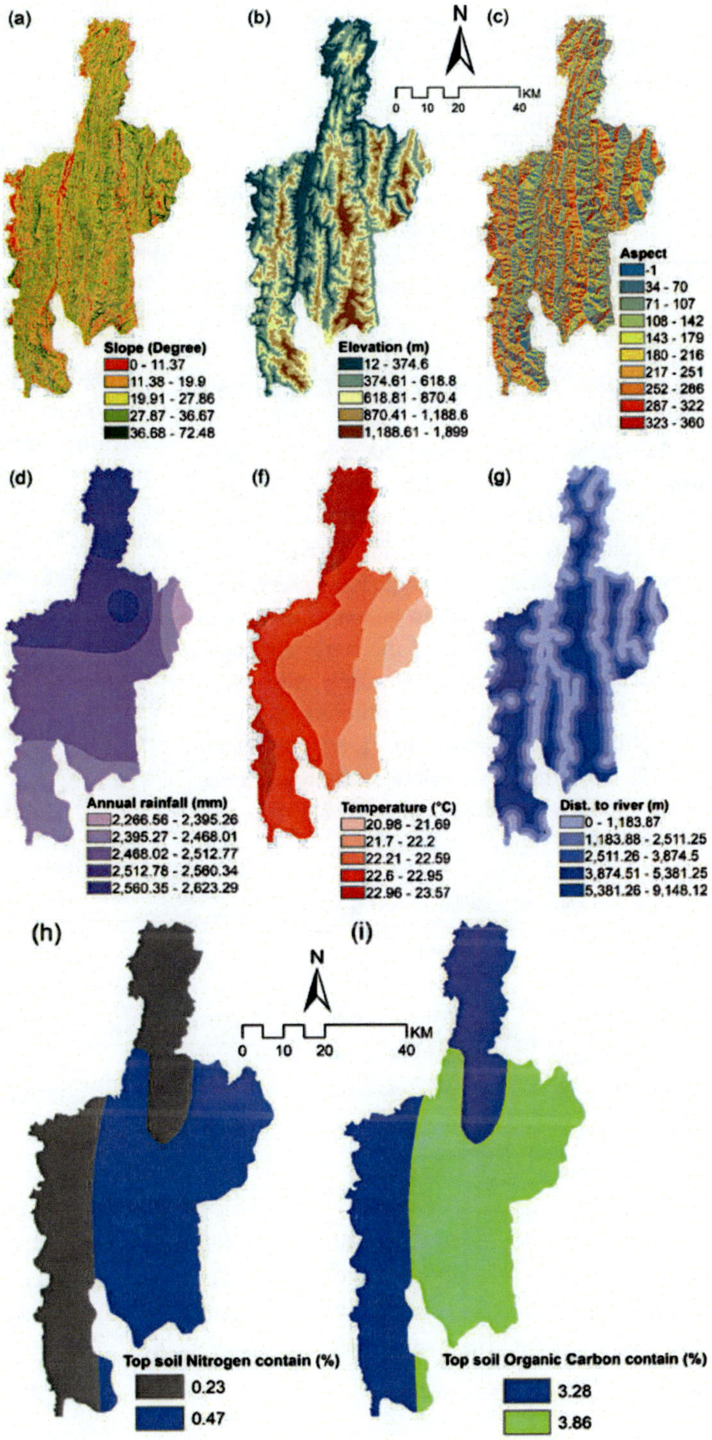

Fig. 9.2 Conditioning factors for land suitability: **a** slope, **b** elevation, **c** aspect, **d** annual rainfall, **e** temperature, **g** distance to river, **h** topsoil nitrogen content, and **i** topsoil organic carbon content

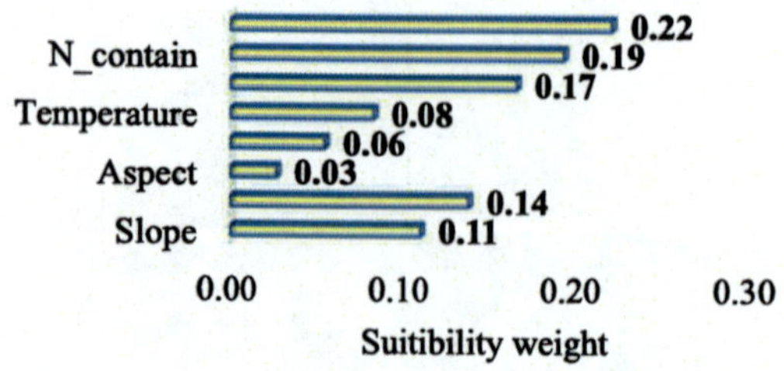

Fig. 9.3 Suitability weight assigned to the conditioning factors

based on K_{ia}, K_{ib}, K_{ic}, and K_i value (Fig. 9.2 and Table 9.5).

As per suitability level, the area has been classified into five suitability classes, like very high suitability area, high suitability area, moderate suitability area, low suitability area, and very low suitability area (Fig. 9.4). Very high suitability areas are found in moderate slope and elevation along the roads of the study area covering 23.74% of the area. Similarly, high suitability area has been found buffering the very susceptibility areas covering areal extension 34.89% forwarded by moderate suitability areas (26.34%), low suitability areas (12.05%), and very low suitability areas (2.98%) (Tables 9.5 and 9.6).

9.4.1 Very High Suitability Zone

The slope of the district ranges between 0° and 72.48° and is diverse in nature. A low slope is found at the valley, and a steep slope is near the hilltop. The very high suitability area belongs to a slope range between 19° and 27°, similar to that in Yingde city in Guangdong provinces, China (Chen et al. 2022). Table 9.7 shows that in the suitability areas, the elevation value ranges between 14 and 1840 m, with an average value of 680 m. In case of the aspect factor, the southeast facing slope is more suitable than other slope-facing areas.

The area's annual rainfall ranges between 2266 and 2623 mm, whereas the very high suitability zone is found where the rainfall average is 2471.7 mm with an average temperature of 22.43 °C. While considering the distance to a river, between 3000 and 4000 m from a river course is highly suitable for tea cultivation. The

Table 9.5 Values of alternative used for interpolation

Latitude	Longitude	Mia	Mib	Mic	Mi
24.38541956230	93.02001609810	2.98781E–05	370.304	0.18768	123.625
24.34050009330	93.03946354440	2.85034E–05	1058.47	0.18071	353.06
24.24046704220	93.01083314370	3.11773E–05	698.984	0.19466	233.221
24.33797902130	93.03934096410	2.66563E–05	3239.64	0.17184	1080.18
24.19829076160	93.00159515250	2.66274E–05	3302.1	0.17171	1101.01
…					…
23.80106352250	92.94626228340	7.67245E–05	28.1397	0.43453	9.62267
23.85143655230	92.93133995280	6.86926E–05	67.2562	0.3922	22.6714
23.83059849870	92.93353552340	2.96358E–05	514.687	0.18646	171.766
23.51487410580	92.97713261540	6.83109E–05	38.7269	0.39018	13.1401
23.52111306630	92.97506832520	6.95475E–05	90.3769	0.39672	30.3935

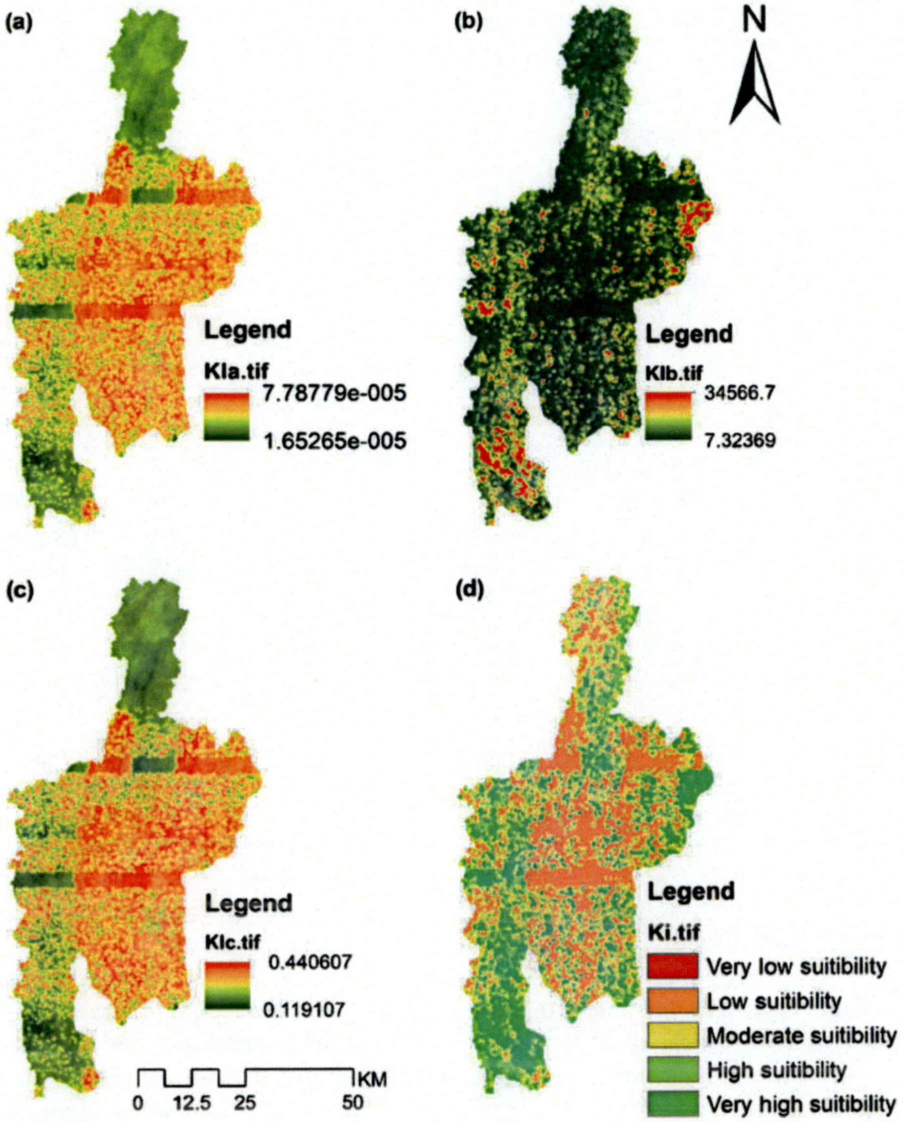

Fig. 9.4 Relative weight, **a** Kia, **b** Kib, **c** Kic, and **d** land suitability zones for tea cultivation

Table 9.6 Areal extension of different suitability class

Suitability classes	Mi class	Area	Area (%)
Very high	4.1–4.3	819.158	23.74
High	4.4–4.4	1203.87	34.89
Moderate	4.5–4.6	908.864	26.34
Low	4.7–4.8	415.727	12.05
Very low	4.9–5.6	102.768	2.98

Table 9.7 Statistical table of investigation factors in tea plantations in very high suitability class

Factors	Min	Max	Mean	SD
Slope	0	71.8545	21.4596	10.044414
Elevation	14	1840	680.142	307.707083
Aspect	− 1	359.799	142.346	104.360116
Annual rainfall	2266.56	2617.88	2471.7	52.527717
Temperature	20.9904	23.5453	22.4355	0.520793
Dist. to river	0	8881.98	3264.79	2021.553544
N_contain	0.23	0.47	0.3249	0.117345
OC_contain	3.28	3.86	3.50933	0.283583

Table 9.8 Statistical table of investigation factors in tea plantations in high suitability class

Factors	Min	Max	Mean	STD
Slope	0	70.5298	22.4567	10.149155
Elevation	16	1898	625.798	303.80085
Aspect	− 1	359.825	165.273	100.304778
Annual rainfall	2271.62	2622.84	2502.13	52.934739
Temperature	21.0185	23.5537	22.5363	0.447531
Dist. to river	0	9148.12	2768.66	1855.54515
N_contain	0.23	0.47	0.32509	0.117387
OC_contain	3.28	3.86	3.50981	0.283684

very high suitability classes are found with high soil nutrient areas for soil factors.

9.4.2 High Suitability Zone

Table 9.8 shows high suitability zone is located at places with an average slope of 22.45° and an average elevation of 625 m. In case of the aspect factor, south facing slopes dominate in this zone. The annual rainfall of this zone is quite higher than very high suitability zones at about 2502.13 mm. With respect to the distance factor, the area between 2000 and 3000 m is mainly affected by soil erosions.

Similar to very high suitability zone, the amount of soil nutrients is also found high in high suitability zones. Here in the study, the moderate, low, and very low suitability classes are considered non-profitable for tea cultivation.

9.5 Validation of the Study Outcome

Initially, the kappa statistics were implemented in the psychological test to compare two independent responses (Cohen 1960). Since then, various fields, including modeling, and mapping, have used the kappa statistic. In the chapter, we applied kappa statistics to compare land suitability zones with *Jhum* land in the study area. The kappa statistics were based on a confusion matrix of two maps, i.e., land suitability and *Jhum* map, using Fig. 9.5.

All the processes have been done using the SCP plugin in QGIS 3.10.7 environment. Here, we used existing farmland that local people used and classified it into five classes according to area extension and production. As a result, the overall accuracy for the present study is 87.27%, representing almost perfect classification.

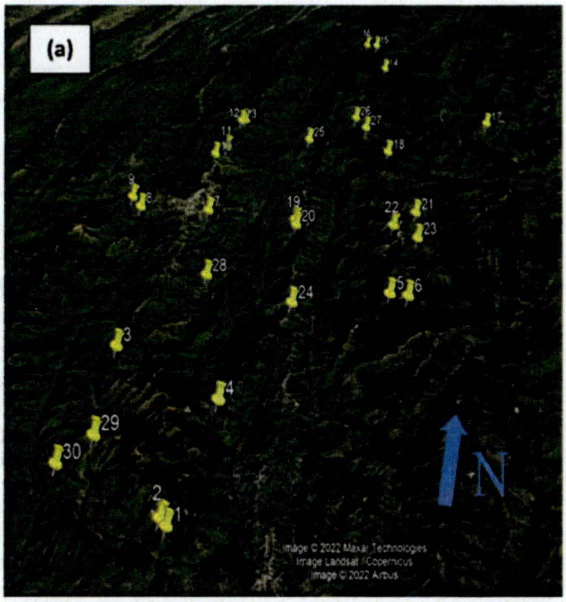

ErrMatrixCode	Reference	Classified	PixelSum
1	1	1	2549
3	1	2	15
6	1	3	17
10	1	4	9
15	1	5	9
2	2	1	88
5	2	2	992
9	2	3	1
19	2	5	123
13	3	3	700
22	3	5	857
12	4	2	405
21	4	4	647
25	5	5	3065

> ERROR MATRIX (pixel count)
> Reference

V_Classified	1	2	3	4	5	Total
1	2549	88	0	0	0	2637
2	15	992	0	405	0	1412
3	17	1	700	0	0	718
4	9	0	0	647	0	656
5	9	123	857	0	3065	4054
Total	2599	1204	1557	1052	3065	9477

> AREA BASED ERROR MATRIX
> Reference

V_Classified	1	2	3	4	5	Area	Wi
1	0.2295	0.0079	0.0000	0.0000	0.0000	[illegible]	0.2374
2	0.0037	0.2451	0.0000	0.1001	0.0000	[illegible]	0.3489
3	0.0062	0.0004	0.2568	0.0000	0.0000	[illegible]	0.2634
4	0.0017	0.0000	0.0000	0.1188	0.0000	[illegible]	0.1205
5	0.0001	0.0009	0.0063	0.0000	0.0225	[illegible]	0.0298
Total	0.2412	0.2543	0.2631	0.2189	0.0225	[illegible]	
Area	[illegible]	[illegible]	[illegible]	[illegible]	[illegible]	[illegible]	
SE	0.0020	0.0043	0.0015	0.0042	0.0002		
SE area	7015334	[illegible]	5347176	[illegible]	693267		
95% CI area	13750055	[illegible]	[illegible]	[illegible]	1358805		
PA [%]	95.1638	96.3852	97.6069	54.2643	100.0000		
UA [%]	96.6629	70.2550	97.4930	98.6280	75.6043		
Kappa hat	0.9560	0.6011	0.9660	0.9824	0.7504		

Overall accuracy [%] = 87.2772
Kappa hat classification = 0.8321

Area unit = metre^2
SE = standard error
CI = confidence interval
PA = producer's accuracy
UA = user's accuracy

Fig. 9.5 a location of sample *Jhum* land used for validation, **b** Kappa matrix has done SCP plugin in QGIS environment

9.6 Conclusions

The district is dominated by Mizo people who have been practicing *Jhum* since historical times and get very little production from such cultivation. The current study is focused on identifying suitable lands for tea cultivation in the Aizawl district of Mizoram. Different environmental criteria, namely elevation, slope, aspect, long-term average rainfall, annual temperature, distance to a river, soil nitrogen content, and soil organic carbon content, are evaluated using CoCoSo MCDM with the help of subjective weights of criteria. Soil nitrogen content is credited the highest weightage, and aspect is credited the lowest weightage. The final integration has been done in the GIS environment. As per the alternative ranking, the area has been classified into five land suitability zones: very high, high, moderate, low, and very low suability. The overall result shows that 23.74% of the total areas are highly suitable for tea cultivation. Although there is the unavailability of all the thematic layers in a digital format required for land suitability analysis, the present study will help government agencies to restart tea cultivation in the study area.

References

Abrishamchi A, Ebrahimian A, Tajrishi M, Mariño MA (2005) Case study: application of multicriteria decision making to urban water supply. J Water Resour Plan Manag 131(4):326–335

Aizawl District Disaster Management Plan (2019) Department of Disaster Management & Rehabilitation, Government of Mizoram, Department of Agriculture, Government of Mizoram, 2020–21

Aslam B, Maqsoom A, Khalil U, Ghorbanzadeh O, Blaschke T, Farooq D et al (2022) Evaluation of different landslide susceptibility models for a local scale in the Chitral District, Northern Pakistan. Sensors22(9):3107

Ashraf N, Ahmad SR, Ahmad A, Javed MA (2020) Assessment of Land Suitability for Tea Cultivation Using Geo-Informatics in the Mansehra and Abbottabad District. Pakistan Journal of Scientific & Industrial Research Series A: Phys Sci 63(1):65–70

Banihashemi SA, Khalilzadeh M, Zavadskas EK, Antucheviciene J (2021) Investigating the environmental impacts of construction projects in time-cost trade-off project scheduling problems with CoCoSo multi-criteria decision-making method. Sustainability 13(19):10922

Barman J, Biswas B (2022) Application of e-TOPSIS for ground water potentiality zonation using morphometric parameters and geospatial technology of Vanvate Lui Basin, Mizoram, NE India. J Geol Soc India 98 (10):1385–1394

Barman J, Biswas B, Das J (2023) Mizoram, the capital of landslide: a review of articles published on landslides in Mizoram, India. In: Das J, Bhattacharya SK (eds) Monitoring and managing multi-hazards. GIScience and Geo-environmental modelling. Springer, Cham. https://doi.org/10.1007/978-3-031-15377-8_6

Basak A, Das J, Rahman ATM, Pham QB (2021) An integrated approach for delineating and characterizing groundwater depletion hotspots in a coastal state of India. J Geol Soc India 97(11):1429–1440. https://doi.org/10.1007/s12594-021-1883-z

Bo LI, Zhang F, Zhang LW, Huang JF, Jin Z-F, Gupta DK (2012) Comprehensive suitability evaluation of tea crops using GIS and a modified land ecological suitability evaluation model. Pedosphere 22 (1):122–130

Chen P, Li C, Chen S, Li Z, Zhang H, Zhao C (2022) Tea cultivation suitability evaluation and driving force analysis based on AHP and geodetector results: a case study of Yingde in Guangdong, China. Remote Sens 14(10):2412

Cohen J (1960) A coefficient of agreement for nominal scales. Educational and psychological measurement, 20(1):37–46

Das AC, Noguchi R, Ahamed T (2020) Integrating an expert system, GIS, and satellite remote sensing to evaluate land suitability for sustainable tea production in Bangladesh. Remote Sens 12(24):4136

Das J, Gayen A, Saha S, Bhattacharya SK (2017) Modelling of alternative crops suitability to tobacco based on analytical hierarchy process in Dinhata subdivision of Koch Bihar district, West Bengal. Model Earth Syst Environ 3(4):1571–1587. https://doi.org/10.1007/s40808-017-0392-y

Eastern Mirror (easternmirrornagaland.com) (accessed on Saturday, 10 Dec 2022)

Gahlod NS, Binjola S, Ravi R, Arya VS (2017) Land-site suitability evaluation for tea, cardamom and rubber using geo-spatial technology in Wayanad district, Kerala. J Appl Nat Sci 9(3):1440–1447

Guitouni A, Martel JM (1998) Tentative guidelines to help choosing an appropriate MCDA method. Eur J Oper Res 109(2):501–521

Han WY, Li X (2018) Ahammed GJ (2018) Stress physiology of tea in the face of climate change. Springer, Singapore

Harris I, Osborn TJ, Jones P et al (2020) Version 4 of the CRU TS monthly high-resolution gridded multivariate climate dataset. Sci Data 7:109. https://doi.org/10.1038/s41597-020-0453-3

Khosravi K, Shahabi H, Pham BT, Adamowski J, Shirzadi A, Pradhan B et al (2019) A comparative assessment of flood susceptibility modeling using multi-criteria decision-making analysis and machine learning methods. J Hydrol 573:311–323

Layomi Jayasinghe S, Kumar L, Sandamali J (2019) Assessment of potential land suitability for tea (*Camellia sinensis* (L.) *O. Kuntze*) in Sri Lanka using a GIS-based multi-criteria approach. Agriculture 9 (7):148

Luo Y, Zhang X, Qin Y, Yang Z, Liang Y (2021) Tourism attraction selection with sentiment analysis of online reviews based on probabilistic linguistic term sets and the IDOCRIW-COCOSO model. Int J Fuzzy Syst 23(1):295–308

Mandal T, Saha S, Das J, Sarkar A (2022) Groundwater depletion susceptibility zonation using TOPSIS model in Bhagirathi river basin, India. Model Earth Syst Environ 8(2):1711–1731. https://doi.org/10.1007/s40808-021-01176-7

Mitra R, Das J (2022) A comparative assessment of flood susceptibility modelling of GIS-based TOPSIS, VIKOR, and EDAS techniques in the Sub-Himalayan foothills region of Eastern India. Environ Sci Pollut Res. https://doi.org/10.1007/s11356-022-23168-5

Mitra R, Saha P, Das J (2022) Assessment of the performance of GIS-based analytical hierarchical process (AHP) approach for flood modelling in Uttar Dinajpur district of West Bengal, India. Geomat Nat Haz Risk 13(1):2183–2226. https://doi.org/10.1080/19475705.2022.2112094

Mukherjee D, Mandal B, Maji A, Biswas B (2017) Impact of various planting dates and suitable nutrient management practices for (*Triticum aestivum* L.) enhanced wheat productivity. Int J Bioresour Sci 4(1):29–34

Panabokke CR, Amarasinghe L, Pathiranage SRW, Wijeratne MA, Amarathunga SLD (2008) Land suitability classification and mapping of tea lands in Ratnapura district. S L J Tea Sci 73(1):1–10

Parthasarathy VA, Chattopadhyay PK, Bose TK (2006) Plantation Crops-I; Naya Udyog, Calcutta, India. ISBN 13: 9788185971971. 38

Peng X, Zhang X, Luo Z (2020) Pythagorean fuzzy MCDM method based on CoCoSo and CRITIC with score function for 5G industry evaluation. Artif Intell Rev 53(5):3813–3847

Rahaman SA, Aruchamy S (2022) Land suitability evaluation of tea (*Camellia sinensis* L.) plantation in Kallar watershed of Nilgiri Bioreserve, India. Geographies 2(4):701–723

Stanujkic D, Popovic G, Zavadskas EK, Karabasevic D, Binkyte-Veliene A (2020) Assessment of progress towards achieving sustainable development goals of the "Agenda 2030" by using the CoCoSo and the Shannon entropy methods: the case of the EU Countries. Sustainability 12(14):5717

The Telegraph—Telegraph Online, Daily Telegraph, Sunday Telegraph—Telegraph (accessed on Saturday, 10 Dec 2022)

Velasquez M, Hester PT (2013) An analysis of multi-criteria decision making methods. Int J Oper Res 10 (2):56–66

Xing W, Zhou C, Li J, Wang W, He J, Tu Y et al (2022) Suitability evaluation of tea cultivation using machine learning technique at town and village scales. Agronomy 12(9):2010

Yazdani M, Zarate P, Kazimieras Zavadskas E, Turskis Z (2019) A combined compromise solution (CoCoSo) method for multi-criteria decision-making problems. Manag Decis 57(9):2501–2519

Zolekar RB, Bhagat VS (2015) Multi-criteria land suitability analysis for agriculture in hilly zone: remote sensing and GIS approach. Comput Electron Agric 118:300–321

Check for updates

10 Integration of Information Technology and Operation Technology in Agriculture Toward Sustainable and Competitive Farming

Jitranjan Sahoo, Manoranjan Dash, and Preeti Y. Shadangi

Abstract

For years, digital technologies have offered many innovative and adaptable solutions for enhancing agricultural productivity. The agriculture 4.0 wave proposes automated farming, which is possible because of physical and digital technology integration, which demands farmers to migrate towards mechanical operations. This convergence of information technology (IT) and operational technology (OT) started an agricultural initiative that will integrate farming data with external parameters. Thus, what should combine artificial intelligence, 5G networks, and different autonomous systems to eliminate errors and thus save energy? This chapter investigates the adoption of IT–OT in the agriculture industry from the farmer's perspective. This research has a unique contribution to bringing innovation to agriculture. It provides a direction to the markets of IT–OT integration technology in developing suitable marketing strategies to improve the adoption in the rural market for sustainable and competitive farming.

Keywords

Agriculture 4.0 · IT–OT · Competitive farming · Sustainable agriculture

10.1 Introduction

Worldwide population is estimated to reach 9.8 billion by 2050, putting significant pressure on existing agricultural land to increase production. To feed the world's population, global production will need to rise by at least 70% from the 2006 levels. However, since arable land cannot expand any further, the solution is to improve the productivity of existing farmland. The agricultural industry can leverage IoT in achieving this and which included sensors, products involved in networking, computing power advanced technology, and new networking enhancement in enriching the interconnections capabilities thus to increase productivity. By collecting and analyzing real-time agricultural data, the farming industry can determine what they require to enhance production efficiency and establish a better IoT market.

The integration of IoT technology has the potential to transform various facets of our daily lives, including home automation, which involves

J. Sahoo (✉) · M. Dash · P. Y. Shadangi
Faculty of Management Sciences,
Siksha O Anusandhan (Deemed to be University),
Bhubaneswar, India
e-mail: jitranjansahoo@gmail.com

M. Dash
e-mail: manoranjandash@soa.ac.in

P. Y. Shadangi
e-mail: preetishadangi@soa.ac.in

J. Das and S. Halder (eds.), *Advancement of GI-Science and Sustainable Agriculture*,
GIScience and Geo-environmental Modelling, https://doi.org/10.1007/978-3-031-36825-7_10

the use of internet-connected appliances, home automation systems, and energy management devices, culminating in the creation of a "smart home" that provides consumers with improved security and energy efficiency. Aiello et al. (2017) explained their study of green economy as an intermediating between sustainability and profitability by enhancing farming operations to be efficient. They proposed an innovative multi-sensor decision fusion methodology. IoT will bring radical change in economic and social well-being and thus present challenges and opportunities for the coming years. Bing(2012) suggested a intelligent system for agriculture for organic melon and fruit production based on IoT, with enabled new technologies i.e. RFID sensors etc. Expert system was integrated to the Iot in order to capture the data for effective decision.

Maheswari et al. (2022) argue that the Internet of things (IoT) has permeated all industries and is projected to create a 4.7 billion US dollar linked agriculture market by 2025, which has greatly impacted the agriculture industry. IoT has various applications in areas such as climate change, environmental monitoring, irrigation, seed and plant monitoring, and decision-making mechanisms, resulting in reduced costs and increased yields. Parmar and Kumar (2022) highlight that IoT technology can be applied to greenhouses, livestock breeding, and agricultural machinery. By using networked sensors, IoT can transform agriculture, industry, and energy production and distribution by increasing information availability along the production value chain. Kacira et al. (2005) and Rajaoarisoa et al. (2012) conducted green house research on sensing and control strategies for sustainable greenhouse production. Ranjit et al. (2022) suggest that the agricultural industry can leverage emerging technologies like IoT to meet the growing demands of the population. IoT can reduce costs, waste, and improve yield quality, making smart agriculture necessary. However, Farooq et al. (2022) indicate that challenges remain in less developed regions, where communication network infrastructure is lacking, and farmers need incentives to invest in expensive IoT systems. Hu and Qian (2011) IoT application was integrated into the crop growth models to make system more intelligent and adaptive and it focuses on the practical deployment with engineering challenges. Jiang and Zhang (2013) and Quy et al. (2022) suggested in order to improve the agriculture planning level they suggested IoT based agricultural service platform which will be used in production, transportation and after sale service processing. Ravikanth et al. (2017) and Rodriguez de la Concepcion et al. (2014) suggested advanced quality system for agriculture and efficient wireless sensor network. The chapter aims to explore the importance and applications of IoT in agriculture, with an emphasis on the challenges facing SMART agriculture.

10.2 Smart Agriculture

The term "smart agriculture" typically refers to the implementation of Internet of things (IoT) solutions in agriculture, and the use of these solutions is increasingly prevalent (Wang et al. 2022). According to BI Intelligence, the number of agriculture IoT devices installed is projected to reach 75 million by 2020, with a yearly growth rate of 20%. Concurrently, the global smart agriculture market is anticipated to triple in size by 2025, with a value of $15.3 billion, compared to just over $5 billion in 2016.

10.2.1 How IoT is Shaping Agriculture

The utilization of technologies and IoT has the capability to revolutionize agriculture in various ways. IoT offers numerous benefits to agriculture such as collecting vast amounts of data using smart agriculture sensors, such as weather patterns, soil quality, crop growth progress, and animal health. This data can be used to monitor the status of the business, staff performance, equipment efficiency, and more. This level of control over internal processes can reduce production risks, allowing for better planning of product distribution. With the ability to anticipate production output accurately, producers can ensure their products remain unsold or misused. Furthermore, increased control over production can aid in cost management and waste reduction, enabling farmers to identify any anomalies in crop growth or livestock health, minimizing the risks of losing their yield. Additionally, the use of smart devices can enhance business efficiency by automating various processes such as irrigation, fertilization, and pest control throughout the production cycle. This leads to better product quality and higher volumes through the maintenance of higher standards of crop quality and growth capacity with the help of automation.

10.2.2 Instances in Agriculture Using IoT

The utilization of IoT sensors and applications has revolutionized agriculture, offering a range of options. One of the most popular gadgets used in smart agriculture is the weather station, equipped with several sensors, scattered throughout the field to collect environmental data. These weather stations can help create a map of climate conditions, identify suitable crops, and facilitate precision farming. The greenhouse automation system also employs a similar principle to source environmental data and adjusts the climate

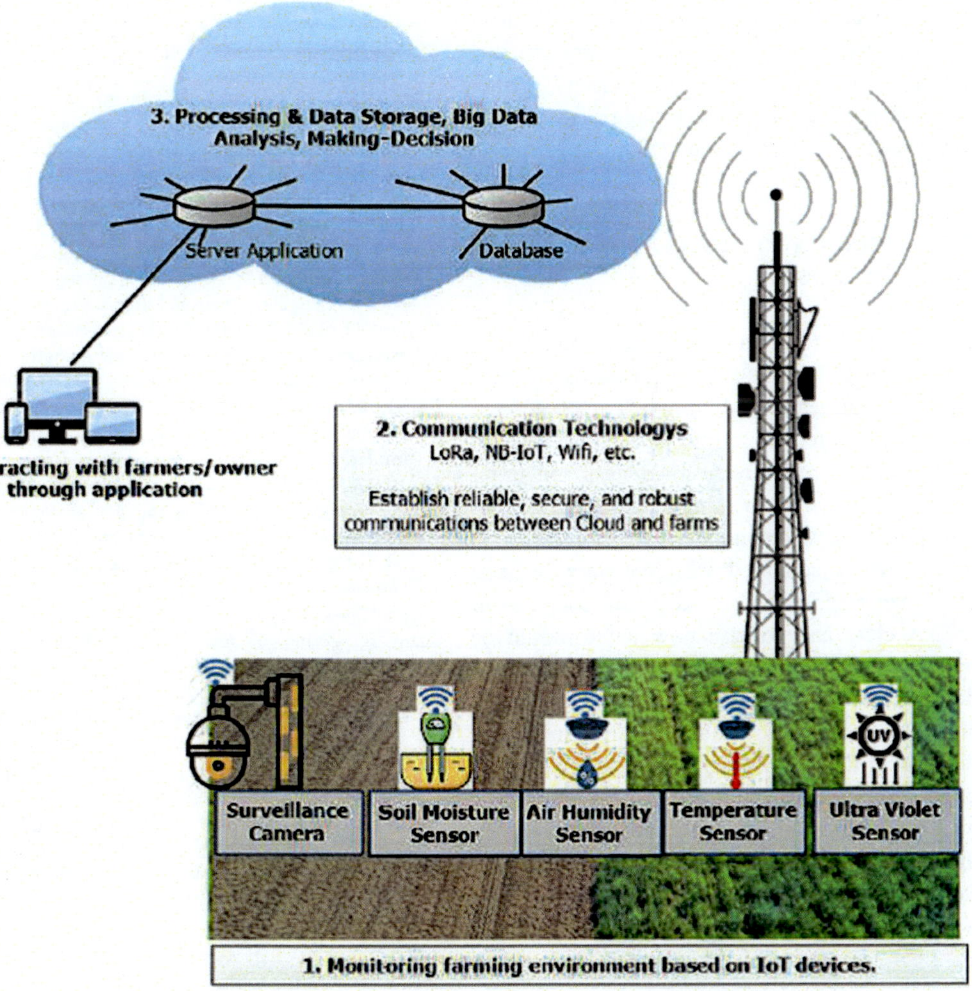

Fig. 10.1 IoT ecosystem for smart agriculture

parameters accordingly. Another type of IoT product used in agriculture is crop management devices, which are placed in the field to gather data related to crop health, such as temperature, precipitation, and leaf water potential. Cattle monitoring and management are another area where IoT sensors can be used to monitor the health and performance of farm animals. End-to-end farm management systems, which include multiple agriculture IoT devices, sensors, and a dashboard with analytical features, represent a more comprehensive approach to smart farming. These systems allow remote farm monitoring and streamline most business operations. Smart farming practices are essential to meet the immediate societal needs while preserving natural resources for future generations. The IoT-enabled smart farms help monitor every process, reduce waste, and improve productivity, particularly in the face of unfavorable climate conditions caused by climate changeFig. 10.1.

10.3 Making IoT Sustainable in Odisha

The adoption of IoT in agriculture is crucial, but many farmers in Odisha lack access to the necessary infrastructure and knowledge. To address this issue, it is recommended to establish internet infrastructure and increase awareness and training programs for farmers and stakeholders. Providing farmers with computers, smartphones, and other devices equipped with software in their native language can also aid in the adoption of IoT. A clear and concise flow chart should be made available to explain how to operate these devices, and regular training and sensitization should be provided. Additionally, involving farmers' children in the training and application of IoT can promote future sustainability. Finally, well-trained service agents should be available in the field to provide support and clarification to farmers.

10.4 Conclusion

The agricultural industry's quest for enhanced productivity and quality has given rise to innovative technologies, one of which is the "Internet of things" (IoT). IoT has the potential to transform agriculture by facilitating precision farming and thereby optimizing production. Autonomous systems would render farms and greenhouses even more precise in their production methods. With IoT's constant evolution, there is a rise in emerging applications and services, enabling farmers to increase resource usage efficiency, maximize profitability, and align production with market demand.

References

Aiello G, Giovino I, Vallone M, Catania P, Argento A (2017) A decision support system based on multisensor data fusion for sustainable greenhouse management. J Clean Prod. https://doi.org/10.1016/j.jclepro.2017.02.197

Bing F (2012) Research on the agriculture intelligent system based on IOT. In: 2012 International Conference on Image Analysis and Signal Processing pp 1–4

Corkery G, Ward S, Kenny C and Hemmingway P (2013) Incorporating smart sensing technologies into the poultry industry. J World's Poult Res 3(4):106–128

Farooq MS, Sohail OO, Abid A, Rasheed SA (2022) Survey on the role of IoT in agriculture for the implementation of smart livestock environment. IEEE 10:9483–9505

Hu X, Qian S (2011) IOT application system with crop growth models in facility agriculture. In Proceedings e 6th international conference on computer sciences and convergence information technology, ICCIT 2011, pp 129–133

Jiang R, Zhang Y (2013) Research of agricultural information service platform based on internet of things. In: 2013 12th International Symposium on Distributed Computing and Applications to Business, Engineering & Science, Kingston upon Thames, UK pp 176–180. https://doi.org/10.1109/DCABES.2013.39

Kacira M, Sase S, Okushima L, Ling PP (2005) Plant response-based sensing for control strategies in sustainable greenhouse production. J Agric Meteorol 61 (1):15–22

Maheshwari R, Vidyarthi M, Vidyarthi P (2022) Significant role of IoT in agriculture for smart farming. In; IoT and AI technologies for sustainable living: a practical handbook, pp 43–55

Parmar M, Kumar R (2022) Overview of IoT in the Agroecosystem. In: Mor RS, Kumar D, Singh A (eds) Agri-Food 4.0 (Advanced Series in Management, Vol 27). Emerald Publishing Limited, Bingley, pp 111–122. https://doi.org/10.1108/S1877-636120220000027008

Quy VK, Hau NV, Anh DV, Quy NM, Ban NT, Lanza S, Muzirafuti A (2022) IoT-enabled smart agriculture: architecture, applications, and challenges. Appl Sci 12 (7):3396

Rajaoarisoa LH, M'Sirdi NK, Balmat JF (2012) Micro-climate optimal control for an experimental greenhouse automation. In: CCCA12, Marseille, France pp 1–6. https://doi.org/10.1109/CCCA.2012.6417903

Ranjit PS, Mahesh GS, Sreenivasa M (2022) Applications of IoT in agriculture. In: Internet of things for agriculture 4.0: impact and challenges. CRC Press, pp 17–34

Ravikanth L, Jayas DS, White White NDG, Fields PG, Sun DW (2017) Extraction of spectral information from hyperspectral data and application of hyperspectral imaging for food and agricultural products. Food Bioprocess Technol 10:1–33. https://doi.org/10.1007/s11947-016-1817-8

Rodriguez de la Concepcion A, Stefanelli R, Trinchero D (2014) A wireless sensor network platform optimized for assisted sustainable agriculture. In: IEEE global humanitarian technology conference (GHTC 2014), pp 159–165. https://doi.org/10.1109/GHTC.2014.6970276

Wang H, Zhao Y, Shao C (2022) IoT for agricultural information generation and recommendation: a deep learning-based approach. Mob Inf Syst 2022 (7378755):1–9. https://doi.org/10.1155/2022/7378755
.

Green Synthesis and Application of Biogenic Nanomaterials as a Blueprint in Mitigation of Abiotic Stress in Crop Plants: A Conceptual Review

11

Saswati Bhattacharya and Jayita Saha

Abstract

Plants being sessile, constantly encounter environmental perturbations that restrict their growth, development, and crop yield. Abiotic stressors like salt, heavy metals, drought, flooding, cold, and elevated temperatures impose heavy yield penalties yearly. The environmental fluctuations resulted in stress conditions impelled the scientific community to focus on developing stratagems to sustain the development and growth of a plant under adverse environmental conditions. Due to the constantly changing global climatic conditions, developing dependable and eco-friendly approaches to overcome the production barrier are of paramount importance. Phytonanotechnology is therefore considered as a viable alternative in mitigating environmental stresses with minimum negative repercussions. Conventional synthesis of nanomaterials (NMs) based on physical and chemical means became matters of concern due to their possible environmental emissions, which could have come up with detrimental effects on the ecosystem. Therefore, the synthesis of NPs from green sources has been projected as a safe and environment-friendly method of nanoparticle (NP) synthesis. The biogenic NPs derived from various organisms like bacteria, algae, fungi, and higher plants evade the use of chemical stabilizers owing to their intrinsic stability, thereby reducing toxic emissions to the environment. Here, we have elaborated the latest developments of NP biogenesis of various sizes and shapes synthesized using the reducing power of secondary metabolites in natural extracts. This chapter also discussed the recent advances in biogenic NPs in ameliorating abiotic stress response, improving plant defense mechanisms, restoring crop yield, and improving growth and development. Therefore, biogenic NPs might come up with a future roadmap for the agricultural community to improve stress-resilience and sustainable development of crop plants.

Keywords

Abiotic stress · Crop plant · Green synthesis · Nanotechnology · Nanoparticle · Tolerance

S. Bhattacharya
Department of Botany, Dr. A.P.J. Abdul Kalam Government College, New Town, Rajarhat, India

J. Saha (✉)
Department of Botany, Rabindra Mahavidyalaya, Champadanga, Hooghly, West Bengal, India
e-mail: ijayita@gmail.com

J. Das and S. Halder (eds.), *Advancement of GI-Science and Sustainable Agriculture*, GIScience and Geo-environmental Modelling, https://doi.org/10.1007/978-3-031-36825-7_11

11.1 Introduction

The term abiotic stress is used to describe the environmental factors that limit plant growth, vigor, and fertility. Stochastic climate change in every corner of the earth is paying a heavy toll on total crop production worldwide. Abiotic stressors brought on by these climatic changes include soil with increased salt and heavy metal content, changing precipitation patterns, water stress in the form of catastrophic floods or droughts, etc. (Saha et al. 2015; Khan et al. 2021a; Kumari et al. 2022). Additionally, global warming has greatly facilitated the salinization of arable land as it has accelerated evaporation, causing unpredictable soil water content due to deluge or drought and changing precipitation patterns. All of these catastrophes mentioned above have a significant adverse effect on global food security (Mafakheri et al. 2021). It is reported that, from 1990 to 2013, in two decades, there has been a 37% increase in the salinity of irrigated farmland (Qadir et al. 2014). The incessant shortfall in precipitation (meteorological drought) associated with elevated evapotranspiration demand results in an agricultural drought that occurs owing to a lack of abundant moisture needed for the natural growth of plants and the completion of their life cycle (Wahab et al. 2022). Moreover, heavy metals are continuously dribbled into the soil due to poor agricultural practices, rapid urbanization, and industrial development. Increased heavy metal contamination in agricultural lands poses major health dangers to humans and restricts crop output (Rehman et al. 2018). Several strategies and approaches have been evolved by researchers from various fields in order to offset the adverse impact of abiotic stress and aid the plant with efficient adaptations.

The eminent physicist Richard Phillips Feynman proposed the idea of nanotechnology in 1959. Prof Norio Taniguchi is credited with coining 'nanotechnology' a decade later (Ansari et al. 2020). In the epoch of agriculture, nanotechnology has emerged as a promising tool to combat abiotic stress without fettering the growth of a plant and crop productivity (Zahedi et al. 2020). Numerous physiological pathways in plants can be modified by NPs, which are also capable of altering gene expression that impedes the development of plants (Chandrika et al. 2018). Nanoparticles (NPs) have extremely small particle sizes between 0.1 and 100 nm. They are significant because of their microscopic size and unique properties compared to the bulk material. The tiny particle size attributes them to the considerably higher surface area-to-volume ratio, magnetism, electrical conductivity, chemical reactivity, physical strength, optical effects, etc., and NPs differ from those of the bulk (Boisseau and Lobaton 2011; Ansari et al. 2020). Recently, plant scientists have proceeded with a sustainable, inexpensive, eco-friendly, and non-toxic method for synthesizing NPs of biological origin compared to chemical or physical methods of NPs' synthesis (Kumari et al. 2022). Plant system responds to NPs based on a dose-dependent manner, as high NP doses confer oxidative injuries to the biomolecules resulting in cellular damage and cell death, while lower concentrations of NPs (at optimal levels) function as the key administrator of plant growth (Chandel et al. 2022). The application of NPs boosts the plant's adaptability toward adverse environmental factors like temperature stress, heavy metal toxicity, salinity, and drought that is mediated by adjusting metabolic pathways and stimulating intrinsic antioxidative defense system to curtail the Reactive Oxygen Species (ROS) generated from stress injury (Khan and Upadhyaya 2019). Following their exogenous administration, an NP influx plays a crucial role in controlling the expression of stress-responsive genes, and the protein they produce channelizes the physiological and biochemical processes leading to plant stress tolerance responses. After sensing the changing environment, the biogenic NPs act as signaling molecules that harmonize the underlying signaling networks to help the plant to reprogram the events from germination to senescence and to acquire stratagem to combat stress factors (Chakraborty et al. 2022).

Various biological sources include bacteria, viruses, actinomycetes, fungi, algae, and other parts of phanerogams, such as roots, shoots, twigs, stems, leaves, fruits, flowers, and seeds, which are widely explored for the green synthesis of NPs. The chapter presents a comprehensive synopsis of recent advances in the biosynthesis of NPs of diverse sizes and shapes, employing the reducing power of different secondary metabolites found in natural extracts. Efforts have been made to enlighten the recent updates in green nanotechnology and the role of biogenic nanoparticles (BNPs) in ameliorating plant abiotic stress by enhancing plants' innate defense mechanisms that restore crop yields and ensure agricultural sustainability.

11.2 Materials and Methods

As this chapter is concerned, the basic sources of data used to formulate the prime objective of the work are to focus the fundamental concepts like 'green synthesis', 'biogenic nanomaterials', and 'abiotic stresses' from various secondary fields or data sources. In the true sense of the term, this chapter has used secondary and tertiary data to incorporate the prime objective of this review kind of research. Firstly, the relevant keywords have been chosen, and thereafter, the significant literature survey has been made on the basis of prime objective of the study. The basic method of the literature survey was online survey from genuine sources dominated in the scientific knowledge world. The thematic sub-paragraph presentations of the concepts are as follows.

11.3 Plant Responses to Abiotic Stresses

According to an approximate estimate by Cramer et al., based on the Food and Agriculture Organization, the USA (FAO) reports, abiotic stressors affect about 96.5% of all rural lands worldwide (Cramer et al. 2011). The majority of the crops shows sensitivity toward abiotic stress. Of the sessile nature of plants and the ubiquity of abiotic stress, plants have developed plenty of tolerance mechanisms at morphological, molecular, and biochemical levels, but this endurance affects their productivity. Plants restrict their vegetative and reproductive development at the onset of stressful situations, using their energy reserves and metabolic precursors to withstand the stress's impacts (Iqbal et al. 2020).

Stress sensing, signaling, and exhaustion are three primary stages of the abiotic stress response of plants (Bhattacharya and Kundu 2020). After sensing stress, a plant can mitigate the stress-induced effects by escape or tolerance. Throughout stress acclimatization, numerous metabolites are produced, and several genes are turned on or off. Many changes have been found at the morpho-anatomical, biochemical, and molecular levels. For example, at the onset of arid conditions, the drought response mechanism is triggered whenever the plants detect a water crisis. Genes related to drought response are switched on, and their expressions result in synthesizing hormones, osmoticums, and other metabolites within the cell to alleviate the stress. Signal molecules like Ca^{2+}, inositol-1, 4, 5-triphosphate (IP3), abscisic acid (ABA), cyclic adenosine 5'-diphosphate ribose (cADPR), nitrogen oxides (NO), etc. are produced in response to drought and bring forth changes at morphological to the molecular level. Coordinated actions of these stress responders affect the host physiology and aid plants in adaptation during stress. The regulatory gene products such as calcium-dependent protein kinases (CDPKs), mitogen-activated protein kinases (MAPKs), HD-zip/bZIP, AP2/ERF, MYB, WRKY, and NAC can alter plant morphology or physiology and help plants to survive in arid conditions by altering signal transduction pathways or by acting as transcription factors to regulate the expression of downstream genes (Yang et al. 2021). An increase in (Na^+) contents in the soil limits the influx of Na^+ in the cell. Moreover, an increase in cytosolic free calcium (Ca^{2+}) concentration has also been observed (Yang et al. 2019). The sequestered action of several plant growth regulators (PGR) plays a pivotal role in combating salinity stress. Salinity also disrupts cellular ionic homeostasis. Unwanted salt accumulation in the cells results in higher production of ROS, which reduces physiological efficiencies (Hashem et al. 2016). Plants can resist heavy metal toxicity by limiting (heavy metal) absorption from the surrounding environment, extracellular and cytoplasmic complexation, and chelation (Yu et al. 2019). High-temperature stress may also add to the production of cellular ROS. The unfavorable temperature or light affects the chloroplast and decreases the functionality of the photosynthetic electron transport chain (PETC). The triplet state of chlorophyll in the photosystem II (PSII) can be induced by extra energy and transmit excitation energy to O_2, creating $_1O^2$ (Szymańska et al. 2017). Prevalence of abiotic stress causes the reduction of photosystem-I (PS-I), which may often result in the production of O_2 and then H_2O_2 (Foyer et al. 2006; Foyer and Noctor 2016). To counter oxidative stress, plants frequently garner a variety of antioxidants. There are many enzymatic antioxidants like catalases (CAT), peroxidase (POX), glutathione reductase (GR), superoxide dismutase (SOD), etc., and some nonenzymatic antioxidants present in a plant cell (Asada 1999; Saha et al. 2015). Redox shuttling across cellular compartments is an important strategy often used by plants to tolerate stress (Geigenberger and Fernie 2014). An organic acid, ascorbate, is crucial for buffering redox systems and transmitting redox equivalents across organelles (Igamberdiev and Bykova 2018). After exposure to stress, ROS is produced in mitochondria and is eventually scavenged by different antioxidant systems, either enzymatic or nonenzymatic. Alternate respiratory pathways significantly reduce ROS generation using the enzyme alternative oxidase (AOX) (Dikshit et al. 2021). The ubiquinone pool of AOXs, which present at the inner mitochondrial membrane, accompanied by ascorbate as a cofactor, reduces O_2 to H_2O and dissipates heat (Saha et al. 2016). The agro-production is inevitably compromised by the inherent in vivo stress response systems outlined above. To sustain crop yield at a higher level, the plant needs to be assisted by additional stress ameliorators that can strengthen the stress responder mechanisms to cope with changing environment with no or minimum compromisation of the crop production.

11.4 Advantages of Using Green NPs

Sustainable advancement is the development that satisfies the demands of the present and balances future generations' potential (Robert et al. 2005). Therefore, limiting the release and accumulation of detrimental chemicals in nature is crucial to create a greener and safer planet for future generations. The majority of chemical procedures used for nanoparticle production are excessively expensive and employ toxic substances that pose some environmental threats (Nath and Banerjee 2013). Techniques with green chemicals have arisen as a novel avenue in the industry of chemical substances in the past twenty years due to the limitations and drawbacks of traditional physical or chemical methods of NP production (Rónavári et al. 2018, 2021). Compared to conventional methodologies, it is more favorable to use microorganisms (bacteria/fungi) or natural materials (fruit juice, polysaccharides, and plant extracts) as they possess abundant hydrogen atoms for NPs manufacturing. Bio-NPs are more biocompatible and have less production cost (Li et al. 2011; El-Sherbiny and Salih 2018). The green synthesis processes also do not require chemical precursors, toxic solvents, or extra-reducing agents (Roy et al. 2019; Naikoo et al. 2021). Choices of an eco-friendly or green solvent, some natural extracts as a suitable reducing agent, and a non-toxic stabilizing substance are the three key prerequisites for the synthesis of nanoparticles (Jadoun et al. 2021).

NPs Characterization:

The synthesized NPs must undergo fine-tuned characterization for their mechanical, physical, and chemical properties before releasing in various fields. Physicochemical parameters of NPs, including their size, surface area, shape, structure, surface morphology, stability, mineral, and elemental decomposition, and purity of NPs can be determined using Ultraviolet–visible spectroscopy (UV–vis), X-Ray Diffraction (XRD), Electron Microscopy, Fourier Transform Infrared Spectroscopy (FTIR), Energy-Dispersive Spectroscopy (EDS), etc. (Dikshit et al. 2021).

11.5 Biogenesis of Green NPs

The biosynthetic pathway involves viruses, microbes (Table 11.1), and higher plants (Table 11.2) to create NPs that are safe, biocompatible, and eco-friendly for biomedical (Razavi et al. 2015) and agricultural applications. The biological organisms employed to produce the NPs are called the 'nano-factories' (Marchev et al. 2020). The mechanism of the production varies from one organism to another (Fariq et al. 2017). Bioproduction of NPs is classified into two categories.

11.5.1 Biosorption

The process involves the association of metal cations present in the surrounding fluid to the organism's cell wall. Stable NPs are formed due to interactions with the cell wall or peptides; the process is energy-independent (Pantidos 2014; Saravanan et al. 2021). The different polysaccharide compounds containing glycoprotein, lipopolysaccharide, etc., are naturally secreted by organisms. The molecules that typically have anionic functional groups can draw cations from polluted or aqueous solutions. In the case of bacteria, the cell wall having peptidoglycan, teichoic acids, liposaccharides, and phospholipids, specifically allow the binding of positive metal ions to negative charges. Chitin, the primary component of the fungal cell wall, also remains involved in the heavy metal complexation and subsequently synthesizes NPs (Wang et al. 2018).

11.5.2 Bioreduction

The process of bioreduction involves the synthesis of biologically stable ions by reducing metal ions chemically. This process utilizes the natural flora and its inert enzymes and can be separated from the polluted environment (Jamkhande et al. 2019). A variety of chemicals found in microbial cells and plant cells, including amides, amines, carbonyl groups, terpenoids, phenolics, flavonoids, alkaloids, proteins, pigments, and other reducing agents, may elicit NP production (Asmathunisha and Kathiresan 2013). Bacteria, fungi, and algae sometimes release compounds with a high propensity to oxidize or reduce metal ions to produce zero-valent/magnetic NPs (Saravanan et al. 2021).

11.6 Sources of Biogenic NPs

11.6.1 Virus: As a Source of NP Production

Viruses' ability to mono-disperse, their NP synthesis's stability and robustness, and chiefly genetic modification are easier on viral platforms. All these characteristics make viruses an attractive source for producing nano-conjugates containing noble metal NPs. A homogeneous shape, size, well-documented structures, and presence of diverse functional groups on their surface make the viral particle, and even often only the protein cage, suitable for regulated production of mono-dispersed NPs (Saratale et al. 2018). The Tobacco Mosaic Virus (TMV) employed for spontaneous synthesis of crystalline, uniform palladium NPs of 1–2 nm size does not require any additional reducing agent (Yang et al. 2013). TiO_2 nanowire production by infecting Escherichia coli with M13 phage has also been reported (Chen et al. 2015). The viral NPs are frequently utilized in imaging, targeted therapies, immune therapeutics, drug administration, and vaccination. However, the production of viral-mediated NPs still has significant drawbacks, including that the virus may not be expressed in a host microbe. Moreover, safety issues must be

Table 11.1 Green synthesis of NPs from microorganisms and cryptogams

Sl. no.	NP type	Name of the organism	Organism type	Shape and size	Production	References
1	Ag	*Bacillus cereus*	Bacteria	4–5 nm; spherical	Intracellular	Ganesh Babu and Gunasekaran, (2009)
2	Ag	*Corynebacterium glutamicum*		5–50 nm, irregular	Extracellular	Sneha et al. (2010)
3	Ag	*Corynebacterium glutamicum*		10–30 nm; spherical	Extracellular	Jo et al. (2016)
4	Ag	*Bacillus methylotrophicus*		10–30 nm; spherical	Extracellular	Wang et al. (2016)
5	Ag	*Phanerochaete Chrysosporium*		34–90 nm; spherical, oval	Extracellular	Saravanan et al. (2018)
6	Ag	*Weissella oryzae*		150 nm; spherical	Extra and intracellular	Singh et al. (2016a)
7	Hg	*Enterobacter sp.*		2–5 nm; spherical	Intracellular	Saratale et al. (2018)
8	CdS	*Escherichia coli*		2–5 nm; elliptical, spherical	intracellular	Saratale et al. (2018)
9	Ag/Au	*Bhargavaea indica*		Anisotropic (Ag) Floral (Au)	Extracellular	Singh et al. (2016b)
10	CdS	*Bacillus amyloliquefaciens*		3–4 nm; Hexagonal/cubic	Extracellular	Singh et al. (2011)
11	Ti	*Lactobacillus sp.*		40–60 nm; spherical	Extracellular	Prasad et al. (2007)
12	Ag	*Streptomyces sp. Strain:LK3*	Actinomycetes	5 nm; spherical	Extracellular	Karthik et al. (2014)
13	Au	*Rhodococcus sp.*		5–15 nm; Mostly spherical	Intracellular	Ahmad et al. (2003)
14	Ag/Au	*Neurospora crassa*	Fungi	Less than 100 nm quasispherical	Extra and intracellular	Castro-Longoria et al. (2011)
15	Ag	*Phomopsis liquidambaris*		18.7 nm spherical	Extracellular	Seetharaman et al. (2018)
16	Ti, Zr, ZrO2	*Fusarium oxysporum*		6–11 nm (Ti, spherical) 3–11 nm (Zr, quasispherical) 3–11 nm (ZrO_2, spherical)	Extra and intracellular	Bansal et al. (2004, 2005)
17	Ag	*Candida utilis NCIM 3469*		20–80 nm; spherical	Extracellular	Waghmare et al. (2015)
18	Ag	*Yarrowia lipolytica NCYC 789*		15 nm; spherical	Extracellular	Apte et al. (2013)
19	Ag	*Penicillium fellutanum*		5–25 nm; spherical	Extracellular	Kathiresan et al. (2009)

(continued)

Table 11.1 (continued)

Sl. no.	NP type	Name of the organism	Organism type	Shape and size	Production	References
20	Se	*Rhodotorula mucilaginosa*	Aquatic yeast	83 nm, 478 nm; spherical and rod shaped	Extra and intracellular	Ashengroph and Tozandehjani (2022)
21	Ag	*Saccharomyces cerevisiae Strain MKY3*	yeast	2–5 nm; hexagonal	Extracellular	Kowshik et al. (2002a)
22	CuO	*Macrocystis pyrifera*	Algae	5–50 nm; spherical	Extracellular	Araya-Castro et al. (2021)
23	Ag	*Macrocystis pyrifera*		50–100 nm; triangular, spherical	Extracellular	Kathiraven et al. (2015)
24	Pd	*Sargassum sp.*		5–10 nm; octahedral	Extracellular	Momeni and Nabipour (2015)
25	Au	*Chlorella vulgaris*		2–10 nm; self-assembled 3D structure	Extracellular	Annamalai and Nallamuthu (2015)
26	Ag	*Ulva compressa*		66.3 nm and 81.8 nm cuboidal	Extracellular	Minhas et al. (2018)
27	Ag	*Gracilaria corticata*		18–36 nm; spherical	Extracellular	Kumar et al. (2013)

Source Compiled by authors

considered, as many viruses are pathogenic, and there is a necessity for further study before large-scale applications can be made (Chakraborty et al. 2022; Saratale et al. 2018).

11.6.2 Bacteria: As a Source of NPs

Microorganisms are frequently used in the synthesis of NPs because of their ease of cultivation, rapid reproduction rate, and ability to thrive under ambient pH, temperature, and pressure (Saravanan et al. 2021). There are numerous microorganisms that produce various inorganic components, either extracellularly or intracellularly. The bacteria and Actinomycetes can convert metal ions to metallic NPs most of the time extracellularly. Most importantly, their proficiency in generating NPs is owing to their higher rate of reproduction in a small scale of time and comparatively easier cultivation (Saratale et al. 2018; Saravanan et al. 2021). The reductase enzyme manufactured by bacteria aids in the bioreduction mechanism and serves as the basis for synthesizing all NPs by bacteria (Saravanan et al. 2021). Due to their resistance to the metal, many bacteria can thrive in environments with high metal ion concentrations. During this process, formation of extracellular complexes occurs due to metal precipitation, altered solubility, toxicity, biosorption, bioaccumulation, efflux systems, and the absence of dedicated metal transport methods (Husseiny et al. 2007; Razavi et al. 2015). The concept of green synthesis of AgNPs from bacteria is an age-old idea. Lactic acid producers like Enterococcus faecium, Lactococcus garvieae, and Pediococcus pentosaceus are used for the synthesis of AgNP in a two-stage procedure. Initially, Ag ions are accumulated inside the cell wall area through biosorption trailed by subsequent metallic reduction and, finally, the production of AgNPs (Sintubin et al. 2009). The evidence of bacterial AgNP production is also found in the Pseudomonas stutzeri

Table 11.2 Green synthesis of NPs from higher plant source

Sr. no.	NP type	Biological source	Parts used	Shape and size	References
1	Ag	Acalypha indica	*Leaf*	20–30 nm; spherical	Krishnaraj et al. (2010)
2	Ag	*Convolvulus arvensis*	Leaf	28 nm; spherical	Hamedi et al. (2017)
3	Se	*Sorghum bicolor*	Leaf	10–40 nm; mostly he–agonal	Djanaguiraman et al. (2018)
4	Cu	*Ginkgo biloba*	Leaf	15–20 nm; spherical	Nasrollahzadeh and Mohammad Sajadi (2015)
5	Fe	Gardenia jasminoides	Leaf	32 nm; rock-like	Naseem and Farrukh (2015)
6	Pb	Cocos nucifera	Leaf	47 nm; spherical	Elango and Roopan (2015)
7	Pd	Catharanthus roseus	Leaf	40 nm; spherical	Kalaiselvi et al. (2015)
8	Ag/TiO_2	Euphorbia prostrata	Leaf	10–15 nm(Ag); 81.7–84.7 nm (TiO_2); spherical	Zahir et al. (2015)
9	S	Rosmarinus officinalis	Leaf	5–80 nm; spherical	Al Banna et al. (2020)
10	Au	Euphorbia fischeriana	Root	20–60 nm; spherical	Zhang et al. (2020)
11	Ag	Chasmanthera dependens	Stem	24.53–98.38 nm; cuboidal	Aina et al. (2019)
12	Ag	Momordica charantia	Stem	27.81 nm; quasispherical	Akinsiku et al. (2018)
13	Au/Ag	*Cibotium barometz roots*	Root	6 nm and 23 nm; spherical	Wang et al. (2017)
14	Ag	Anogeissus latifolia		5.5–5.9 nm; spherical	Kora et al. (2012)
15	Ag	*Salvia leriifolia*	Root	27 nm; spherical	Baghayeri et al. (2018)
16	Ag	*Beta vulgaris*	Root	5–100 nm; spherical	Bin-Jumah et al. (2020)
17	Ag	*Berberis vulgaris*	Root	30–70 nm; spherical	Behravan et al. (2019)
18	Ag	*Zingiber officinale*	Root	10 nm; spherical	Judith Vijaya et al. (2017)
19	ZnO	*Sphagneticola trilobata*	Root	65–80 nm; irregular	Shaik et al. (2020)
20	Ag	*Jatropha curcas*	latex	20–40 nm; spherical to uneven	Bar et al. (2009)
21	Ag	*Nelumbo nucifera*	Seeds	12.9 nm; quasispherical	He et al. (2018)
22	Fe_2O_3	*Punica granatum*	Seed	25–55 nm; semispherical	Bibi et al. (2019)
23	Fe_3O_4	*Borassus flabellifer*	Seed coat	35 nm; hexagonal	Sandhya and Kalaiselvam (2020)

(continued)

Table 11.2 (continued)

Sr. no.	NP type	Biological source	Parts used	Shape and size	References
24	Zn	Artocarpus gomezianus	Fruit	Less than 20 nm; spherical	Suresh et al. (2015)
25	Au	Lycium chinense	Fruit	20–100 nm; polydispersed agglomerated particles	Chokkalingam et al. (2019)
26	ZnO	Musa acuminata	Fruit skin	30–80 nm; of various shapes	Abdullah et al. (2020)
27	Ag	Ferulago macrocarpa	Flower	14–25 nm; spherical	Azarbani and Shiravand (2020)
28	Ag	Nyctanthes arbortristis	Flower	5–20 nm; spherical to oval	Gogoi et al. (2015)
29	Ag	Spartium junceum	Flower	15–25 nm; spherical	Nasseri et al. (2019)

Source Compiled by authors

AG259 strain derived from silver mines (Jorge de Souza et al. 2019). The silver NP production is effectively mediated by a Cupriavidus strain obtained from urban soil (Ameen et al. 2020). Samadi et al. isolated the same from Proteus mirabilis PTCC 1710, obtained from photographic waste. These NPs are spherical with a size range from 10 to 20 nm confirmed through TEM imaging (Samadi et al. 2009). When exposed to precursor ions, Lactobacillus bacteria, frequently found in buttermilk, aid in forming submicron-sized crystals of silver, gold, and gold–silver alloy NPs (Nair and Pradeep 2002). Escherichia fergusonii, Shigella sp., Enterobacter cloacae, Klebsiella sp., Bacillus, and Paenibacillus sp. are all common soil bacteria capable of synthesizing silver NPs with spherical to hexagonal forms with an average size range of 10–20 nm (Pourali and Yahyaei 2016). Johnston et al. showed the ability of the bacteria, Delftia acidovorans to produce pure gold NPs. He explained that delftibactin, a small non-ribosomal peptide, aids in fabricating Au-NPs (Johnston et al. 2013). Geobacillus stearothermophilus, Pseudomonas fluorescens, and Staphylococcus epidermidis are also widely used for the green production of Au-NPs, spherical within the size range between 5 and 90 nm (Shukla and Iravani 2018).

Cd Sulfide NPs can be synthesized from Rhodopseudomonas palustris when the culture is supplemented with 1 mM $CdSO_4$ solution for 72 h at 30 °C temperature (Sweeney et al. 2004; Bai et al. 2009). Further investigations have indicated that the cytoplasmic enzyme cysteine desulfhydrase is responsible for synthesizing CdS nanocrystals, and the protein thus secreted is stabilized and produces CdS NPs (Bai et al. 2009). Bacteria in heavy metal-contaminated alpine areas are used to synthesize zero-valent palladium (Pd0) NPs (Schlüter et al. 2014). The hydrogenase enzyme in Escherichia coli enables the bacteria to synthesize Pd0 NPs (Lloyd et al. 1998). The pH variations in Aeromonas hydrophila suspension culture augmented with zinc oxide triggers oxidoreductase enzymes of different pH sensitivities, leading to the form Zn-NPs (Jayaseelan et al. 2012). Due to the potential for tellurite resistance, bacteria like Escherichia sp. WYS and Raoultella sp. obtained from wastewater can produce tellurite NPs (Nguyen et al. 2019). The production of Cadmium Sulfide (CdS) nanocrystals from Rhodopseudomonas palustris, a photosynthetic bacterium, suggested the synthesis of the nanocrystals which is mediated by the intracellular enzyme, Ce S-lyase, present in the cytoplasm of the bacteria (Jayabalan et al. 2019).

11.6.3 Actinomycetes as a Source of NPs

The actinomycetes, a new source of biological NP, showed monodispersity stability and significant biocidal activity toward various pathogens (Golinska et al. 2014). The actinomycetes demand special attention over other microbial organisms because of their saprophytic behavior and the synthesis of several bioactive chemicals and extracellular enzymes (Chakraborty et al. 2022). The reductase enzyme produced by Streptomyces sp. plays a crucial role in reducing metal salts and thus plays a pivotal role in the green synthesis of Ag, Zn, and Cu NPs (Karthik et al. 2014). Genera like Thermomonospora, Nocardia, Streptomyces, and Rhodococcus have been investigated for their role in producing Au-NPs (El-Batal et al. 2015). The cytoplasmic membranes and mycelial surface are the sites of NP production in actinomycetes (Ahmad et al. 2003). The synthesis of NPs is caused by an electrostatic interaction between the negatively charged carboxylate groups found in the mycelial cell wall enzymes and positively charged silver ions (Abdeen et al. 2014).

11.6.4 Algae as a Source of NPs

The World Health Organization (WHO) affirmed that proteins found in algae could play a crucial role in medications and other nutraceutical industries. Marine ecology is mainly made up of an algal population. In algal cells, proteins, carbohydrates, minerals, and other bioactive (polyphenols, tocopherols, chlorophyll, and other pigments) function as reducing and stabilizing agents (Jacob et al. 2021). Algae may make NPs both extracellularly and intracellularly. The process of NP synthesis with algae includes the preparation of algal extract and the suspension of the metal precursor. The key factors that control nanomaterial production are ambient temperature, pH, extract concentration, precursor, time, etc. (Aboelfetoh et al. 2017). Encapsulation of Klebsormidium flaccidum with a silica gel suspension resulted in the color transformation of the chloroplast from green to purple, indicating a reduction of gold salt within the entrapped cells. Further, Transmission Electron Microscopic analysis affirmed the presence of NADPH and NADPH reductase enzyme that facilitates NP production (Sicard et al. 2010). There are reports of the synthesis of AgNPs by algae, such as Chaetomorpha linum, Padina gymnosperm, and Sargassum wightii (Kannan et al. 2013; Singh et al. 2012). After 120 h of incubation, Spirulina platensis biomass produced globular AgNPs (size range 7–16 nm) by extracellular reduction of $AgNO_3$ at 37 °C temperature and at a pH of 5.6 (Govindaraju et al. 2008). Investigations on Chlorella vulgaris extract have revealed the formation of triangular and hexagonal single-crystalline gold nanoplates (Xie et al. 2007). The reports indicate a single-step synthesis of ZnO-NPs by S. muticum (Namvar et al. 2015). The mean size of these ZnO particles is 42 nm, and the polysaccharide in the alga plays a vital role in NP synthesis (Namvar et al. 2015). Bifurcaria bifurcate a Phaeophyceae member is reported to produce CuO-NPs (Abboud et al. 2014). Arya et al. have prepared AgNPs of 40–90 nm size range from Botryococcus braunii green algae and characterized its catalytic property in synthesizing benzimidazoles (Arya et al. 2019). In a recent study with a Rhodophycean member, Portieria hornemannii has been shown to produce AgNPs (Fatima et al. 2020). Spherical nanostructures of Ag are obtained from an aqueous extract of the marine Phaeophyceae member, Macrocystis pyrifera (Araya-Castro et al. 2021).

11.6.5 Fungi as a Source of NPs

Instead of bacteria, mycosynthesis of NP can be a good alternative for stable NP synthesis. Most fungi have significant metabolites with increased biomagnification potential and straightforward downstream processing, making them simple to cultivate for effective, economic NP synthesis (Alghuthaymi et al. 2015; Singh et al. 2016c). Furthermore, fungi are more competent in uptaking and ingesting metals than bacteria and show a better tolerance, notably in terms of the property of the metal salts to bind with the cell wall of the fungal biomass and resulting in a higher yield of NPs (Castro-Longoria et al. 2011). Three plausible routes for metal NP mycosynthesis have been proposed: electron shuttle quinones, nitrate reductase activity, and sometimes the both (Alghuthaymi et al. 2015). Fungal enzymes like nitrate reductases and α-NADPH-dependent reductases from Fusarium oxysporum and Penicillium sp. play a crucial role in NP synthesis (Anil Kumar et al. 2007).

Several studies have demonstrated the application of yeasts for synthesizing metallic NPs because of their greater resistance to toxic metals (Saratale et al. 2018). MKY3, a yeast strain resistant to silver, can produce AgNPs extracellularly of 2–5 nm average size (Kowshik et al. 2002b). Bhainsa et al. have investigated the extracellular production of AgNPs from Aspergillus fumigatus, and the size ranges between 5 and 25 nm (Ahmad et al. 2002; Bhainsa and D'Souza 2006). AgNPs can also be synthesized using Pleurotus and Phoma sp., as well as an edible mushroom Volvariella volvacea (Zhou et al. 2009). Mono- and bimetallic Au or Ag NPs are effectively produced by the non-pathogenic filamentous fungus Neurospora crassa (Bansal et al. 2005). The production of ultra-small-sized copper and copper oxide (CuO) NPs is possible using the white-rot fungus Stereum hirsutum (Cuevas et al. 2015). Phomopsis liquidambar has produced AgNPs (Seetharaman et al. 2018). Ag nanocrystal synthesis has been conducted using the ligninolytic fungi Trametes trogii. In addition to the supplementation of the fungal filtrate with $AgNO_3$, the role of pH is also very crucial. The NP production is optimum at alkaline pH of 13 (Kobashigawa et al. 2019). Wanarska and Maliszewska (2019) have reported that Penicillium cyclopium may produce metallic silver NPs. The cell wall proteins and saccharides are responsible for synthesizing metallic AgNPs, even from the perished fungal biomass. FTIR mycelium analysis confirmed that the fungal mycelia's saccharides or proteins are responsible for the biomineralization of Ag. These biomolecules are present in the cell-free extract, and bioreduction occurs when silver ions are incubated with polypeptides that contain thiol groups (Wanarska and Maliszewska 2019). Besides AgNPs, there are ample fungi capable of synthesizing Au-NPs. Their higher tolerance to metals and bioaccumulation ability contribute unique advantages to the filamentous fungi over other microbes (Saravanan et al. 2021). The fungal phytopathogen, Fusarium oxysporum synthesizes spherical or hexagonal gold nanoparticles with an optimal size of 20 nm (Naimi-Shamel et al. 2019). Trichoderma viride has been employed to establish a rapid and environmentally friendly technique for synthesizing Au-NPs by synthesizing secondary metabolites and denatured proteins even at a high temperature of 100 °C. The NPs thus produced serve as biocatalysts and convert 4-nitrophenol to 4-aminophenol, a novel alternative in green bioremediation (Mishra et al. 2014). Extract of Flammulina velutipes produced by boiling tiny fragments of mushroom in distilled water and incubating for 30 min with chloroauric acid triggers the nucleation process and, ultimately, Au-NP formation indicated by the production of deep violet color (Rabeea et al. 2020). CuO-NP can be obtained from Trichoderma asperellum by supplementing the mycelia-free aqueous solution with copper nitrate. The hydryl ion of water reacts with copper to form Cu-hydroxide. The cellular enzymes and proteins convert this copper hydroxide into CuO-NPs (Saravanakumar et al., 2019). Saccharomyces cerevisiae has been employed in synthesizing MnO-NP production (Salunke et al. 2015). There are reports of the utilization of S. cerevisiae in the production of cadmium telluride quantum dots that are systemically biocompatible (Kowshik et al. 2002a).

11.6.6 Higher Plant as a Source of NPs

However, most microorganism-based NP syntheses are slower and have a moderate-to-poor yield. The NP recovery necessitates additional downstream processing. Furthermore, challenges associated with microorganism-based NP manufacturing involve complex processes such as collecting microbiological samples, isolation, proper culturing, and maintenance (Singh et al. 2016c). In the green synthesis of NPs, phytonanotechnology has opened a new arena of eco-friendly, convenient, efficient, stable, rapid, and cost-effective fabrication of NPs. Extraction of different plant parts, like roots, shoots, fruit, etc., is used for making NPs (Table 11.2). AgNP biosynthesis may be accomplished most simply by reducing Ag^+ and fusing it with biomolecules like vitamins, polysaccharides, saponins, amino acids, proteins, terpenes, phenols, alkaloids, etc. (Tolaymat et al. 2010). An intriguing work by Bar et al. demonstrated a concise green synthesis pathway for AgNPs from silver salts like $AgNO_3$ utilizing a Jatropha curcas leaf extract. Consequently, in 4 h, fairly homogeneous (10–20 nm) AgNPs have been produced (Bar et al. 2009). A study with Acalypha indica leaf extracts has shown achievable plant-based AgNP production (Krishnaraj et al. 2010). Crystalline Au-NPs of almost uniform size and shape are obtained from the dehydrated roots of Euphorbia fischeriana (Zhang et al. 2020). Both Au and Ag NPs can be derived efficiently from the pulverized dried roots of Cibotium barometz (Wang et al. 2017). Elephantopus scaber and Salvia leriifolia leaf extracts are reported to produce AgNPs (Baghayeri et al. 2018). Zn is an important micronutrient indispensable for plant growth, development, and maturation; its dearth can be detrimental (Ibrahim et al. 2016). The better agronomic output of Abelmoschus esculentus has been achieved using ZnO-NPs made from Citrus medica fruit peels (Keerthana et al. 2021). CuO-NPs can also often have plenty of effects on crop plants. Utilizing Cuscuta reflexa leaf extract, the Cu-NPs can be reduced from Cu^{+2} ions to Cu-NPs. The graphene oxide/MnO_2 nanocomposites are used to immobilize the Cu-NPs (Naghdi et al. 2018). Pt and Pd are both expensive and highly valued in the production of NP (Jadoun et al. 2021). An aqueous solution of [Pd $(OAc)_2$] is stirred for one hour at 60 °C temperature along with a methanolic leaf extract of Catharanthus roseus. The decoction comprises a mixture of eight compounds containing -OH groups responsible for reducing the metal ion to metallic NPs. A change in the color of the solution indicated the formation of Pd-NPs (Kalaiselvi et al. 2015). Due to their unique morphologies and surface chemistry, titanium oxide NPs is of considerable interest (Jadoun et al. 2021). Annona squamosa leaf extract and an aqueous solution of TiO_2 salt when combined together at room temperature produce spherical TiO_2-NPs (Roopan et al. 2012). Cucurbita pepo seed extract is used for green synthesis of TiO_2-NPs of uniform spherical size (Abisharani et al. 2019). Resource distribution, availability and acquisition of nutrients, biogeochemical processes, interactions between microbes and roots, and spatiotemporal heterogeneity coupled with soil complexes are all aided by the structural dynamics and morphology of the roots (Erktan et al. 2018). Hence, roots are considered a very important site for the green synthesis of NPs. AgNP synthesis has been reported from root extracts from Beta vulgaris (Bin-Jumah et al. 2020), Borassus aethiopum (Danbature et al. 2020), Zingiber officinale (Judith Vijaya et al. 2017), etc. There are great uses of the plant stem and shoot extract as a source of different NP productions. There are reports of the extraction of AgNPs from the decoction of the stem of Coleus aromaticus (Vanaja et al. 2013), Salvadora persica (Tahir et al. 2015), and Momordica charantia (Akinsiku et al. 2018). The stem extracts of Leucas lavandulifolia are used to generate spherical selenium (Se) NPs. Se-NPs are synthesized when Se ions are reduced in the presence of diverse phytoconstituents, such as polyphenols and water-soluble heterocyclic compounds (Kirupagaran et al. 2016). The floral parts of a plant are always rich in pigments like chlorophyll, xanthophyll, carotenoids, anthocyanins, etc., along with other bioactive compounds like phenols and flavonoids add to the color and aroma of the flower. The floral extracts of Ferulago macrocarpa (Azarbani and Shiravand 2020) and Cuscuta reflexa (Shaik et al. 2020) are widely used in AgNP production.

11.7 Role of Green NPs in Abiotic Stress Response

11.7.1 Salt Stress

Salt stress is an ominous factor that jeopardizes growth and crop yield worldwide. Soil salinization has resulted from the gradual worsening of environmental conditions, climate change, and poor irrigation practices, affecting approximately 20% of the earth's irrigated agricultural lands. Salinity adversely affects the molecular, physiological, and biochemical processes, limiting quantitative and qualitative food production. The onset of salt stress leads to immediate osmotic and ionic imbalances that elevate the build-up of ROS and toxic ions with a concomitant oxidative burst instigating lipid peroxidation, distortion of biomolecules, and ultimately resulting in cell death.

Biogenic NPs are considered an eco-friendly and low-cost method that can be used as an antidote to increasing salt stress resilience in crop plants. Recent findings revealed that seed priming with ZnO-NPs (5 and 10 mg/L), synthesized from *Agathosma betulina,* significantly improved shoot lengths, fresh weights, anatomical structure, lowered Na^+/K^+ ratio (1.53 and 0.58) mending element distribution, reduced biomolecules damage by declining oxidative stress in Sorghum bicolor under the treatment of 400 mM NaCl salt (Rakgotho et al. 2022). Bio-synthesized gold NPs (Au-NPs) recuperated the wheat plant (*Triticum aestivum*) from salt stress by modifying the K^+/Na^+ ratio, nitrogen assimilation, chlorophyll content, stomatal dynamism, controlled reactive oxygen/nitrogen species accumulation (Wahid et al. 2022). Se-NPs can be synthesized from selenium dioxide as a precursor molecule using a reducing/capping agent derived from the aqueous flower extract of *Allamanda cathartica.* They exhibited restored germination percentage (31%), shoot length (92%), root length (78%), and total chlorophyll content (49%) in *Brassica campestris* (TS-36 variety) under 200 mM NaCl stress (Sarkar and Kalita 2022a). In another study, they have shown that biosynthesized Se-NPs (30 mg L^{-1}) from fresh grape aqueous extract enhanced SOD, CAT, Ascorbate peroxidase (APX), and POX activities by 41.20%, 64.10%, 63.06%, and 70.43%, respectively, phenolic and flavonoid content by 98.88% and 86.90%, respectively, and 61.89% free radical scavenging activity in mustard plants (TS-36 variety) grown hydroponically under 200 mMNaCl (Sarkar and Kalita 2022b). In addition to that, the significant increase in the seed germination percentage (39.66%), root and shoot length (75% and 60.64%, respectively), dry and fresh weight per plant (41.2% and 22.11%, respectively), water content percentage (1.02%), chlorophyll content (81.92%), carbohydrate content (24.65%), and protein content (79.14%) has also been observed. Hydroponically applied bio-Se-NPs produced using a leaf extract of barley plants drastically compensate for the adverse effect of salt stress (100 mM) in *Hordeum vulgare* that is correlated with declined malondialdehyde (MDA) level, significant accumulation of phenolic content, metabolic adjustment, and rising shoot dry weight (Habibi and Aleyasin 2020). Application of 120 mg L-1 Silica NPs (Si-NPs), as a foliar spray, obtained from rice husk has been investigated in three rice varieties (Pokkali, KDML105, and IR29) under salinity and observed stress alleviation by inducing net photosynthesis rate, lowering H_2O_2 content, increased activity of peroxidase, catalase, and ascorbate peroxidase (Larkunthod et al. 2022). Under a two-year field trial study with saline irrigation water, the application of Green nanosilica (GNS) as a foliar spray obtained from plant biowaste boosted fruit output, decreased leaf Na^+ concentrations, increased nutrient absorption, antioxidant, and osmoregulatory (such as proline and total sugar content), and imparted salt avoidance in *Musa* sp. (Ding et al. 2022). In maize plants, the restorative effect from salt stress is confirmed through the application of biologically synthesized copper NPs (Cu-NPs, 100 mg kg^{-1}) from *Klebsiella pneumonia* (strain NST2). The green Cu-NPs, when mixed with saline soil, recovered root and shoot length, fresh weight, and dry weight, neutralized the lipid peroxidation and oxidative damage by stimulating antioxidants and demoting the cellular

ROSlevels and Na + and Cl- content when compared to control (Noman et al. 2021). The biogenesis of titanium dioxide NPs (TiO_2-NPs, 40 mg/L) using the extract of *Buddleja asiatica* leaf has shown a substantial increase in dry weight, plant height, fresh weight, shoot, and root length, RWC, leaf count per plant, chlorophyll a and b, and total chlorophyll contents, thus proved to be favorable to augment agronomic growth and physiological attributes of two wheat varieties (Faisalabad-08 and NARC-11) under salinity (Mustafa et al. 2021). The salt tolerance potential was scored by applying a foliar spray of zinc oxide nanoparticles (ZnO-NPs) derived from *Moringa* leaf extract onto *Vicia faba* (cultivars: Giza-716 and Sakha 4). The results revealed significant enhancement in various plant growth parameters, photosynthetic pigments, proline and phenol levels, ions (Na^+, K^+, Ca^{2+}, and Zn^{2+}), pigments (chlorophyll and carotenoids), as well as enzyme activities (polyphenol oxidase—PPO, ascorbate peroxidase—APX, and catalase—CAT), when compared to control plants (Ragab et al. 2022; Mogazy and Hanafy 2022). The ameliorative role of 100 mg L^{-1} green Se-NPs synthesized by *Bacillus cereus* TAH examined in wheat seed germination under a salt-exposed environment has resulted in 25, 25, 39.4, and 11% enrichment of germination percent, mean germination time, vigor index, and germination rate index, respectively (Ghazi et al. 2022).The Se-NPs' treatment under high Ec values of 14 ds m^{-1} in a gnotobiotic sand system divulged marked increases by 22.8, 24.9, 19.2, and 20% of the shoot and root length, fresh and dry weight, respectively, compared to controls. Biogenic Se-NPs and ZnO-NPs out of an aqueous extract of the leaves of *Moringa* mitigated the negative effect of salinity on the growth, yield, antioxidant activity, and phytoconstituents accumulation in garlic plants (Sids 40) which have been associated with increased total phenolic and flavonoids compounds, enhanced H_2O_2 scavenging potential of antioxidants (El-Saber 2021). The positive effect of silver nanoparticles (AgNPs), fabricated using *Rosmarinus officinalis,* markedly elevated cellular levels of photosynthetic pigments, accumulated soluble sugars, proline, and soluble proteins, decreased H_2O_2 and malondialdehyde content with a concomitant increase in enzymatic and nonenzymatic antioxidants, thus bestowed with salt tolerance on *Linum usitatissimum* (Khalofah et al. 2021). Furthermore, exposure to various concentrations of AgNPs on *Triticum aestivum* and *Lathyrus sativus* stimulated fresh and dry weight, root and shoot length, seed sprouting, soluble sugar, total chlorophyll content, proline content, and antioxidant enzymes during salinity (Abasi et al. 2022). A comparative study has been done between green zinc NPs (Zn-GNPs) synthesized using the leaf extract of *Sorghum bicolor* L. and chemically synthesized Zn-CNPs produced by the co-precipitation method. It has been found that Zn-GNPs (0.3%) more efficiently triggered the salt tolerance response on the growth of Okra (*Abelmoschus esculentus*) under saline environment and resulted in significant enhancement of the fresh and dry weight of shoot and root, chlorophyll contents, and antioxidant activity (Zafar et al. 2021).

11.7.2 Drought Stress

Temperature dynamics, global warming, rainfall inconsistencies, shifts in monsoon patterns, and light intensity has resulted in inescapable stresses to the plant kingdom. Drought stress begins to manifest without prior indication, thereby impeding plant growth and crop yield and damaging plant morphological, physiological, biochemical, and molecular attributes, leading to reduced photosynthetic ability and energy production (Seleiman et al. 2021). Low photosynthesis is immediately followed by stomatal closure, membrane damage, disrupted various enzymes activities, impaired ATP synthesis, altered water relations, and reduced water retention capacity in plants that ultimately end up with the plant striving for more water and consequently aggravating permanent wilting (Farooq et al. 2012).

Drought adaptation through green NPs' applications became the most promising and environment-friendly strategy to compensate for the damaging effect of water stress through maintaining cellular homeostasis and is considered of great potential in agriculture. Recent studies explored the healing attributes of many Biogenic NPs to relieve the dangerous effects of drought conditions. For example, foliar spray of green ZnO-NPs (25 and 50 mg/L) on tomato plants significantly increased shoot and root biomass, shoot dry weight (2–2.5-fold), ascorbic acid, and free phenols, raised the activity of SOD, APX, and CAT, decreased MDA and H_2O_2 contents, thereby minimizing drought-induced oxidative stress in response to severe drought conditions (25%) compared to ZnO-NPs-untreated plants (El-Zohri et al. 2021). Green synthesized ZnO-NP-II (ZnO-NP-II; size = 75 nm) from *Lawsonia inermis* extract showed greater healing efficacy and post-stress recovery potential over chemical method (ZnO-NP-I; size = 100 nm) to mitigate the hazards brought on by water stress in rice (*Oryza sativa*) Kopilee cultivar seedlings under the hydroponic system (Shome et al. 2022). They have reached this conclusion by measuring the increase in fresh and dry mass, root and shoot length and decrease in H_2O_2 and O^2 contents due to augmentation of CAT, GPX, SOD, and GR activity. Moreover, foliar application of 100 ppm green ZnOx NPs synthesized using the leaf extract of *Camellia sennesis* encouraged the defensive system of *Coriander sativum* that helps them to withstand water stress by enhancing chlorophyll and proline content, CAT, SOD, and APX activity, declining MDA content, thus, restore agronomic attributes for high crop productivity (Khan et al. 2021b). The plants' defensive condition demonstrated the positive benefits of biosynthesizing Se-NPs produced from the buds of *Allium sativum* (Ikram et al. 2020). Application of 30 mg/L green Se-NPs exogenously on drought-tolerant (V1) and drought-susceptible (V2) wheat varieties at the trifoliate stage stemmed marked induction in plant height, shoot and root length, fresh weight, dry weight, leaf area, number, and length under water-deficient conditions. Investigations have been done with biogenic iron oxide NPs (Fe_3O_2-NPs) produced from ginger (*Zingiber officinale*) and cumin (*Cuminum cyminum*) seed extracts on wheat plants (Noor et al. 2022). The effectiveness of ginger-derived Fe_3O_2-NPs (0.6 mM) and cumin seed-derived Fe_3O_2-NPs (1.2 mM) in improving drought resistance in wheat plants has been demonstrated. These NPs have been found to stimulate germination and increase survival percentage. They also promote the accumulation of chlorophyll a, b, and carotenoids, as well as soluble sugars, proline content, and total iron content in both roots and shoots. Additionally, these NPs enhance the activity of SOD, peroxidase, and ascorbate peroxidase (APX) enzymes, while reducing lipid peroxidation and electrolyte leakage under drought stress, in comparison to the non-treated control group. Furthermore, 50 mg/L of biosynthesized magnetite (Fe_3O_4) NPs from leaves of *Polyalthia longifolia* when applied on two different varieties of *Trigonella foenum-graecum* (Afg 1 and Afg 3) contributed to better plant growth parameters, enhanced production of photosynthetic pigments, and overall photosynthetic performance, thus lessening drought stress response (Bisht et al. 2022). On the other hand, Fe_3O_4-NPs (both capped as well as bare) synthesized from a marine alga *Chaetomorpha antennina* (green algae) employed as nano-fertilizer on drought-stressed *Setaria italica* plants exhibited enhancement of overall plant growth (Sreelakshmi et al. 2021). They also suggested that iron uptaken by the plants facilitated the accumulation of photoassimilates, iron, chlorophyll, and soluble sugar content, thus confirming its ameliorative role in overcoming drought stress. Additionally, nanopriming and foliar application of phyto-synthesized FeONPs (100 ppm) using leaf extract of *Prosopis cineraria* at the seedling development stage (20 days) and tillering stage (30 days) of wheat (*Triticum aestivum*) plant encouragingly influenced all the morphological and yield attributes in the wheat crop and bestowed with drought resilience under rain-fed conditions in comparison to control (Singh et al. 2022).

11.7.3 Heavy Metal Stress

Heavy metals (HM), for instance, Manganese (Mn), Nickel (Ni), Iron (Fe), Copper (Cu),

Cadmium (Cd), Cobalt (Co), Mercury (Hg) Zinc (Zn), and arsenic (As) are conventional elements which have been accumulated air, water, and soil through several anthropogenic activities, industrial trash. Although some of these metals function as micronutrients during normal plant growth and physiological processes, their surplus accrual in the environment compels the plant to reach a high risk (Ghori et al. 2019). Under prolonged toxic exposure in HM, plants experienced excessive oxidative stress for the excessive production of ROS that eventually showed detrimental effects on cellular components such as membranes, nucleic acids, proteins, pigments, and homeostasis between ROS and enzymatic or nonenzymatic antioxidants.

Several studies have been made to understand the fundamental mechanism of action of cheap and environmentally sound NPs that have been attributed as the savior for alleviating the toxic effect of HM in crop plants (Mathur et al. 2022; García-Ovando et al. 2022). AgNPs and CuO-NPs derived from the leaf extract of *Catharanthus roseus* have been utilized to wipe out chromium and cadmium, thus improving the present circumstances of metalloid pollution in soil and water environments (Verma and Bharadvaja 2022). Disk-shaped biogenic hydroxyapatite (HAP) nanoparticles (NPs), synthesized using *Aspergillus niger* fermentation broth, has applied to the habitat of mung-bean seedlings under Cd stress. HAP NPs' exposure in the presence of Cd stress demonstrated a drastic reduction of Cd content in the stem, a remarkable decline in H_2O_2 and MDA content, and a significant increase of SOD, CAT, and APX, thereby eliminating the oxidative stress due to Cd stress by blocking Cd translocation in mung-bean seedlings and restore seedling growth (Shen et al. 2022). External application of magnesium oxide nanoparticles (MgO-NPs) synthesized from an *Enterobacter* sp. strain RTN2 drastically improved rice plant resilience by minimizing the ill-effect of As-contaminated soil. Biogenic MgO-NPs (200 mg kg^{-1}) have resulted in significant induction of the plant biomass, antioxidant enzymatic contents, and reduction in ROS, As uptake, and translocation compared to the control treatment. It is also suggested to be used to formulate a nano-fertilizer for the sustainable production of rice (Ahmed et al. 2021). Synthesis of silicon nanoparticles (Bio-Si-NPs) has been performed using potassium silica fluoride substrate and the identified *Aspergillus tubingensis* AM11 strain and employed in 5.0 mmol/L concentration as a foliar spray on *Phaseolus vulgaris* to investigate its shielding effect under heavy metalscontaminated saline soil (El-Saadony et al. 2021). They observed notable convalescence in terms of plant growth and production, pigment contents (chlorophylls and carotenoids), rate of transpiration and net photosynthesis, stomatal conductance, relative water content (RWC), total soluble sugars, free proline contents, N, P, K, Ca^{2+}, K^+/Na^+ contents, and antioxidative enzymes' activities (peroxidase, CAT, APX, SOD). The remarkable decline in electrolyte leakage, MDA, H_2O_2, $O_2^{\bullet-}$, Na^+, Pb, Cd, and Ni contents in leaves and pods of *Phaseolus vulgaris* compared to control has also been discerned. The easing effect of green copper NPs (Cu-NPs) derived from a native strain of *Klebsiella pneumoniae* to relieve oxidative stress in wheat plants due to chromium (Cr) stress has been studied (Noman et al. 2020a, b). Soil fortification with bio-Cu-NPs (25 and 50 mg kg^{-1}) significantly enhances plant growth and biomass, along with cellular antioxidants, whereas it reduced the ROS levels and translocation of Cr from soil to plant body by immobilizing the Cr in the soil when compared to control. Additionally, they have examined the healing role of the Cu-NPs (100 mg kg^{-1}), synthesized from a bacterium *Shigella flexneri* SNT22 on wheat plants under cadmium (Cd) toxicity and recorded similar responses like increased plant length (44.4%), shoot dry weight (28.26%), nitrogen contents (41.60%), phosphorus contents (58.79%), P, N, Ca^{2+}, K^+, Ca^{2+}/Na^+, and K^+/Na^+ contents with a concomitant reduction in Na^+ content and acropetal Cd translocation (49.62%) in a wheat plant (Noman et al. 2020b). Further reports indicated that treatment with Bio-FeO-NPs (100 mg/L) produced from leaf extract of *Adiantum lunulatum* facilitated the rice plants to withstand the adverse effect owing to As stress by increasing percentage of seed germination, the growth, and vigor of the seedlings, reducing uptake and translocation of As, a decline in oxidative stress (Chatterjee et al. 2021). Combined application of the *Bacillus* sp. strain ZH16 with biogenic MoNPs considerably decreased plants' arsenic (As) translocation by 30.3% in As-contaminated agricultural soils. Furthermore, Bio-MoNPs exhibited biocompatibility with bacterial strain and stimulated indole-3-acetic acid synthesis, ACC deaminase activity, and phosphate solubilization, promoting the morphological parameters, nutrients content, and ionic balance of wheat plants under As-spiked conditions (Ahmed et al. 2022).

11.7.4 Other Abiotic Factors

Besides the salinity, drought, and heavy metal stress, green synthesized NPs have also acted as protective shields for many crop plants against many other abiotic factors. There are reports of application of 75 mg/l biosynthesized AgNPs generated using plant extract of *Moringa oleifera* on wheat plants at the trifoliate stage under heat stress in a range of 35–40 °C for three h/day for about three days (Iqbal et al. 2019). This investigation evidenced marked increase in plant root length (5.4%), shoot length (26.1%), root number (7.5%), plant fresh weight (2%), plant dry weight (0.60%), leaf area (33.8%), leaf number (4.8%), fresh leaf weight (0.15%), and leaf dry weight (0.18%) as compared over control under heat stress. Foliar application and soil irrigation of biogenic NPs, obtained from green tea (20 mg per plant), have been used as an enticing substitute that prohibits toxicity and injury brought on by oxidative stress as evidenced by fall in APX and CAT levels, rise in R-S-nitrosothiols (RSNOs), phenolic and total flavonoid content in lettuce (*Lactuca sativa*) (Kohatsu et al. 2021). Growth, physiology, antioxidant defense, and yield parameters' studies revealed that foliar administration of 25 mg/L biogenic AgNPs has been proved effective as a model anti-ozonant ethylene urea (EDU) against a phytotoxic pollutant, Tropospheric ozone (O_3) in two wheat cultivars (HD-2967 & DBW-17) at both vegetative and reproductive stages, and assigned as a promising ozone protectant agent (Kannaujia et al. 2022).

11.8 Challenges of Using Biogenic NPs

Recent advancement in the environmentally friendly synthesis of NPs and their possible applications in agricultural fields have become a major promising area with an endeavor to sustain crop yield and productivity worldwide. Enormous evidence has demonstrated their economic, non-toxic, and eco-friendly nature, compared to that of the conventional method of chemical synthesis of NPs. However, some challenges encumber the global synthesis of biogenic NPs and subsequent utilization. For instance, reactants (plant extract, microorganism inoculum, chemical compounds as oxidizing and reducing agents) and physical parameters of the reaction with reduced reaction time need to be optimized elaborately for controlling the shape, size, yield, and stability of the NPs. The physicochemical properties of NPs seek more characterization studies to guarantee quality assurance. Prioritization is required to scale up the commercial green synthesis of the NPs. Additional toxicological dissections are necessitated for the sprawling use of biogenic NPs on plants and animals. Moreover, large-scale production of biocompatible NPs from endemic flora is gaining huge encumbrance due to the interruption of raw material collection. In the coming future, it is essential to address all these limitations with momentous care to come up with a wide spectrum of utilizing biogenic NPs in various fields.

11.9 Conclusion

Global climate change, with steady temperature rise, water and land resources' scarcity due to anthropogenic activity, non-rhythmic seasonal

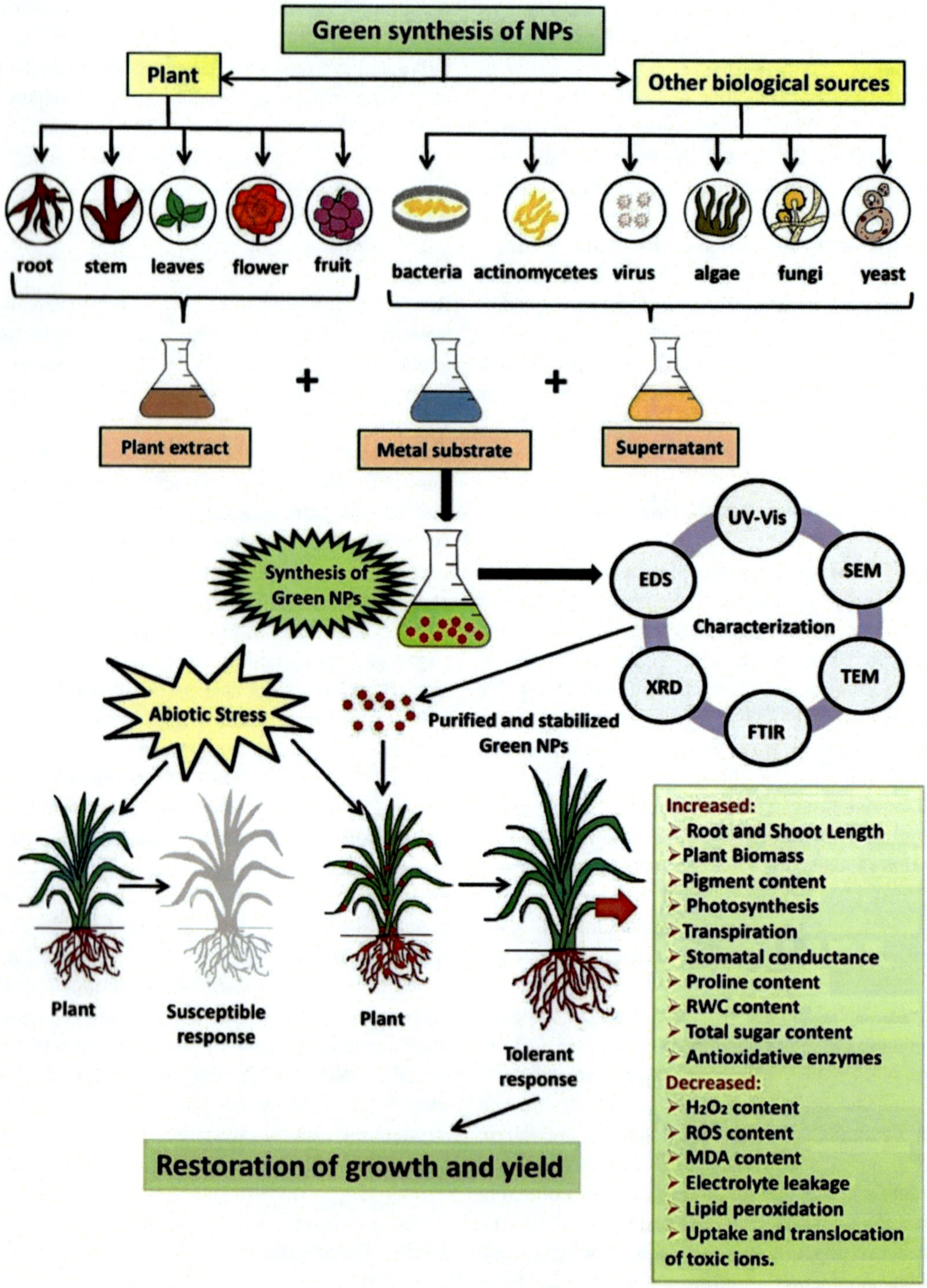

Fig. 11.1 Pictorial illustration of green synthesis of NPs from various biological sources and its amelioration impact on plants under abiotic stress conditions

oscillations, and other abiotic stressors challenge universal agricultural methods. This challenge necessitates the development of several sustainable and eco-friendly solutions to minimize the detrimental effect of stressors on field crops. Green nanotechnology is therefore considered as a viable alternative to traditional ones for alleviating abiotic stresses with minimum toxic environmental emissions. Green NP synthesis has evolved as a safe, low-cost, and environmentally friendly technique of NP synthesis. These bioinspired noble metals, due to their intrinsic stability, inertness, ease of accessibility, and environmentally benign nature, can be employed as reliable administrators of stress mitigation in crop plants, lowering the negative impacts of abiotic stressors. This chapter explored the potential applications of a wide range of biological sources as a precursor of biogenic NPs (Fig. 11.1). This chapter has also explored possible applications of the potential usages of a wide range of biological sources as a precursor of biogenic NPs (Fig. 11.1). In addition to that the most recent evidence exhibiting the positive impact of biogenic NPs restoring growth and production of crop plants in response to diverse abiotic stressors (Fig. 11.1) has been elaborated with citation of recent research trends. We also emphasized how the application of green synthesized NPs is accompanied by stress tolerance by acting as key players governing the underneath physiological and biochemical attributes like seedling germination, growth parameters, mineral uptake, photosynthetic efficiency, antioxidant defense system, accumulation of compatible solutes and ROS, membrane damage, etc. However, there are certain limitations of the green synthesis of NPs that should be dealt with the future researchers. Moreover, in the era of nanotechnology, the biogenesis of NPs can be served as an imperative development that demands global application in the upcoming years, encouraging the future agricultural community to improve the stress-resilience and sustainability of crop plants.

References

Abasi F, Raja NI, Mashwani ZUR, Amjad MS, Ehsan M, Mustafa N, Haroon M, Proćków J (2022) Biogenic silver nanoparticles as a stress alleviator in plants: a mechanistic overview. Molecules 27:3378. https://doi.org/10.3390/molecules27113378

Abboud Y, Saffaj T, Chagraoui A, El Bouari A, Brouzi K, Tanane O, Ihssane B (2014) Biosynthesis, characterization and antimicrobial activity of copper oxide nanoparticles (CONPs) produced using brown alga extract (Bifurcaria bifurcata). Appl Nanosci 4:571–576

Abdeen S, Geo S, Praseetha PK, Dhanya RP (2014) Biosynthesis of silver nanoparticles from Actinomycetes for therapeutic applications

Abdullah FH, Abu Bakar NHH, Abu Bakar M (2020) Low temperature biosynthesis of crystalline zinc oxide nanoparticles from Musa acuminata peel extract for visible-light degradation of methylene blue. Optik 206:164279. https://doi.org/10.1016/j.ijleo.2020.164279

Abisharani JM, Devikala S, Kumar R, Dinesh AM, Kamaraj P (2019) Green synthesis of TiO_2 Nanoparticles using Cucurbita pepo seeds extract. Mater Today Proc 14:302–307. https://doi.org/10.1016/j.matpr.2019.04.151

Aboelfetoh EF, El-Shenody RA, Ghobara MM (2017) Eco-friendly synthesis of silver nanoparticles using green algae (Caulerpa serrulata): reaction optimization, catalytic and antibacterial activities. Environ Monit Assess 189:349. https://doi.org/10.1007/s10661-017-6033-0

Ahmad A, Mukherjee P, Mandal D et al (2002) Enzyme mediated extracellular synthesis of CdS nanoparticles by the Fungus, *Fusarium oxysporum*. J Am Chem Soc 124:12108–12109. https://doi.org/10.1021/ja027296o

Ahmad A, Senapati S, Khan MI, Kumar R, Ramani R, Srinivas V, Sastry M (2003) Intracellular synthesis of gold nanoparticles by a novel alkalotolerant actinomycete, Rhodococcus species. Nanotechnology 14:824. https://doi.org/10.1088/0957-4484/14/7/323

Ahmed T, Noman M, Manzoor N, Shahid M, Hussaini KM, Rizwan M, Ali S, Maqsood A, Li B (2021) Green magnesium oxide nanoparticles-based modulation of cellular oxidative repair mechanisms to reduce arsenic uptake and translocation in rice (Oryza sativa L.) plants. Environmental Pollution 288:117785. https://doi.org/10.1016/j.envpol.2021.117785

Ahmed T, Noman M, Rizwan M, Ali S, Ijaz U, Nazir MM, ALHaithloul HAS, Alghanem SM, Abdulmajeed AM, Bin Li (2022) Green molybdenum nanoparticles-mediated bio-stimulation of Bacillus sp. strain ZH16 improved the wheat growth by managing in planta nutrients supply, ionic homeostasis and arsenic accumulation. J Hazard Mater 423:127024. https://doi.org/10.1016/j.jhazmat.2021.127024

Aina DA, Owolo O, Lateef A, Aina FO, Hakeem AS, Adeoye-Isijola M, Adediji I (2019) Biomedical applications of *Chasmanthera dependens* stem extract mediated silver nanoparticles as antimicrobial, antioxidant, anticoagulant, thrombolytic, and larvicidal agents. Karbala Int J Modern Sci 5(2):2. https://doi.org/10.33640/2405-609X.1018

Akinsiku AA, Dare EO, Ajanaku KO, Adekoya JA, Ayo-Ajayi J (2018) Green synthesized optically active organically capped silver nanoparticles using stem extract of African cucumber (Momordica charantia). J Mater Environ Sci 3:902–908

Al Banna LS, Salem NM, Jaleel GA, Awwad AM (2020) Green synthesis of sulfur nanoparticles using Rosmarinusofficinalis leaves extract and nematicidal activity against Meloidogyne javanica. Chem Int 6(3): 137–143

Alghuthaymi MA, Almoammar H, Rai M, Said-Galiev E, Abd-Elsalam KA (2015) Myconanoparticles: synthesis and their role in phytopathogens management. Biotechnol Biotechnol Equip 29:221–236. https://doi.org/10.1080/13102818.2015.1008194

Ameen F, AlYahya S, Govarthanan M, ALjahdali N, Al-Enazi N, Alsamhary K, Alshehri WA, Alwakeel SS, Alharbi SA (2020) Soil bacteria Cupriavidus sp. mediates the extracellular synthesis of antibacterial silver nanoparticles. J Mol Struct 1202:127233. https://doi.org/10.1016/j.molstruc.2019.127233

Anil Kumar S, Abyaneh MK, Gosavi SW, Kulkarni SK, Pasricha R, Ahmad A, Khan MI (2007) Nitrate reductase-mediated synthesis of silver nanoparticles from $AgNO_3$. Biotechnol Lett 29:439–445. https://doi.org/10.1007/s10529-006-9256-7

Annamalai J, Nallamuthu T (2015) Characterization of biosynthesized gold nanoparticles from aqueous extract of Chlorella vulgaris and their anti-pathogenic properties. Appl Nanosci 5:603–607. https://doi.org/10.1007/s13204-014-0353-y

Ansari M, Shahzadi K, Ahmed S (2020) Nanotechnology: a breakthrough in agronomy. In: Javad S (ed) Nanoagronomy. Springer International Publishing, Cham, pp 1–21

Apte M, Sambre D, Gaikawad S, Joshi S, Bankar A, Kumar AR, Zinjarde S (2013) Psychrotrophic yeast Yarrowia lipolytica NCYC 789 mediates the synthesis of antimicrobial silver nanoparticles via cell-associated melanin. AMB Expr 3:32. https://doi.org/10.1186/2191-0855-3-32

Araya-Castro K, Chao T-C, Durán-Vinet B, Cisternas C, Ciudad G, Rubilar O (2021) Green Synthesis of copper oxide nanoparticles using protein fractions from an aqueous extract of brown algae Macrocystis pyrifera. Processes 9:78. https://doi.org/10.3390/pr9010078

Arya A, Mishra V, Chundawat TS (2019) Green synthesis of silver nanoparticles from green algae (Botryococcus braunii) and its catalytic behavior for the synthesis of benzimidazoles. Chem Data Collect 20:100190. https://doi.org/10.1016/j.cdc.2019.100190

Asada K (1999) The water-water cycle in chloroplasts: scavenging of active oxygens and dissipation of excess photons. Annu Rev Plant Physiol Plant Mol Biol 50:601–639. https://doi.org/10.1146/annurev.arplant.50.1.601

Ashengroph M, Tozandehjani S (2022) Optimized resting cell method for green synthesis of selenium nanoparticles from a new Rhodotorula mucilaginosa strain. Process Biochem 116:197–205. https://doi.org/10.1016/j.procbio.2022.03.014

Asmathunisha N, Kathiresan K (2013) A review on biosynthesis of nanoparticles by marine organisms. Colloids Surf, B 103:283–287. https://doi.org/10.1016/j.colsurfb.2012.10.030

Azarbani F, Shiravand S (2020) Green synthesis of silver nanoparticles by Ferulago macrocarpa flowers extract and their antibacterial, antifungal and toxic effects. Green Chem Lett Rev 13:41–49. https://doi.org/10.1080/17518253.2020.1726504

Baghayeri M, Mahdavi B, Hosseinpor-Mohsen Abadi Z, Farhadi S (2018) Green synthesis of silver nanoparticles using water extract of Salvia leriifolia: antibacterial studies and applications as catalysts in the electrochemical detection of nitrite. Appl Organomet Chem 32:e4057. https://doi.org/10.1002/aoc.4057

Bai HJ, Zhang ZM, Guo Y, Yang GE (2009) Biosynthesis of cadmium sulfide nanoparticles by photosynthetic bacteria Rhodopseudomonas palustris. Colloids Surf, B 70:142–146. https://doi.org/10.1016/j.colsurfb.2008.12.025

Bansal V, Rautaray D, Ahmad A, Sastry M (2004) Biosynthesis of zirconia nanoparticles using the fungus Fusarium oxysporum. J Mater Chem 14:3303–3305. https://doi.org/10.1039/B407904C

Bansal V, Rautaray D, Bharde A, Ahire K, Sanyal A, Ahmad A, Sastry M (2005) Fungus-mediated biosynthesis of silica and titania particles. J Mater Chem 15:2583–2589. https://doi.org/10.1039/B503008K

Bar H, Bhui DKr, Sahoo GP, Sarkar P, De SP, Misra A (2009) Green synthesis of silver nanoparticles using latex of Jatropha curcas. Colloids Surf A Physicochem Eng Asp 339:134–139. https://doi.org/10.1016/j.colsurfa.2009.02.008

Behravan M, Hossein Panahi A, Naghizadeh A, Ziaee M, Mahdavi R, Mirzapour A (2019) Facile green synthesis of silver nanoparticles using Berberis vulgaris leaf and root aqueous extract and its antibacterial activity. Int J Biol Macromol 124:148–154. https://doi.org/10.1016/j.ijbiomac.2018.11.101

Bhainsa KC, D'Souza SF (2006) Extracellular biosynthesis of silver nanoparticles using the fungus Aspergillus fumigatus. Colloids Surf, B 47:160–164. https://doi.org/10.1016/j.colsurfb.2005.11.026

Bhattacharya S, Kundu A (2020) Sugars and sugar polyols in overcoming environmental stresses. In: Protective chemical agents in the amelioration of plant abiotic stress. John Wiley & Sons, Ltd, pp 71–101

Bibi I, Nazar N, Ata S, Sultan M, Ali A, Abbas A, Jilani K, Kamal S, Sarim FM, Khan MI, Jalal F,

Iqbal M (2019) Green synthesis of iron oxide nanoparticles using pomegranate seeds extract and photocatalytic activity evaluation for the degradation of textile dye. J Market Res 8:6115–6124. https://doi.org/10.1016/j.jmrt.2019.10.006
Bin-Jumah M, AL-Abdan M, Albasher G, Alarifi S (2020) Effects of green silver nanoparticles on apoptosis and oxidative stress in normal and cancerous human hepatic cells in vitro. Int J Nanomedicine 15:1537–1548. https://doi.org/10.2147/IJN.S239861
Bisht S, Sharma V, Kumari N (2022) Biosynthesized magnetite nanoparticles from Polyalthia longifolia leaves improve photosynthetic performance and yield of Trigonella foenum-graecum under drought stress. Plant Stress 5:100090. https://doi.org/10.1016/j.stress.2022.100090
Boisseau P, Loubaton B (2011) Nanomedicine, nanotechnology in medicine. C R Phys 12:620–636. https://doi.org/10.1016/j.crhy.2011.06.001
Castro-Longoria E, Vilchis-Nestor AR, Avalos-Borja M (2011) Biosynthesis of silver, gold and bimetallic nanoparticles using the filamentous fungus Neurospora crassa. Colloids Surf, B 83:42–48. https://doi.org/10.1016/j.colsurfb.2010.10.035
Chakraborty S, Singh A, Roychoudhury A (2022) Biogenic nanoparticles and generation of abiotic stress-resilient plants: a new approach for sustainable agriculture. Plant Stress 6:100117. https://doi.org/10.1016/j.stress.2022.100117
Chandel SS, Shree B, Mondal K, Mali S (2022) Mechanism of nanoparticles-mediated alleviating biotic and abiotic stresses in agricultural crops: recent advances and future perspectives. In: Rajput VD, Verma KK, Sharma N, Minkina T (eds) The role of nanoparticles in plant nutrition under soil pollution: nanoscience in nutrient use efficiency. Springer International Publishing, Cham, pp 117–139
Chandrika KP, Singh A, Tumma MK, Yadav P (2018) Nanotechnology prospects and constraints in agriculture. In: Dasgupta N, Ranjan S, Lichtfouse E (eds) Environmental nanotechnology, vol 1. Springer International Publishing, Cham, pp 159–186
Chatterjee A, Mridha D, Banerjee J, Chanda S, Ray K, Acharya K, Das M, Roychowdhury T, Sarkar J (2021) Green synthesis of iron oxide nanoparticles and their ameliorative effect on arsenic stress relief in Oryza sativa seedlings. Biocatal Agric Biotechnol 38:102207. https://doi.org/10.1016/j.bcab.2021.102207
Chen M, Xu N, Cao X, Zhou K, Chen Z, Wang Y, Liu C (2015) Facilitated transport of anatase titanium dioxides nanoparticles in the presence of phosphate in saturated sands. J Colloid Interface Sci 451:134–143. https://doi.org/10.1016/j.jcis.2015.04.010
Chokkalingam M, Singh P, Huo Y, Soshnikova V, Ahn S, Kang J, Mathiyalagan R, Kim YJ, Yang DC (2019) Facile synthesis of Au and Ag nanoparticles using fruit extract of Lycium chinense and their anticancer activity. J Drug Deliv Sci Technol 49:308–315. https://doi.org/10.1016/j.jddst.2018.11.025
Cramer GR, Urano K, Delrot S, Pezzotti M, Shinozaki K (2011) Effects of abiotic stress on plants: a systems biology perspective. BMC Plant Biol 11:163. https://doi.org/10.1186/1471-2229-11-163
Cuevas R, Durán N, Diez MC, Tortella GR, Rubilar O (2015) Extracellular biosynthesis of copper and copper oxide nanoparticles by Stereum hirsutum, a native white-rot fungus from Chilean forests. J Nanomaterials 16(57):57. https://doi.org/10.1155/2015/789089
Danbature WL, Shehu Z, Yoro M, Adam MM (2020) Nanolarvicidal Effect of green synthesized Ag-Co Bimetallic nanoparticles on Culex quinquefasciatus mosquito. Adv Biol Chem 10:16. https://doi.org/10.4236/abc.2020.101002
Dikshit PK, Kumar J, Das AK, Sadhu S, Sharma S, Singh S, Gupta PK, Kim BS (2021) Green synthesis of metallic nanoparticles: applications and limitations. Catalysts 11:902. https://doi.org/10.3390/catal11080902
Ding Z, Zhao F, Zhu Z, Ali EF, Shaheen SM, Rinklebe J, Eissa MA (2022) Green nanosilica enhanced the salt-tolerance defenses and yield of Williams banana: a field trial for using saline water in low fertile arid soil. Environ Exp Bot 197:104843. https://doi.org/10.1016/j.envexpbot.2022.104843
Djanaguiraman M, Belliraj N, Bossmann SH, Prasad PVV (2018) High-temperature stress alleviation by selenium nanoparticle treatment in grain sorghum. ACS Omega 3:2479–2491. https://doi.org/10.1021/acsomega.7b01934
Elango G, Roopan SM (2015) Green synthesis, spectroscopic investigation and photocatalytic activity of lead nanoparticles. Spectrochim Acta Part A Mol Biomol Spectrosc 139:367–373. https://doi.org/10.1016/j.saa.2014.12.066
El-Batal AI, Mona SS, Al-Tamie MSS (2015) Biosynthesis of gold nanoparticles using marine Streptomyces cyaneus and their antimicrobial, antioxidant and antitumor (in vitro) activities. J Chem Pharm Res 7:1020–1036
El-Saadony MT, Desoky E-SM, Saad AM, Eid RSM, Selem E, Elrys AS (2021) Biological silicon nanoparticles improve Phaseolus vulgaris L. yield and minimize its contaminant contents on a heavy metals-contaminated saline soil. J Environ Sci 106:1–14. https://doi.org/10.1016/j.jes.2021.01.012
El-Saber M (2021) Effect of biosynthesized Zn and Se nanoparticles on the productivity and active constituents of garlic subjected to saline stress. Egypt J Desert Res 71:99–128. https://doi.org/10.21608/ejdr.2021.81793.1085
El-Sherbiny IM, Salih E (2018) Green synthesis of metallic nanoparticles using biopolymers and plant extracts. In: Green metal nanoparticles. John Wiley & Sons, Ltd, pp 293–319
El-Zohri M, Al-Wadaani NA, Bafeel SO (2021) Foliar sprayed green zinc oxide nanoparticles mitigate drought-induced oxidative stress in tomato. Plants 10:2400. https://doi.org/10.3390/plants10112400

Erktan A, McCormack ML, Roumet C (2018) Frontiers in root ecology: recent advances and future challenges. Plant Soil 424:1–9. https://doi.org/10.1007/s11104-018-3618-5

Fariq A, Khan T, Yasmin A (2017) Microbial synthesis of nanoparticles and their potential applications in biomedicine. J Appl Biomed 15:241–248. https://doi.org/10.1016/j.jab.2017.03.004

Farooq M, Hussain M, Wahid A, Siddique KHM (2012) Drought Stress in plants: an overview. In: Aroca R (ed) Plant responses to drought stress: from morphological to molecular features. Springer, Berlin, Heidelberg, pp 1–33

Fatima R, Priya M, Indurthi L, Radhakrishnan V, Sudhakaran R (2020) Biosynthesis of silver nanoparticles using red algae Portieria hornemannii and its antibacterial activity against fish pathogens. Microb Pathog 138:103780. https://doi.org/10.1016/j.micpath.2019.103780

Foyer CH, Noctor G, Verrier P (2006) Photosynthetic carbon–nitrogen interactions: modelling inter-pathway control and signalling. In: Annual plant reviews Volume 22: Control of primary metabolism in plants. John Wiley & Sons, Ltd, pp 325–347

Foyer CH, Noctor G (2016) Stress-triggered redox signalling: what's in pROSpect? Plant, Cell Environ 39:951–964. https://doi.org/10.1111/pce.12621

Ganesh Babu MM, Gunasekaran P (2009) Production and structural characterization of crystalline silver nanoparticles from Bacillus cereus isolate. Colloids Surf, B 74:191–195. https://doi.org/10.1016/j.colsurfb.2009.07.016

García-Ovando AE, Ramírez Piña JE, Esquivel Naranjo EU, Cervantes Chávez JA, Esquivel K (2022) Biosynthesized nanoparticles and implications by their use in crops: Effects over physiology, action mechanisms, plant stress responses and toxicity. Plant Stress 6:100109. https://doi.org/10.1016/j.stress.2022.100109

Geigenberger P, Fernie AR (2014) Metabolic control of redox and redox control of metabolism in plants. Antioxid Redox Signal 21:1389–1421. https://doi.org/10.1089/ars.2014.6018

Ghazi AA, El-Nahrawy S, El-Ramady H, Ling W (2022) Biosynthesis of nano-selenium and its impact on germination of wheat under salt stress for sustainable production. Sustainability 14:1784. https://doi.org/10.3390/su14031784

Ghori N-H, Ghori T, Hayat MQ, Imadi SR, Gul A, Altay V, Ozturk M (2019) Heavy metal stress and responses in plants. Int J Environ Sci Technol 16:1807–1828. https://doi.org/10.1007/s13762-019-02215-8

Gogoi N, Babu PJ, Mahanta C, Bora U (2015) Green synthesis and characterization of silver nanoparticles using alcoholic flower extract of Nyctanthes arbortristis and in vitro investigation of their antibacterial and cytotoxic activities. Mater Sci Eng, C 46:463–469. https://doi.org/10.1016/j.msec.2014.10.069

Golinska P, Wypij M, Ingle AP, Gupta I, Dahm H, Rai M (2014) Biogenic synthesis of metal nanoparticles from actinomycetes: biomedical applications and cytotoxicity. Appl Microbiol Biotechnol 98:8083–8097. https://doi.org/10.1007/s00253-014-5953-7

Govindaraju K, Basha SK, Kumar VG, Singaravelu G (2008) Silver, gold and bimetallic nanoparticles production using single-cell protein (Spirulina platensis) Geitler. J Mater Sci 43:5115–5122. https://doi.org/10.1007/s10853-008-2745-4

Habibi G, Aleyasin Y (2020) Green synthesis of Se nanoparticles and its effect on salt tolerance of barley plants. Int J Nano Dimension 11(2):145–157

Hamedi S, Shojaosadati SA, Mohammadi A (2017) Evaluation of the catalytic, antibacterial and anti-biofilm activities of the Convolvulus arvensis extract functionalized silver nanoparticles. J Photochem Photobiol, B 167:36–44. https://doi.org/10.1016/j.jphotobiol.2016.12.025

Hashem A, Abd_Allah EF, Alqarawi AA, Al-Huqail AA, Shah MA (2016) Induction of osmoregulation and modulation of salt stress in acacia gerrardii Benth. by arbuscular mycorrhizal fungi and bacillus subtilis (BERA 71). BioMed Res Int 2016:e6294098. https://doi.org/10.1155/2016/6294098

He Y, Li X, Zheng Y, Wang Z, Ma Z, Yang Q, Yao B, Zhao Y, Zhang H (2018) A green approach for synthesizing silver nanoparticles, and their antibacterial and cytotoxic activities. New J Chem 42:2882–2888. https://doi.org/10.1039/C7NJ04224H

Husseiny MI, El-Aziz MA, Badr Y, Mahmoud MA (2007) Biosynthesis of gold nanoparticles using Pseudomonas aeruginosa. Spectrochim Acta Part A Mol Biomol Spectrosc 67:1003–1006. https://doi.org/10.1016/j.saa.2006.09.028

Ibrahim RK, Hayyan M, AlSaadi MA, Hayyan A, Ibrahim S (2016) Environmental application of nanotechnology: air, soil, and water. Environ Sci Pollut Res 23:13754–13788. https://doi.org/10.1007/s11356-016-6457-z

Igamberdiev AU, Bykova NV (2018) Role of organic acids in the integration of cellular redox metabolism and mediation of redox signalling in photosynthetic tissues of higher plants. Free Radical Biol Med 122:74–85. https://doi.org/10.1016/j.freeradbiomed.2018.01.016

Ikram M, Raja NI, Javed B, Mashwani Z-R, Hussain M, Hussain M, Ehsan M, Rafique N, Malik K, Sultana T, Akram A (2020) Foliar applications of bio-fabricated selenium nanoparticles to improve the growth of wheat plants under drought stress. Green Process Synth 9:706–714. https://doi.org/10.1515/gps-2020-0067

Iqbal M, Raja NI, Mashwani Z-U-R, Hussain M, Ejaz M, Yasmeen F (2019) Effect of silver nanoparticles on growth of wheat under heat stress. Iran J Sci Technol Trans Sci 43:387–395. https://doi.org/10.1007/s40995-017-0417-4

Iqbal S, Waheed Z, Naseem A (2020) Nanotechnology and abiotic stresses. In: Javad S (ed) Nanoagronomy. Springer International Publishing, Cham, pp 37–52

Jacob JM, Ravindran R, Narayanan M, Samuel SM, Pugazhendhi A, Kumar G (2021) Microalgae: a prospective low cost green alternative for nanoparticle synthesis. Curr Opin Environ Sci Health 20:100163. https://doi.org/10.1016/j.coesh.2019.12.005

Jadoun S, Arif R, Jangid NK, Meena RK (2021) Green synthesis of nanoparticles using plant extracts: a review. Environ Chem Lett 19:355–374. https://doi.org/10.1007/s10311-020-01074-x

Jamkhande PG, Ghule NW, Bamer AH, Kalaskar MG (2019) Metal nanoparticles synthesis: an overview on methods of preparation, advantages and disadvantages, and applications. J Drug Deliv Sci Technol 53:101174. https://doi.org/10.1016/j.jddst.2019.101174

Jayabalan J, Mani G, Krishnan N, Pernabas J, Devadoss JM, Jang HT (2019) Green biogenic synthesis of zinc oxide nanoparticles using Pseudomonas putida culture and its In vitro antibacterial and anti-biofilm activity. Biocatal Agric Biotechnol 21:101327. https://doi.org/10.1016/j.bcab.2019.101327

Jayaseelan C, Rahuman AA, Kirthi AV, Marimuthu S, Santhoshkumar T, Bagavan A, Gaurav K, Karthik L, Rao KVB (2012) Novel microbial route to synthesize ZnO nanoparticles using Aeromonas hydrophila and their activity against pathogenic bacteria and fungi. Spectrochim Acta Part A Mol Biomol Spectrosc 90:78–84. https://doi.org/10.1016/j.saa.2012.01.006

Jo JH, Singh P, Kim YJ, Wang C, Mathiyalagan R, Jin C-G, Yang DC (2016) Pseudomonas deceptionensis DC5-mediated synthesis of extracellular silver nanoparticles. Artif Cells Nanomed Biotechnol 44:1576–1581. https://doi.org/10.3109/21691401.2015.1068792

Johnston CW, Wyatt MA, Li X, Ibrahim A, Shuster J, Southam G, Magarvey NA (2013) Gold biomineralization by a metallophore from a gold-associated microbe. Nat Chem Biol 9:241–243. https://doi.org/10.1038/nchembio.1179

Jorge de Souza TA, Rosa Souza LR, Franchi LP (2019) Silver nanoparticles: an integrated view of green synthesis methods, transformation in the environment, and toxicity. Ecotoxicol Environ Saf 171:691–700. https://doi.org/10.1016/j.ecoenv.2018.12.095

Judith Vijaya JJ, Jayaprakash N, Kombaiah K, Kaviyarasu K, Kennedy LJ, Ramalingam RJ, Al-Lohedan HA, Mansoor-Ali VM, Maaza M (2017) Bioreduction potentials of dried root of Zingiber officinale for a simple green synthesis of silver nanoparticles: antibacterial studies. J Photochem Photobiol B: Biology 177:62–68. https://doi.org/10.1016/j.jphotobiol.2017.10.007

Kalaiselvi A, Roopan SM, Madhumitha G, Ramalingam C, Elango G (2015) Synthesis and characterization of palladium nanoparticles using Catharanthus roseus leaf extract and its application in the photocatalytic degradation. Spectrochim Acta Part A Mol Biomol Spectrosc 135:116–119. https://doi.org/10.1016/j.saa.2014.07.010

Kannan RRR, Arumugam R, Ramya D, Manivannan K, Anantharaman P (2013) Green synthesis of silver nanoparticles using marine macroalga Chaetomorpha linum. Appl Nanosci 3:229–233. https://doi.org/10.1007/s13204-012-0125-5

Kannaujia R, Singh P, Prasad V, Pandey V (2022) Evaluating impacts of biogenic silver nanoparticles and ethylenediurea on wheat (Triticum aestivum L.) against ozone-induced damages. Environ Res 203:111857. https://doi.org/10.1016/j.envres.2021.111857

Karthik L, Kumar G, Kirthi AV, Rahuman AA, Bhaskara Rao KV (2014) Streptomyces sp. LK3 mediated synthesis of silver nanoparticles and its biomedical application. Bioprocess Biosyst Eng 37:261–267. https://doi.org/10.1007/s00449-013-0994-3

Kathiraven T, Sundaramanickam A, Shanmugam N, Balasubramanian T (2015) Green synthesis of silver nanoparticles using marine algae Caulerpa racemosa and their antibacterial activity against some human pathogens. Appl Nanosci 5:499–504. https://doi.org/10.1007/s13204-014-0341-2

Kathiresan K, Manivannan S, Nabeel MA, Dhivya B (2009) Studies on silver nanoparticles synthesized by a marine fungus, Penicillium fellutanum isolated from coastal mangrove sediment. Colloids Surf, B 71:133–137. https://doi.org/10.1016/j.colsurfb.2009.01.016

Keerthana P, Vijayakumar S, Vidhya E, Punitha VN, Nilavukkarasi M, Praseetha PK (2021) Biogenesis of ZnO nanoparticles for revolutionizing agriculture: a step towards anti-infection and growth promotion in plants. Ind Crops Prod 170:113762

Khalofah A, Kilany M, Migdadi H (2021) Phytostimulatory influence of comamonas testosteroni and silver nanoparticles on Linum usitatissimum L. under salinity stress. Plants 10:790. https://doi.org/10.3390/plants10040790

Khan Z, Upadhyaya H (2019) Chapter 15—Impact of nanoparticles on abiotic stress responses in plants: an overview. In: Tripathi DK, Ahmad P, Sharma S, Chauhan DK, Dubey NK (eds) Nanomaterials in plants, algae and microorganisms. Academic Press, Cambridge, pp 305–322

Khan MIR, Palakolanu SR, Chopra P, Rajurkar AB, Gupta R, Iqbal N, Maheshwari C (2021a) Improving drought tolerance in rice: ensuring food security through multi-dimensional approaches. Physiol Plant 172:645–668. https://doi.org/10.1111/ppl.13223

Khan MT, Ahmed S, Shah AA, Noor Shah A, Tanveer M, El-Sheikh MA, Siddiqui MH (2021b) Influence of Zinc oxide nanoparticles to regulate the antioxidants enzymes, some osmolytes and agronomic attributes in Coriandrum sativum L. grown under water stress. Agronomy 11:2004. https://doi.org/10.3390/agronomy11102004

Kirupagaran R, Saritha A, Bhuvaneswari S (2016) Green synthesis of selenium nanoparticles from leaf and stem extract of Leucas lavandulifolia Sm. and their application. J Nanosci Technol, pp 224–226

Kobashigawa JM, Robles CA, Martínez Ricci ML, Carmarán CC (2019) Influence of strong bases on the synthesis of silver nanoparticles (AgNPs) using the ligninolytic fungi Trametes trogii. Saudi J Biol Sci 26:1331–1337. https://doi.org/10.1016/j.sjbs.2018.09.006

Kohatsu MY, Lange CN, Pelegrino MT, Pieretti JC, Tortella G, Rubilar O, Batista BL, Seabra AB, de Jesus TA (2021) Foliar spraying of biogenic CuO nanoparticles protects the defence system and photosynthetic pigments of lettuce (Lactuca sativa). J Clean Prod 324:129264. https://doi.org/10.1016/j.jclepro.2021.129264

Kora AJ, Beedu SR, Jayaraman A (2012) Size-controlled green synthesis of silver nanoparticles mediated by gum ghatti (Anogeissus latifolia) and its biological activity. Org Med Chem Lett 2:17. https://doi.org/10.1186/2191-2858-2-17

Kowshik M, Ashtaputre S, Kharrazi S, Vogel W, Urban J, Kulkarni SK, Paknikar KM (2002a) Extracellular synthesis of silver nanoparticles by a silver-tolerant yeast strain MKY3. Nanotechnology 14:95. https://doi.org/10.1088/0957-4484/14/1/321

Kowshik M, Deshmukh N, Vogel W, Urban J, Kulkarni SK, Paknikar KM (2002b) Microbial synthesis of semiconductor CdS nanoparticles, their characterization, and their use in the fabrication of an ideal diode. Biotechnol Bioeng 78:583–588. https://doi.org/10.1002/bit.10233

Krishnaraj C, Jagan EG, Rajasekar S, Selvakumar P, Kalaichelvan PT, Mohan N (2010) Synthesis of silver nanoparticles using Acalypha indica leaf extracts and its antibacterial activity against water borne pathogens. Colloids Surf, B 76:50–56. https://doi.org/10.1016/j.colsurfb.2009.10.008

Kumar P, Senthamil Selvi S, Govindaraju M (2013) Seaweed-mediated biosynthesis of silver nanoparticles using Gracilaria corticata for its antifungal activity against Candida spp. Appl Nanosci 3:495–500. https://doi.org/10.1007/s13204-012-0151-3

Kumari S, Khanna RR, Nazir F, Albaqami M, Chhillar H, Wahid I, Khan MIR (2022) Bio-synthesized nanoparticles in developing plant abiotic stress resilience: a new boon for sustainable approach. Int J Mol Sci 23:4452. https://doi.org/10.3390/ijms23084452

Larkunthod P, Boonlakhorn J, Pansarakham P, Pongdontri P, Thongbai P, Theerakulpisut P, Larkunthod P, Boonlakhorn J, Pansarakham P, Pongdontri P, Thongbai P, Theerakulpisut P (2022) Synthesis and characterization of silica nanoparticles from rice husk and their effects on physiology of rice under salt stress. Chilean J Agric Res 82:412–425. https://doi.org/10.4067/S0718-58392022000300412

Li X, Xu H, Chen Z-S (2011) Chen G (2011) Biosynthesis of nanoparticles by microorganisms and their applications. J Nanomaterials 8(1–8):16. https://doi.org/10.1155/2011/270974

Lloyd JR, Yong P, Macaskie LE (1998) Enzymatic recovery of elemental palladium by using sulfate-reducing bacteria. Appl Environ Microbiol 64:4607–4609. https://doi.org/10.1128/AEM.64.11.4607-4609.1998

Mafakheri M, Kordrostami M, Al-Khayri JM (2021) Abiotic stress in plants: socio-economic consequences and crops responses. In: Al-Khayri JM, Ansari MI, Singh AK (eds) Nanobiotechnology: mitigation of abiotic stress in plants. Springer International Publishing, Cham, pp 1–28

Marchev AS, Yordanova ZP, Georgiev MI (2020) Green (cell) factories for advanced production of plant secondary metabolites. Crit Rev Biotechnol 40:443–458. https://doi.org/10.1080/07388551.2020.1731414

Mathur S, Singh D, Ranjan R (2022) Remediation of heavy metal(loid) contaminated soil through green nanotechnology. Front Sustain Food Syst 6:932424. https://doi.org/10.3389/fsufs.2022.932424

Minhas FT, Arslan G, Gubbuk IH, Akkoz C, Ozturk BY, Asıkkutlu B, Arslan U, Ersoz M (2018) Evaluation of antibacterial properties on polysulfone composite membranes using synthesized biogenic silver nanoparticles with Ulva compressa (L.) Kütz. and Cladophora glomerata (L.) Kütz. extracts. Int J Biol Macromol 107:157–165. https://doi.org/10.1016/j.ijbiomac.2017.08.149

Mishra A, Kumari M, Pandey S et al (2014) Biocatalytic and antimicrobial activities of gold nanoparticles synthesized by *Trichoderma* sp. Bioresour Technol 166:235–242. https://doi.org/10.1016/j.biortech.2014.04.085

Mogazy AM, Hanafy RS (2022) Foliar spray of biosynthesized zinc oxide nanoparticles alleviate salinity stress effect on Vicia faba plants. J Soil Sci Plant Nutr 22:2647–2662. https://doi.org/10.1007/s42729-022-00833-9

Momeni S, Nabipour I (2015) A simple green synthesis of palladium nanoparticles with Sargassum alga and their electrocatalytic activities towards hydrogen peroxide. Appl Biochem Biotechnol 176:1937–1949. https://doi.org/10.1007/s12010-015-1690-3

Mustafa N, Raja NI, Ilyas N, Ikram M, Mashwani Z-R, Ehsan M (2021) Foliar applications of plant-based titanium dioxide nanoparticles to improve agronomic and physiological attributes of wheat (Triticum aestivum L.) plants under salinity stress. Green Process Synth 10:246–257. https://doi.org/10.1515/gps-2021-0025

Naghdi S, Sajjadi M, Nasrollahzadeh M, Rhee KY, Sajadi SM, Jaleh B (2018) Cuscuta reflexa leaf extract mediated green synthesis of the Cu nanoparticles on graphene oxide/manganese dioxide nanocomposite and its catalytic activity toward reduction of nitroarenes and organic dyes. J Taiwan Inst Chem Eng 86:158–173. https://doi.org/10.1016/j.jtice.2017.12.017

Naikoo GA, Mustaqeem M, Hassan IU, Awan T, Arshad F, Salim H, Qurashi A (2021) Bioinspired and green synthesis of nanoparticles from plant extracts with antiviral and antimicrobial properties: A critical review. J Saudi Chem Soc 25:101304. https://doi.org/10.1016/j.jscs.2021.101304

Naimi-Shamel N, Pourali P, Dolatabadi S (2019) Green synthesis of gold nanoparticles using Fusarium oxysporum and antibacterial activity of its tetracycline conjugant. J De Mycologie Médicale 29:7–13. https://doi.org/10.1016/j.mycmed.2019.01.005

Nair B, Pradeep T (2002) Coalescence of nanoclusters and formation of submicron crystallites assisted by lactobacillus strains. Cryst Growth Des 2:293–298. https://doi.org/10.1021/cg0255164

Namvar F, Azizi S, Ahmad MB, Shameli K, Mohamad R, Mahdavi M, Tahir PM (2015) Green synthesis and characterization of gold nanoparticles using the marine macroalgae Sargassum muticum. Res Chem Intermed 41:5723–5730. https://doi.org/10.1007/s11164-014-1696-4

Naseem T, Farrukh MA (2015) Antibacterial activity of green synthesis of iron nanoparticles using lawsonia inermis and gardenia jasminoides leaves extract. J Chem 2015:e912342. https://doi.org/10.1155/2015/912342

Nasrollahzadeh M, Mohammad Sajadi S (2015) Green synthesis of copper nanoparticles using Ginkgo biloba L. leaf extract and their catalytic activity for the Huisgen [3+2] cycloaddition of azides and alkynes at room temperature. J Colloid Interface Sci 457:141–147. https://doi.org/10.1016/j.jcis.2015.07.004

Nasseri MA, Shahabi M, Allahresani A, Kazemnejadi M (2019) Eco-friendly biosynthesis of silver nanoparticles using aqueous solution of Spartium junceum flower extract. Asian J Green Chem 3:382–390. https://doi.org/10.22034/ajgc.2018.144365.1099

Nath D, Banerjee P (2013) Green nanotechnology—a new hope for medical biology. Environ Toxicol Pharmacol 36:997–1014. https://doi.org/10.1016/j.etap.2013.09.002

Nguyen VK, Choi W, Ha Y, Gu Y, Lee C, Park J, Jang G, Shin C, Cho S (2019) Microbial tellurite reduction and production of elemental tellurium nanoparticles by novel bacteria isolated from wastewater. J Ind Eng Chem 78:246–256. https://doi.org/10.1016/j.jiec.2019.06.006

Noman M, Ahmed T, Hussain S, Niazi MBK, Shahid M, Song F (2020a) Biogenic copper nanoparticles synthesized by using a copper-resistant strain Shigella flexneri SNT22 reduced the translocation of cadmium from soil to wheat plants. J Hazard Mater 398:123175. https://doi.org/10.1016/j.jhazmat.2020.123175

Noman M, Shahid M, Ahmed T, Tahir M, Naqqash T, Muhammad S, Song F, Abid HMA, Aslam Z (2020b) Green copper nanoparticles from a native Klebsiella pneumoniae strain alleviated oxidative stress impairment of wheat plants by reducing the chromium bioavailability and increasing the growth. Ecotoxicol Environ Saf 192:110303. https://doi.org/10.1016/j.ecoenv.2020.110303

Noman M, Ahmed T, Shahid M, Niazi MBK, Qasim M, Kouadri F, Abdulmajeed AM, Alghanem SM, Ahmad N, Zafar M, Ali S (2021) Biogenic copper nanoparticles produced by using the Klebsiella pneumoniae strain NST2 curtailed salt stress effects in maize by modulating the cellular oxidative repair mechanisms. Ecotoxicol Environ Saf 217:112264. https://doi.org/10.1016/j.ecoenv.2021.112264

Noor R, Yasmin H, Ilyas N, Nosheen A, Hassan MN, Mumtaz S, Khan N, Ahmad A, Ahmad P (2022) Comparative analysis of iron oxide nanoparticles synthesized from ginger (Zingiber officinale) and cumin seeds (Cuminum cyminum) to induce resistance in wheat against drought stress. Chemosphere 292:133201. https://doi.org/10.1016/j.chemosphere.2021.133201

Pantidos N (2014) Biological synthesis of metallic nanoparticles by bacteria, fungi and plants. J Nanomed Nanotechnol 5:1. https://doi.org/10.4172/2157-7439.1000233

Pourali P, Yahyaei B (2016) Biological production of silver nanoparticles by soil isolated bacteria and preliminary study of their cytotoxicity and cutaneous wound healing efficiency in rat. J Trace Elem Med Biol 34:22–31. https://doi.org/10.1016/j.jtemb.2015.11.004

Prasad K, Jha AK, Kulkarni AR (2007) Lactobacillus assisted synthesis of titanium nanoparticles. Nanoscale Res Lett 2:248–250. https://doi.org/10.1007/s11671-007-9060-x

Qadir M, Quillérou E, Nangia V, Murtaza G, Singh M, Thomas RJ, Drechsel P, Noble AD (2014) Economics of salt-induced land degradation and restoration. Natural Resour Forum 38:282–295. https://doi.org/10.1111/1477-8947.12054

Rabeea MA, Owaid MN, Aziz AA, Jameel MS, Dheyab MA (2020) Mycosynthesis of gold nanoparticles using the extract of Flammulina velutipes, Physalacriaceae, and their efficacy for decolorization of methylene blue. J Environ Chem Eng 8:103841. https://doi.org/10.1016/j.jece.2020.103841

Ragab SM, Turoop L, Runo S, Nyanjom S (2022) Nanoparticle Treatments based on zinc oxide and moringa oleifera Leaf extracts alleviate salinity stress in Faba Bean (Vicia faba L.). J Agric Chem Environ 11:42–65. https://doi.org/10.4236/jacen.2022.111004

Rakgotho T, Ndou N, Mulaudzi T, Iwuoha E, Mayedwa N, Ajayi RF (2022) Green-synthesized zinc oxide nanoparticles mitigate salt stress in sorghum bicolor. Agriculture 12:597. https://doi.org/10.3390/agriculture12050597

Razavi M, Salahinejad E, Fahmy M et al (2015) Green chemical and biological synthesis of nanoparticles and their biomedical applications. In: Basiuk VA, Basiuk EV (eds) Green processes for nanotechnology: from inorganic to bioinspired nanomaterials. Springer International Publishing, Cham, pp 207–235. https://doi.org/10.1007/978-3-319-15461-9_7

Rehman K, Fatima F, Waheed I, Akash MSH (2018) Prevalence of exposure of heavy metals and their impact on health consequences. J Cell Biochem 119:157–184. https://doi.org/10.1002/jcb.26234

Robert KW, Parris TM, Leiserowitz AA (2005) What is sustainable development? Goals, indicators, values, and practice. Environ Sci Policy Sustain Dev 47:8–21. https://doi.org/10.1080/00139157.2005.10524444

Rónavári A, Igaz N, Gopisetty MK, Szerencsés B, Kovács D, Papp C, Vágvölgyi C, Boros IM, Kónya Z, Kiricsi M, Pfeiffer I (2018) Biosynthesized silver and gold nanoparticles are potent antimycotics against opportunistic pathogenic yeasts and dermatophytes. Int J Nanomedicine 13:695–703. https://doi.org/10.2147/IJN.S152010

Rónavári A, Igaz N, Adamecz DI, Szerencsés B, Molnar C, Kónya Z, Pfeiffer I, Kiricsi M (2021) Green Silver and gold nanoparticles: biological synthesis approaches and potentials for biomedical applications. Molecules 26:844. https://doi.org/10.3390/molecules26040844

Roopan SM, Bharathi A, Prabhakarn A, Abdul Rahuman A, Velayutham K, Rajakumar G, Padmaja RD, Lekshmi M, Madhumitha G (2012) Efficient phytosynthesis and structural characterization of rutile TiO_2 nanoparticles using Annona squamosa peel extract. Spectrochim Acta Part A Mol Biomol Spectrosc 98:86–90. https://doi.org/10.1016/j.saa.2012.08.055

Roy A, Bulut O, Some S, Kumar Mandal A, Deniz Yilmaz M (2019) Green synthesis of silver nanoparticles: biomolecule-nanoparticle organizations targeting antimicrobial activity. RSC Adv 9:2673–2702. https://doi.org/10.1039/C8RA08982E

Saha B, Borovskii G, Panda SK (2016) Alternative oxidase and plant stress tolerance. Plant Signal Behav 11:e1256530. https://doi.org/10.1080/15592324.2016.1256530

Saha J, Brauer EK, Sengupta A, Popescu SC, Gupta K, Gupta B (2015) Polyamines as redox homeostasis regulators during salt stress in plants. Front Environ Sci 3

Salunke BK, Sawant SS, Lee SI, Kim BS (2015) Comparative study of MnO_2 nanoparticle synthesis by marine bacterium *Saccharophagus degradans* and yeast *Saccharomyces cerevisiae*. Appl Microbiol Biotechnol 99:5419–5427. https://doi.org/10.1007/s00253-015-6559-4

Samadi N, Golkaran D, Eslamifar A, Jamalifar H, Fazeli MR, Mohseni FA (2009) Intra/extracellular biosynthesis of silver nanoparticles by an autochthonous strain of proteus mirabilis isolated fromphotographic waste. J Biomed Nanotech 5:247–253. https://doi.org/10.1166/jbn.2009.1029

Sandhya J, Kalaiselvam S (2020) Biogenic synthesis of magnetic iron oxide nanoparticles using inedible borassus flabellifer seed coat: characterization, antimicrobial, antioxidant activity and in vitro cytotoxicity analysis. Mater Res Express 7:015045. https://doi.org/10.1088/2053-1591/ab6642

Saratale RG, Karuppusamy I, Saratale GD, Pugazhendhi A, Kumar G, Park Y, Ghodake GS, Bharagava RN, Banu JR, Shin HS (2018) A comprehensive review on green nanomaterials using biological systems: recent perception and their future applications. Colloids Surf, B 170:20–35. https://doi.org/10.1016/j.colsurfb.2018.05.045

Saravanakumar K, Shanmugam S, Varukattu NB, MubarakAli D, Kathiresan K, Wang M-H (2019) Biosynthesis and characterization of copper oxide nanoparticles from indigenous fungi and its effect of photothermolysis on human lung carcinoma. J Photochem Photobiol, B 190:103–109. https://doi.org/10.1016/j.jphotobiol.2018.11.017

Saravanan M, Arokiyaraj S, Lakshmi T, Pugazhendhi A (2018) Synthesis of silver nanoparticles from Phenerochaete chrysosporium (MTCC-787) and their antibacterial activity against human pathogenic bacteria. Microb Pathog 117:68–72. https://doi.org/10.1016/j.micpath.2018.02.008

Saravanan A, Kumar PS, Karishma S, Vo D-VN, Jeevanantham S, Yaashikaa PR, George CS (2021) A review on biosynthesis of metal nanoparticles and its environmental applications. Chemosphere 264:128580. https://doi.org/10.1016/j.chemosphere.2020.128580

Sarkar RD, Kalita MC (2022a) Green synthesized Se nanoparticle-mediated alleviation of salt stress in field mustard, TS-36 variety. J Biotechnol 359:95–107. https://doi.org/10.1016/j.jbiotec.2022.09.013

Sarkar RD, Kalita MC (2022b) Se nanoparticles stabilized with Allamanda cathartica L. flower extract inhibited phytopathogens and promoted mustard growth under salt stress. Heliyon 8:e09076. https://doi.org/10.1016/j.heliyon.2022b.e09076

Schlüter M, Hentzel T, Suarez C, Koch M, Lorenz WG, Böhm L, Düring R-A, Koinig KA, Bunge M (2014) Synthesis of novel palladium(0) nanocatalysts by microorganisms from heavy-metal-influenced high-alpine sites for dehalogenation of polychlorinated dioxins. Chemosphere 117:462–470. https://doi.org/10.1016/j.chemosphere.2014.07.030

Seetharaman PK, Chandrasekaran R, Gnanasekar S, Chandrakasan G, Gupta M, Manikandan DB, Sivaperumal S (2018) Antimicrobial and larvicidal activity of eco-friendly silver nanoparticles synthesized from endophytic fungi Phomopsis liquidambaris. Biocatal Agric Biotechnol 16:22–30. https://doi.org/10.1016/j.bcab.2018.07.006

Seleiman MF, Al-Suhaibani N, Ali N, Akmal M, Alotaibi M, Refay Y, Dindaroglu T, Abdul-Wajid HH, Battaglia ML (2021) Drought stress impacts on plants and different approaches to alleviate its adverse effects. Plants 10:259. https://doi.org/10.3390/plants10020259

Shaik AM, David Raju M, Rama Sekhara Reddy D (2020) Green synthesis of zinc oxide nanoparticles using aqueous root extract of Sphagneticola trilobata Lin and investigate its role in toxic metal removal, sowing germination and fostering of plant growth. Inorg Nano-Metal Chem 50:569–579. https://doi.org/10.1080/24701556.2020.1722694

Shen Y, Li J, Zhang S, Jiang X, Liang J, Li T, Guo R, Guan W, Yang L (2022) Cd stress alleviation in mung-bean seedlings with biogenic hydroxyapatite nanoparticles as ecofriendly remediation agents. Environ Sci: Nano 9:3844–3858. https://doi.org/10.1039/D2EN00536K

Shome S, Tewari S, Bhattacharya MK, Panda SK, Upadhyaya H (2022) Phytofunctionalized ZnO nanoparticles ameliorate water stress and its recovery in Oryza sativa L. Acta Physiol Plant 44:137. https://doi.org/10.1007/s11738-022-03477-5

Shukla AK, Iravani S (2018) Green synthesis, characterization and applications of nanoparticles. Elsevier

Sicard C, Brayner R, Margueritat J, Hémadi M, Couté A, Yéprémian C, Djediat C, Aubard J, Fiévet F, Livage J, Coradin T (2010) Nano-gold biosynthesis by silica-encapsulated micro-algae: a "living" bio-hybrid material. J Mater Chem 20:9342–9347. https://doi.org/10.1039/C0JM01735C

Singh BR, Dwivedi S, Al-Khedhairy AA, Musarrat J (2011) Synthesis of stable cadmium sulfide nanoparticles using surfactin produced by Bacillus amyloliquifaciens strain KSU-109. Colloids Surf, B 85:207–213. https://doi.org/10.1016/j.colsurfb.2011.02.030

Singh R, Misra V, Mudiam MKR, Chauhan LKS, Singh RP (2012) Degradation of γ-HCH spiked soil using stabilized Pd/Fe0 bimetallic nanoparticles: pathways, kinetics and effect of reaction conditions. J Hazard Mater 237–238:355–364. https://doi.org/10.1016/j.jhazmat.2012.08.064

Singh P, Kim YJ, Wang C, Mathiyalagan R, Yang DC (2016a) Weissella oryzae DC6-facilitated green synthesis of silver nanoparticles and their antimicrobial potential. Artif Cells Nanomed Biotechnol 44:1569–1575. https://doi.org/10.3109/21691401.2015.1064937

Singh P, Kim YJ, Wang C, Mathiyalagan R, Yang DC (2016b) Microbial synthesis of flower-shaped gold nanoparticles. Artif Cells Nanomed Biotechnol 44:1469–1474. https://doi.org/10.3109/21691401.2015.1041640

Singh P, Kim Y-J, Zhang D, Yang D-C (2016c) Biological synthesis of nanoparticles from plants and microorganisms. Trends Biotechnol 34:588–599. https://doi.org/10.1016/j.tibtech.2016.02.006

Singh A, Basnal N, Shukla G, Chaudhary N, Singh S, Gaurav SS (2022) Evaluation of efficacy of Phyto-synthesized iron oxide nanoparticles in contributing drought resilience in wheat (Triticum aestivum L.). Nanotechnology 33:485101. https://doi.org/10.1088/1361-6528/ac8c48

Sintubin L, De Windt W, Dick J, Mast J, van der Ha D, Verstraete W, Boon N (2009) Lactic acid bacteria as reducing and capping agent for the fast and efficient production of silver nanoparticles. Appl Microbiol Biotechnol 84:741–749. https://doi.org/10.1007/s00253-009-2032-6

Sneha K, Sathishkumar M, Mao J, Kwak IS, Yun Y-S (2010) Corynebacterium glutamicum-mediated crystallization of silver ions through sorption and reduction processes. Chem Eng J 162:989–996. https://doi.org/10.1016/j.cej.2010.07.006

Sreelakshmi B, Induja S, Adarsh PP, Rahul HL, Arya SM, Aswana S, Haripriya R, Aswathy BR, Manoj PK, Vishnudasan D (2021) Drought stress amelioration in plants using green synthesised iron oxide nanoparticles. Mater Today Proc 41:723–727. https://doi.org/10.1016/j.matpr.2020.05.801

Suresh D, Shobharani RM, Nethravathi PC, Pavan Kumar MA, Nagabhushana H, Sharma SC (2015) Artocarpus gomezianus aided green synthesis of ZnO nanoparticles: Luminescence, photocatalytic and antioxidant properties. Spectrochim Acta Part A Mol Biomol Spectrosc 141:128–134. https://doi.org/10.1016/j.saa.2015.01.048

Sweeney RY, Mao C, Gao X, Burt JL, Belcher AM, Georgiou G, Iverson BL (2004) Bacterial Biosynthesis of cadmium sulfide nanocrystals. Chem Biol 11:1553–1559. https://doi.org/10.1016/j.chembiol.2004.08.022

Szymańska M, Schneider M, Chateau-Smith C, Nezelof S, Vulliez-Coady L (2017) Psychophysiological effects of oxytocin on parent–child interactions: a literature review on oxytocin and parent–child interactions. Psychiatry Clin Neurosci 71:690–705. https://doi.org/10.1111/pcn.12544

Tahir K, Nazir S, Li B, Khan AU, Khan ZUH, Ahmad A, Khan FU (2015) An efficient photo catalytic activity of green synthesized silver nanoparticles using Salvadora persica stem extract. Sep Purif Technol 150:316–324. https://doi.org/10.1016/j.seppur.2015.07.012

Tolaymat TM, El Badawy AM, Genaidy A, Scheckel KG, Luxton TP, Suidan M (2010) An evidence-based environmental perspective of manufactured silver nanoparticle in syntheses and applications: a systematic review and critical appraisal of peer-reviewed scientific papers. Sci Total Environ 408:999–1006. https://doi.org/10.1016/j.scitotenv.2009.11.003

Vanaja M, Rajeshkumar S, Paulkumar K, Gnanajobitha G, Malarkodi C, Annadurai G (2013) Phytosynthesis and characterization of silver nanoparticles using stem extract of Coleus aromaticus. Int J Mater Biomater Appl 3:1–4

Verma A, Bharadvaja N (2022) Plant-mediated synthesis and characterization of silver and copper oxide nanoparticles: antibacterial and heavy metal removal activity. J Clust Sci 33:1697–1712. https://doi.org/10.1007/s10876-021-02091-8

Waghmare SR, Mulla MN, Marathe SR, Sonawane KD (2015) Ecofriendly production of silver nanoparticles using Candida utilis and its mechanistic action against pathogenic microorganisms. Biotech 5:33–38. https://doi.org/10.1007/s13205-014-0196-y

Wahab A, Abdi G, Saleem MH, Ali B, Ullah S, Shah W, Mumtaz S, Yasin G, Muresan CC, Marc RA (2022) Plants' physio-biochemical and phyto-hormonal responses to alleviate the adverse effects of drought stress: a comprehensive review. Plants 11:1620. https://doi.org/10.3390/plants11131620

Wahid I, Rani P, Kumari S, Ahmad R, Hussain SJ, Alamri S, Tripathy N, Khan MIR (2022) Biosynthesized gold nanoparticles maintained nitrogen metabolism, nitric oxide synthesis, ions balance, and stabilizes the defense systems to improve salt stress tolerance in wheat. Chemosphere 287:132142. https://doi.org/10.1016/j.chemosphere.2021.132142

Wanarska E, Maliszewska I (2019) The possible mechanism of the formation of silver nanoparticles by Penicillium cyclopium. Bioorg Chem 93:102803. https://doi.org/10.1016/j.bioorg.2019.02.028

Wang C, Kim YJ, Singh P, Mathiyalagan R, Jin Y, Yang DC (2016) Green synthesis of silver nanoparticles by Bacillus methylotrophicus, and their antimicrobial activity. Artif Cells Nanomed Biotechnol 44:1127–1132. https://doi.org/10.3109/21691401.2015.1011805

Wang D, Markus J, Wang C, Kim Y-J, Mathiyalagan R, Aceituno VC, Ahn S, Yang DC (2017) Green synthesis of gold and silver nanoparticles using aqueous extract of Cibotium barometz root. Artif Cells Nanomed Biotechnol 45:1548–1555. https://doi.org/10.1080/21691401.2016.1260580

Wang L, Liu X, Lee D-J, Tay J-H, Zhang Y, Wan C-L, Chen X-F (2018) Recent advances on biosorption by aerobic granular sludge. J Hazard Mater 357:253–270. https://doi.org/10.1016/j.jhazmat.2018.06.010

Xie J, Lee JY, Wang DIC, Ting YP (2007) Identification of active biomolecules in the high-yield synthesis of single-crystalline gold nanoplates in algal solutions. Small 3:672–682. https://doi.org/10.1002/smll.200600612

Yang C, Choi CH, Lee CS, Yi H (2013) A facile synthesis–fabrication strategy for integration of catalytically active viral-palladium nanostructures into polymeric hydrogel microparticles *via* replica molding. ACS Nano 7(6):5032–5044. https://doi.org/10.1021/nn4005582

Yang Y, Wu Y, Ma L, Yang Z, Dong Q, Li Q, Ni X, Kudla J, Song C, Guo Y (2019) The Ca^{2+} Sensor SCaBP3/CBL7 Modulates plasma membrane H+-ATPase activity and promotes alkali tolerance in arabidopsis. Plant Cell 31:1367–1384. https://doi.org/10.1105/tpc.18.00568

Yang X, Lu M, Wang Y, Wang Y, Liu Z, Chen S (2021) Response mechanism of plants to drought stress. Horticulturae 7:50. https://doi.org/10.3390/horticulturae7030050

Yu G, Ma J, Jiang P, Li J, Gao J, Qiao S, Zhao Z (2019) The mechanism of plant resistance to heavy metal. IOP Conf Ser: Earth Environ Sci 310:052004. https://doi.org/10.1088/1755-1315/310/5/052004

Zafar S, Hasnain Z, Aslam N, Mumtaz S, Jaafar HZ, Wahab PEM, Qayum M, Ormenisan AN (2021) Impact of Zn nanoparticles synthesized via green and chemical approach on okra (Abelmoschus esculentus L.) growth under salt stress. Sustainability 13:3694. https://doi.org/10.3390/su13073694

Zahedi SM, Karimi M, Teixeira da Silva JA (2020) The use of nanotechnology to increase quality and yield of fruit crops. J Sci Food Agric 100:25–31. https://doi.org/10.1002/jsfa.10004

Zahir AA, Chauhan IS, Bagavan A, Kamaraj C, Elango G, Shankar J, Arjaria N, Roopan SM, Rahuman AA, Singh N (2015) Green synthesis of silver and titanium dioxide nanoparticles using euphorbia prostrata extract shows shift from apoptosis to G0/G1 arrest followed by necrotic cell death in leishmania donovani. Antimicrob Agents Chemother 59:4782–4799. https://doi.org/10.1128/AAC.00098-15

Zhang T, Dang M, Zhang W, Lin X (2020) Gold nanoparticles synthesized from Euphorbia fischeriana root by green route method alleviates the isoprenaline hydrochloride induced myocardial infarction in rats. J Photochem Photobiol, B 202:111705. https://doi.org/10.1016/j.jphotobiol.2019.111705

Zhou W, He W, Zhong S et al (2009) Biosynthesis and magnetic properties of mesoporous Fe_3O_4 composites. J Magn Magn Mater 321:1025–1028. https://doi.org/10.1016/j.jmmm.2008.10.007

Part II

Agro-ecology, People and GI-Science

12 Employment Potential of Sericulture for Underprivileged Section: Assessment of Value Chain Analysis in Bangladesh

Md. Kamruzzaman, Md. Abdullah Al Mamun, Jayanta Das, Kamruzzaman, and G. M. Monirul Alam

Abstract

Value chain development has a significant impact on employment generation and poverty reduction. The value chain analysis is essential for an industry's successful establishment. The chapter aims to perform a SWOT analysis of the value chain of the sericulture industry in Bangladesh. We conducted 48 in-depth interviews from two districts for this study. Results reveal that mulberry leaves, silkworm eggs, cocoons, raw silk, silk yarn, and fabric are the main products in the value chain in the study area. Supply of agricultural goods, producing cocoons, reeling, weaving, dying, and printing are the major input functions in the value chain. Selling cocoons, raw silk, and silk yarn at the distribution functions plays a vital role in the value chain. Mulberry farmers get about 30% of profits from one *Bigha* farm each year. Due to low prices, farmers are not interested in cocoon production. Unfavorable climatic conditions are another drawback behind the decreasing production of cocoons. Farmers get only about 25% of profits using traditional reeling, which is much lower than the modern method. As a result, the local weaving industry depends on imported silk yarn, contributing to declining local product demand.

Keywords

Sericulture · Value chain · Silkworm production · Problems · Bangladesh

12.1 Introduction

The development of a country depends on its participation in the process of industrialization, which largely depends on global supply chains (Gereffi and Lee 2012). Chain refers to a vertical relationship between producers and buyers. It also indicates a movement of products to the consumers (Meaton et al. 2013). The efficient and timely distribution of products across the

Md. Kamruzzaman (✉) · Md. A. Al Mamun
Institute of Bangladesh Studies, University of Rajshahi, Rajshahi, Bangladesh
e-mail: mkzaman@ru.ac.bd

Md.A. Al Mamun
e-mail: shuzon.mamun@gmail.com

J. Das
Department of Geography, Rampurhat College, Rampurhat, Birbhum 731224, India
e-mail: jayanta.das@gmail.com

Kamruzzaman
Social Science Research Council, Ministry of Planning, Government Republic of Bangladesh, Dhaka, Bangladesh
e-mail: ssrc.planningdivision@yahoo.com

G. M.M. Alam
Department of Agribusiness, Bangabandhu Sheikh Mujibur Rahman Agricultural University, Gazipur, Bangladesh
e-mail: gmalam@bsmrau.edu.bd

J. Das and S. Halder (eds.), *Advancement of GI-Science and Sustainable Agriculture*, GIScience and Geo-environmental Modelling, https://doi.org/10.1007/978-3-031-36825-7_12

supply chain is the key issue in international trade (Gereffi and Lee 2012). The value chain transforms a conception into a product or service and delivers it to final consumers through different stages (Kaplinsky and Morris 2001). The process leads to the final utilization of a product or service (Sturgeon 2001; Ponte et al. 2019).

One can understand the potential impact of value chain development on poverty reduction using value chain analysis which includes governance, coordination, policies, and operations to create links and reliance between actors in the chain. It also specifies the existence and evolution of the relationship between the actors (Rosales et al. 2017). The global value chain focuses on creating values for goods and services throughout economic activities (Gereffi 2011, 2013). It upgrades the distribution channel by establishing the structural connectivity between the input and output processes among farms and countries (Lee and Gereffi 2015; Akram 2015). It also helps to understand the creation, capture, and sustenance of values in different industries (Gereffi and Lee 2016).

Sericulture is the mixture of growing plants and rearing insects that produce silk (Ruiz et al. 2020). Tesfa et al. (2014) and Sime and Siraj (2020) have studied the status, opportunities, and constraints of the production systems and marketing channels of sericulture in Ethiopia. Similarly, Jantakat and Tangjaturasopon (2012) have pointed out the barriers (e.g., weak sales and marketing systems with insufficient data) in the value chain process for silk production in the Nakhonchaiburin zone, Thailand. The chapter has also emphasized silk production and marketing. Angadi et al. (2013) have drawn attention to expanding the new applications of the waste of mulberry plants, which may bring additional income and establish the sericulture industry as more cost-effective. Shukla (2012) has addressed the cost of farming mulberry plants and the profits from the cultivation and silkworm rearing in the Udaipur district in Rajasthan, India. He has focused on human labor in the expensive sericulture industry. Eswarappa (2011) has discussed the role of Community-Based Organizations (CBOs) in developing the sericulture sector in Andhra Pradesh, India. Sujatha et al. (2015) have investigated the impact of socioeconomic factors on the adoption of new technologies in the sericulture industry in Andhra Pradesh, India, such as education, experience, cocoon yield, and cocoon price. They have found a constructive impact on decisions of farmer regarding the issue. Porrasa et al. (2017) have studied the Hilsa value chain that provides data and indications on developments and profit-making of Hilsa fish in Bangladesh. Hossain et al. (2019) have investigated the market dynamics of sunflowers in coastal Bangladesh. They have identified the shortage of quality seeds of sunflowers and their price fluctuation. The whole value chain of sunflowers, which includes collectors, processors, wholesalers, and retailers, gets more profit than farmers. Uddin et al. (2018) have assessed the value chain of fishes like Pangas and Tilapia and analyzed the governance of the market players in Bangladesh. They have indicated the negative scenarios of fishes like Pangas and Tilapia production due to poor market infrastructure and increasing fodder costs.

However, none of the studies looked at the strength, weaknesses, opportunities, and threats for further development of the value chain of the sericulture industry in Bangladesh. The ultimate aim of this chapter is to figure out the complete value chain of sericulture in Bangladesh and to intend to generate employment opportunities. Furthermore, the study aimed at analyzing the value chain to (i) understand the main features of value chains of sericulture; (ii) identify and measure the weaknesses and threats along the value chain; (iii) identify existing profitability in the sericulture industry; and (iv) suggest the strengths and opportunities of this sector.

12.2 Methods

12.2.1 Study Area Selection

Sericulture is widely produced in Bagha, Charghat, Paba, and Godagari Upazilas of Rajshahi district, Bholahat, Shibganj, Nachol, and Gomostapur of Chapainawabganj district, and some selected areas of Thakurgaon and Panchagorh districts in Bangladesh. To achieve the predetermined objectives, we have purposively selected sericulture production areas: Bagha and Charghat Upazilas from Rajshahi district and Bholahat Upazila from Chapainawabganj district because sericulture production is much higher in these areas than that of other areas of Bangladesh. We have considered Bholahat Upazila of Chapainawabganj district as the most important production and trading area of sericulture for our analytical research on the value chain of sericulture (Figs. 12.1 and 12.2).

Fig. 12.1 Sericulture cultivated areas in Bangladesh

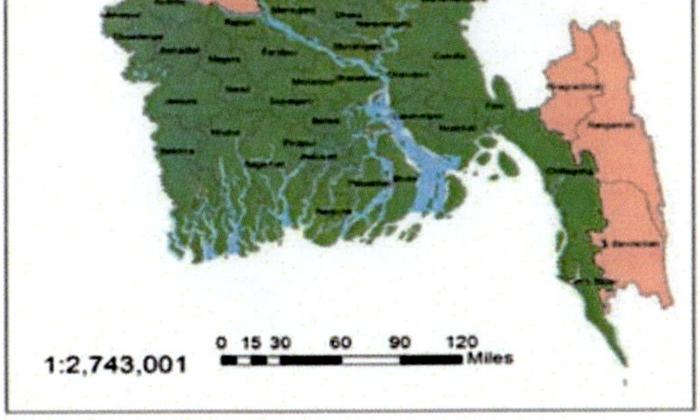

Fig. 12.2 Map showing the study areas

12.2.2 Sample Size Selection

The sample size for the in-depth interview was 48. Among the selected samples, 36 were from experienced farmers, two from modern reelers (who have dealt with reeling industry), three from traditional reelers from Bholahat Upazila, two from the weaving industry, two from traditional weavers, and three officers from Bangladesh Sericulture Research and Training Institute (BSRTI) (one from each wing namely cocoon/silkworm, reeling and farming research wings, respectively). Key informants were selected from farmers, agriculture extension officials, researchers, owners of the reeling and weaving industry (both traditional and modern), and traders. The total number of key informants was 18; among them, six were from farmers, four from agriculture extension officials (local and head office), two researchers from BSRTI, two from reeling industries (traditional and modern), two from weaving industries (traditional and modern), and two from traders.

12.2.3 Data Sources and Collection Tools

We collected secondary data from different sources such as the Bangladesh Sericulture Development Board (BSDB) and Bangladesh Sericulture Research and Training Institute, different research articles published in different journals, annual reports of BSDB, and privately owned silk industries and websites. A literature survey was conducted using document and content analysis for secondary data collection. We collected primary data from the respondents of the study area through in-depth interviews, FGD, and Key-Informant Interviews (KII). The face-to-face interviews were conducted with structured and semi-structured questionnaires to collect the required data. For cost margin analysis, relevant information regarding fixed cost, variable cost, revenue, etc., was collected through a structured questionnaire. We also collected comprehensive data from key informants using face-to-face interviews. In addition, we conducted three focus group discussions, two in two Upazilas consisting of farmers and sericulture extension officials and one at Rajshahi City consisting of officials from BSRTI and BSDB, modern and traditional reelers and weavers.

12.2.4 Data Analyzing Tools

We used descriptive statistics to analyze the collected quantitative data using Statistical Package for Social Sciences (IBM SPSS) version 22.0 software. The qualitative data were interpreted using an inductive reasoning process based on research objectives and findings of the quantitative data. Nguyen and Eiligman (2010) used an analytical framework that illustrates the functions and links between each process of value chain.

(A) Analytical framework of the value chain

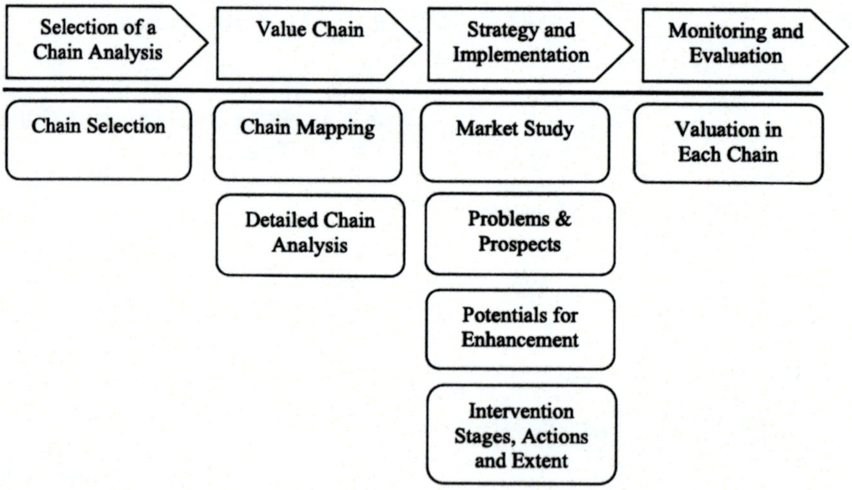

(B) Analytical framework for valuation of sericulture chain

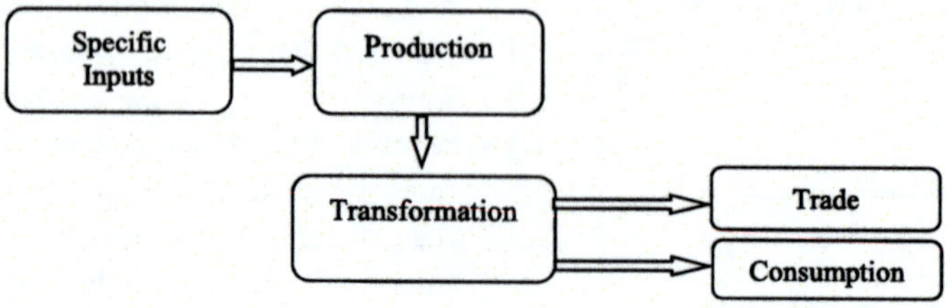

We used the above analytical framework to develop functional chain mapping, identifying categories of actors in value chains and their relations, constraints of each stage of value chains as a whole, and providing some preliminary suggestions for further promotion of sericulture in Bangladesh. Furthermore, we used a contingent valuation method for valuation measurement to explore the economic prospects of sericulture.

12.3 Results and Discussion

12.3.1 Main Products in the Sericulture Value Chain

A value chain identifies customers' demands and adds value to the products or services of industries (Jantakat and Tangjaturasopon 2012). A long value chain and structure are the sericulture industry's characteristics that include several products with economic value (Buhroo et al. 2018). It starts from silk reeling and spinning in the agriculture sector and ends up in the textile and apparel companies. The supply chain creates value for various products (e.g., cocoons, silk yarn, silk textiles, and apparel) of the sericulture industry (Hui 2010).

Mulberry Leaves

Mulberry leaves are the only source to rear silkworms (Rohela et al. 2020). A mulberry orchard usually has a 15–20 years life cycle of optimal production. Mulberry leaves can be harvested six months after plantation. It depends on the variety of mulberry species and weather. At maturity, one *bigha* land produces around 2000–2500 kg of leaves at the field level (BSRTI 2014).

Silkworm Egg

The fluid state of proteins in a worm is known as silkworm (Saikia 2011). The first stage of silk production is the laying of silkworm eggs in a controlled environment with appropriate instruments, such as an aluminum box. After that, it is ensured that they are free from diseases. The female silkworm lays 300–400 eggs simultaneously, called a cluster of eggs. Then the tiny eggs are incubated for up to seven days until they hatch into larvae. In the case of a natural environment, hatching time depends on season and weather, as practiced in Bangladesh. Generally, farmers buy silkworm eggs from BSDB nurseries at a minimum price. However, they cannot use modern technologies. Farmers need more space for rearing in a disease-free environment, rather than using their living room or nearby living room. Farmers usually use bamboo-made coarse mats (*Chatai*). Conceptually, it is called the first instar. In Bangladesh, white and yellow species having different resistance abilities are usually used for rearing silkworms. The yellow species can resist hot weather conditions (like the weather condition in Rajshahi), while the white species can cope with the cold weather in the northern part of Bangladesh (like the weather condition in Panchagarh). Larva, which has a special type of protein, consumes mulberry leaves that produce silk. However, the finest silk yarn depends on the quality of leaves and management at different stages.

Cocoon

Once hatched, farmers in Bangladesh put larvae in a locally bamboo-made *Chatai* and feed them huge amounts of chopped young mulberry leaves. This stage is called the second instar. In this way, silkworms complete five instars to produce a cocoon. Usually, in the last instar, silkworms consume about 80% of the full feeds of their life cycle. After the fifth instar, the mature larvae are transferred into bamboo-made round shape casing (locally called *Chandrika*) for spinning to produce silk, and these larvae transform into cocoons within 24–48 h. The silkworm rotates its body to construct a cocoon and produces about a kilometer of silk filament, and the length depends on the species. The quality and length of filament depend on the quality of round shape casing and the timing of shifting from *Chatai* to *Chandrika*. In Bangladesh, farmers use bamboo-made casing, which is unsuitable for spinning. Good quality cocoon is the main benchmark for producing high quality silk (Halder 1999). The cocoon must be boiled or dried within 2-3 days to prepare for reeling. Otherwise, the larva inside the cocoon will turn into a live moth and the moth comes out by cutting the cocoon shell. These cocoons will not be used for reeling.

Raw Silk

The cocoon is treated with hot air or steam and boiling water before producing the raw silk. The silk is then unbound from the cocoon by softening the sericin. Then, the silk delicately and carefully unwind from 4 to 8 cocoons together at a time, sometimes with a slight twist to create a single strand. One kg of raw silk yarn generally requires 12–15 kg of cocoons, which depends on seasonal production (cycles) in the study areas.

However, one kg of raw silk normally requires only 7.6 kg of cocoons in other countries like China, Vietnam, etc.

Dead Larva

Once the cocoon is boiled in hot water and raw silk is reeled, the dead larva inside the cocoon can be served as fish and dairy feed and even considered as organic fertilizer in Bangladesh. However, farmers are usually reluctant to sell the dead larva due to market scarcity. We found little demand of dead larva in local market. Each kilogram of dead larva's sell price is only BDT 4–5 which is not profitable due to its high processing cost. Sometimes farmers use it as poultry feed; otherwise, they throw away this dead larva.

Silk Yarn

Raw silk still contains sericin and fibroin. After washing in soap and boiling water, the silk yarn is left soft, lustrous, and up to 30% lighter. The silk yarn is twisted into a strong strand for weaving or knitting. Creating the silk yarn is called 'throwing,' which prevents the thread from splitting into its constituent fibers.

Silk Fabric

Weavers use silk yarn to produce final products after dying and designing it. In some cases, designing parts is very important after weaving. In Bangladesh, silk products are mostly three pieces, scarves, handbags, *sharee*, etc., for women and *panjabi*, shirts, ties, etc., for men and silk clothes, tablecloths, silk photo frames, etc., for general use, which usually sell in domestic as well as international markets. In the study area of Bholahat, some reelers make silk fabrics and silk brocade using *motka* silk (yarn made from silk waste) and sell it in different places in Bangladesh.

12.3.2 The Analytical Framework of the Value Chain

There are different actors and stages involved in the sericulture industry. Value chain analysis with detailed descriptions of these actors and stages is shown in Fig. 12.3.

12.3.3 Description of Value Chain Actors

Table 12.1 shows the description of value chain actors at different stages in the study area.

12.3.4 State of Supporting Institutions

The sericulture industry in Bangladesh suffers from a lack of institutional assistance (Krishnan and Gurung 2015). It needs more support from meso-level bodies such as local authorities, business associations, agro-forestry institutes, or trade promotion agencies. Due to the absence of specific policy from those institutions, developing of sericulture subsector is generally mentioned as an objective in the local socioeconomic development plans of BSDB. However, more improvement needs to be made due to a requirement for more funding sources and technologies. In some places within the study areas, such as Mirganj, Bagha, and Bholahat, farmers get very modest financial and technical support from local BSDB offices. Besides, some constraints remain at the macro- and micro-levels. The main limitations identified at the macro- and micro-levels are illustrated in Fig. 12.4.

12.3.5 Analysis of the Value Chain

The calculation and presentation of value addition for sericulture are based on the following facts (Table 12.2).

The life cycle of the mulberry orchard is 20–25 years for optimal leaf production. After three years, about 30% of profit gets from one *Bigha* per year in the mulberry garden (Table 12.3). Initially, leaf production is minimal; after three years, it will increase.

However, at present, farmers are uninterested in mulberry gardening due to the limited market of mulberry leaves. The production of the present mulberry leaves variety is lower than the production, predicted by BSRTI. Consequently, farmers have decreased silkworm rearing. So,

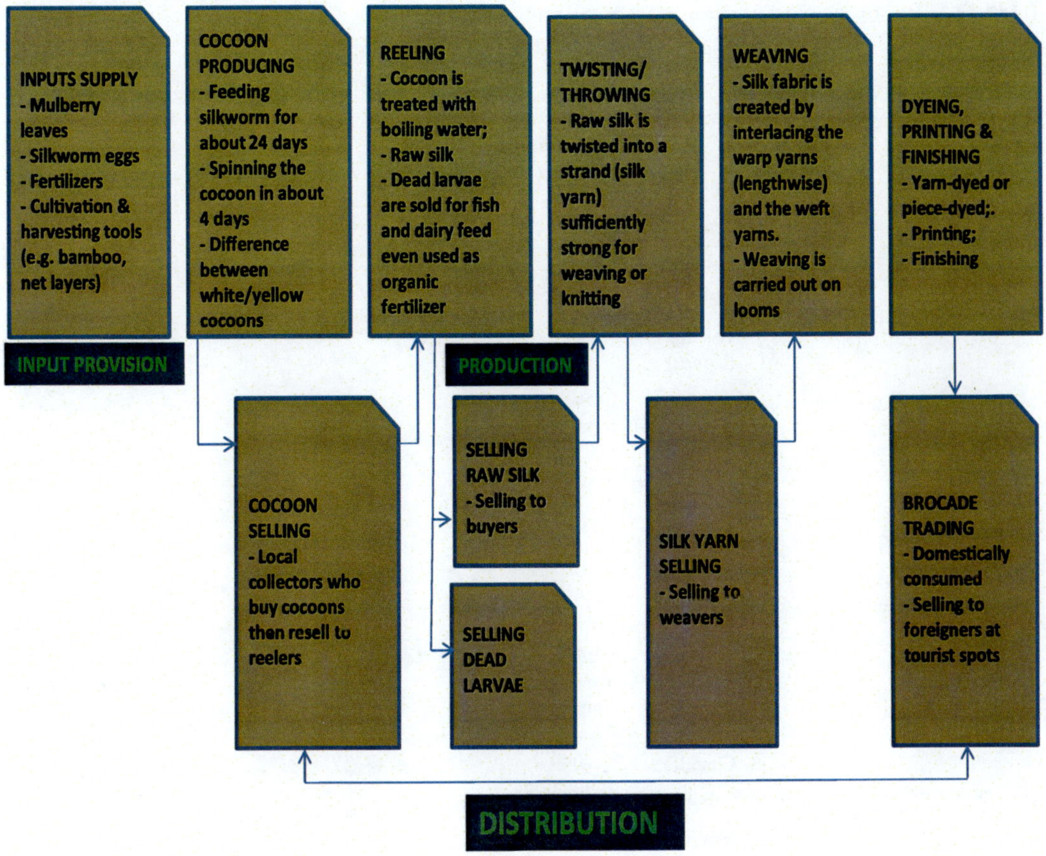

Fig. 12.3 Sericulture value chain functions

mulberry production might be profitable if the silkworm rearing and price of cocoon increase.

Table 12.4 shows that silkworm rearing is a profitable business in Bangladesh, as found in the study areas. From 100 Disease-free Layer: in sericulture, one egg, which means a cluster of eggs laid by a silkworm (DFL) rearing in a cycle, we found 55 kg cocoons, but it varies from farmer to farmer, area to area, and also from cycle to cycle (*Bondh*). We found more than nearly 30% of profits from one cycle in 100 DFL eggs. From our observation and FGD during the field survey, farmers face different problems in selling cocoons, such as monopoly pricing, lack of modern technology, and insufficient and disease-free quality leaves. Mulberry orchards are converted into mango orchards, and existing mulberry orchards are affected and polluted by different pesticides used in different crop production, especially in mango cultivation in the study area. Therefore, it takes work to motivate farmers toward silkworm cultivation. Current cocoon production needs to be increased in the study areas. In addition, the price of cocoons could be more cost-effective, discouraging the farmers from rear the silkworm. The yield of mulberry leaves and silkworm production are hampered by various risk factors, such as hot and humid weather, lack of quality eggs and mulberry leaves.

A. **Reeling**

See Tables 12.5 and 12.6.

Two types of reeling are found in the study areas. In Bholahat, some farmers are still follow the traditionally reeling, but in the other two

Table 12.1 Description of value chain actors in the study area at present

Location		Inputs supply	Output	Reeling/twisting	Weaving/finishing	Distributions
Rajshahi	Bagha	Silkworm	Cocoon production	No reelers	No weavers	Local buyers from farmers
		– Eggs				
		– *Chatai*				
		– *Chandrica*				
		– Labor				
		– Medicine				
		– Feeding (leaves)				
		Mulberry	Leave production			
		– Sapling				
		– Fertilizer				
		– Pesticides				
		– Land				
		– Labor				
		– Fencing				
	Charghat	Silkworm	Mulberry plantation and rearing silkworms for Cocoon production	No reelers	No weavers	Local buyers from farmers
		– Eggs				
		– Chatai				
		– Chandrica				
		– Labor				
		– Medicine				
		– Feeding (leaves)				
		Mulberry				
		– Sapling				
		– Fertilizer				
		– Pesticides				
		– Land				
		– Labor				
		– Fencing				
Chapainawabganj	Bholahat	Silkworm	Cocoon production	Most of the cocoon farmers are involved in reeling	Insufficient supply of silk yarn for weaving and brocade products	Local buyers for selling to others, produce silk yarn and Andy silk cloths
		– Eggs				
		– *Chatai*				
		– *Chandrica*				
		– Labor				
		– Medicine				
		– Feeding (leaves)				
		Mulberry				
		– Sapling				

(continued)

Table 12.1 (continued)

Location		Inputs supply	Output	Reeling/ twisting	Weaving/finishing	Distributions
		– Fertilizer				
		– Pesticides				
		– Land				
		– Labor				
		– Fencing				

Source Field survey

places, there are no traditional as well as modern reelers. Farmers usually use *Kathghai*, traditional wooden equipment, for reeling. Farmers are getting about 25% of profits through this method, but in modern reeling, it is higher (40.90%), as shown in the above table. The high quality of silk yarn produced through modern reeling, results in a higher price than traditionally produced silk. In the sericulture value chain, reeling is a more profitable stage than mulberry cultivation and silkworm rearing. Though mulberry cultivation is less profitable, it is inevitable in the sericulture value chain. However, it is unfortunate that farmers are abundant with cocoons rearing due to insufficient feed for silkworm rearing, causing an extreme decrease of the cocoons which are the raw materials of silk yarn.

B. **Weaving**

See Tables 12.7 and 12.8.

Weaving and designing are closely related to sericulture finish products. Nowadays, weaving is found only in the BISIC area of Rajshahi City. However, we found a few traditional weavers in Bholahat of Chapainawabganj district, which was once the epicenter of sericulture. Recently, these weavers are producing only silk *Thaan* (an entire sheet of woven fabric) without dying and designing, and this *Thaan* sells into the modern weaving industries at Rajshahi and Dhaka for further dying and designing into a finished product. Modern weaving industries in Rajshahi mainly use imported silk yarn due to an insufficient supply of local yarn; imported silk yarn is more cost-effective than local yarn due to government tax policy. In the case of dying and designing, it is more profitable than other stages of the sericulture value chain. Firstly, there are no risk factors in dying and designing. There are a few dying and designing industries in Rajshahi (only two are running in full swing). It creates a monopoly market for the traditional and modern silk *Thaan*. On the one hand, limited yarn and weaving industries create a monopoly market. In addition, the government yarn and weaving industry has been closed since 2002. Consequently, it pushes farmers to abandon silkworm rearing.

12.3.6 SWOT Analysis

A. Strength

- BSDB and BSRTI head offices are located in Rajshahi;
- There is sufficient infrastructure such as office space, research and training institute, and sericulture nursery in different Upazila levels in Bangladesh under BSDB;
- Available skilled human resources of BSDB and BSRTI;
- Household traditions of sericulture in this region;
- Favorable environment;

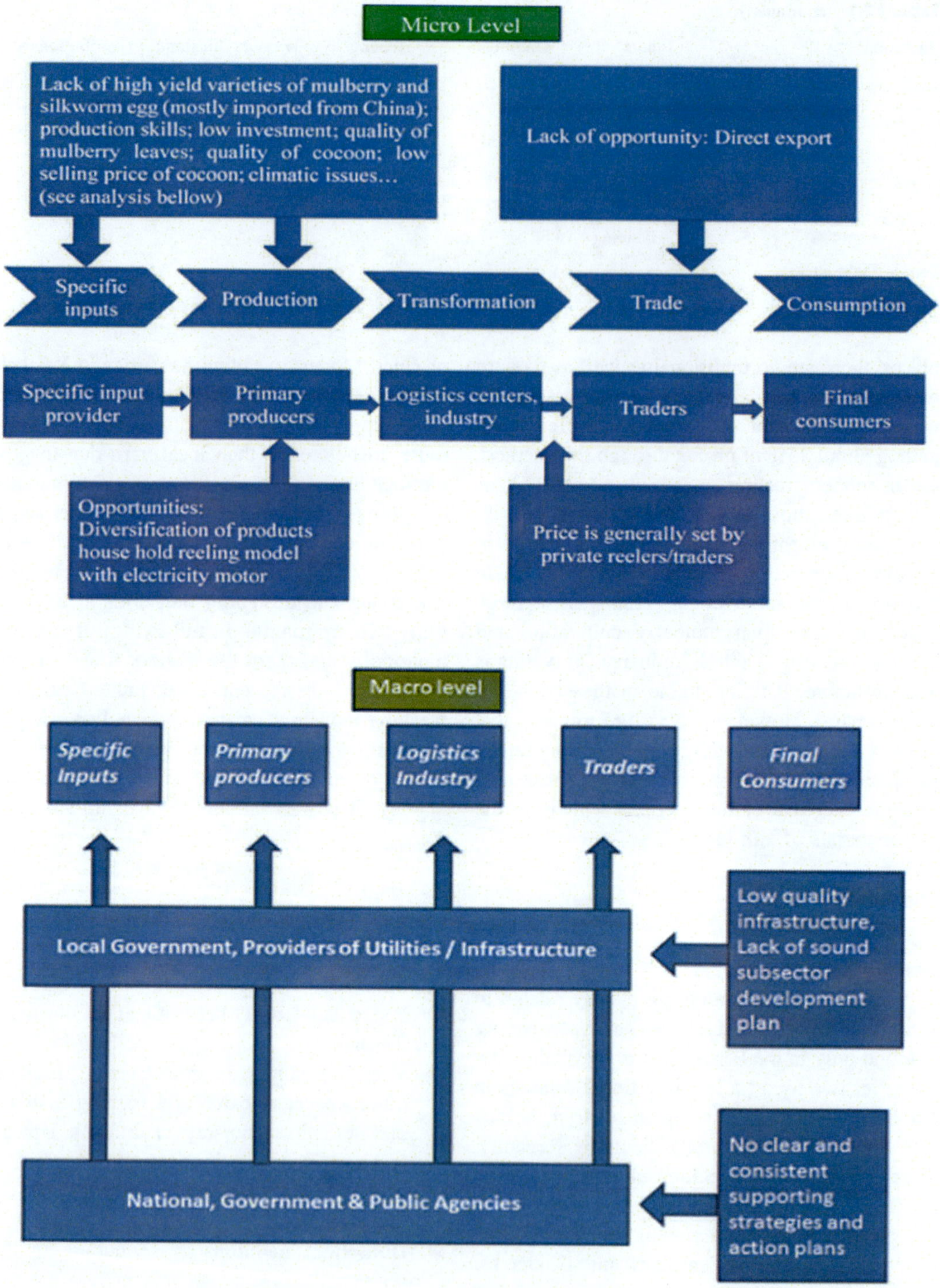

Fig. 12.4 Critical points at macro- and micro-level support institutions

Table 12.2 Mulberry cultivation cost analysis (from starting up to three years/*Bigha*)

Type of inputs	Unit price	Amount	Types of outputs	Total amount	Unit price	Amount
Land (using price/*Bigha*)	8000.00	24,000.00	Leaves	3200/kg	Tk. 5/kg	16,000.00
Land preparation	2500.00	2500.00	Fuel	Lump-sum		8000.00
Sapling	Tk. 50/100	750.00	Mix crop	Lump-sum	Tk. 5000.00/ year	15,000.00
Planting	Tk.1500/*Bigha*	1500.00	Fruit			
Fertilizer	three years/ *Bigha*	4800.00				
Insecticides	three years/*Bigha*	1000.00				
Irrigation	three years/*Bigha*	2500.00				
Weeding/pruning	three years/*Bigha*	2000.00				
Routing care						
Total amount		39,000.00		Total Amount		39,000.00

Source Field survey

Table 12.3 Mulberry cultivation cost analysis (after three years/*Bigha*) in a year

Type of inputs	Unit price	Amount	Types of outputs	Total amount	Unit price	Amount
Land (using price/Bigha)	8000.00	8000.00	Leaves	2500.00	Tk. 5/kg	12,500.00
Land preparation	00	00.00	Fuel	Lump-sum		4000.00
Sapling	00	00.00	Mix crop	Lump-sum	Tk. 0.00/year	00.00
Planting	Tk. 00/*Bigha*	00.00	Fruit			
Fertilizer	Three years/*Bigha*	1200.00				
Insecticides	One year/*Bigha*	500.00				
Irrigation	One year/*Bigha*	1000.00				
Weeding/pruning	One year/*Bigha*	1000.00				
Routing care		1000.00				
Total amount		12,700.00		Total amount		16,500.00
Profit (16,500.00–12,700.00) = 3800/12700				29.92%		

Source Field survey

- Traditional marketing chain;
- Huge amount of unemployed women in this region.

B. Weaknesses

- Inappropriate state policy (including import, tax, etc.) of sericulture development;
- Financial constraints;
- Unfavorable administrative systems;
- Inadequate HYV and local weather suitability species of silkworm and mulberry;
- Lack of modern technologies in research as well as at the cultivation level;
- Lack of equilibrium between backward and forward industries and production systems;
- Insufficient private entrepreneurship in sericulture.

C. Opportunities

- Generation of home employment opportunities for women in rural households that will empower them as income generators;

Table 12.4 Silkworm rearing per 100 DFL in one cycle

Type of inputs	Amount /No	Unit price	Total amount	Types of outputs	Unit price in Tk	Total amount
Eggs	100	2.10	210.00	Cocoon 55 kg	250.00	13,750.00
Chatai/Dala	32	100	3200/8 = 400.00	Litter		500.00
Chandrika	32	200	6400/8 = 800.00	Dead larva		500.00
$CaCO_3$	3 kg	20	60.00			
Net	60 Gauge	70	4200/12 = 350.00			
Jute, paper, etc.	–	100	100.00			
Leaves	1000	5	5000.00			
Labor	21	200.00	4200.00			
Total			11,120.00	Total		14,750.00
Profit (14,750.00–11,120.00) = Tk. 3630.00			32.64%			

Source Field survey

Table 12.5 Traditional reeling (*Kathghai*), 3.5 kg silk yarn/day

Type of inputs	Amount	Unit price	Total amount	Types of outputs	Unit price (Tk.)	Total amount
Process of cocoon	Lump-sum	1500.00	1000.00	Fine silk yarn 3.5 kg	3500.00	12,250.00
Cocoon	40 kg	250.00	10,000.00	*Matka* silk yarn lump-sum	2000.00	2000.00
Labor	1	700.00	700.00	Dead larva	Lump-sum	1000.00
Total			12,200.00	Total		15,250.00
Profit (15,250.00–12,200.00) = Tk. 3050.00			25.00%			

Source Field survey

Table 12.6 Modern reeling per 1 kg silk yarn

Type of inputs	Amount /No	Unit price	Total amount	Types of outputs	Unit price (Tk.)	Total amount
Process of cocoon	Lump-sum	250.00	250.00	Fine silk yarn 1 kg	4000.00	4000.00
Cocoon	10	250.00	2500.00	Matka silk yarn lump-sum	500.00	500.00
Labor	1	400.00	400.00	Dead larva	Lump-sum	150.00
Other material costs	6	150.00	150.00			
Total			3300.00	Total		4650.00
Profit (4350.00.00–3300.00) = Tk. 1350.00			40.90%			

Source Field survey

Table 12.7 Weaving (modern)

Inputs	Amount	Unit price	Processing cost	Total amount	Output	Unit price	Total amount
Silk yarn	1 kg	4000	3600	7600	4 *Sharees*	2250	9000
Total				7600	Total		9000
Profit (9000–7600) = Tk.1400				18.42%			

Source Field survey

Table 12.8 Designing (*Sharee*)

Inputs	Unit price	Designing cost	Total amount	Output	Unit price	Total amount
Silk cloth	2250	1000–10,000	3250–12,250	Finish product	4550–17,150	4550–17,150
Profit (4550–17,150) − (3250–12,250) = 1300–4900				40.00%		

Source Field survey

- Opportunity for additional income, especially for rural poor;
- Diversification of sericulture production by introduction of modern technologies;
- Restoring the moribund heritage of Rajshahi silk;
- Can be fulfilled the silk yarn of the country evading import dependency;
- Increase export and earning foreign currencies;
- Introducing controlled silkworm rearing systems.

D. Threats

- Government policy and planning;
- Unfavorable weather for existing species for increasing production cycles;
- Shifting trends of cultivating mango rather than mulberry.

12.4 Conclusion

The sericulture industry is an important sector for employing rural people. This study has made a SWOT analysis of the value chain of the sericulture industry in Bangladesh. It has found a big gap in the value chain of silk production. Leaves produced from mulberry cultivation need more support in the value chain. Mulberry cultivation needs long-term investment to make a profit. However, the current demand for mulberry leaves has decreased results in shifting to other crops from mulberry to get more profits. Besides, the silkworm is reared only four times a year due to unfavorable environment. Cocoon is the only output produced by the farmers but they need to get the proper price. As a result, cocoon production is gradually decreasing and it is in a vulnerable situation. The sericulture industry in Bangladesh is household-based and medium or large-scale enterprises have yet to be developed. Thus, crucial action is required to develop the silk industry in an underdeveloped country like Bangladesh. However, there is potential for environment-friendly sericulture production in Bangladesh. Thus, equilibrium development of backward and forward linkages is needed evading import dependency.

References

Akram S (2015) Analysis of the silk value chain in Pakistan. Int J Modern Trends Eng Res 2(8):223–236

Angadi BSR, Reddy M, Sivaprasad V (2013) Scope of product diversification and value creation in Indian sericulture industry. J Eng Comput Appl Sci (JEC&AS) 2(5):2319–5606

BSRTI (2014) Annual report 2014. Bangladesh Sericulture Research Institute, Rajshahi

Buhroo ZI, Bhat MA, Malik MA, Kamili AS, Ganai NA, Khan IL (2018) Trends in development and utilization of sericulture resources for diversification and value

addition. Int J Entomol Res 6(1):27–47. https://doi.org/10.33687/entomol.006.01.2069
Eswarappa K (2011) Developmental initiatives and sericulture in a south Indian village. South Asia Res 31 (3):213–229. https://doi.org/10.1177/026272801103100302
Gereffi G (2011) Global value chains and international competition. Antitrust Bull 56(1):37–56
Gereffi G (2013) Global value chains in a post-Washington consensus world. Rev Int Polit Econ. https://doi.org/10.1080/09692290.2012.756414
Gereffi G, Lee J (2016) Economic and social upgrading in global value chains and industrial clusters: why governance matters. J Bus Ethics 133:25–38. https://doi.org/10.1007/s10551-014-2373-7
Gereffi G, Lee J (2012) Why the world suddenly cares about global supply chains. J Supply Chain Manage 48 (3). https://doi.org/10.1111/j.1745-493X.2012.03271.x
Halder SR (1999) Viability of sericulture programme of BRAC: results of a cost-benefit analysis. Bangladesh J Agric Econs XXII 2:99–116
Hossain MI, Afroz S, Das M, Haque MM, Islam MS, Lim-Camacho L (2019) Value chain analysis of sunflower in coastal areas of Amtali upazila of Barguna district. J Bangladesh Agric Univ 17(2):244–250. https://doi.org/10.3329/jbau.v17i2.41989
Hui N (2010) Changes in silk production and trade structure in China after the 1980s. In: 4th Asian rural sociology association (ARSA) international conference, Legazpi City, Philippines, pp 223–235
Ponte S, Gereffi G, Raj-Reichert G (eds) (2019) Introduction to the Handbook on Global Value Chains. Edward Elgar Publishing
Jantakat C, Tangjaturasopon A (2012) Barriers of value chain for development of silk product in Nakhonchaiburin Zone, Thailand. In: 2012 International conference on economics, business innovation, Singapore, vol 38, pp 109–112
Kaplinsky R, Morris M (2001) A handbook for value chain research. International Development Research Center, Ottawa, Canada
Krishnan R, Gurung TR (2015) Sericulture scenario in SAARC region: a re-emerging industry for poverty alleviation in SAARC region synthesis. In: Gurung TR, Bokhtiar SM, Kumar D (eds) Sericulture scenario in SAARC region: a re-emerging industry for poverty alleviation in SAARC region. SAARC Agricultural Centre, Dhaka, pp 1–9
Lee J, Gereffi G (2015) Global value chains, rising power farms and economic and social upgrading. Crit Perspect Int Bus 11(3/4):319–339. https://doi.org/10.1108/cpoib-03-2014-0018
Meaton J, Abebe B, Wood A (2013) Forest spice development: the use of value chain analysis to identify opportunities for the sustainable development of Ethiopian Cardamom (Korerima). Sustain Dev 23 (1):1–15. https://doi.org/10.1002/sd.1563
Nguyen T, Eiligman A (2010) Value chain study for sericulture in PHU THO, HOA BINH, HOA and NGHE AN, Viet Nam. Joint Progamme on Green Production and Trade to Increase Income and Employment Opportunities for the Rural Poor
Porrasa I, Mohammed EY, Ali L, Ali MS, Hossain MB (2017) Power, profits and payments for ecosystem services in Hilsa fisheries in Bangladesh: A value chain analysis. Marine Policy 84(60–68). http://dx.doi.org/https://doi.org/10.1016/j.marpol.2017.06.031
Rohela GK, Shukla P, Muttanna KR, Chowdhury SR (2020) Mulberry (*Morus* spp.): an ideal plant for sustainable development. Trees For People (2):100011. https://doi.org/10.1016/j.tfp.2020.100011
Rosales RM, Pomeroy R, Calabio IJ, Batong M, Cedo K, Escara N, Facunla V, Gulayan A, Narvadez M, Sarahadil M, Sobreveg MA (2017) Value chain analysis and small-scale fisheries management. Mar Policy 83:11–12. https://doi.org/10.1016/j.marpol.2017.05.023
Ruiz A, Caballero B, Martínez Y, Vega R, Valdés A, Pérez MDC (2020) Analysis of the energy balance in the *Morus alba-Bombyx mori* system in Cuba's sericulture. Int J Agric Econ 5(1):30–35. https://doi.org/10.11648/j.ijae.20200501.14
Saikia JN (2011) Supply chain linkages and constraints in natural silk sector of Assam: a study of Muga and Eri silk. Int J Multidisc Manage Stud 1(3):167–194
Shukla R (2012) Economics of rainfed sericulture-a study in the district of Udaipur in Rajasthan, India. Bangladesh J Agric Res 37(1):49–54
Sime D, Siraj Z (2020) Sericulture in Ethiopia: Production status, opportunities, challenges and potential areas. A review. J Entomol Zool Stud 8(6):1–10
Sturgeon TJ (2001) How do we define value chains and production networks? IDS Bull 32(3):9–18
Sujatha B, Reddy PL, Babu MAS, Reddy BAP, Kumar S, Naik SS (2015) Socioeconomic factors influencing the adoption levels of new Sericulture technologies by different farming groups in Anantapur District of Andhra Pradesh. Int J Agric Ext 3(2):149–153
Tesfa A, Ejigu K, Yetayew A, Assefa H (2014) Assessment of value chain of sericulture products in Amhara region, Ethiopia. Int J Environ Eng Nat Resour 1 (2):61–69
Uddin MT, Goswami A, Rahman MS, Dhar AR, Khan MA (2018) Value chain of pangas and tilapia in Bangladesh. J Bangladesh Agric Univ 16(3):503–512. https://doi.org/10.3329/jbau.v16i3.39448

13 Development of Objective-Based Multi-criteria Decision-Making Approach in Crop Suitability Assessment for Maize Production Using GIS

Rajib Mitra, Amit Sarkar, Golap Hossain, Dipesh Roy, Goutam Mandal, Jayanta Das, and Deepak Kumar Mandal

Abstract

Farmers are constantly adopting new tactics to ensure significant yield production, providing sustainable income in economic globalization and climatic unpredictability. Evaluation of crop suitability is the first and foremost requirement for increasing yield production and enables cultivators to utilize the land appropriately for different crops. GIS integration with MCDM is an effective and widely used method for land suitability studies. This study has prepared the suitability map for maize cultivation of the Dinhata sub-division of Koch Bihar district in India by integrating one of the famous objective-based MCDM techniques (i.e., entropy weighting method) and GIS tools. A total of twelve criteria, i.e., elevation, slope, TWI, drainage density, mNDWI, nitrogen, phosphorus, potassium, organic carbon, rainfall, temperature, and LULC, have been selected for this study. From the result, it was estimated that 16.86% area is highly suitable (S1) for maize cultivation. Almost half (47.01%) area is estimated as moderately suitable (S2), and only 5.99% of the area is estimated as not suitable (N1 and N2) for maize cultivation. Therefore, this study identifies that maize is one of the dominant crops in the study area. This kind of research is always convenient for assessing suitable sites of different crop production, primarily in the agriculturally leading areas, providing more scope to the policymakers to identify the land's limitations and ensure sustainable agriculture in the region.

Keywords

Land suitability · Remote sensing · Entropy weighting method (EWM) · Dinhata sub-division · Sustainability

13.1 Introduction

The growing population is naturally having a huge impact on agricultural land and other natural resources. This situation is much direr in developing countries like India, Bangladesh, Pakistan, and Myanmar. Due to this growing population, wetlands, pastures, forests, and farmlands are transforming into new settlements or built-up areas (Akıncı et al. 2013). As the proportion of land related to agricultural is limited globally, the pressure on these agricultural

R. Mitra (✉) · A. Sarkar · G. Hossain · D. Roy · G. Mandal · D. K. Mandal
Department of Geography and Applied Geography, University of North Bengal, Siliguri, West Bengal, India
e-mail: rajibmitrageo@gmail.com

J. Das
Department of Geography, Rampurhat College, Birbhum, Rampurhat 731224, India

J. Das and S. Halder (eds.), *Advancement of GI-Science and Sustainable Agriculture*, GIScience and Geo-environmental Modelling, https://doi.org/10.1007/978-3-031-36825-7_13

lands is also increasing. As a result, the health and productivity of this agricultural land are deteriorating from time to time to fulfill the rising demand of the excessive population (Cowie et al. 2018), which drives the land to be improperly used (Elaalem et al. 2011). Therefore, proper land use planning has been a crucial task (in recent times) so that these natural resources are appropriately handed over to future generations, and they can use these natural resources in a well-planned and sustainable manner (Ahmed et al. 2016).

Land suitability assessment is an important undertaking in land use planning. Suitability analysis is performed to determine the suitability of a piece of land for specific uses. (FAO 1985). Land use suitability analysis is the process of determining a specific fitness of land area for certain services such as cultivation, forestry, recreation, and the degree of suitability (Malczewski 2004). Analyzing a piece of land's suitability can help determine its intrinsic and possible capabilities for various goals (Bandyopadhyay et al. 2009). It also depends on the crop's requirements and the land's properties (Mustafa et al. 2011). It is very important and should be done to consider the local demands and conditions (Kihoro et al. 2013). By evaluating existing and potential capabilities, land suitability analysis can assist in developing new tactics to boost agricultural productivity (Pramanik 2016). It can also assist in identifying priority locations for prospective management or policy actions, such as land or soil restoration projects (Yalew et al. 2016). Hence, utilizing modern techniques, it is necessary to identify suitable farmland and carry out land use planning to conduct a rational analysis and appraisal of the soil and land resources (Özkan et al. 2020).

It is acknowledged that there is no set standard for the factors to be considered when evaluating the agricultural potential of a piece of land (Akıncı et al. 2013). Data of large variety and quantity, including climatic attributes such as rainfall and temperature; intestinal soil properties such as temperature, texture, moisture, depth, aeration, fertility, and salinity and exotic soil conditions such as slope, accessibility, and flooding are analyzed in the assessment of land suitability for agricultural purpose (Wang 1994; Joerin et al. 2001; Kahsay et al. 2018). Therefore, suitability for a particular use is assessed by contrasting its requirements with the traits and attributes of various land components. The decision-making process for determining land suitability is quite intricate and entails a variety of decision indicators and criteria. Due to these requirements, multivariate criteria decision-making is used to choose the best crop from the currently grown ones. On the contrary, to conduct a multi-criteria decision-making (MCDM) approach, several techniques are helpful in decision-making, viz., Weighted Linear Combination (WLC) method, Entropy Weighting Method (EWM), Analytical Hierarchy Process (AHP), Analytic Network Process (ANP) method and the Technique for Order of Preference by Similarity to Ideal Solution (TOPSIS). Entropy weighting technique is a popular MCDM technique, which helps decision-makers to make the best possible choice while taking into account various qualitative, quantitative, and sometimes conflicting factors. (Malekinezhad et al. 2021). Shannon first developed the idea of entropy, widely employed across the fields, i.e., engineering, economics, sciences, finance, spectral analysis, language modeling, and social sciences (Shannon 1948). The entropy-based technique uses initial criteria value information to calculate weights in an objective manner (Chen et al. 2018). In order to identify the relative significance of every criterion in relation to the choice and outcome, most MCDM methods determine the weight of the criteria before starting to solve the problems (Nyimbili and Erden 2020). Instead of relying on expert opinion, entropy weights can assess the amount of meaningful information offered by the index based on the dissimilarity between feature values (Taheriyoun et al. 2010). The main benefit of the Entropy technique over subjectively based weighting methods lies in the fact that it removes anthropogenic involvement from the weighting procedure, allowing for the objective determination of the criteria weights (Ding et al. 2017). These weights measured using the entropy method are considered as objective weights and it implies that the greater the entropy, the smaller the relative weight will be and vice versa. The above-said statement signifies lesser contextual information since some may be missed or incomprehensible (Zou et al. 2006).

Every day, more pressure is placed on agricultural land. Hence, site suitability analysis for cultivating a particular crop has recently been crucial in understanding suitable areas for particular crop production. Not all types of crops can be cultivated in any area; a suitable area for the production of particular crops is very important. There are numerous numbers of researches have been done across the country and the world. Recently, the entropy weighting method (EWM) has been used in GIS-based MCDM to study migration modeling (Rashid 2018), waste landfill sites (Ding et al., 2017), landslide susceptibility zonation mapping (Jaafari et al. 2014), flood risk assessments (Haghizadeh et al. 2017). Some research work has also been done for water quality evaluation (Zou et al. 2006), assessment of the suitability for urban emergency facilities (Nyimbili and Erden 2020), the spatial distribution of crop areas (You and Wood 2006) using entropy weighting method. A literature review indicates that using the EWM under a GIS environment has just been implied for crop suitability evaluation in the study area. The general land evaluation for a specific crop using integrated GIS and EWM in the Dinhata sub-division of Koch Bihar district has yet to be studied. Therefore, in this research work, the EWM approach has been applied following the GIS environment to evaluate land suitability for maize production in the Dinhata sub-division of Koch Bihar district, West Bengal, based on a survey of the literature and expert opinions. The wide and important criteria developed and employed in this study were essential for obtaining the best choice output. The professional knowledge about the suitability of a particular location provided by this study broadens the understanding of sustainable maize planning and serves as a crucial guide for achieving the best maize production program. Hence, the outcome of this research would set a benchmark to delineate the suitable areas to cultivate maize in the study area.

13.2 Materials and Methods

13.2.1 Study Area

Koch Bihar, situated in the north-eastern part of West Bengal, is considered the most important district for agriculture in Sub-Himalayan North Bengal region. The economy of the district is primarily centered on agriculture. The research area Dinhata sub-division of the Koch Bihar district is located between 25° 57′ 24″ and 26° 14′ 06″ north latitudes and 89° 15′ 17″ to 89° 37′ 36″ east longitudes (Fig. 13.1). The Dinhata sub-division is located in the southern part of the district. It cover 764.84 km^2, or around 14.85% of the entire geographical area of the district. The sub-division is divided into the Dinhata-I, Dinhata-II, and Sitai administrative blocks. The majority of the population in the sub-division is related to agriculture, which provides a significant part of their income. Nearly 60% of the district is suitable for paddy production, and other agricultural products of this area are maize, jute, mustard, tobacco, wheat, potato, barley, sugarcane, ginger, garlic, vegetable crop, etc. Maize (Zia Mais) cultivation has increased in recent trends in the district as a whole, and Dinhata sub-division has accounted for 36% of the total maize production within the district, which is 1571 MT on 645 hectors of agricultural land in Dinhata sub-division (District Statistical Handbook 2013). Since agriculture is the district's dominant sector, the current study is intended to analyze the crop suitability assessment for maize production in the Dinhata sub-division of Koch Bihar District.

13.2.2 Data Sources and Thematic Layer Construction

Several parameters have been used to analyze crop suitability for maize production. Each of these parameters significantly impacts the

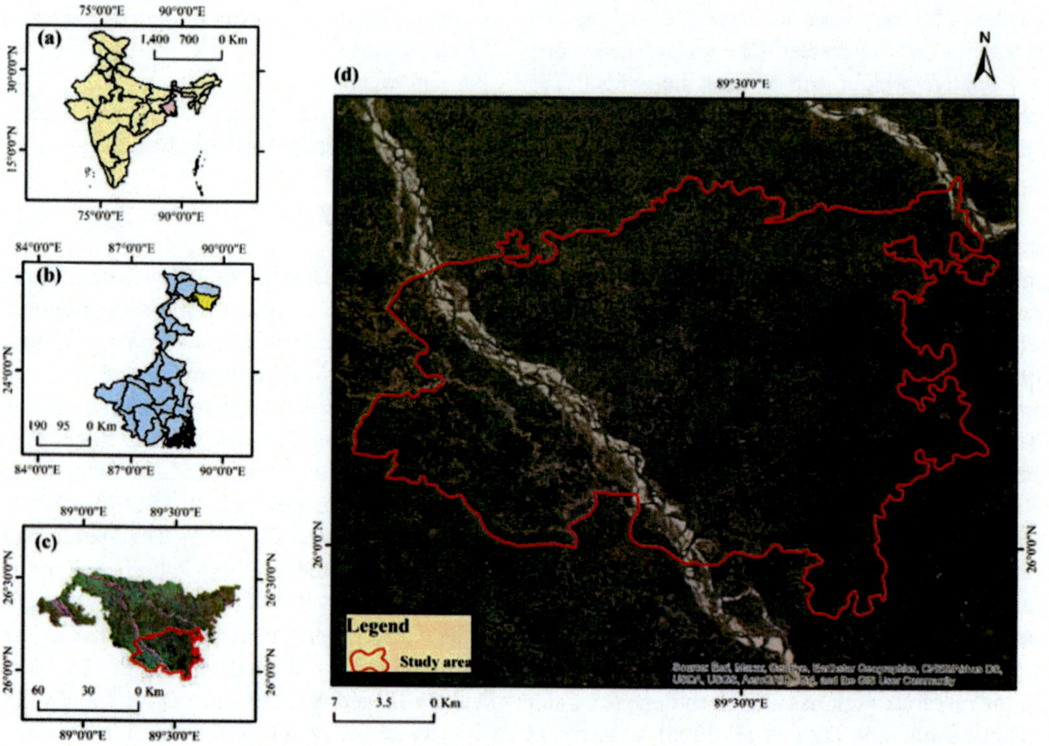

Fig. 13.1 Location map of the study area a India, b West Bengal, c Koch Bihar district, and d Dinhata sub-division

mapping of crop suitability for the production of maize. Finding the appropriate parameters requires a thorough review of the literature, as well as field observations and the advice of experts in the study area. The source of the many applied parameters of the crop suitability for maize production is shown in Table 13.1. To create thematic layers in the GIS platform, the study initially developed a spatial database of parameters related to the suitability of crop cultivation for maize production. The Sentinel-2 and SRTM DEM (30 m) data were downloaded, and these types of data processing in ArcGIS 10.4.2 software. The SRTM DEM (30 m) has been used to map the elevation, slope, TWI, and drainage density. The CRU TS of the UK's National Center for Atmospheric Science (NCAS) website's climatic data (temperature and rainfall grid data) was assembled and used to calculate the study area's annual rainfall and mean temperature. For preparing the several pedological factor map of the study area, the data of NBSS and LUP has been collected and then compiled by the researchers.

13.2.3 Description of EWM

The concept of the entropy weighting method was first proposed by Shannon (Shannon 1948), and it is widely employed across the fields (i.e., engineering, economics, sciences, finance, spectral analysis, language modeling, and social sciences). It is a system or informational uncertainty measurement derived from probability theory and applied to the source data (attribute values of alternatives). To solve most Multiple Criteria Decision-Making (MCDM)-related problems, it is necessary to determine the weights of criteria in order to identify their relative importance compared to other criteria that influence decision alternatives and outcomes. Entropy weights can be used to determine the amount of valuable information provided by an index, based on the

Table 13.1 Data sources used to crop suitability evaluation for maize production

Clusters	Criterion	Descriptions source	Resolution/scale	Type of GIS data
Topography	Elevation	Derived from ASTER DEM and prepared the thematic layer using ArcGIS Retrieved from: https://search.earthdata.nasa.gov/search	30 m	Raster
	Slope			
	Topographic Wetness Index			
Hydrology	Drainage Density	Derived from ASTER DEM and prepared the thematic layer using ArcGIS Retrieved from: https://search.earthdata.nasa.gov/search	30 m	Raster
	mNDWI	Using Sentinel-2 Retrieved from: https://scihub.copernicus.eu	10 m	Raster
Climate	Rainfall	Thiessen polygons were constructed to locate the influence area Retrieved from: https://crudata.uea.ac.uk/cru/data/hrg/	0.5° × 0.5° grid	NetCDF
	Temperature			
Pedology	Nitrogen	Maps on nitrogen content (kg/ha), phosphorous content (kg/ha), potassium content (kg/ha), and organic carbon content (%) for the Koch Bihar district have been collected and then compiled for the Dinhata sub-division by the researchers from NBSS and LUP	1 km x 1 km	Raster
	Phosphorous			
	Potassium			
	Organic Carbon			
Anthropogenic	LULC	Using Sentinel-2 Retrieved from: https://scihub.copernicus.eu	10 m	Raster

differences between attribute values, without relying on subjective input from experts or decision-makers. The primary benefit of the entropy approach over subjectively based weighting methods is that it excludes human intervention from the weighting process, allowing for the objective determination of the criteria weights (Ding et al. 2017; Taheriyoun et al. 2010). The term' objective weights' refers to the weights determined by the entropy approach, and it means that the higher the entropy, the lower the value differences between the objects being evaluated, and consequently, the smaller the relative weight, and vice versa. This denotes a lesser level of information because some can be lost or unintelligible (Zou et al. 2006; Munier et al. 2019). The following steps can be used to calculate the objective weights using the Shannon entropy approach (Hwang and Yoon 1981; Malczewski and Rinner 2015; Alinezhad and Khalili 2019):

Step 1: Construction of a decision matrix '*D*,' which displays the performance of *m* viable options with respect to *n* evaluation attributes (criterion).

$$\text{Let } D = \begin{bmatrix} X_{11} & X_{12} & \cdots & X_{1m} \\ X_{21} & X_{22} & \cdots & X_{2m} \\ X_{31} & X_{31} & \cdots & X_{3m} \\ X_{41} & X_{42} & \cdots & X_{4m} \end{bmatrix} \quad (13.1)$$

be a decision matrix.

Step 2: Normalization of Indices: The original value of data is normalized, and the dimensional effects are eliminated. The normalization has performed using the following equations:

$$P(X_{ij}) = \frac{X_{ij} - \min\{X_{ij}\}}{\max\{X_{ij}\} - \min\{X_{ij}\}}$$
$$P(X_{ij}) = \frac{\max X_{ij} - \{X_{ij}\}}{\max\{X_{ij}\} - \min\{X_{ij}\}} \quad (13.2)$$

where X_{ij} is related to the value of the j th object (j = 1, 2, 3, …, n) on the ith (i = 1, 2, 3, …, m).

Step 3: To assess a system with n object and m indices, the Rényi entropy of order α value pi for ith index can be defined using Eq. 13.3, which is also called Shannon entropy:

$$H_a(X) = \frac{1}{1-a}\log\left(\sum_{i=1}^{n} f_i^a\right) \quad (13.3)$$

where $f_{ij} = \frac{P_{ij}}{\sum_{i=1}^{m} P_{ij}}, a \geq 0 \& a \neq 1$

13.2.4 Integration of GIS with EWM

This chapter's main procedure incorporates the entropy weighting MCDM method with GIS. Here we integrated the method with ArcGIS software. Firstly, the values of the prepared thematic layers were extracted by using the 'Create Fishnet' tool of the Data Management tool of ArcGIS. The area of the grid of fishnets was 1kmX1km, and a total of 737 fishnets were generated in the study area. After extracting each layer's values of 737 fishnets, EWM was computed in an MS Excel sheet. The beneficial and non-beneficial criteria were identified based on their effects on the selected crops of the study area. Table 13.2 represents the categorization of the criteria for EWM.

The evaluation matrix for CSA is represented in Table 13.3. In the evaluation matrix, the maximum value is 3410.64, and the minimum value is −4. After obtaining the values of the evaluation matrix, the normalized vector was computed (Table 13.4). Then the weights were computed for each layer (Table 13.5), and finally, based on the weights, the suitability map of the Dinhata sub-division was prepared in ArcGIS. The 'IDW' technique was utilized in the present study to prepare the map. The methodological framework of the research has been presented in Fig. 13.2.

Table 13.2 Categorization of the criteria for EWM

Criteria type	Criteria
Beneficial	Potassium
Beneficial	Phosphorus
Beneficial	Nitrogen
Beneficial	Organic carbon
Non-beneficial	Drainage density
Non-beneficial	Elevation
Non-beneficial	mNDWI
Beneficial	TWI
Beneficial	Temperature
Non-beneficial	Slope
Beneficial	Rainfall
Non-beneficial	LULC

Table 13.3 Evaluation matrix for CSA using EWM

Criteria Fishnet (F) No.	K	P	N	OC	DD	E	mNDWI	TWI	T	S	R	LULC
F1	1	3	2	1	0.420578	31	−0.07096	7.82229	24.0211	0.681932	3049.5	1
F2	1	2	3	3	0.584131	26	0.035544	16.4966	24.2229	1.07815	3017.1	5
F3	1	2	3	3	0.412963	28	−0.15459	5.43168	24.2555	7.40565	3013.12	3
F4	1	2	1	2	0.429064	35	−0.12699	6.68046	24.0509	4.2646	3086.8	2
F5	1	3	1	2	0.273743	41	−0.06168	7.23296	24.0488	2.45735	3082.61	2
…	…	…	…	…	…	…	…	…	…	…	…	…
F733	3	3	1	1	0.209255	22	0.157364	10.3591	24.5856	0	3268.15	4
F734	3	3	3	3	0.037615	25	−0.11628	13.2047	24.3809	4.42396	3220.74	2
F735	3	3	1	1	0.168723	30	−0.05511	5.86628	24.5922	4.81088	3267.85	2
F736	3	3	2	1	0.021258	27	−0.13219	10.1249	24.5608	0.340978	3229.57	4
F737	3	2	3	3	0	23	−0.03479	12.5564	24.5549	0	3224.37	2

N.B. *K* potassium, *P* phosphorus, *N* nitrogen, *OC* organic carbon, *DD* drainage density, *E* elevation, *mNDWI* modified normalized difference water index, *TWI* topographic wetness index, *T* temperature, *S* slope, *R* rainfall, and *LULC* land use land cover

Table 13.4 Normalized vector for CSA using EWM

Criteria Fishnet (F) No	K	P	N	OC	DD	E	mNDWI	TWI	T	S	R	LULC
F1	0	0.666667	0.5	0	0.482323	0.636364	0.360664	0.195913	0.025863	0.056058	0.145575	0
F2	0	0.333333	1	1	0.669887	0.545455	0.576479	0.784735	0.32241	0.088629	0.068919	1
F3	0	0.333333	1	1	0.47359	0.581818	0.191184	0.033635	0.370316	0.608777	0.059503	0.5
F4	0	0.333333	0	0.5	0.492055	0.709091	0.247126	0.118404	0.069655	0.350569	0.233823	0.25
F5	0	0.666667	0	0.5	0.313931	0.818182	0.379469	0.155908	0.066569	0.202005	0.22391	0.25
…	…	…	…	…	…	…	…	…	…	…	…	…
F733	1	0.666667	0	0	0.239976	0.472727	0.823337	0.368114	0.8554	0	0.662881	0.75
F734	1	0.666667	1	1	0.043137	0.527273	0.268826	0.561277	0.554592	0.363669	0.550713	0.25
F735	1	0.666667	0	0	0.193493	0.618182	0.392768	0.063136	0.865099	0.395475	0.662171	0.25
F736	1	0.666667	0.5	0	0.024379	0.563636	0.236586	0.352217	0.818957	0.02803	0.571604	0.75
F737	1	0.333333	1	1	0	0.490909	0.433947	0.51727	0.810287	0	0.559302	0.25

N.B. *K* potassium, *P* phosphorus, *N* nitrogen, *OC* organic carbon, *DD* drainage density, *E* elevation, *mNDWI* modified normalized difference water index, *TWI* topographic wetness index, *T* temperature, *S* slope, *R* rainfall, and *LULC* land use land cover

13.3 Results

13.3.1 Description of the Selected Criteria

In this total, twelve criteria have been assigned to compute the EWM in the GIS platform. The criteria are topographic, hydrologic, climatic, pedogenic, and anthropogenic types and their description is given below.

13.3.1.1 Topographic Criteria

In the case of evaluating the crop suitability for maize in the Dinhata sub-division, three topographic factors have been analyzed: elevation, slope, and Topographic Wetness Index (TWI) (Fig. 13.3a, b, d). The study area is

Table 13.5 Weights of the criteria for CSA using EWM

Criteria	K	P	N	OC	DD	E	mNDWI	TWI	T	S	R	LULC
Fishnet (F) No												
F1	0	0	−0.34657	0	−0.3435	−0.1188	−0.2546	−0.2648	−0.0698	−0.3607	−0.043	0
F2	0	−0.366204	0	0	−0.3598	−0.1785	−0.3635	−0.1456	−0.3643	−0.3484	−0.0671	0
F3	0	−0.366204	0	0	−0.3645	−0.1924	−0.3517	−0.2261	−0.0711	−0.3522	−0.0794	0
F4	0	−0.27031	−0.34657	−0.34657359	−0.188	−0.3306	−0.3678	−0.3561	−0.3649	−0.3528	−0.1689	0
F5	0	0	−0.34657	0	−0.3224	0	−0.3594	−0.2439	−0.3162	−0.3195	−0.1791	0
…	…	…	…	…	…	…	…	…	…	…	…	…
F733	−0.346574	−0.27031	0	0	−0.3678	−0.2774	−0.1922	−0.2473	−0.3477	−0.3142	−0.3639	−0.3466
F734	−0.346574	−0.27031	−0.34657	0	−0.2937	−0.3065	−0.3678	−0.3669	−0.3173	0	−0.3648	−0.3466
F735	−0.346574	0	0	0	−0.2569	−0.3542	−0.3671	−0.2736	−0.361	−0.1002	−0.3651	−0.3466
F736	−0.346574	0	0	0	−0.3486	−0.2973	−0.1557	−0.3023	−0.3546	−0.1002	−0.3653	−0.3466
F737	−0.346574	0	−0.34657	0	−0.3369	−0.2876	−0.3403	−0.3587	−0.2986	0	−0.3671	−0.3466

Here, *K* potassium, *P* phosphorus, *N* nitrogen, *OC* organic carbon, *DD* drainage density, *E* elevation, *mNDWI* modified normalized difference water index, *TWI* topographic wetness index, *T* temperature, *S* slope, *R* rainfall, and *LULC* land use land cover

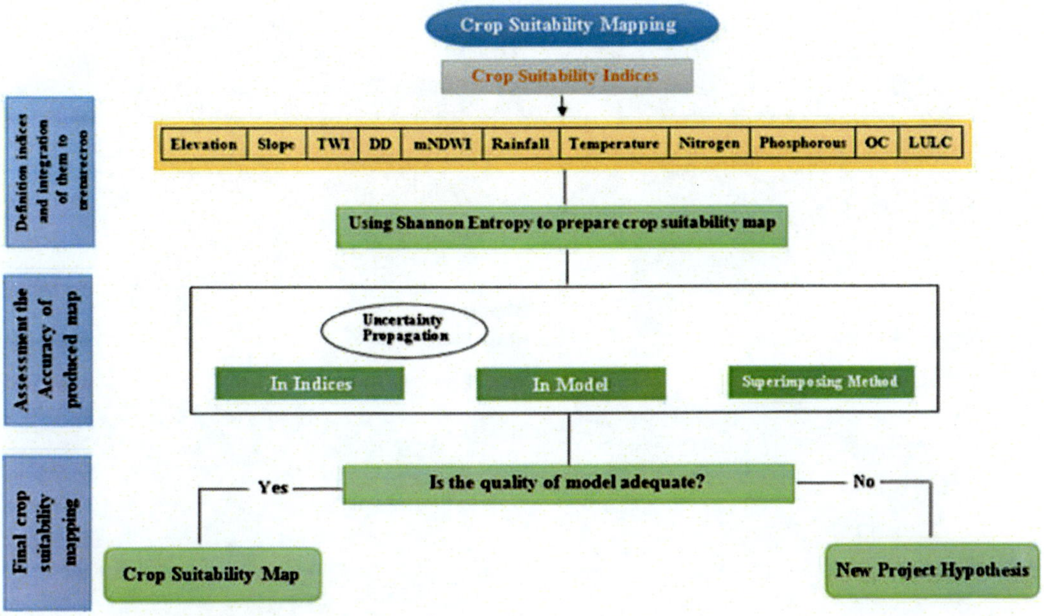

Fig. 13.2 Methodological framework of the current research

generally flat, with only small undulations and a slight southerly slope. The rivers that originate in the northern Himalayas flood low-lying areas during the monsoon season, but locations that are only slightly higher in elevation remain above the water line. The highest elevation of the study area is about 70 m from the mean sea level, and the lowest elevation is 2 m. The relative height is 68 m. Higher elevation is found in the north-western portion, and lower elevation is found in the southern part of the Dinhata sub-division. No hilly tracks exist in this sub-division, so the average slope is moderate to very low. The average slope ranges from 0 to 24.34. TWI has been computed for the present study based on Mitra et al. (2022) and Mitra and Das (2022). TWI is frequently used to measure the influence of topography on hydrological processes. High values in TWI represent low and flat places, whereas low values represent higher elevated lands. The topographic condition of the Dinhata sub-division is relatively flat, as evidenced by Topographic Wetness Index (TWI) values, which ranges from 4.18 to 22.46.

13.3.1.2 Hydrologic Criteria

Hydrological parameters are very significant elements for evaluation in the field of agricultural study. Two important hydrological factors have been considered for current research: drainage density (DD) and Modified Normalized Difference Water Index (mNDWI) (Fig. 13.3c, e). A vast network of river channels runs through the sub-division from the north-west to the south-east. DD has been calculated based on the study by Roy et al. (2022). Higher drainage density indicates high suitability for maize cropping; similarly, lower drainage density indicates lower suitability for maize cropping in the Dinhata sub-division. As maize has been cultivated in the rabi session in West Bengal, river water is utilized for irrigation purposes in the study area. So, DD is considered one of the most significant factors for maize cultivation.

13.3.1.3 Climatic Criteria

The precipitation data has been used to analyze the yearly rainfall of the study area, which is one of the necessary conditions for maize production.

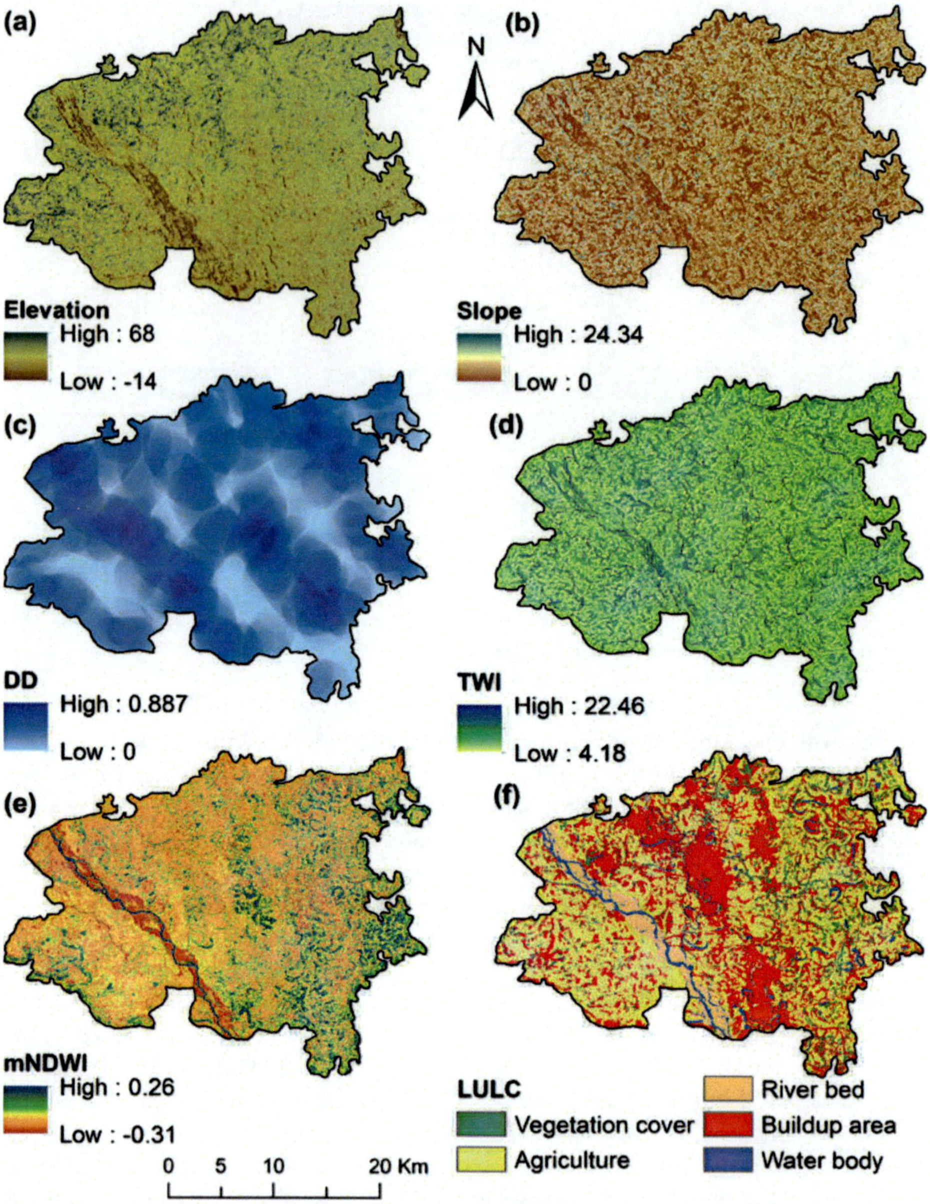

Fig. 13.3 Criteria for maize crop suitability assessment **a** Elevation, **b** Slope, **c** Drainage density (DD), **d** Topographic wetness index (TWI), **e** Modified normalized difference water index (mNDWI), and **f** Land use land cover (LULC)

However, the entire parts of India cultivate maize only in the rainy season except in West Bengal. In West Bengal, maize is cultivated in the winter season, meaning the annual rainfall does not directly affect this particular crop production. The maximum rainfall is observed from June to September. The annual rainfall in the Dinhata sub-division varies from 2981 to 3411 mm (Fig. 13.4f). It is also observed that the maximum annual rainfall has been found in the southern portion, and minimum rainfall is found in the northern part of the Dinhata sub-division. Rainfall is the most significant factor when choosing a site to cultivate maize crops.

Similarly, any region's mean annual temperature directly affects the land suitability of maize crops. The cultivation of maize is best suited to warm temperatures rather than colder temperatures. The research region, around 30% of the Dinhata sub-division, is perfectly suited for the production of maize. The center and northern portions of the study area are the most suitable places for maize cultivation. In the study area, the annual temperature varies from 23.94 to 24.31 °C (Fig. 13.4e). The maximum temperature is observed in the southern portion, and the

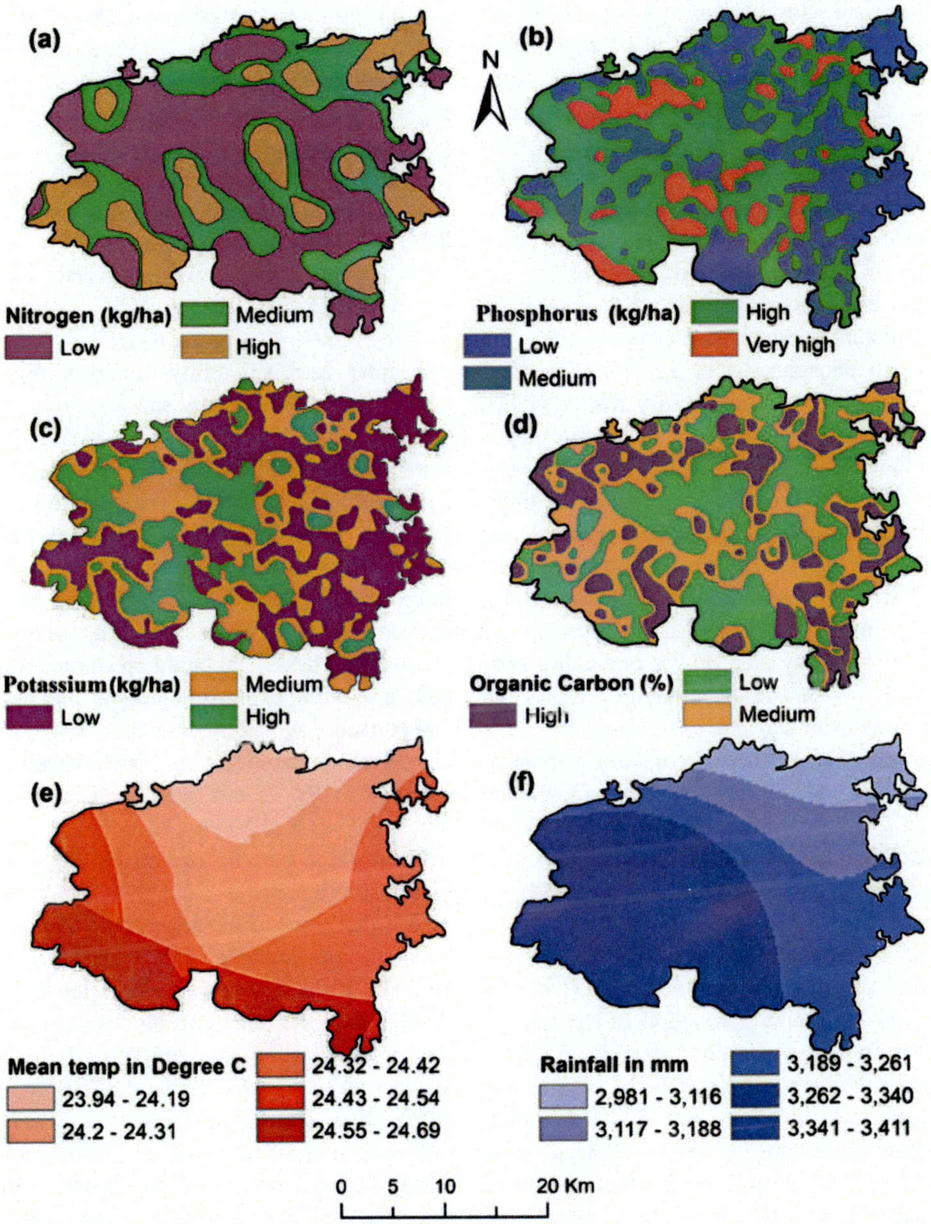

Fig. 13.4 Criteria for maize crop suitability assessment **a** Nitrogen, **b** Phosphorous, **c** Potassium, **d** Organic carbon content in the soil, **e** Mean temperature, and **f** Rainfall

minimum temperature is found in the northern portion of the study area.

13.3.1.4 Pedogenic Criteria

Maize has been grown successfully in different soil types, from loamy sand to clay loam (Williams et al. 2016). Soils that possess high water holding capacity, a suitable amount of nitrogen, potassium, and phosphorous, and a good content of organic matter are deemed ideal for achieving high maize productivity. In the Dinhata sub-division, a high nitrogen concentration in soil is found in the southern portions, and a low concentration is observed mainly in the central part (Fig. 13.4a). The primary element that plants need in order to grow more effectively is nitrogen. Strong growth and dark green color are promoted by sufficient nitrogen (Cheema et al. 2010). Another important pedogenic factor is the presence of phosphorous in the soil. The plant height, grain weight per cob, dry matter yields, and grain yield all rise after phosphorus application (Khan et al. 2005). The study area clearly shows that the western portion is a high-concentration zone of phosphorous, and the eastern part is a lower-concentration area (Fig. 13.4b). Similarly, potassium also plays a vital role in growing maize cultivation in any region. In the study area, lower potassium content in soil is found in the eastern portions, and high potassium in soil has been observed in the western part (Fig. 13.4c). The central portion of the study area has been found to have a low level of organic carbon, and a high-concentration of organic carbon has been observed in the northern and southern parts of the region (Fig. 13.4d).

13.3.1.5 Anthropogenic Criteria

Regarding anthropogenic criteria, LULC is crucial in growing maize production in any region. Side by side, to prepare the LULC map for the Dinhata sub-division of Koch Bihar district, the supervised classification method (Maximum Likelihood algorithm) has been utilized. Five types of LULC classes were identified, viz., vegetation cover, agricultural land, river bed, build-up area, and water bodies, to analyze the land suitability for maize cultivation in this study area. Out of all five categories of LULC, the agricultural field is the dominant land use category observed in this area. In this investigation, the build-up areas and vegetation cove are considered unsuitable for cultivation. As maize is a winter season cropping system in West Bengal, water from several water bodies has been used to grow this crop. LULC classes are also responsible for groundwater recharge (Mitra and Roy 2022); therefore, they have a direct and indirect role in selecting suitable land for cultivation. Figure 13.3f clearly depicted the LULC categories of the Dinhata sub-division.

13.3.2 Description of the Crop Suitability Assessment (CSA)

The final land suitability zonation for maize cultivation in the Dinhata sub-division was generated using the EWM. Figure 13.5 shows the suitability zonation map with five separate zones such as high suitability (S1), moderate suitability (S2), low suitability (S3), currently not suitable (N1), and permanently not suitable (N2). Among the twelve criteria for constructing the final maize crop suitability zonation, seven were identified as beneficial based on studied literature and agricultural expertise opinions, and five were categorized as non-beneficial criteria. The 'moderate' suitability zone contains most of the areas (47.01%) of this study area, followed by the 'high' zone (30.14%), 'low' zone (16.86%), 'currently not suitable' zone (4.85%), and 'permanently not suitable' zone (1.14%). It is clearly visible that most of the areas of the Dinhata sub-division are suitable for maize production. Dinhata-II block is more suitable for maize cultivation than the other two blocks in the Dinhata sub-division. High suitability zone for maize production has been found in Bhetaguri, Singimari, Uttar Baishguri, Dighattari, Kumarganj, Shyamsing, and Sadiatar Kuthi. Conversely, low suitable areas are observed in Nagarsitai, Khamar Sitai, Bholachatra, Pirpalasiti, Bharali, Jhuripara, Dinhata, Choto Atiabari, and Karishal. In the study area, it is also found that central portions

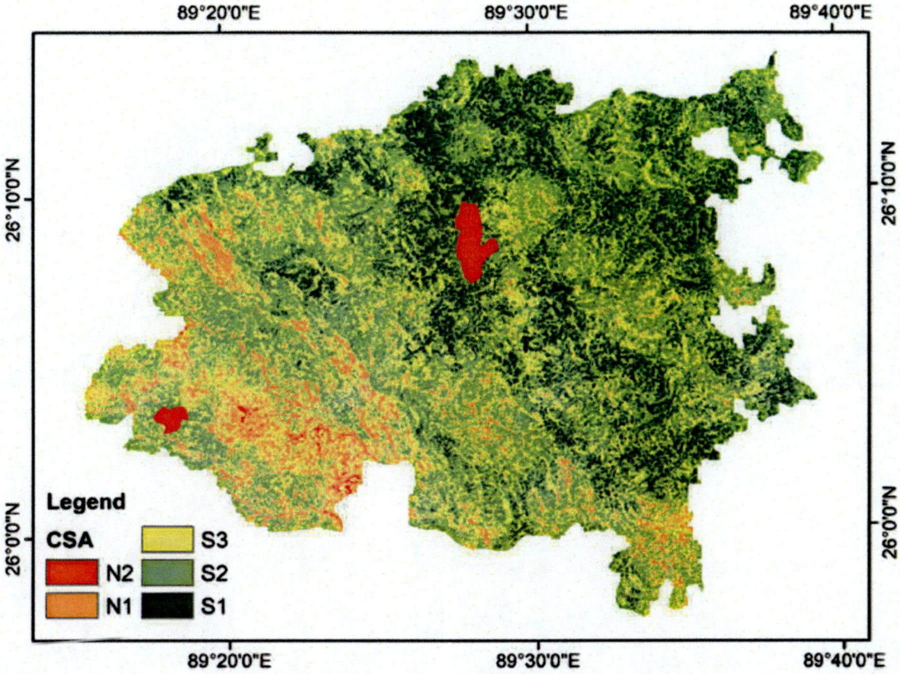

Fig. 13.5 Crop suitability zonation for maize production based on the FAO method (1985)

are moderately suitable areas. Table 13.6 depicts the distribution of areas in different suitability types based on EWM.

Direct field visits and photographic evidence validated this GIS-based entropy model's accuracy and robustness. The researchers collected several photographs using handheld GPS during the field visit. Subsequently, the application of the current model to preparing the suitability zonation for maize cultivation in the Dinhata sub-division is considered almost perfectly accurate. Figure 13.6 manifest the field photographs of four popular maize cultivation area of the Dinhata sub-division. The field photographs were collected at different times based on the growth of maize in this region. The field survey was carried out in 2022, and information on local

Table 13.6 Distribution of areas of different suitability types in the Dinhata sub-division

Suitability types		Area in sq. km	Area in percentage
	N2	8.46	1.14
	N1	35.99	4.85
	S3	223.57	30.14
	S2	371.71	47.01
	S1	125.11	16.86

Fig. 13.6 Maize cultivation in different places of Dinhata sub-division **a** Gitaldaha, **b** Petla, **c** Ataiabari, and **d** Nayarhat

farmers' attitudes toward problems and prospects in the production of maize has been noted.

13.4 Discussion

In India, maize is the 3rd most important cereal crop, representing almost 4% of the world's maize area and 2% of total production. In India, nearly 32 million tons of maize were produced in an area of 9.9 million hectares in 2020–21. States like Karnataka, Madhya Pradesh, Maharashtra, Terengganu, Rajasthan, Uttar Pradesh, Bihar, Tamil Nadu, Gujarat, etc., are predominantly produced maize crops. The area under maize cultivation has increased gradually due to its commercial importance in West Bengal over the last two decades. In 2020 nearly 1.64 million tons of maize were produced in West Bengal; recently, the state Government has decided to raise maize cultivation by 33%. In West Bengal, Koch Bihar is a predominant agriculture district where various cereal crops are produced. The selected study area is the Dinhata sub-division, where maize is produced in 36% of the district's total area.

This study used topographic, soil, climatic, and anthropogenic factors to measure crop suitability classes of maize crops (known as Makai) in the Dinhata sub-division of Koch Bihar district. Some researchers have emphasized and integrated GIS with entropy weighting models and included variable weights to prepare final suitability maps because GIS provides flexibility and accuracy in land use institutions. Our findings show that incorporating the entropy method with GIS can help policymakers and planners make appropriate and wise decisions. It is very important to estimate the range and class of crop suitability in this region for maize cultivation.

Maize has recently been adopted as a cereal crop in the study area. The changing cropping pattern has resulted in the adaptation of various new cereal crops viable in regional settings. Therefore, evaluating various factors related to sustainable agricultural practices is essential. The maize is a Kharif season crop in north India that requires adequate, well-distributed rainfall, sandy loam to silt textured soil with good drainage capacity. Nitrogen, organic carbon, phosphorous, and potassium are the most important nutrients for maize crops, followed by rainfall, drainage density, temperature, and slope. In the study region, about 17% of the land is highly suitable, and 47% is moderately suitable due to the abundance of previously mentioned criteria.

The previous study by Das et al. (2017) integrated the GIS-based subjective AHP model for assessing alternative crops in replacement of tobacco cultivation. However, the GIS-based objective entropy weighting method has been assigned in this chapter. The investigation by Mitra and Sarkar (2018) provides the general land suitability evaluation of the Dinhata-I block using the FAO method. The current study's importance is manifest in the first-time employment of the MCDM EWM and cross-verification of the model using field visits. After evaluating this study, it suggested that other MCDM and machine learning methods can be adopted to estimate the land suitability for maize crops more accurately. The comparative assessment of land suitability can be done using several models in the same study area. That will help the policymakers and agricultural planners in the management of lands.

13.5 Conclusion

In this chapter, we used spatial analytic methods to find the regions that would be suitable for growing maize. We assessed land suitability for maize cultivation in the 764.84 sq km area of the Dinhata sub-division of Koch Bihar district, West Bengal, by identifying 12 different attributes. These results conclude that potassium, phosphorus, nitrogen, organic carbon, topographic moisture index, topographic, and rainfall were the most important beneficial factors in maize productivity across land units. In contrast, drainage density, elevation, modern common difference index, slope, and LULC are non-beneficial factors for maize cultivation in the study area. The study identified the lands near Nayarhat, Gidaldaha, Bhetaguri, Singimari, Atharabari, and Uttar Baishguri as highly suitable sites for maize crop production based on favorable environmental and biophysical factors. At the same time, the agricultural lands near Nagarsitai, Khamarsitai, Bholachatua, Dinhata, and Karishal are shown to be not suitable for maize cultivation. A closer look at the percentage coverage of land suitable for maize cultivation in the study area indicates a very high variation among different parts of the selected subdivision. For example, a larger portion (47.01%) of the Dinhata sub-division (middle west, south, and east) is classified as 'high suitable,' while only 1.14% (north-west, west) of the study area is classified in the 'not suitable' category. Only 16.84% share of the study area was revealed as 'very high' suitable land for maize cultivation especially found in the middle, north, and north-eastern portion of the study area. The results of the entropy weighting method are verified by field inspection and presented as quite simple and transparent for use by decision-makers. The results showed that the decision problem solution obtained with the entropy weighting method was much more accurate and less sensitive. According to experts, this entropy weighting technique can identify criteria weights with varying relative importance for accurately assessing productivity and appropriate agricultural management, especially while facing several complicated parameters.

References

Ahmed GB, Shariff ARM, Balasundram SK, bin Abdullah AF (2016) Agriculture land suitability analysis evaluation based multi criteria and GIS approach. In: IOP conference series: earth and environmental science, vol 37, no 1, p 012044. IOP Publishing

Akıncı H, Özalp AY, Turgut B (2013) Agricultural land use suitability analysis using GIS and AHP technique.

Comput Electron Agric 97:71–82. https://doi.org/10.1016/j.compag.2013.07.006

Alinezhad A, Khalili J (2019) New methods and applications in multiple attribute decision making (MADM). Springer, Berlin/Heidelberg, Germany, ISBN 9783030150082

Bandyopadhyay S, Jaiswal RK, Hegde VS, Jayaraman V (2009) Assessment of land suitability potentials for agriculture using a remote sensing and GIS based approach. Int J Remote Sens 30(4):879–895. https://doi.org/10.1080/01431160802395235

Cheema MA, Farhad W, Saleem MF, Khan HZ, Munir A, Wahid MA, Rasul F, Hammad HM, Cheema M, Farhad W (2010) Nitrogen management strategies for sustainable maize production. Crop Environ 1(1):49–52

Chen ZM, Yeh YL, Chen TC (2018) Assessment of a regional flood disaster indicator via an entropy weighting method. Nat Hazard Rev 19(2):05018002

Cowie AL, Orr BJ, Castillo Sanchez VM, Chasek P, Crossman ND, Erlewein A, Welton S (2018) Land in balance: the scientific conceptual framework for land degradation neutrality. Environ Sci Policy 79:25–35. https://doi.org/10.1016/j.envsci.2017.10.011

Das J, Gayen A, Saha S, Bhattacharya SK (2017) Modelling of alternative crops suitability to tobacco based on analytical hierarchy process in Dinhata Subdivision of Koch Bihar district, West Bengal. Model Earth Syst Environ 3(4):1571–1587. https://doi.org/10.1007/s40808-017-0392-y

Ding X, Chong X, Bao Z, Xue Y, Zhang S (2017) Fuzzy comprehensive assessment method based on the entropy weight method and its application in the water environmental safety evaluation of the Heshangshan drinking water source area, Three Gorges Reservoir area, China. Water 9:329

District Statistical Handbook (2013) Government of West Bengal. Retrieved from: http://wbpspm.gov.in/publications/District%20Statistical%20Handbook

Elaalem M, Comber A, Fisher P (2011) A comparison of fuzzy AHP and Ideal point methods for evaluating land suitability. Trans GIS 15(3):329–346. https://doi.org/10.1111/j.1467-9671.2011.01260.x

FAO (1985) Guidelines: land evaluation for irrigated agriculture. FAO Soils Bull 55

Haghizadeh A, Siahkamari S, Haghiabi AH, Rahmati O (2017) Forecasting flood-prone areas using Shannon's entropy model. J Earth Syst Sci 126(3). https://doi.org/10.1007/s12040-017-0819-x

Hwang CL, Yoon K (1981) Methods for multiple attribute decision making. In: Multiple attribute decision making. Springer, Berlin/Heidelberg, Germany, pp 58–191

Jaafari A, Najafi A, Pourghasemi HR, Rezaeian J, Sattarian A (2014) GIS-based frequency ratio and index of entropy models for landslide susceptibility assessment in the Caspian forest, northern Iran. Int J Environ Sci Technol 11(4):909–926. https://doi.org/10.1007/s13762-013-0464-0

Joerin F, Thériault M, Musy A (2001) Using GIS and outranking multi-criteria analysis for land-use suitability assessment. Int J Geogr Inf Sci 15(2):153–174. https://doi.org/10.1080/13658810051030487

Kahsay A, Haile M, Gebresamuel G, Mohammed M (2018) GIS-based multi-criteria model for land suitability evaluation of rainfed teff crop production in degraded semi-arid highlands of Northern Ethiopia. Model Earth Syst Environ. https://doi.org/10.1007/s40808-018-0499-9

Khan MA, Abid M, Hussain N, Masood MU (2005) Effect of phosphorous levels on growth and yield of maize (Zea mays L.) cultivars under saline conditions. Int J Agric Biol 3:511–514

Kihoro J, Bosco NJ, Murage H (2013) Suitability analysis for rice growing sites using a multi-criteria evaluation and GIS approach in great Mwea region, Kenya. Springerplus 2(1):265. https://doi.org/10.1186/2193-1801-2-265

Malczewski J (2004) GIS-based land-use suitability analysis: a critical overview. Prog Plan 62(1):3–65. https://doi.org/10.1016/j.progres

Malczewski J, Rinner C (2015) Multicriteria decision analysis in geographic information science. Springer, New York, NY, USA

Malekinezhad H, Sepehri M, Pham QB, Hosseini SZ, Meshram SG, Vojtek M, Vojteková J (2021) Application of entropy weighting method for urban flood hazard mapping. Acta Geophys 69(3):841–854. https://doi.org/10.1007/s11600-021-00586-6

Mitra R, Das J (2022) A comparative assessment of flood susceptibility modelling of GIS-based TOPSIS, VIKOR, and EDAS techniques in the Sub-Himalayan foothills region of Eastern India. Environ Sci Pollut Res. https://doi.org/10.1007/s11356-022-23168-5

Mitra R, Sarkar P (2018) Assessment of land suitability through conventional method: a micro level analysis in Koch Bihar District, West Bengal. Res Rev J 3(9):123–1232

Mitra R, Saha P, Das J (2022) Assessment of the performance of GIS-based analytical hierarchical process (AHP) approach for flood modelling in Uttar Dinajpur district of West Bengal, India. Geomat Nat Haz Risk 13(1):2183–2226. https://doi.org/10.1080/19475705.2022.2112094

Mitra R, Roy D (2022) Delineation of groundwater potential zones through the integration of remote sensing, geographic information system, and multi-criteria decision-making technique in the sub-Himalayan foothills region, India. Int J Energy Water Resour, pp 1–21. https://doi.org/10.1007/s42108-022-00181-5

Munier N, Hontoria E, Jiménez-Sáez F (2019) Strategic approach in multi-criteria decision making. Springer International Publishing, Cham, Switzerland

Mustafa AA, Singh M, Sahoo RN, Ahmed N, Khanna M, Sarangi A, Mishra AK (2011) Land suitability analysis for different crops: a multi criteria decision making approach using remote sensing and GIS. Researcher 3(12):61–84

Nyimbili PH, Erden T (2020) A hybrid approach integrating entropy-AHP and GIS for suitability assessment of urban emergency facilities. ISPRS Int J Geo Inf 9 (7):419. https://doi.org/10.3390/ijgi9070419

Özkan B, Dengiz O, Turan İD (2020) Site suitability analysis for potential agricultural land with spatial fuzzy multi-criteria decision analysis in regional scale under semi-arid terrestrial ecosystem. Sci Rep 10 (1):1–18

Pramanik MK (2016) Site suitability analysis for agricultural land use of Darjeeling district using AHP and GIS techniques. Model Earth Syst Environ 2(2). https://doi.org/10.1007/s40808-016-0116-8

Rashid MFA (2018) Capabilities of a GIS-based multi-criteria decision analysis approach in modelling migration. GeoJournal. https://doi.org/10.1007/s10708-018-9872-5

Roy D, Das S, Mitra R (2022) An application of geospatial-based multi-criteria decision-making technique to identify landslide susceptibility zones in the Ragnu Khola River Basin of Darjeeling Himalayan region, India. Appl Geomatics, pp 1–19. https://doi.org/10.1007/s12518-022-00468-6

Shannon CE (1948) A mathematical theory of communication. Bell Syst Tech J 27:379–423

Taheriyoun M, Karamouz M, Baghvand A (2010) Development of an entropy-based fuzzy eutrophication index for reservoir water quality evaluation. J Environ Heal Sci Eng 7:1–14

Wang F (1994) The use of artificial neural networks in a geographical information system for agricultural land-suitability assessment. Environ Plan A Econ Space 26 (2):265–284. https://doi.org/10.1068/a260265

Williams A, Hunter MC, Kammerer M, Kane DA, Jordan NR, Mortensen DA, Smith RG, Snapp S, Davis AS (2016) Soil water holding capacity mitigates downside risk and volatility in US rainfed maize: time to invest in soil organic matter? PLoS ONE 11(8): e0160974

Yalew SG, van Griensven A, Mul ML, van der Zaag P (2016) Land suitability analysis for agriculture in the Abbay basin using remote sensing, GIS and AHP techniques. Model Earth Syst Environ 2(2). https://doi.org/10.1007/s40808-016-0167-x

You L, Wood S (2006) An entropy approach to spatial disaggregation of agricultural production. Agric Syst 90(1–3):329–347. https://doi.org/10.1016/j.agsy.2006.01.008

Zou Z-H, Yi Y, Sun JN (2006) Entropy method for determination of weight of evaluating indicators in fuzzy synthetic evaluation for water quality assessment. J Environ Sci 18:1020–1023

14 Declining Groundwater Level and Its Impact on Irrigation and Agro-production

Shekhar Singh, Dheeraj Mohan Gururani, Anil Kumar, Yogendra Kumar, Manoj Singh Bohra, and Priyanka Mehta

Abstract

Over 80% population of India uses groundwater for household purposes and irrigation water demand. Due to the option of easy availability and minimal initial investment, groundwater is the primary water resource for several industries, including irrigation. More than 90% of annual groundwater withdrawal for irrigation explains the significant reliance on groundwater. Such high dependency, non-monitored, and injudicious use has led to the overexploitation of aquifers across the country, resulting in the depletion of the groundwater level and negatively impacting crop production. To estimate this negative impact more clearly, we managed to run a state-level multiple regression analysis by utilizing agricultural census, groundwater evaluation, and gridded weather data sources for a specific period to understand the impact of groundwater on grain cultivation in India. The chapter clearly shows a link between decreased agricultural productivity and a decline in groundwater level. To protect and the optimum use of vital groundwater resources, thorough research at the block or tehsil level can give more information needed to make decisions on issues that could have long-term environmental and ecological effects.

Keywords

Groundwater depletion · Irrigation · Yield agro-production · Water resource

14.1 Introduction

Our planet is known as the 'Blue Planet' as a sizable section of its surface is covered with water and manifests as water vapor in the atmosphere and groundwater below the ground's surface. Given that the earth is a closed system, very little substance, including water, ever leaves the planet or enters the atmosphere; the water on the planet billions of years ago is still there. The hydrologic cycle is how the earth purifies and refills the water supply. Although there is much water on the planet, but just a small portion (nearly 0.3%) is useable by people. The remaining 99.7% is

S. Singh (✉) · A. Kumar · M. S. Bohra
Department of Soil and Water Conservation Engineering, G.B. Pant University of Agriculture and Technology, Pantnagar, Udham Singh Nagar 263145, Uttarakhand, India
e-mail: shekharsingh861@gmail.com

D. M. Gururani
Department of Civil Engineering, Institute of Meerut, MIET College, 250005, Meerut, Uttar Pradesh, India

Y. Kumar · P. Mehta
Department of Irrigation and Drainage Engineering, G.B. Pant University of Agriculture and Technology, Pantnagar, Udham Singh Nagar 263145, Uttarakhand, India

J. Das and S. Halder (eds.), *Advancement of GI-Science and Sustainable Agriculture*, GIScience and Geo-environmental Modelling, https://doi.org/10.1007/978-3-031-36825-7_14

dispersed across the earth's seas, soils, ice caps, and atmosphere; even so, a large portion of the usable 0.3% is inaccessible, and most of the water humans' use is provided by rivers (Gleick 1993). The visible bodies of water are referred to as surface waters. Aquifer moisture is how most fresh water is kept underground. A river can continue to flow even without precipitation because groundwater can feed the streams. Both surface and groundwaters are helpful to humans (Alizade Govarchin Ghale et al. 2018).

In contrast to other mineral resources, groundwater receives its yearly replenishment from meteoric precipitation, making it the most abundant and precious resource on earth. One of the biggest worries that have emerged in the modern world is groundwater depletion. There is a risk that groundwater will be drained as the population grows and the demand for water rises (Das et al. 2021b). Compared to surface water, groundwater moves more slowly. Consequently, refilling or recharging also requires a lot more time. Although the loss of groundwater is not immediately apparent to us, it also significantly lowers the water table. Despite being hidden, the groundwater is used for various things, including domestic, industrial, and agricultural needs. Urbanization is a major factor in groundwater depletion (Das et al. 2020a). It has led to significant groundwater depletion due to industrial, domestic, and agricultural uses. The groundwater is drained when water from underground is removed in excess. In numerous regions of the world, groundwater is starting to run out, despite being a crucial source of irrigation; in India, where more than 60% of irrigation is accomplished with groundwater combined, and where the water level has decreased by an average of greater than 8 m during the 1980s (Rodell et al. 2009; Siebert et al. 2010; Aeschbach-Hertig and Gleeson 2012; Sekhri 2012; Saha et al. 2018). Since the aquifers pump to increase groundwater withdrawal capacities, groundwater frameworks have increased above twenty million (Mukherji et al. 2013). Thanks to laws encouraging heavy subsidized, unbundled, and unmanaged power, agricultural workers have primarily used groundwater when they need it in the majority of India. Given the grim situation of groundwater depletion in India, it is imperative to comprehend how it affects the nation's food output. Studies show that groundwater depletion has already impacted agricultural productivity and will continue to do so in the future, resulting in decreased crop output and a heightened sensitivity to the adverse impact of climate change. Change in groundwater depth has become one of the main challenges today's dates, which has also attracted many researchers to it (Singh et al. 2021; Gururani et al. 2023).

The present chapter is concerned to the Delhi and Haridwar regions. As Delhi is being affected by urbanization and Haridwar is an industrial area, the groundwater changes were sure to fluctuate. Trend analysis for groundwater depletion was performed monthly and seasonally using Mann–Kendall (MK) and Theil–Sen's estimator (TSE). It enables comparing data over a specific time frame and detecting upward and downward trends and handling any missing values that may be present in the dataset. The MK equation is often used in statistical analysis to detect a significant monotonic trend in a variable over time. If the variable exhibits a monotonic upward or downward trend, it means that it increases or decreases consistently over time without fluctuating randomly or irregularly.

14.2 Study Area and Data Acquisition

The study is done for the Delhi and Haridwar district of Uttarakhand, India, with an aerial extent of 1484 km^2 and 12.3 km^2, respectively, and a detailed location of the research area is depicted in Fig. 14.1. The groundwater data from the Gravity Recovery and Climate Experiment (GRACE) satellite from 2003 to 2015 were obtained from the JPL website by NASA. The study area is located at 28° 36′ 36″ N 77° 13′ 48″E and 29.945°N 78.163°E, an area suffering from groundwater depletion. There have historically been seven cities connected to the Delhi region. The earliest, Indraprastha, is mentioned in the Sanskrit epic Mahabharata (400 BCE and 200 CE), where a city is described as located on a mound beside the Yamuna River. Haridwar has been referred to in the scriptures under several different names, including Kapilasthana, Gangadwara, and Mayapuri. The Char Dham Badrinath, Kedarnath, Gangotri, and Yamunotri which are four main centers of pilgrimage in Uttarakhand could be reached from there.

14.3 Material and Methods

14.3.1 Previous Study

The groundwater levels' trends from 1996 to 2018 for the Delhi Metropolitan Region (DMR) were observed and analyzed. From Central Groundwater Board, the station-level data at a seasonal scale for 256 stations were used for geospatial analysis. The results revealed a downward trend in groundwater, except in floodplains of the Yamuna River, caused by rapid population growth along with the excessive groundwater extraction (Roy et al. 2020). Alterations to the climate as well as human-caused factors like modifications to land use were the major cause of groundwater depletion in northern India. They used observations and statistical analyses (parametric and non-parametric) to present the changes in groundwater storage at various spatial scales. The study suggested the unsustainable extraction of groundwater, frequent dryness, and rising temperature as the reasons for groundwater depletion in Delhi, Punjab, and Haryana (Panda et al. 2021). The groundwater unsustainability was induced due to some identified factors in India. The study evaluated the groundwater condition by considering observations of 5988 monitoring well and applying reliability, resilience, and vulnerability analysis. The study indicated that rainfall deficiency over the years had been the most responsible for the decline in groundwater (Nair and Indu 2021). The groundwater storage depletion in the nearby Gangetic aquifers is related to groundwater base flow reduction and Ganges River depletion. The study employed GRACE data, surface water data, groundwater modeling, and isotope and hydrogeochemical analyses (Mukherjee et al. 2018). The depletion trend for Tamil Nadu state in India by considering the 11-year data between 2002 and 2012 revealed that the depletion rate in groundwater was around 8% higher than the groundwater recharge rate and suggested sustainable agriculture practices for its improvement (Chinnasamy and Agoramoorthy 2015). The research on groundwater levels over time and space in Punjab also revealed the impact of state policies and decisions on groundwater depletion from 2000 to 2019 (Sidhu et al. 2021a, b). The subsidence in Delhi NCR due to groundwater depletion is analyzed in this chapter, which is considered a period of 26 months from 2011 to 2013 and used Terra SAR-X and persistent scatterers' interferometry. Over-exploitation of groundwater and rapid construction activities were the primary causes of the subsidence as per the result of the study (Malik et al. 2019). The temporal variation in groundwater over nine districts of Delhi was assessed by different techniques. Furthermore, this study considered the period from 1996 to 2019 and employed three distinct approaches to analyze the data: Innovative Trend Analysis (ITA), MK, and TSE methods (Sarma and Singh 2021).

Moreover, a study in Karnal district of Haryana was conducted to identify patterns in pre- and post-monsoon groundwater depths by the MK and TSE methods, and they employed time series modeling to forecast changes in respective seasons. Additionally, the study uncovered a decrease in the groundwater depth between 1974 and 2010, with a noteworthy decline between 2001 and 2010 (Patle et al. 2015). The non-parametric MK and TSE techniques were applied to evaluate the annual trends of groundwater storage changes (GWSC) and terrestrial water storage (TWS) in India. The results indicated a significant reduction in GWSC and TWS for Northern India. These variations are attributed to annual rainfall variations due to global warming (Vissa et al. 2019). Twenty wells of Shilabati river basin in West Bengal were chosen to assess the trend in seasonal groundwater levels between

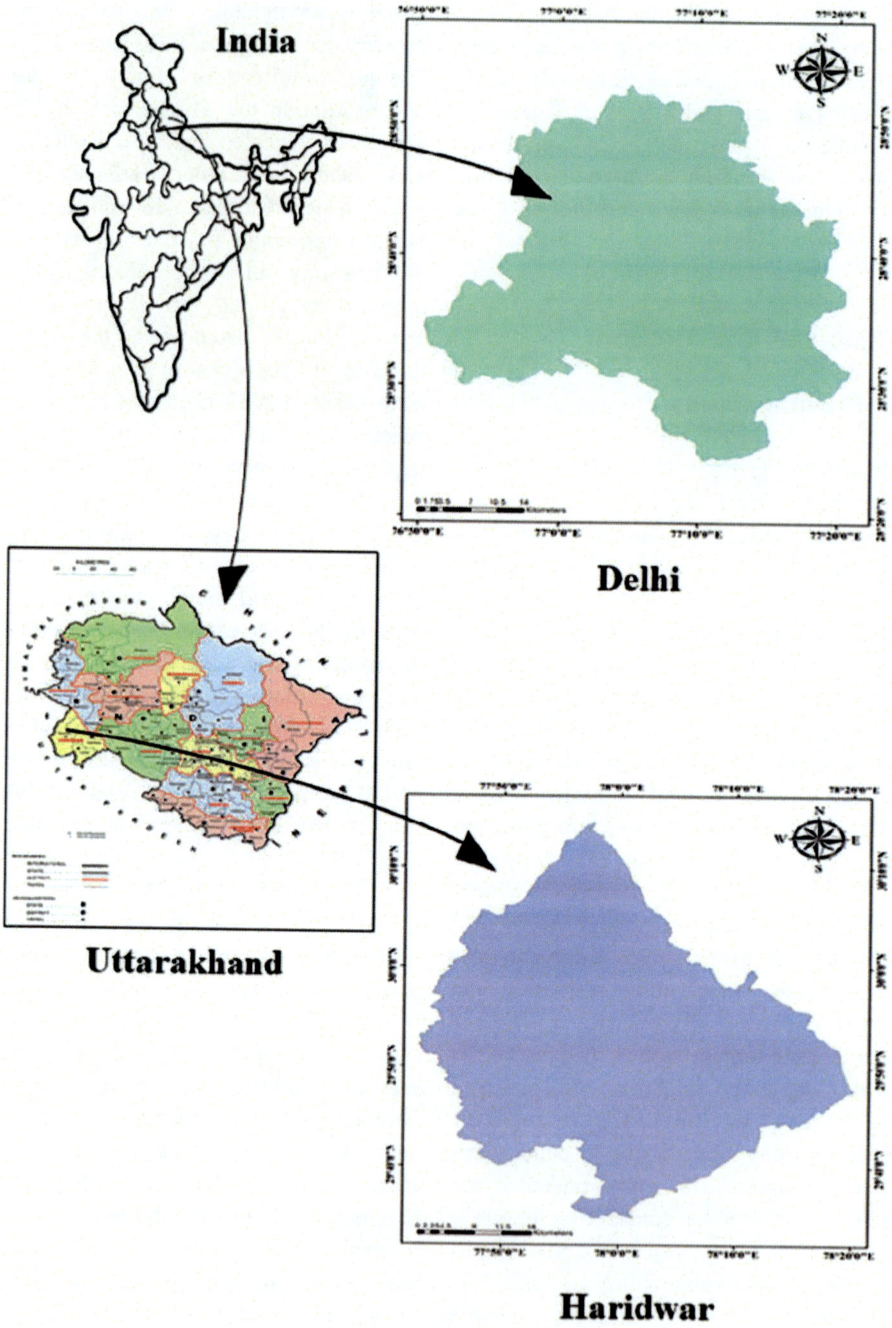

Fig. 14.1 Map depicting the study area

1996 and 2018 which was evaluated by employing statistical tests such as the MK test. Nearly about 60% wells observed exhibited a decrease in their water levels, especially in the post-monsoon season. These wells are primarily situated close to agricultural land, where extensive groundwater extraction from submersible pumping wells has been noted (Halder et al. 2020). In the Malwa region of Punjab, GWL data for 90 observation wells/piezometers were analyzed over 21 years, from 1997 to 2018. The Modified Mann–Kendall (MMK) test and TSE were employed to analyze the trend for the GWL changes in each well. As per the trend analysis, most of the Malwa region's wells significantly declined their GWL between 1997 and 2018 (Sahoo et al. 2021). From 2002 to 2016, high-land irrigation by wells and tube wells led to a steep groundwater decline in the Northwest and Northeastern India. Additionally, rapid urbanization with high runoff and less groundwater table recharge have contributed to groundwater depletion throughout India (Cao and Roy 2020).

14.3.2 Present Study

The present study employed the MK and TSE methods to assess the monthly and seasonal trends consisting of 14 years of GRACE satellite data from 2002 to 2015. This study primarily concentrates on the problem of depleting groundwater and its impact on irrigation and agro-production, and the methodologies of the current study are described as follows.

14.3.3 Mann–Kendall Test

The MK test (Mann 1945; Kendall 1975) is a non-parametric test that determines whether a trend is linear or nonlinear even when the data do not follow any particular distribution (Das and Bhattacharya 2018). Moreover, it addresses the dataset's missing values. The formula for the test statistics is given below.

$$S = \sum_{i=1}^{n-1} \sum_{j=i+1}^{n} \operatorname{sgn}(P_j - P_i), \tag{14.1}$$

$$\operatorname{sgn}(P_j - P_i) = \begin{Bmatrix} +1 & if & (P_j > P_i) \\ 0 & if & (P_j = P_i) \\ -1 & if & (P_j < P_i) \end{Bmatrix}, \tag{14.2}$$

where P_1, P_2, P_3,…, P_n rainfall time series of n dataset and 'sgn' is acknowledged as signum function. Equation 14.3 describes the variance of the MK test when the dataset length exceeds 10 (Xu et al. 2007; Gan 1998; Pandey et al. 2023; Kumar et al. 2023).

$$\operatorname{Var}(s) = \frac{1}{18}\left[n(n-1)(2n+5) - \sum_{g=1}^{q} t_g(t_g-1)(2t_g+5)\right], \tag{14.3}$$

where q is the number of tied groups, t_g is the amount of data in the tied groups, and 'Var' is the variance of the MK test. The following formula estimates (Eq. 14.4) the standard normal statistic Z_{cal} for circumstances where $n > 10$ (Hirsch et al. 1982; Gan 1998; Das et al. 2021a).

$$Z_{cal} = \begin{Bmatrix} \frac{S-1}{\sqrt{\operatorname{Var}(S)}} & \text{if} & (S<0) \\ 0 & \text{if} & (S=0) \\ \frac{S-1}{\sqrt{\operatorname{Var}(S)}} & \text{if} & (S>0) \end{Bmatrix}. \tag{14.4}$$

When the value Z_{cal} is obtained, and comparison is made with the Standard Normal Distribution (Z) table, and if the value of Z_{cal} is higher than the standard Z-value, the null hypothesis (Ho) is excluded, indicating that there is a statistically significant trend in the dataset; otherwise, H_o is accepted, indicating that the trend is not statistically significant (Das et al. 2019; Singh and Kumara 2021).

14.3.4 Theil–Sen Estimator

Theil–Sen estimator (TSE) is employed for calculating the strength or magnitude of the existing trend within the dataset, which was proposed by

Theil (1950) and Sen (1968). The positive β values are marked as an upward trend, whereas the negative β values are designated as a downward trend (Hirsch et al. 1982; Xu et al. 2010; Pandey et al. 2023; Kumar et al. 2023; Das et al. 2020b). The following equation is used for computing the strength of the trend:

$$\beta = \text{Median}\left(\frac{X_i - X_j}{j - i}\right) \text{ For all } i \leq jj \quad (14.5)$$

where β is known as slope estimator and value of $1 < i < j < n$.

14.4 Results and Discussion

In the present chapter, the MK (5% significance level) and TSE tests were employed to examine the impact of groundwater on irrigation and agricultural production during 2002–2015. The MK test involves comparing the Z_{cal} value with the standard Z-value (±1.96), and if the Z_{cal} is larger than the $Z_{1-\alpha/2}$, it suggests that null hypothesis cannot be supported, and a statistically substantial trend appears in the data. The results of the MK and TSE tests are presented in Table 14.1, where all the Z_{cal} values (ranging from −3.2335 to −4.4536 throughout all months and seasons) are bigger than ± 1.96 or falling in the rejection range (Yadav et al. 2014; Guntu et al. 2020; Seenu and Jayakumar 2021; Kumar et al. 2023). Based on the aforementioned result, it can be inferred that the significant decreasing trend has been detected.

When groundwater data were examined on a monthly basis, a significant declining trend (5% significance level) was observed for each month, with magnitudes ranging from −5.14 to −6.69 mm/year. In contrast, when groundwater data were analyzed on a seasonal basis, the techniques (MK and TSE) found a similar significant

Table 14.1 Trend and its strength detection by MK and TSE

Time interval	Delhi			Haridwar		
	*Z_{cal}-value	**TSE	Trend	*Z-value	**TSE	Trend
January	−3.8436	−5.94	Yes	−3.9656	−5.09	Yes
February	−3.8436	−6.25	Yes	−4.4536	−5.32	Yes
March	−3.8436	−5.14	Yes	−3.4775	−4.06	Yes
April	−4.0876	−5.27	Yes	−3.8436	−4.38	Yes
May	−4.1563	−5.41	Yes	−3.2335	−4.05	Yes
June	−3.9656	−6.02	Yes	−3.5995	−4.92	Yes
July	−4.2096	−5.99	Yes	−3.3555	−4.73	Yes
August	−3.7215	−6.22	Yes	−3.4775	−4.43	Yes
September	−4.3316	−6.03	Yes	−4.3316	−5.20	Yes
October	−4.0876	−6.25	Yes	−3.8436	−5.11	Yes
November	−3.9656	−6.69	Yes	−4.0876	−5.24	Yes
December	−4.0876	−6.33	Yes	−4.3316	−5.17	Yes
Pre-monsoon	−4.3316	−15.61	Yes	−3.8436	−12.75	Yes
Post-monsoon	−4.0876	−19.52	Yes	−4.2096	−14.96	Yes
Winter	−4.0876	−12.17	Yes	−4.3316	−10.63	Yes
Monsoon	−4.4536	−24.62	Yes	−3.9656	−18.61	Yes

* 5% Significance level, (−) decreasing trend, ** Magnitude of trend in mm/year

trend of decreasing water levels, but with greater magnitudes ranging from −12.17 to −24.62 mm/year. This indicates that the rate of decline was more pronounced during certain seasons. The negative Z_{cal} value (as shown in Table 14.1) demonstrates a statistically substantial negative trend in the groundwater data, indicating a decline in water levels. This trend is consistent with the prevalence of groundwater extraction in the study area.

Excessive groundwater consumption, reduction in rainfall amount, and unfavorable climatic conditions in affected areas could all contribute to a negative or declining trend in groundwater. This negative groundwater pattern hinders the growth of the study region and lowers agriculture production, putting food availability at risk.

14.5 Conclusions

The most popular non-parametric MK test was employed to identify trends in the groundwater data at a 5% significance level. Additionally, TSE is used to determine the magnitude of the trend. The study's findings demonstrated that the groundwater levels in the Haridwar and Delhi regions have drastically decreased since a strong negative trend persisted during the whole data period (2003–2015). The findings of this study also enable us to conclude that groundwater extraction has been discovered to be much more than groundwater replenishment for the Delhi and Haridwar regions. Additionally, excessive groundwater consumption significantly reduces its availability, which affects agricultural production. On behalf of this study, the authors advised that the local authorities may employ conjunctive water use to decrease overexploitation of groundwater, which in turn increases agro-production yields, and the consumptive water consumption in that area can help to reduce the groundwater problems up to some extent. Additionally, it is necessary to urge local farmers to use groundwater wisely.

References

Aeschbach-Hertig W, Gleeson T (2012) Regional strategies for the accelerating global problem of groundwater depletion. Nat Geosci 5(12):853–861

Alizade Govarchin Ghale Y, Altunkaynak A, Unal A (2018) Investigation anthropogenic impacts and climate factors on drying up of Urmia Lake using water budget and drought analysis. Water Resour Manag 32(1):325–337. https://doi.org/10.1007/s11269-017-1812-5

Cao Y, Roy SS (2020) Spatial patterns of seasonal level trends of groundwater in India during 2002–2016. Weather 75(4):123–128

Chinnasamy P, Agoramoorthy G (2015) Groundwater storage and depletion trends in Tamil Nadu State, India. Water Resour Manag 29(7):2139–2152

Das J, Bhattacharya SK (2018) Trend analysis of long-term climatic parameters in Dinhata of Koch Bihar district, West Bengal. Spat Inf Res 26(3):271–280. https://doi.org/10.1007/s41324-018-0173-3

Das J, Mandal T, Saha P (2019) Spatio-temporal trend and change point detection of winter temperature of North Bengal, India. Spat Inf Res 27(4):411–424. https://doi.org/10.1007/s41324-019-00241-9

Das J, Rahman AS, Mandal T, Saha P (2020a) Challenges of sustainable groundwater management for large scale irrigation under changing climate in Lower Ganga River basin in India. Groundw Sustain Dev 11(1–11):100449. https://doi.org/10.1016/j.gsd.2020.100449

Das J, Rahman AS, Mandal T, Saha P (2021a) Exploring driving forces of large-scale unsustainable groundwater development for irrigation in lower Ganga River basin in India. Environ Dev Sustain 23(5):7289–7309. https://doi.org/10.1007/s10668-020-00917-5

Das J, Mandal T, Rahman AS, Saha P (2021b) Spatio-temporal characterization of rainfall in Bangladesh: an innovative trend and discrete wavelet transformation approaches. Theoret Appl Climatol 143(3):1557–1579. https://doi.org/10.1007/s00704-020-03508-6

Das J, Mandal T, Saha P, Bhattacharya SK (2020b) Variability and trends of rainfall using non-parametric approaches: a case study of semi-arid area. MAUSAM 71(1):33–44. https://doi.org/10.54302/mausam.v71i1.4

Gan TY (1998) Hydroclimatic trends and possible climatic warming in the Canadian Prairies. Water Resour Res 34(11):3009–3015

Gleick PH (1993) Water in crisis, vol 100. Oxford University Press, New York

Guntu RK, Rathinasamy M, Agarwal A, Sivakumar B (2020) Spatiotemporal variability of Indian rainfall using multiscale entropy. J Hydrol 587:124916. https://doi.org/10.1016/j.jhydrol.2020.124916

Gururani DM, Singh S, Joshi H, et al (2023) Assessment of Groundwater Prospects Zones Using RS, GIS, and MIF Methods. In Surface and Groundwater Resources Development and Management in Semi-arid Region: Strategies and Solutions for Sustainable Water Management. Cham: Springer International Publishing. pp 317–335

Halder S, Roy MB, Roy PK (2020) Analysis of groundwater level trend and groundwater drought using Standard Groundwater Level Index: a case study of an eastern river basin of West Bengal, India. SN Appl Sci 2:507

Hirsch RM, Slack JR, Smith RA (1982) Techniques of trend analysis for monthly water quality data. Water Resour Res 18(1):107–121

Kendall MG (1975) Rank correlation methods. Griffin, London, UK

Kumar U, Singh DK, Panday SC et al (2023) Spatio-temporal trend and change detection of rainfall for Kosi River basin, Uttarakhand using long-term (115 years) gridded data. Arab J Geosci 16(3):173

Malik K, Kumar D, Perissin D (2019) Assessment of subsidence in Delhi NCR due to groundwater depletion using Terra SAR-X and persistent scatterers interferometry. Imaging Sci J 67(1):1–7

Mann HB (1945) Nonparametric tests against trend. Econometrica, pp 245–259

Mukherjee A, Bhanja SN, Wada Y (2018) Groundwater depletion causing reduction of baseflow triggering Ganges river summer drying. Sci Rep 8(1):1–9

Mukherji A, Rawat S, Shah T (2013) Major insights from India's minor irrigation censuses 1986–87 to 2006–07. Econ Polit Weekly, pp 115–124

Nair AS, Indu J (2021) Assessment of groundwater sustainability and identifying factors inducing groundwater depletion in India. Geophys Res Lett **48**(3)

Panda DK, Ambast SK, Shamsudduha M (2021) Groundwater depletion in northern India Impacts of the sub-regional anthropogenic land-use, socio-politics and changing climate. Hydrol Process **35**(2)

Pandey BW, Negi VS, Anand S et al (2023) Estimation of anomalies and temporal temperature and precipitation trends in The Cryospheric Himalayan Highland Region (CHHR), Uttarkashi, Uttarakhand, India. Mausam 74(1):29–42

Patle GT, Singh DK, Sarangi A et al (2015) Time series analysis of groundwater levels and projection of future trend. J Geol Soc India 85(2):232–242

Rodell M, Velicogna I, Famiglietti JS (2009) Satellite-based estimates of groundwater depletion in India. Nature 460:999–1002

Roy SS, Rahman A, Ahmed S et al (2020) Alarming groundwater depletion in the Delhi Metropolitan Region a long-term assessment. Environ Monit Assess 192(10):1–14

Saha D, Marwaha S, Mukherjee A (eds) (2018) Groundwater resources and sustainable management issues in India. In: Clean and sustainable groundwater in India. Berlin Springer

Sahoo S, Swain S, Goswami A et al (2021) Assessment of trends and multi-decadal changes in groundwater level in parts of the Malwa region, Punjab. Groundw Sustain Deve, India, p 14

Sarma R, Singh SK (2021) Temporal variation of groundwater levels by time series analysis for NCT of Delhi, India. In: Advances in water resources and transportation engineering. Springer, Singapore, pp 191–203

Seenu PZ, Jayakumar KV (2021) Comparative study of innovative trend analysis technique with Mann-Kendall tests for extreme rainfall. Arab J Geosci 14 (7). https://doi.org/10.1007/s12517-021-06906-w

Sekhri S (2012) India Policy forum vol 9, ed S Shah et al. Sage, New Delhi

Sidhu BS, Sharda R, Singh S (2021a) Spatio-temporal assessment of groundwater depletion in Punjab. Groundw Sustain Deve, India, p 12

Sidhu BS, Sharda R, Singh S (2021b) Spatio-temporal assessment of groundwater depletion in Punjab, India. Groundw Sustain Deve 12:100498

Siebert S, Burke J, Faures JM et al (2010) Groundwater use for irrigation a global inventory. Hydrol Earth Syst Sci 14:1863–1880

Singh S, Kumara S (2021) Non-Parametric Trend Analysis in South-East Regions of Uttarakhand, India. Int J Earth Sci Knowl Appl 3(3):301–304

Singh S, Kumara S, Kumar V (2021) Analysis of groundwater quality of Haridwar Region by application of Nemerow pollution index method. Indian J Ecol 48:1149–1154

Sen PK (1968) Estimates of the regression coefficient based on Kendall's tau. J Amer Statist Assoc 63 (324):1379–1389

Theil H (1950) A rank-invariant method of linear and polynomial regression analysis. Indag Math 12(85): 173

Vissa NK, Anandh PC, Behera MM et al (2019) ENSO-induced groundwater changes in India derived from GRACE and GLDAS. J Earth Syst Sci 128(5):1–9

Xu ZX, Li JY, Liu CM (2007) Long-term trend analysis for major climate variables in the Yellow River basin. Hydrol Process 21:1935–1948

Xu Z, Liu Z, Fu G, Chen Y (2010) Trends of major hydro climatic variables in the Tarim River basin during the past 50 years. J Arid Environ 74(2):256–267. https://doi.org/10.1016/j.jaridenv.2009.08.014

Yadav R, Tripathi SK, Pranuthi G, Dubey SK (2014) Trend analysis by Mann-Kendall test for precipitation and temperature for thirteen districts of Uttarakhand. J Agrometeorol 16(2):164–171

Check for updates

15 Impact of Shifting Cultivation and Changing Land Use on the Hydrology of Iril Watershed, Manipur

Rebati Sinam

Abstract

Shifting cultivation is the most prevalent agricultural system in the hill slopes of Manipur. Forested hilly slopes occupy about 90% of the geographical area of Manipur. The present study analyses the pattern of land use land cover (LULC) for three decadal periods of 2001, 2011, and 2021 of an ungauged watershed Iril River, Manipur. The study assesses the change in the forest type concerning the effect of shifting cultivation and permanent agriculture in the hilly slopes. The impact of LULC change on surface runoff and streamflow is examined using the Soil and Water Assessment Tool (SWAT) model. 'Dense Forest' has a loss of about 24%, while 'Scrubland, Shrubs, and Mixed Forest' has a gain of about 55% from 2001 to 2021. It is observed that most dense forest areas are converted to moderately dense and open forest areas. This transformation is attributed to the prevalence of shifting cultivation and forest fires in the watershed. Overall accuracy assessment of the classification stands at 88.24%, 91.62%, and 92.55% for 2001, 2011, and 2021, respectively. SWAT Model simulation of the Iril watershed shows a validation performance of Nash–Sutcliffe Efficiency of 0.68 and R^2 of 0.69, and percent bias of −2.6%. Results showed that the LULC of the watershed does influence the output of surface runoff and streamflow, but the proportion is different. For a Dense Forest change of about 24–26%, the change in surface runoff is only 2–2.35%, and the change in streamflow is negligible (less than 1%).

Keywords

Iril river · SWAT model · Land use land cover (LULC) · Manipur · Model simulation · Shifting cultivation

15.1 Introduction

Precipitation is considered the primary factor influencing the nature and behavior of runoff. However, it is known that the amount of runoff depends on other factors like the soil type, slope, antecedent moisture content, land use land cover (LULC), etc., of the catchment. Forest and green vegetation influence the runoff volume by delaying or intercepting the falling rainwater. Urbanization, characterized by a great extent of impermeabilization and concretization, leads to flooding and higher runoff. Curve Number (CN) represents the relationship between LULC and hydrograph which Soil Conversation Service

R. Sinam (✉)
Centre for the Study of Regional Development, School of Social Sciences, Jawaharlal Nehru University, New Delhi, India
e-mail: rebatisinam0@gmail.com

J. Das and S. Halder (eds.), *Advancement of GI-Science and Sustainable Agriculture*, GIScience and Geo-environmental Modelling, https://doi.org/10.1007/978-3-031-36825-7_15

developed, later modified by National Resources Conservation Service, and has been incorporated in many rainfall-runoff models like HEC-HMS, SWAT, GLEAMS, VIC, EPA-SWMM, etc. (Khare et al. 2017; Setti et al. 2020). Different authors and scholars have studied the impact of LULC on hydrology due to its importance for sustainable water management and resource conservation (Letha et al. 2011; Sajikumar and Remya 2014; Guzha et al. 2018; Manaschintan and Nalini 2020; Umukiza et al. 2021). Though the impact of LULC on hydrology is widely acknowledged, it takes work to draw generalizations due to variability in local conditions (Umukiza et al. 2021). The different rate of population growth, economic development, land utilization, and resource exploration creates various LULC changes. While some region takes decades to change their forest proportion, others take only a few years to achieve substantial change. Deforestation and the rapid transformation of deforested land into urban areas or cropland is the most common land use change. This transformation is mainly responsible for causing significant changes in runoff characteristics (Garg et al. 2019). Given the increasing incidents of flash floods, inundation, water pollution, and inconsistent rainfall pattern, the study of the impact of LULC on hydrology is essential for any future planning.

The advancement in science and technology and mathematical operations helps create probable and possible scenarios of LULC change. The use of Geographic Information Systems (GIS) and Remote sensing has facilitated the study of land use changes with more ease and higher accessibility. One can execute the impact assessment studies using different time series land use data (Baker and Miller 2013). Most relationship study at the watershed level uses mathematical models like rainfall-runoff, physical, conceptual, and black box models. Hydrological models are simulation models that produce an output or series of outputs in response to an input or series of inputs (Dwarakish and Ganasri 2015). In the rainfall-runoff model, the inputs are a function of the watershed, like area, elevation, channel network geometry, soil, land use, precipitation, temperature, solar radiation, etc., which upon execution, generates a time series of streamflow as output. This model can be used to examine the impact of LULC on the amount of streamflow. In the Soil and Water Assessment Tool (SWAT) model, the output is not restricted to streamflow but extended to a wide range of variables like evaporation, organic content, mineral content, sediment loads, etc. Using probable future land use scenarios; one can determine the corresponding expected streamflow (Anand et al. 2020). However, it is to be noted that the accuracy of the output is dependent on the input data's robustness and the parameters' selection. Moreover, a sufficient and consistent series of observed data are required to validate the results with high accuracy. To achieve near-reality level output, proper calibration of the simulated data, parameterization, sensitivity analysis, and validation is quintessential in hydrological modeling. However, these steps are often complicated for an ungauged catchment with little or no data. Notwithstanding the limitations, studies have been done by several authors to model ungauged catchments and assess the impact of land use change on the hydrology (Gitau and Chaubey 2010; Sajikumar and Remya 2014; Mishra et al. 2017; Sisay et al. 2017; Kumar et al. 2018; Patil and Nataraja 2020; Tanksali and Soraganvi 2021).

Manipur is one of the hilly states in the Northeastern part of India. The state has almost 75% of its geographical area under forest cover. About 5.97% is classed as a very dense forest, 39.05% as moderately dense forest, and 54.98% as open forest as per forest canopy density class. In terms of physiographical structure, the state has a central flat plain surrounded by low-lying rugged hills interspersed with narrow valleys. The region enjoys a wet tropical climate with heavy rainfall ranging between 120 and 270 cm. The average temperature ranges between 1–38 °C. Major rivers include the Imphal river, Iril river, Nambul River, Barak River, Thoubal River, and Nambol River. Imphal and Nambul flow through the heart of Imphal City. Iril river is one of the south-flowing rivers on the eastern side with high discharge potential and flows for a distance of 144 km up to its confluence with Imphal river at Lilong. Flooding and overbanking of rivers is an annual event during the heavy monsoon season.

The state has about 40% tribal population who mainly dwells in hilly areas. Due to the increase in population growth, much of the area is converted into urban built-up land and cropland. Shifting cultivation and forest fires are common practices, one of the state's major problems. According to the Forest Survey of India (2019) report, about 4.48% of the state's total forest area is 'extremely prone' to forest fires, 33.13% are 'very highly fire prone,' and 35.85% are 'highly fire prone area'. Forest fires are one of the reasons for mass deforestation in the state. The changes in forest and land use cover are bound to impact the hydrology of the river basins.

Shifting cultivation is a much-contested issue concerning its environmental sustainability. This unique cultivation method commences with clearing forest cover at the slopes of hills, then letting it dry and burning the stump and stubbles, especially during winter and early spring (Shaw et al. 2022). The cropping season will start during spring, and the plot is used for one year up to three years at the maximum. After this, the plot is left fallow for many years, subjected to the natural process of forest succession. The fallow years are usually 20–30 years in olden times but owing to the increase in population and increasing demands, the fallow period has been reduced to 7–11 years (Bhuyan 2019). The above said concept is also called Jhum Cycle. Studies have shown that the extent of shifting cultivation in North East India has reduced over the years with socioeconomic diversification, higher education, and rising opportunities. Thong et al. (2019) observed decreasing trend in shifting cultivation in the Champhai of Mizoram. The main drivers are increasing education levels, high employment in government service, and environmental awareness. Concerning the detrimental effects of Shifting Cultivation, there is a sharp division between two groups of scholars—one critical of this practice while the other defending the same. Most scholars have reiterated the harmful effects of shifting cultivation on the environment. It is commonly viewed as the main driver of forest fragmentation, soil loss/erosion, and changes in the biophysical environment (Shaw et al. 2022). However, Nath et al. (2022) assert that shifting cultivation is a 'complex and misunderstood form of land use' and recommend a more in-depth understanding of the system in a local context to avoid any unscientific generalization on its relation to an environmental problem. As in the case of Northeast India, Shifting Cultivation is rooted in its cultural and ethnic heritage (Pandey et al. 2022). The biggest challenge to this conundrum is that though the system has been practiced since thousands of years ago, there seems to be no major environmental concern, but why does this practice gain sudden momentum in today's environmental context? Various scholars have studied and contested the sustainability of this cultivation system (Khumbah 2022; Nath et al. 2022; Borah et al. 2022). Thong et al. (2020) studied the forest recovery of shifting cultivation during the fallow period. They found that there are naturally occurring species that can help accelerate the process of forest succession in the Jhum Fallows. Borah et al. (2022) underscore the importance of this practice for preserving bird diversity and maintaining carbon stocks. Some studies resound that shifting cultivation is a carbon-neutral system (Yuen et al. 2013) that has the least effect on hydrology (Ziegler et al. 2009), prevents soil erosion (Valentin et al. 2008), rather enhances soil organic contents (Bruun et al. 2006).

Despite the contestation behind the practice of Jhum cultivation, its undeniable impacts on the surrounding environment cannot be undermined. The main argument should rather concern the extent, magnitude, and direction of its impact on biodiversity, soil, hydrology, land, economics, and the people. It is also to be noted that Manipur has been experiencing rapid change in its land use dynamics due to various developmental and infrastructural activities. This is bound to impact the hydrological behavior of the region. Putting this in the hindsight, the main objective of the chapter is to examine the decadal land use changes, evaluate the SWAT model's output, and study the potential impact of LULC on runoff characteristics of the forested region of the Iril river watershed using the SWAT model.

15.2 Study Area

The area of interest for the present study is the Iril watershed, drained by the Iril river and located between 24° 48′ N–25° 24′ N latitude and 93° 56′ E–94° 18′ E longitude in the state of Manipur, India. Figure 15.1 gives the location of the study area. Iril river has a length of 141 km up to the Outlet point at Moirang Kampu. Iril river is locally known as the 'River of Blood.' Iril river starts from Poumai Naga Village of Lakhamai, flowing through small towns like Saikul, Lamlai, Top, and Irilbung joins the Imphal River at Lilong. The region is predominantly covered with high to moderately thick forest, open forest, and Scrubland. The elevation ranges between 779–2440 m above sea level. The watershed covers parts of Manipur's Senapati, Kangpokpi, Ukhrul, and Imphal East districts. Iril watershed constitutes the sub-basin of the Imphal-Manipur River Basin in the upper catchment. It has an area of 1574.38 km^2 with a watershed boundary of 278.76 km.

For land use classification, Landsat data is used. The primary data required for SWAT Modeling are DEM, land use data, soil, meteorological data, and observed discharge data. The data sources used in the present analysis are tabulated in Table 15.1. The dataset required for soil has been downloaded through the link https://swat.tamu.edu/data/india-dataset/.

15.3 Methodology

15.3.1 LULC Classification

Landsat images used in the study are radiometrically corrected. For radiometric correction, the procedure given by USGS is followed. For this purpose, Landsat level 1 data can be rescaled to top of atmosphere (TOA) reflectance by using the radiometric rescaling formula as prescribed

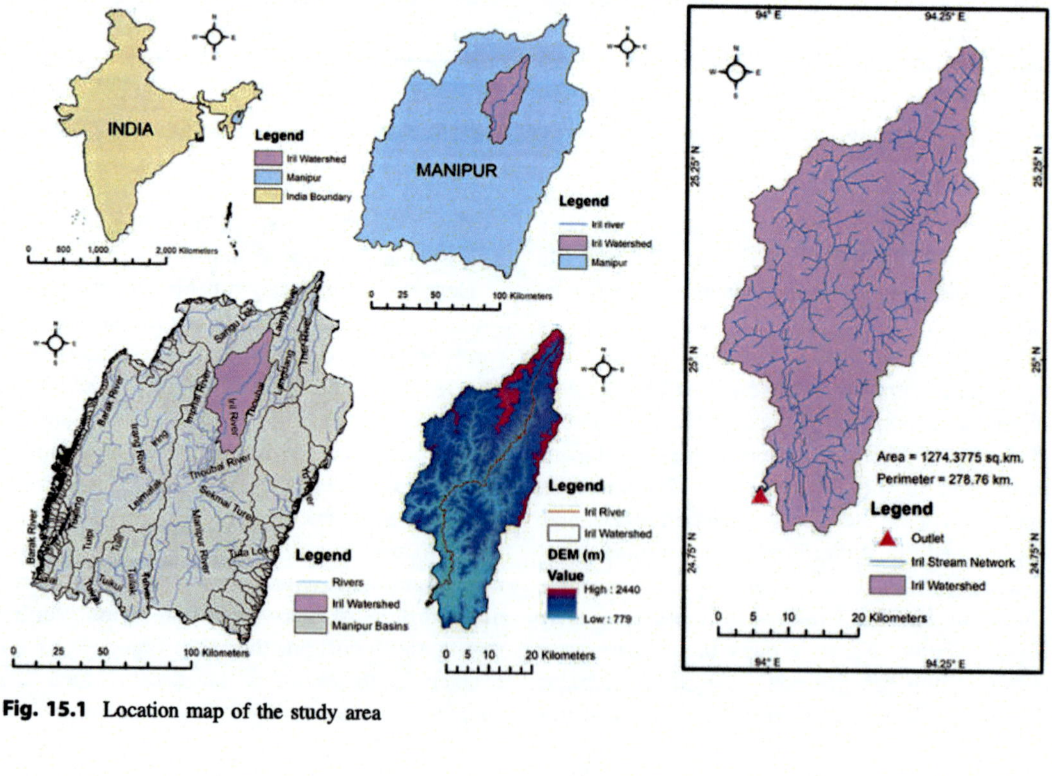

Fig. 15.1 Location map of the study area

Table 15.1 Data sources

S. N.	Type	Period	Sources
1	Elevation (DEM) 30 m × 30 m		Shuttle Radar Topography Mission (SRTM), NASA (https://www.earthexplorer.usgs.gov)
2	Soil data	–	Digital soil map, Food and Agricultural Organization (FAO) – (soil_HWSD_FAO)
3	Land use data	2001, 2011, 2021	Landsat imagery (https://www.earthexplorer.usgs.gov)
4	Rainfall data gridded 0.25° × 0.25°	1981–2020	Indian Meteorological Department (IMD), Pune (https://www.imdpune.gov.in/Clim_Pred_LRF_New/Gridded_Data_Download.html)
5	Temperature 1° × 1°	1981–2020	
6	Daily discharge (Moirang Kampu station-iril river)	1999–2003	Loktak Development Authority (LDA), Manipur

by USGS provided in the metadata of the dataset file. The formula of the conversion can be found at the link https://www.usgs.gov/core-science-systems/nli/landsat/using-usgs-landsat-level-1-data-product.

1. After the radiometric correction, the corrected bands are stacked into composite bands. All analysis is done using ArcGIS 10.8 and Erdas Imagine 2014 software.
2. From the composite band's file, the Iril watershed is extracted using the clip function in ArcGIS.
3. The resultant image is then processed in Erdas Imagine for image classification.
4. The image is combined into different bands for better visual interpretation and accurate selection of training samples. For instance, band 5, band 4, and band 3 are combined for False Color Composite (FCC) for Landsat 8 (2021) image.
5. The image pixels are divided into five LULC classes, as given below in Table 15.2, along with a description.
6. Supervised classification is employed in Erdas Imagine software, and the maximum likelihood classification algorithm classifies the image.
7. The rectification of the classification error is done by recoding the classes after cross-examining with the respective feature using google earth and the natural color Landsat imagery.
8. An accuracy assessment is conducted after completing the LULC classification to check the accuracy level. As used by other studies, kappa analysis and error matrix of accurate measurement is executed in this study. Google Earth images and pan-sharpened images of Landsat 5 and 8 natural color images are used as reference images. Dividing the total number of correctly classified points or pixels by the total number of reference points or pixels gives the overall accuracy while dividing the number of correctly classified pixels in each category by either the total number of pixels in the corresponding column (Producer's accuracy) or row (User's accuracy) gives the individual class accuracy.

15.3.2 SWAT Model Description

The SWAT model is a continuous-time hydrological model that is a semi-distributed, process-based river basin model developed to evaluate the impacts of alternative management decisions on water resources and non-point-source pollution in large river basins—the developmental history of SWAT spans over three decades. SWAT model has been used extensively in many studies (Mishra et al. 2007; Rostamian et al. 2008; Setegn et al. 2009; Zhang et al. 2008; Schmalz and Fohrer 2009; Githui et al. 2009;

Table 15.2 Description of the type of LULC classes

S. N.	Class	Description
1	Waterbodies and Wetland	Rivers, lakes, ponds, flooded vegetation, flooded cultivated area, fisheries, and waterlogged fields
2	Built-up land	Any concrete area, residential, settlement, and man-made structures
3	Cropland	All cultivated areas with no stagnant water with or without standing crops
4	Scrubland, shrubs, and mixed forest	This category includes all areas with scanty vegetation, scrubs, shrubs, barren land, fallow, dispersed forest area, open forest, mixed forest, disserted Jhum Cultivation, etc.
5	Dense forest	All those areas with a thick canopy cover more than 70% with a height of more than 2 m

Easton et al. 2010; Thampi et al. 2010; Shi et al. 2011; Abbaspour et al. 2015; Molina-Navarro et al. 2017; Hallouz et al. 2019; Oo et al. 2020; Hu et al. 2020; Tufa and Sime 2021; Akoko et al. 2021; Malik et al. 2022).

SWAT operates on a 'daily time step and is primarily designed to project the impact of land use and management on water, sediment, and agricultural chemical yields in the ungauged watershed' (Arnold et al. 2012). The model is capable of long-time simulation as it is computationally efficient. However, it requires various input data for its different components. Some important components are weather, soil, land use, plant growth, nutrients, pesticides, hydrology, etc. SWAT mainly functions by dividing the watershed into multiple smaller sub-watersheds, again divided into 'hydrologic response units (HRUs)' comprising homogenous land use, soil, and slope characteristics. The main driving force behind SWAT processes is water balance, which impacts plant growth, sediment transport, nutrients, pesticides, and pathogens. Water balance is dependent on a hydrologic cycle which is climatic driven and involves a wide range of moisture and energy inputs like precipitation, temperature, wind or solar radiation, etc. SWAT uses these observed data and generates simulated data like canopy storage, surface runoff, lateral flow, water yield, evaporation, groundwater recharge, and return flows. SWAT has a wide range of applications. The details of SWAT can be referred to from SWAT documentation available online.

In this present chapter, SWAT is used to simulate runoff in the context of understanding the impact of LULC on runoff characteristics. The steps involved in the SWAT modeling are given below:

1. Collection of Input Data: The input data used in the chapter include Digital Elevation Model, Soil information, Land use data, Meteorological data (Rainfall, Temperature, Solar Radiation, Wind Speed, Relative Humidity), and Observed Discharge data. The sources of these input data are given in the Data sources section.
2. Software Installation: ARCSWAT version 2012 is installed and integrated with ArcMap 10.5. Software is downloaded free of cost from the swat.tamu.edu website.
3. Pre-processing of Input data: The spatial data (DEM, Soil Map, Land Use Map) are reprojected to WGS_1984_UTM_Zone_46N corresponding to the location of Iril watershed. All the raster files are resampled to 30 × 30 m spatial resolution, even though SWAT has no problem processing raster with varying resolutions. The soil classes and parameters are according to the classification provided by Food and Agricultural Organization. For meteorological data, daily Rainfall (0.25° × 0.25°) and daily Temperature (1.0° × 1.0°) data as provided by IMD Pune are used. The software simulates the other required meteorological data (Solar radiation, Wind Speed, Relative Humidity).

For Land Use Land Cover (LULC), three LULCs are classified for 2001, 2011, and 2021. There are five LULC classes: Water bodies and Wetlands, Built-Up Land, Cropland, Scrubland, Shrubland, and Mixed Forest Dense Forest. These classes are reclassified to corresponding classes per the SWAT database for a land cover class given in the text file 'CropLu' in the SWAT database. For Waterbodies and Wetlands, it is assigned WATR (water), Built-Up Land is assigned URHD (Residential-High Density), Cropland is assigned AGRL (Agricultural Land-Generic), Scrubland, Shrubland, and Mixed Forest is assigned FRST (Forest-Mixed), and Dense Forest is assigned FRSE (Forest-Evergreen).

4. Processing of the model: The model run involves five basic steps:
   a. Project Set-Up
   b. Watershed Delineator: DEM set-up → Stream Definition → Stream Network → Outlet and Inlet Definition → Watershed outlet selection, and definition → Calculation of sub-basin parameters. As the stream network created by DEM does not conform to the real ground, the actual river network is burned in the DEM in the burn option. The outlet point is chosen at Moirang Kampu (24.82° N, 93.98° E) and the watershed delineated corresponds to an area of 1274.4 km^2.
   c. HRU Analysis: Land use data, → Soil data → Slope → Overlay. For slope, five slope classes are created 0–10, 10–20, 20–30, 30–40, and 40–9999 classes Imphal Watershed. In HRU definition, the threshold for land use, soil, and slope percentage are kept at 0–0–0%.
   d. Write input Tables: Weather station → Write SWAT input Tables. Daily Rainfall and Temperature are added here. Gridded Rainfall and Temperature (Table 15.3) obtained from Indian Meteorological Department are used.

5. SWAT simulation: Run SWAT → Run SWAT output. The simulation period is from 1/1/1995 till 12/31/2020 with a warm-up period of 4 years.
6. Post-Processing: The model run's result is in the scenario folder in the project file within the 'TableOut' and 'TxtInOut' folders. The output flows are then analyzed in Microsoft excel, compared, and examined.

After successful model building, model calibration and validation with the actual data are required to achieve realistic output results (Arnold et al. 2012). Calibration is adjusting the model input parameters to achieve the best simulation match with the actual observation. It is an effort to parameterize better to reduce prediction uncertainty. Calibration involves selecting the parameter values by comparing model predictions for a given set of assumed conditions with observed data for the same conditions. Calibration and validation are done by dividing the observed data into two datasets for each purpose.

Several statistical tests are available to determine the fitness between the simulated and observed data for judging the SWAT predictions. The statement mentioned above includes a coefficient of determination (R^2), Nash Sutcliffe Efficiency (NSE), root mean square error (RMSE), nonparametric test, *t*-test, objective functions, autocorrelation, etc. (Arnold et al. 2012). The formulae for the calculation of some of the common objective functions are given below:

Nash–Sutcliffe Efficiency (NSE)

$$\mathrm{NSE} = 1 - \frac{\sum_{i=1}^{n}(O-S)^2}{\sum_{i=1}^{n}(O-A)^2}.$$

RSR is the ratio of root mean square error to standd deviation of observed data

$$\mathrm{RSR} = \frac{\mathrm{RMSE}}{\mathrm{SD}} = \frac{\sqrt{\sum_{i=1}^{n}(O-S)^2}}{\sqrt{\sum_{i=1}^{n}(O-A)^2}}.$$

PBIAS, or Percent Bias, is given by

Table 15.3 Information on rainfall and temperature data used in the SWAT model

Rainfall grid point	Name	Lat	Long	Elevation (m)
1	pcp74_110	24.75	93.75	1402
2	pcp74_111	24.75	94	792
3	pcp74_112	24.75	94.25	1222
4	pcp75_110	25	93.75	1354
5	pcp75_111	25	94	828
6	pcp75_112	25	94.25	1074
7	pcp76_110	25.25	93.75	1201
8	pcp76_111	25.25	94	1185
9	pcp76_112	25.25	94.25	1037
Temperature grid point				
1	tmp18_27	24.5	93.5	585
2	tmp18_28	24.5	94.5	186
3	tmp19_27	25.5	93.5	499
4	tmp19_28	25.5	94.5	1354

$$\text{PBIAS} = \frac{\sum_{i=1}^{n}(O - S) \times 100}{\sum_{i=1}^{n}(O)}.$$

where O = Observed Discharge, S = Simulated Discharge, A = Mean observed Discharge, N = Number of observations.

The statistical ranges of these tests and their interpretation are given in Table 15.4.

In this present chapter, since there is insufficient observed data at the outlet as the rivers are ungauged, calibration and validation are impossible for Iril watershed. There is observed data from 1999 to 2003, which is needed for both calibration and validation. Therefore, this data is used to validate the simulated data. The model is run multiple times to achieve the near-accurate level of the observed data based on the physical reasoning of the author by changing various input factors. For instance, in the case of Iril watershed, the slope classes and HRU definition are changed multiple times and validated again to achieve the most satisfactory result. The fitness

Table 15.4 Performance classification of objective function

Objective functions	Value range	Performance classification
R^2	$0.7 < R^2 < 1$ $0.6 < R^2 < 0.7$ $0.5 < R^2 < 0.6$ $R^2 < 0.5$	Very good Good Satisfactory Unsatisfactory
NSE	$0.75 < \text{NSE} \leq 1.00$ $0.65 < \text{NSE} \leq 0.75$ $0.50 < \text{NSE} \leq 0.65$ $\text{NSE} \leq 0.50$	Very good Good Satisfactory Unsatisfactory
RSR	$0.00 \leq \text{RSR} \leq 0.50$ $0.50 \leq \text{RSR} \leq 0.60$ $0.60 \leq \text{RSR} \leq 0.70$ $0.70 \leq \text{RSR}$	Very good Good Satisfactory Unsatisfactory
PBIAS	$\text{PBIAS} \leq \pm 10$ $\pm 10 \leq \text{PBIAS} \leq \pm 15$ $\pm 15 \leq \text{PBIAS} \leq \pm 25$ $\text{PBIAS} \geq \pm 25$	Very good Good Satisfactory Unsatisfactory

of the simulation is tested using the four objective functions listed above. Using an excel template file 'SWAT Calibration Helper v1.0' created by Shrestha (2016) and observed in SWAT CUP Version 2012, the objective functions are generated.

15.4 Results and Discussion

15.4.1 Accuracy Assessment

Accuracy assessment is an important post-image classification procedure to validate the result of classification. It is very important to understand the results related to change detection and its subsequent application for land management, urban planning, and decision-making (Deng et al. 2016). The most common accuracy assessment method used by almost all LULC change analysts is the error matrix, confusion matrix, or kappa analysis. For the selection of the referenced training sites, stratified random sampling is recommended by most authors (Congalton 1991). This chapter applies a combination of user-defined stratified random sampling concerning Google Earth imagery and pan-sharpened images of Landsat 5 and 8. The result of the accuracy assessment is given in Table 15.5. The kappa coefficient gives the difference between the classified map's true agreement and chance agreement to the reference data. Lillesand et al. (2004) assert that for accurate and good classification, kappa should be higher than a value of 0.80; for moderate accuracy, it must be between 0.40 and 0.80, and any value lesser than 0.40 is of poor classification. As per the USGS classification scheme, an overall accuracy of 85% is the standard value for a satisfactory result (Anderson et al. 1976).

The overall accuracy for 2001, 2011, and 2021 is 85.24%, 91.62%, and 92.55%, respectively, which can be considered a fairly classified image. The kappa coefficient for the three periods is 0.84, 0.88, and 0.90 for 2001, 2011, and 2021 respectively. The limitation of overall accuracy and kappa statistics is that they did not denote each class's accuracy level. Therefore, the producer's accuracy and the user's accuracy give an estimate of the level of accuracy of each class. The producer's accuracy for Dense Forest, Cropland and Scrubland, Shrubs, and Mixed Forest is high in all three periods. Dense Forest has an accuracy value of 98.6%, 97.3%, and 94.6% in 2001, 2011, and 2021 respectively. Similar is the case for the classification of Scrubland, Shrubs, and Mixed Forest, with an accuracy of 81.5% (2001), 96.2% (2011), and 100% (2021), while Cropland has an accuracy of 96.6% (2001), 96.6% (2011), and 100% (2021). Water bodies and Wetland has been classified with the least accuracy. It had only 60% accuracy in 2001 and 69.6% and 68% in 2011 and 2021, respectively. This may be because the classifier needed help tracing the river channel's path in the watershed while the reference data included it. Most referenced water pixels have been classified as Cropland or Built-up land. The images used in the classification are from February, the lean season in the Iril watershed. During the lean season, most rivers dry up and can give a different spectral signature in the sensors. This error is partially rectified by recoding the class in Erdas Imagine software. However, the classification

Table 15.5 Accuracy assessment of land use land cover classification, Iril watershed

Year	Producer's accuracy					User's accuracy					Overall accuracy	Kappa statistics
	WB	BU	CL	SL	DF	WB	BU	CL	SL	DF		
2001	60.0	86.7	96.6	81.5	98.6	100	100	84.8	95.7	81.9	88.24	0.84
2011	69.6	69.2	96.6	96.2	97.3	100	90	84.8	94.3	91.1	91.62	0.88
2021	68.0	81.8	100.0	100.0	94.6	100	60	93.5	89.1	100.0	92.55	0.90

* *WB* Waterbodies and wetland; *BU* Built-up land; *CL* Cropland; *SL* Scrubland, shrubs, and mixed forest; *DF* Dense forest

error must be addressed owing to user errors. Built-up land has an accuracy of 86.7% in 2001, 69.2% in 2011, and 81.8% in 2021. The user's accuracy determines how accurate the existing map is. Waterbodies and wetlands have been accurately classified in all the periods, with an accuracy of 100% in all three periods. Dense Forest has a user accuracy of 81.9% (2001), 91.1% (2011), and 100% (2021), while Scrubland, Shrubs, and Mixed Forest has a user accuracy of 95.7% (2001), 94.3% (2011), and 89.1% (2021). Built-up land has a user's accuracy of 100% (2001), 90% (2011), and 60% (2021), while Cropland has a user's accuracy of 84.8% (2001), 84.8% (2011) and 93.5% (2021). From the result of the accuracy assessment, the classification has been done satisfactorily.

15.4.2 LULC Classification

The result of LULC classification is given in Fig. 15.2 and Table 15.6. As of 2021, Dense forest occupies 47.92% of the total area of Iril watershed, 41.53% by scrubland, shrubs, and mixed forest, 8.89% by cropland, 1.03% by built-up land and 0.63% by waterbodies and wetland. Only built-up land shows a consistent increase over the last three decades. The built-up land area was 7.98 km^2 in 2001, which increased to 8.81 km^2 in 2011 and then to 13.17 km^2 in 2021, a total of 64.95% increase from 2001 to 2011. The rest of the classes show a fluctuating trend. Waterbodies and Wetland showed an increase in 2011 from the 2001 proportion but is reduced again in 2021. From 2001 to 2021 state, it shows an increase of 1.53%. Cropland has more or less occupied the same areal coverage and shows a slight increase of 1.16% from 2001 to 2021. Scrubland, shrubs, and mixed forests show a decrease of 7.51% from 2001 to 2011 but a drastic increase of 68.45% from 2011 to 2021, giving an overall increase of 55.80% in 2021 from 2001 state. The opposite trend is found in the case of dense forest. It shows an increase of 3.12% from 2001 to 2011 but a drastic reduction of 26.6% from 2011 to 2021, giving an overall areal decrease of 24.31% from 2001 to 2021. Manipur forests can be divided into very dense, moderately dense, open, and scrub. It has been found that most very dense forests are converted to moderately dense or open forests.

This drastic change in forest land can be due to the problem of shifting cultivation, selective logging, biotic pressures, and forest fires prevalent in the watershed (Reimeingam 2017). As mentioned earlier, deforestation and forest fire are common issues in the watershed. Most of the forest land is trimmed out to transform into Jhum land and is left bare after harvest. Renewing the deserted land takes years and is often occupied by smaller trees and scrubs, which may be one reason. The alternating increase and decrease of Dense Forest are a common phenomenon of the watershed owing to the wide presence of hill dwellers whose main occupation is forest produce and Jhum cultivation. One limitation of this chapter is that a ground truthing of the classified image could not be performed, and the accuracy assessment is based on Google Earth images of the current year.

15.4.3 Impact of LULC on Runoff Characteristics

The mean daily observed flow of Iril watershed is 27.96 cumecs with a standard deviation of 26.89 cumecs, while the simulated flow is 27.24 with a standard deviation of 23.91 cumecs. The plot of the average monthly streamflow is presented in Fig. 15.3 for the simulation period 1999–2003. The performance (validation) of the simulation is tested using the coefficient of determination (R^2), Nash–Sutcliffe efficiency (NSE), root mean square error-observations standard deviation ratio (RSR), and percent bias (PBIAS), and the result is given in Table 15.7.

R^2 estimates how well the prediction model replicates the variance of observed data. It can range from 0 to 1, where a value of 0 means no correlation between the simulated and observed data, and a value of 1 means perfect correlation. In the case of Iril watershed, the model has an R^2

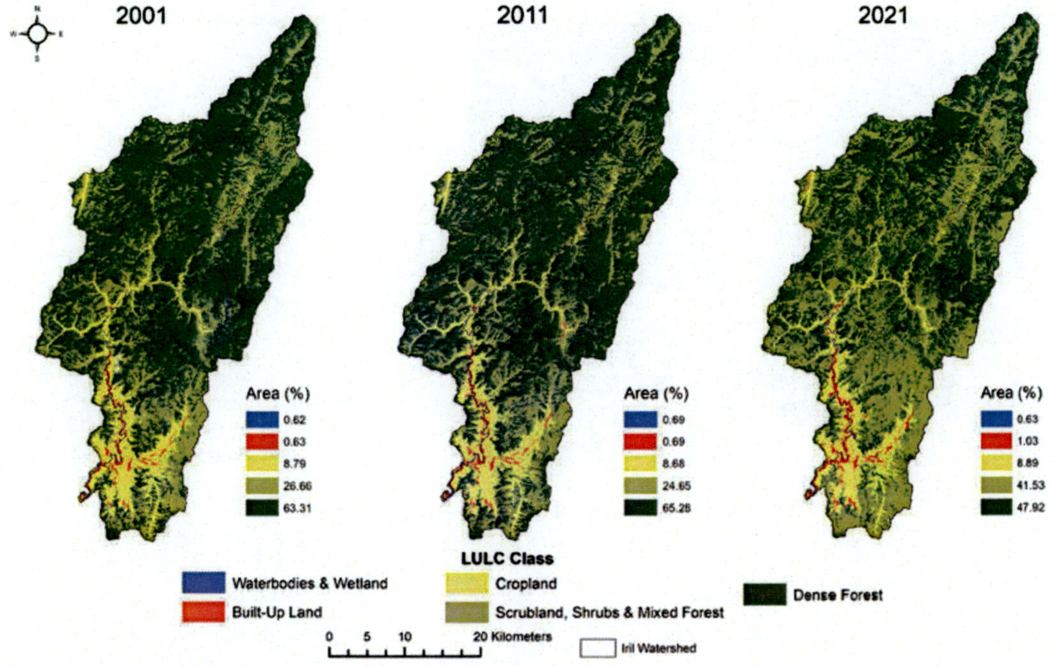

Fig. 15.2 Land use land cover classes of Iril watershed, 2001–2011

Table 15.6 Change in land use land cover class of Iril watershed

Class	2001		2011		2021		Change (%)		
	in km^2	in %	in km^2	in %	in km^2	in %	2001–2011	2011–2021	2001–2021
Waterbodies and wetland	7.92	0.62	8.80	0.69	8.04	0.63	11.13	−8.63	1.53
Built-up land	7.98	0.63	8.81	0.69	13.17	1.03	10.42	49.39	64.95
Cropland	111.99	8.79	110.62	8.68	113.29	8.89	−1.22	2.41	1.16
Scrubland, shrubs, and mixed forest	339.71	26.66	314.19	24.65	529.25	41.53	−7.51	68.45	55.80
Dense forest	806.77	63.31	831.96	65.28	610.62	47.92	3.12	−26.60	−24.31

value of 0.69, which falls in the 'good' category of performance with a Karl Pearson correlation coefficient of 0.83.

Nash–Sutcliffe efficiency measures how well the observed values match with the simulated output along a 1:1 regression line having a slope of 1. Its values range from $-\infty$ to 1, where 1 indicates a perfect match between the simulated and observed values. A model with NSE values ≤ 0 means that the observed data mean is a more accurate predictor than the simulated output. In streamflow, NSE values less than or equal to 0.50 are considered unsatisfactory, and the output is deemed erroneous. A value greater than 0.50 is expected for the result of the model to be in an acceptable range. In the case of the Iril subwatershed, the NSE values stand at 0.68, which falls in 'good' performance. It has been reported that R^2 and NSE are usually biased toward high flows, due to which it has been suggested to use other efficiency test measures (Arnold et al. 2012). In the present chapter, RSR

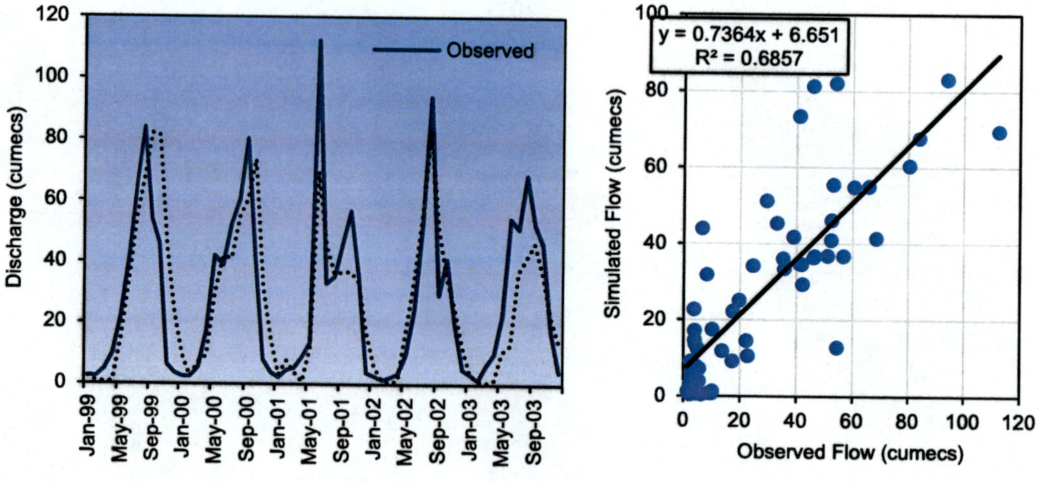

Fig. 15.3 Validation of SWAT model simulation, Iril river watershed

Table 15.7 Validation of the simulation

Name of watershed	Objective function	Result	Performance
Iril watershed	NSE	0.68	Good
	R^2	0.69	Good
	RSR	0.56	Good
	PBIAS	-2.6%	Very good

and PBIAS are calculated as well. RSR is the ratio of root mean square error and standard deviation of the observed data. It ranges from 0 to ∞ where a value of 0 means a lower RMSE and better model simulation performance. For accepting a model simulation, it is expected to have an RSR value less than or equal to 0.7; beyond this value, the model is unsatisfactory. Iril watershed model has an RSR of 0.56, which is 'good.' PBIAS measures the mean tendency of the simulated data to be larger or smaller than the corresponding observed values. An optimal value of 0.0 PBIAS means an accurate model simulation. To be acceptable, the value of PBIAS must be lesser than or equal to ± 25%. Iril river model has a PBIAS of − 2.6%, a 'very good' model performance category. With this statistical result, we can infer that SWAT successfully simulated the Iril subwatershed's streamflow with satisfactory performance. With the assumption that SWAT can generate acceptable streamflow for the given watershed, the model is run using the LULC class image of 2011 and 2021 and keeping all other inputs constant to observe the likely change in the output with different land use input data.

The SWAT model is run using the land use images of 2011 and 2021, and the surface runoff output is compared. The average annual precipitation is 1394.4 mm, with a potential evapotranspiration of about 959 mm. From the SWAT output, it is found that the average curve number for 2001 is 78.54, 2011 is 78.48, and 2021 is 78.9. The average surface runoff for 2001 is 319.71 mm, for 2011 is 318.86 mm, and for 2021, it is 326.61 mm. The result is summarized in Table 15.8. It is apparent from the Table 15.8, that there is a minute change in the hydrological components of the watershed over the three periods. The surface runoff saw a minimal decrease from 319.71 mm in 2001 to 318.86 mm in 2011, which is only 0.27% but increased to 326.61 mm, which is a change of 2.43% from that of 2011. This pattern is similar to the

Table 15.8 Comparison of hydrological components of Iril watershed for all the years

Hydrological component (mm)	LULC 2001	LULC 2011	LULC 2021	Change (%)		
				2001–2011	2011–2021	2001–2021
Precipitation	1394.4	1394.4	1394.4	0	0	0
PET	959.3	959.4	959.3	0	0	0
Evaporation and transpiration	481.1	480.9	484.2	−0.04	0.69	0.64
Average curve number	78.54	78.48	78.9	−0.08	0.54	0.46
Surface runoff	319.71	318.86	326.61	−0.27	2.43	2.16
Lateral flow	119.41	119.63	117.89	0.18	−1.45	−1.27
Return flow	427.51	427.86	419.42	0.08	−1.97	−1.89
Revap from shallow aquifer	19.04	19.03	19.04	−0.05	0.05	0.00
Infiltration uptake/soil moisture redistribution	470.45	470.81	461.93	0.08	−1.89	−1.81
Recharge to deep aquifer	23.52	23.54	23.1	0.09	−1.87	−1.79

watershed's Scrubland, Shrubs, and Mixed Forest trend. The percentage change in Scrubland, Shrubs, and Mixed Forest from 2001 to 2011 is 7.51%, and from 2011 to 2021, the change is 68.45%. The proportion of change in the LULC classes is observed to differ from the proportion of change in their respective hydrological component.

Figure 15.4 gives the difference in average monthly surface runoff (mm) and streamflow (cumecs) for all three periods. The change observed is minimal in most months and less than 10%, except for February between 2011 and 2021. The result in surface runoff is different from that of stream discharge or reach flow. The average stream flow in 2001 was 429.45 cumecs, which slightly increased to 429.63 cumecs (0.04% gain) and decreased to 427.97 cumecs (0.39% loss) in 2021 concerning the preceding year. However, the percentage changes are very low and less than 1%. The wet season (May–October) contributes about 91% of the total surface flow and 83% of the streamflow, as shown in Tables 15.9 and 15.10.

From the present chapter, it is observed that change in the LULC of the watershed does influence the output of surface runoff and stream flow. However, it is observed that LULC is one of many controlling factors in the watershed. The change observed is very low. The percentage of change in land use is not proportional to the percentage change in surface runoff. For a Dense Forest change of about 24–26%, the change in surface runoff is less than 3%, and the change in streamflow is negligible (less than 1%). In many studies, the impact of urbanization on runoff is found to be substantial, but in the present chapter, the proportion of urbanization is around 1% which is not enough to cause a visible change in the hydrology even if there is a consistent decadal positive trend.

15.5 Conclusion

The chapter analyzed the change in the land use pattern of Iril watershed in Manipur. Five classes are identified, and changes observed for three years (2001, 2011, 2021) spanning over three decades. It is observed that built-up land has shown a consistently increasing trend over the years. However, the other classes showed a fluctuating trend. The most important classes in this watershed are Dense Forest and Scrubland, Shrubs, and Mixed Forest, which occupy more than 80% of the total geographical area. LULC classification is done using Landsat imagery and supervised classification. Accuracy assessment of the classification is validated by kappa statistics, user's and producer's accuracy. Overall accuracy

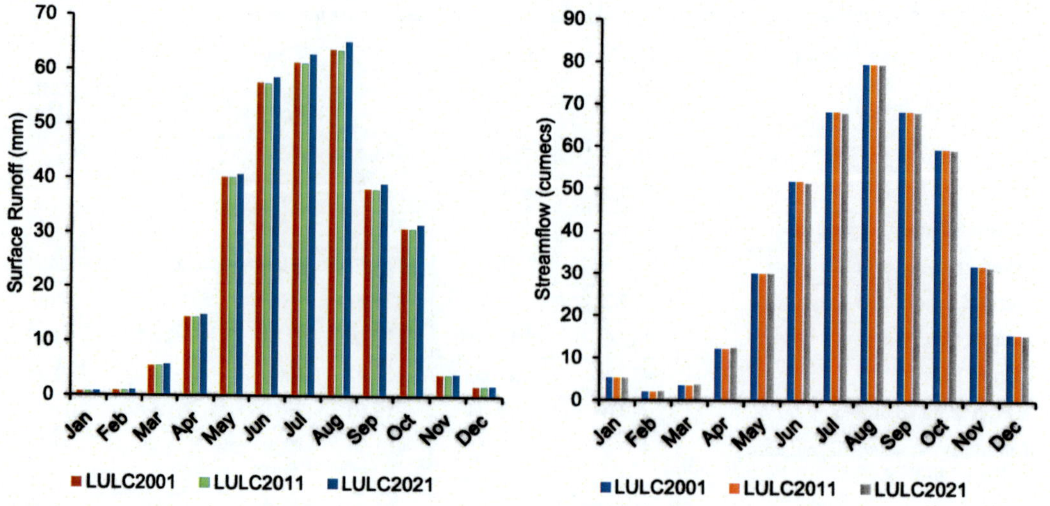

Fig. 15.4 Comparison of simulated average monthly surface runoff and streamflow of 2001, 2011, and 2021

Table 15.9 Comparison of simulated average monthly surface runoff for all the three periods

Month	LULC2001	LULC2011	LULC2021	Change (%)		
	(mm)			2001–11	2011–21	2001–21
January	0.75	0.75	0.79	0.00	5.33	5.33
February	0.95	0.94	1.04	−1.05	10.64	9.47
March	5.53	5.50	5.77	−0.54	4.91	4.34
April	14.48	14.44	14.91	−0.28	3.25	2.97
May	40.21	40.16	40.69	−0.15	1.34	1.19
June	57.64	57.51	58.59	−0.23	1.88	1.65
July	61.38	61.20	62.82	−0.29	2.65	2.35
August	63.78	63.60	65.17	−0.28	2.47	2.18
September	38.14	38.03	39.12	−0.29	2.87	2.57
October	30.93	30.84	31.61	−0.29	2.50	2.20
November	4.03	4.02	4.13	−0.25	2.74	2.48
December	1.92	1.92	1.99	0.00	3.65	3.65
Total	319.74	318.90	326.63	−0.26	2.42	2.15
Wet season	292.08	291.33	298.00	−0.26	2.29	2.03
(%)	91.35	91.35	91.23	–	–	–
Dry season	27.66	27.57	28.63	−0.33	3.84	3.51
(%)	8.65	8.65	8.77	–	–	–

* Wet season (May–October); Dry season (November–April)

assessment of the classification stood at 88.24%, 91.62%, and 92.55% for 2001, 2011, and 2021, respectively. The impact of LULC on hydrological response is examined using SWAT modeling. Due to the lack of sufficient data, the model output is only validated. SWAT Model simulation of the Iril watershed showed a validation performance of NSE of 0.68 and R^2 of

Table 15.10 Comparison of simulated average monthly stream flow for all the three periods

Month	LULC2001	LULC2011	LULC2021	Change (%)		
	(cumecs)			2001–11	2011–21	2001–21
January	5.36	5.36	5.29	0.12	−1.28	−1.17
February	2.13	2.13	2.16	−0.17	1.45	1.28
March	3.62	3.61	3.76	−0.42	4.27	3.83
April	12.34	12.33	12.51	−0.07	1.43	1.35
May	30.09	30.11	30.03	0.07	−0.28	−0.21
June	51.92	51.98	51.52	0.11	−0.88	−0.77
July	68.47	68.51	68.19	0.06	−0.47	−0.40
August	79.71	79.71	79.58	0.01	−0.17	−0.16
September	68.49	68.51	68.31	0.03	−0.30	−0.27
October	59.53	59.55	59.32	0.03	−0.38	−0.35
November	32.00	32.02	31.67	0.08	−1.12	−1.04
December	15.80	15.81	15.65	0.06	−1.04	−0.98
Total	429.45	429.63	427.97	0.04	−0.39	−0.34
Wet season	358.20	358.37	356.94	0.05	−0.40	−0.35
(%)	83.41	83.41	83.40	–	–	–
Dry season	71.25	71.26	71.03	0.02	−0.32	−0.30
(%)	16.59	16.59	16.60	–	–	–

0.69, RSR of 0.56, and PBIAS of −2.6%. From the chapter, the influence of LULC change on runoff is observed, but the change is minimal. The percentage of change in land use is not proportional to the percentage change in surface runoff. For a Forest change of about 24–26%, the change in surface runoff is less than 3%, and the change in streamflow is negligible (less than 1%). It is also noted that land use is one of many controlling factors in the case of streamflow. The watershed being predominantly a forested region, the change in forest type caused the least impact on runoff-streamflow behavior. Despite a positive and high percentage of built-up land, its influence on runoff seems negligible as built-up land occupies less than one percent of the watershed. This study is useful for understanding the relationship between land use change and hydrology. The study explores the usage of mathematical models in simulating real time hydrological behavior in different conditional scenarios.

References

Abbaspour KC, Rouholahnejad E, Vaghefi S, Srinivasan R, Yang H, Kløve B (2015) A continental-scale hydrology and water quality model for Europe: calibration and uncertainty of a high-resolution large-scale SWAT model. J Hydrol 524:733–752. https://doi.org/10.1016/j.jhydrol.2015.03.027

Akoko G, Le TH, Gomi T, Kato T (2021) A review of SWAT model application in Africa. Water 13:1–20

Anand V, Oinam B, Parida BR (2020) Uncertainty in hydrological analysis using multi-GCM predictions and multi-parameters under RCP 2.6 and 8.5 scenarios in Manipur River basin, India. J Earth Syst Sci 129. https://doi.org/10.1007/s12040-020-01492-z

Anderson JR, Hardy EE, Roach JT, Witmer RE (1976) A land use and land cover classification system for use with remote sensor data

Arnold JG, Moriasi DN, Gassman PW, White MJ (2012) SWAT: model use, calibration, and validation. Trans ASABE 55:1491–1508

Baker TJ, Miller SN (2013) Using the soil and water assessment tool (SWAT) to assess land use impact on water resources in an East African watershed. J Hydrol 486:100–111. https://doi.org/10.1016/j.jhydrol.2013.01.041

Bhuyan R (2019) A review note on shifting cultivation in northeast India amidst changing perceptions. Dhaulagiri J Soc Anthropol 13:90–95

Borah JR, Gilroy JJ, Evans KL, Edwards DP (2022) The value of shifting cultivation for biodiversity in Northeast India. Divers Distrib 00:1–14. https://doi.org/10.1111/ddi.13605

Bruun TB, Mertz O, Elberling B (2006) Linking yields of upland rice in shifting cultivation to fallow length and soil properties. Agric Ecosyst Environ 113(1–4):139–149

Congalton RG (1991) A review of assessing the accuracy of classifications of remotely sensed data. Remote Sens Environ 46:35–46

Deng JS, Wang K, Deng YH, Qi GJ (2016) PCA-based land-use change detection and analysis using multitemporal and multisensor satellite data. Int J Remote Sens 29:4823–4838. https://doi.org/10.1080/01431160801950162

Dwarakish GS, Ganasri BP (2015) Impact of land use change on hydrological systems: a review of current modeling approaches. Cogent Geosci 1:1–18. https://doi.org/10.1080/23312041.2015.1115691

Easton ZM, Fuka DR, White ED, Collick AS, Ashagre BB, Mccartney M, Awulachew SB (2010) A multi basin SWAT model analysis of runoff and sedimentation in the Blue Nile, Ethiopia. Hydrol Earth Syst Sci 14:1827–1841. https://doi.org/10.5194/hess-14-1827-2010

Garg V, Nikam BR, Thakur PK, Aggarwal SP, Gupta PK, Srivastav SK (2019) Human-induced land use land cover change and its impact on hydrology. Hydro Res 1:48–56. https://doi.org/10.1016/j.hydres.2019.06.001

Gitau MW, Chaubey I (2010) Regionalization of SWAT model parameters for use in ungauged watersheds. Water 2:849–871. https://doi.org/10.3390/w2040849

Githui F, Mutua F, Bauwens W (2009) Estimating the impacts of land-cover change on runoff using the soil and water assessment tool (SWAT): case study of Nzoia catchment, Kenya. Hydrol Sci Sci Hydrol 54:899–908. https://doi.org/10.1623/hysj.54.5.899

Guzha AC, Rufino MC, Okoth S, Jacobs S, Nóbrega RLB (2018) Impacts of land use and land cover change on surface runoff, discharge and low flows: evidence from East Africa. J Hydrol Reg Stud 15:49–67. https://doi.org/10.1016/j.ejrh.2017.11.005

Hallouz F, Meddi M, Mahé G, Alirahmani S (2019) Modeling of discharge and sediment transport through the SWAT model in the basin of Harraza (Northwest of Algeria). Water Sci 32:79–88. https://doi.org/10.1016/j.wsj.2017.12.004

Hu S, Fan Y, Zhang T (2020) Assessing the effect of land use change on surface runoff in a rapidly urbanized city: a case study of the central area of Beijing. Land 9:1–15

Khare D, Patra D, Mondal A, Kundu S (2017) Impact of land use/land cover change on runoff in the catchment of a hydro power project. Appl Water Sci 7:787–800. https://doi.org/10.1007/s13201-015-0292-0

Khumbah AL (2022) Barom nu: the traditional practice of rice-breeding in shifting agriculture by the Inpui tribe of Manipur, North-East India. Indian J Tradit Knowl 21:191–197

Kumar N, Singh SK, Singh VG, Dzwairo B (2018) Investigation of impacts of land use/land cover change on water availability of tons river basin, Madhya Pradesh, India. Model Earth Syst Environ 4:295–310. https://doi.org/10.1007/s40808-018-0425-1

Letha J, Nair BT, Chand BA (2011) Effect of land use/land cover changes on runoff in a river basin: a case study. Water Resour Manag 145:139–149. https://doi.org/10.2495/WRM110121

Lillesand T, Kiefer R, Chipman J (2004) Remote sensing and image interpretation, 5th edn., p 146

Malik MA, Dar AQ, Jain MK (2022) Modelling streamflow using the SWAT model and multi-site calibration utilizing SUFI-2 of SWAT-CUP model for high altitude catchments, NW Himalaya's. Model Earth Syst Environ 8:1203–1213. https://doi.org/10.1007/s40808-021-01145-0

Manaschintan SB, Nalini SS (2020) Effect of land use—land cover change on runoff characteristics in Mumbai City. In: Ghosh JK, da Silva I (eds) Springer, Singapore, pp 183–192

Mishra A, Froebrich J, Gassman PW (2007) Evaluation of the SWAT model for assessing sediment control structures in a small watershed in India. Trans ASABE 50:469–478

Mishra H, Mario D, Shakti D, Mukesh S, Kumar S, Anjelo S, Denis F, Kumar R (2017) Hydrological simulation of a small ungauged agricultural watershed Semrakalwana of Northern India. Appl Water Sci 7:2803–2815. https://doi.org/10.1007/s13201-017-0531-7

Molina-navarro E, Andersen HE, Nielsen A, Thodsen H, Trolle D (2017) Environmental modelling and software the impact of the objective function in multi-site and multi-variable calibration of the SWAT model. Environ Model Softw 93:255–267. https://doi.org/10.1016/j.envsoft.2017.03.018

Nath AJ, Reang D, Sileshi GW (2022) The shifting cultivation juggernaut: an attribution problem. Glob Challenges 2200051:1–5. https://doi.org/10.1002/gch2.202200051

Oo HT, Zin WW, Cho C, Kyi T (2020) Analysis of streamflow response to changing climate conditions using SWAT model. Civ Eng J 6:194–209

Pandey DK, Dubey SK, De HK (2022) Indigenous people's attachment to traditional agroforestry system: an empirical insight from Churachandpur, Northeast India. Indian J Tradit Knowl 21:1–7

Patil NS, Nataraja M (2020) Effect of land use land cover changes on runoff using hydrological model: a case study in Hiranyakeshi watershed. Model Earth Syst Environ 6:2345–2357. https://doi.org/10.1007/s40808-020-00808-8

Reimeingam M (2017) Shifting cultivation in Manipur: land, labour and environment. J Rural Develop 36(1):97–119. https://doi.org/10.25175/jrd/2017/v36/i1/112705

Rostamian R (2008) Application of a SWAT model for estimating runoff and sediment in two mountainous basins in central Iran. Hydrol Sci J J Des Sci Hydrol—Hydrol SCI J 53:977–988. https://doi.org/10.1623/hysj.53.5.977

Sajikumar N, Remya RS (2014) Impact of land cover and land use change on runoff characteristics. J Environ Manage 1–9. https://doi.org/10.1016/j.jenvman.2014.12.041

Schmalz B, Fohrer N (2009) Advances in geosciences comparing model sensitivities of different landscapes using the ecohydrological SWAT model. Adv Geosci 21:91–98

Setegn S, Yohannes F, Researcher I, Chowdary VM, Remote N, Centre S, Mal BC, Chancellor V (2009) Impact of land use/land cover transformations on Alemaya lake, Ethiopia. J Indian Water Resour Soc 29:40–45

Setti S, Maheswaran R, Sridhar V, Barik KK, Merz B, Agarwal A (2020) Inter-comparison of gauge-based gridded data, reanalysis and satellite precipitation product with an emphasis on hydrological modeling. Atmosphere (basel) 11:1–26. https://doi.org/10.3390/atmos11111252

Shaw AK, Natori Y, Edake S (2022) Introduction: shifting cultivation in Northeast India in the context of South and Southeast Asia. A tradition in transition: understanding the role of shifting cultivation for sustainable development of Northeast India. TERI, pp 2–14

Shi P, Chen C, Srinivasan R, Zhang X, Cai T, Fang X, Qu S, Chen X, Li Q (2011) Evaluating the SWAT model for hydrological modeling in the Xixian watershed and a comparison with the XAJ Model. Water Resour Manag 25:2595–2612. https://doi.org/10.1007/s11269-011-9828-8

Shrestha PK (2016) SWAT Calibration helper v1.0 microsoft excel worksheet

Sisay E, Halefom A, Khare D, Singh L, Worku T (2017) Hydrological modelling of ungauged urban watershed using SWAT model. Model Earth Syst Environ 3:693–702. https://doi.org/10.1007/s40808-017-0328-6

Tanksali A, Soraganvi VS (2021) Assessment of impacts of land use/land cover changes upstream of a dam in a semi-arid watershed using QSWAT. Model Earth Syst Environ 7:2391–2406. https://doi.org/10.1007/s40808-020-00978-5

Thampi SG, Raneesh KY (2010) Influence of scale on SWAT model calibration for streamflow in a river basin in the humid tropics. Water Resour Manag 24:4567–4578. https://doi.org/10.1007/s11269-010-9676-y

Thong P, Sahoo UK, Thangjam U, Pebam R (2020) Pattern of forest recovery and carbon stock following shifting cultivation in Manipur, North-East India. PLoS ONE 15:1–20. https://doi.org/10.1371/journal.pone.0239906

Thong P, Sahoo UK, Pebam R, Thangjam U (2019) Changing trends of shifting cultivation and its drivers in Champhai, Northeast India. Indian J Hill Farming 32:1–4

Tufa FG, Sime CH (2021) Stream flow modeling using SWAT model and the model performance evaluation in Toba sub-watershed, Ethiopia. Model Earth Syst Environ 7:2653–2665. https://doi.org/10.1007/s40808-020-01039-7

Umukiza E, Raude J., Wandera SM, Petroselli A, Gathenya JM (2021) impacts of land use and land cover changes on peak discharge and flow volume in Kakia and Esamburmbur. Hydrology 8. https://doi.org/10.3390/hydrology8020082

Valentin C, Agus F, Alamban R, Boosaner A, Bricquet JP, Chaplot V, Vadari T (2008) Runoff and sediment losses from 27 upland catchments in Southeast Asia: impact of rapid land use changes and conservation practices. Agric Ecosyst Environ 128(4):225–238

Yuen JQ, Ziegler AD, Webb EL, Ryan CM (2013) Uncertainty in below-ground carbon biomass for major land covers in Southeast Asia. For Ecol Manage 310:915–926

Zhang X, Srinivasan R, Van LM (2008) Multi-site calibration of the SWAT model for hydrologic modeling. Trans ASABE 51:2039–2049

Ziegler AD, Bruun TB, Guardiola-Claramonte M, Giambelluca TW, Lawrence D, Lam NT (2009) Environmental consequences of the demise in Swidden cultivation in montane mainland Southeast Asia: hydrology and geomorphology. Hum Ecol 37(3):361–373

16 GIS-Based Road Network Connectivity Assessment and Its Impact on Agricultural Characteristics Using Graph Theory: A Block-Level Study in the Hill Area of Darjeeling District, West Bengal

Surajit Paul, Debasish Roy, and Bipul Chandra Sarkar

Abstract

The topological structure of connectivity and accessibility is one of the important elements that reveal the shape of any region. The road network is a key component of rural development, as it promotes access to economic and social services, increased agricultural and non-agricultural productivity, generating employment which expands rural growth opportunities and overall income leading to poverty reduction. The knowledge of the topographical characteristics of road networks has improved due to the recent development of the Geographic Information System (GIS). The present chapter uses graph theory to examine the accessibility and connectivity of the road network in five hilly blocks of the Darjeeling district named Darjeeling-Pulbazar, Jorebunglow-Sukiapokhri, Kurseong, Mirik, and Rangli-Rangliot. Additionally, the connection index, the detour index, the associated number, and aggregate transport score have all been used to classify the degree of network accessibility. The outcome reveals that under the study area, there is a serious need to identify the improvement of agricultural characteristics through connectivity and network accessibility between the hilly blocks of the Darjeeling district. It is also observed that most of the accessible villages are centrally placed and close to urban areas.

Keywords

Darjeeling district · GIS · Connectivity · Accessibility · Aggregate transport score (ATC)

16.1 Introduction

The purpose of transportation is to make it easier for people to access commodities, services, and knowledge. To lead a fruitful economic and social life, people require access to a wide range of products, services, and information (Sahitya and Prasad 2020). Urban and rural locations, as well as developing and developed nations, have quite different transportation patterns. Research has shown that rural transportation in developing nations has unique characteristics. People move around in rural areas for a variety of subsistence, social, and economic reasons (Holl 2007). Most transportation is done on foot and occurs mostly within and outside the village, far from the road system. Subsequently, access is the ultimate goal

S. Paul (✉) · D. Roy
Department of Geography and Applied Geography, University of North Bengal, Siliguri, West Bengal, India
e-mail: rs_surajit@nbu.ac.in

B. C. Sarkar
Ananda Chandra College, University of North Bengal, Siliguri, West Bengal, India

J. Das and S. Halder (eds.), *Advancement of GI-Science and Sustainable Agriculture*, GIScience and Geo-environmental Modelling, https://doi.org/10.1007/978-3-031-36825-7_16

to reach the essential services, goods, and resources that rural populations require to live decent, productive lives on a social and economic level (Páez et al. 2012). The essential needs of rural residents are water, food, and firewood, and the social and economic welfare components of rural life are such as health and education as well; their basic needs like agriculture, domestic animals, and home industries are all connected to transportation (Nagne and Gawali 2013). The freight movement during production makes transportation an essential part of the agro-industrial complex. Due to its multifaceted functions and significance in maintaining relationships and creating integrations of people, goods, and services, transportation has become a necessary component of modern life (Umoren et al. 2009). Accelerated infrastructure investment in rural areas is required to provide employment and open new business opportunities (Sharma and Ram 2023; Sarkar 2013). All of these ultimately result in a higher standard of living and lessen the vulnerability of the rural poor (Levy 1996). Rural transportation and communication infrastructure development promotes rural economic development by granting access to facilities like education, healthcare, marketing, etc. Rural connectivity is a crucial element of rural development and has a considerable positive impact on the socioeconomic development of rural residents. Investments in rural roads have been proven to raise rural residents out of poverty. Better roads can open up chances for economic growth and poverty reduction in various ways. Improved roads increase farm and non-farm production by increasing the availability of pertinent inputs and lowering input costs. The above said background is made possible by easier access to markets and technology (Binswanger et al. 1993).

Rural and urban areas live on a continuum connected by many sectorial and spatial links (McGranahan et al. 2004; Seto et al. 2012). Urban populations are connected to rural environments through flows of people, goods, and information, which we refer to as rural–urban connectedness (Maity et al. 2021; Seto et al. 2012; Elmqvist et al. 2013; Djurfeldt 2015). Markets enable rural agricultural production in one location to benefit other distant places by facilitating commerce between rural activities and cities and larger regions (Verburg et al. 2011). Farmers near urban markets can more readily purchase agricultural inputs, access services like financing and insurance, and trade their products which can contribute to both a rise in agricultural production and an increase in agricultural specialization (Masters et al. 2013). The degree of spatial difference that can be overcome is called accessibility. It describes the various means of exchange for both individuals and businesses. Urban and regional science have a long history of addressing the issue of accessibility. People may be influenced to convert land if agricultural land or markets are more accessible (Nagendra et al. 2004). Farmers can cut travel times and transportation costs to market towns by expanding the network or making road improvements (such as paving unpaved roads with paved ones) (Dorosh et al. 2018). They can also improve vehicular access to agricultural land and markets throughout the year (Verburg et al. 2011). Thus, farmers frequently boost crop production along established highways to increase agricultural productivity, ultimately changing the land's cover (Hafner 1971; Dorosh et al. 2018). Due to this, agricultural advancements and road connectivity were attempted to be linked in this chapter. The chapter aims to determine whether the expansion of the rural road network will improve or cause any modifications to the agricultural features of the studied area.

16.2 Study Area

The hilly area of Darjeeling district (after the separation of the newly formed Kalimpong district on February 14, 2017) is geographically separated among hills and plains area and located in the northern section of the state of West Bengal. The Darjeeling Himalayan Region is part of the Indian Himalayan Region and is located in the Eastern Himalayas (IHR). Geographically, the Darjeeling Himalayan area is a part of the Outer Himalayas and Lesser Himalayas, often known as the Shivalik Range. The Darjeeling district is the only district in West Bengal with hills and mountains that are a part of the mighty Himalayas, where the Darjeeling Himalayan Region (the study area) was politically located. From a locational perspective, it is between 26° 27′ 10″ and 27° 13′ 05″ N (latitude) and 87° 59′ 30″ and 88° 53′ 00″ E (longitude) (Fig. 16.1). The district is strategically crucial since, in addition to Indian Territory, it shares a border with three other countries. Three countries —Bangladesh (South-East), Nepal (West), and Bhutan—as well as one district of West Bengal state—Jalpaiguri (South-East) and Siliguri sub-division (South)—as well as two states—Bihar (South-West) and Sikkim (North)—encircle the study area (i.e., the Darjeeling Himalayan Region). In addition, due to its unusual geographic location, the district's natural boundaries are shaped by several rivers, including the Mechi, Mahananda, Teesta, Jaldhaka, Rangeet, and Rammam. Darjeeling's geology spans formations from the Archaean to the Pleistocene Sub-Recent and Recent, with an average elevation of 2045 m from mean sea level (District Statistical Handbook 2011). Darjeeling district has a highly extensive road network that almost completely covers the whole area despite its hilly terrain and high altitude. Several State Highways (SH-12, SH-12A) and National Highways NH77 pass through the hill areas of the Darjeeling district, which has a well-built road network in addition to several medium-sized, small-sized, and secondary road networks. The PWD, DCAHC, Siliguri Mahakuma Parishad, and Prime Minister Gram Sadak Yojana manage a road network that is more than 4100 km long. In addition, the Urban Local Bodies (ULB) also have distinct road maintenance. The hilly areas of Darjeeling district cover an area of 1406 km^2 and come under five community development blocks of Darjeeling district named Darjeeling-Pulbazar, Jorebunglow-Sukiapokhri, Kurseong, Mirik, and Rangli-Rangliot. Due to its varied geomorphological circumstances, district Darjeeling has a variety of agro-climatic regions. The highest altitudes are unsuited for crop cultivation since they are nearly always covered in snow. Tea, cinchona, and rubber plantations predominately employ the lower hills (District Statistical Handbook 2013).

16.3 Data Source

The transportation of agricultural goods and services is heavily dependent on the transportation system, which in turn encourages the development of the social and economic spheres. As a result, the transportation system is crucial for any area of agriculture development. Better road systems boost accessibility and mobility, greatly reducing time and travel expenses (Sreelekha et al. 2016). Agricultural developments like access to agricultural fields, machinery, fertilizer, and trading of products directly or indirectly depend on improving the road network (Patarasuk 2013). The recent study used some secondary data sources to achieve its goal, including block-level maps of the Darjeeling district from the Census of India, in 2011 and validation of the map using a Google Earth image. Population-related data like population density, decadal growth rate, etc., were acquired from the Darjeeling District Census Handbook, and transportation-related data like road length, road map, etc., were gathered from an open street map and the Darjeeling District Statistical Handbook, 2014. All agricultural data were collected from the Ministry of Agriculture 2017 (Census of Agriculture 2017). The nodes and arcs were meticulously counted to calculate various indices. Measures such as the aggregate transport score, detour index, associated number, and road density index were calculated in Microsoft Excel 2013 and ArcGIS 10.5v to determine the connectivity and accessibility of five blocks of the study area.

16.4 Methodology

Several network indices based on graph theory are used to assess the usability and effectiveness of the network analysis (Demšar et al. 2008). A graph comprises a collection of nodes or vertices linked by a collection of edges (Arif et al. 2020). The edges are lines that connect two of the corresponding vertices, and the vertices are dots positioned at the intersection of two or more edges (Derrible 2011). Alpha (α), beta (β), and gamma (γ) index were first introduced in this discipline by Garrison and Marble in 1962, 1964, and 1965. Kansky (1963) attempts to relate the topology of the road network with economic development by using a variety of metrics, including the cyclomatic number, network diameter, and alpha, beta, and gamma index (Derrible 2009). A variety of metrics, including the alpha index, beta index, gamma index, network density, cyclomatic number, and aggregate transportation score, are used by many researchers to assess the degree of connectivity (Levinson 2012; Nagne 2013). The current chapter relies on a few chosen graph theory measures. A base map for determining network indices is provided in Fig. 16.1, the hill area of Darjeeling district (study area) with a total of 2041 edges and 1921 nodes.

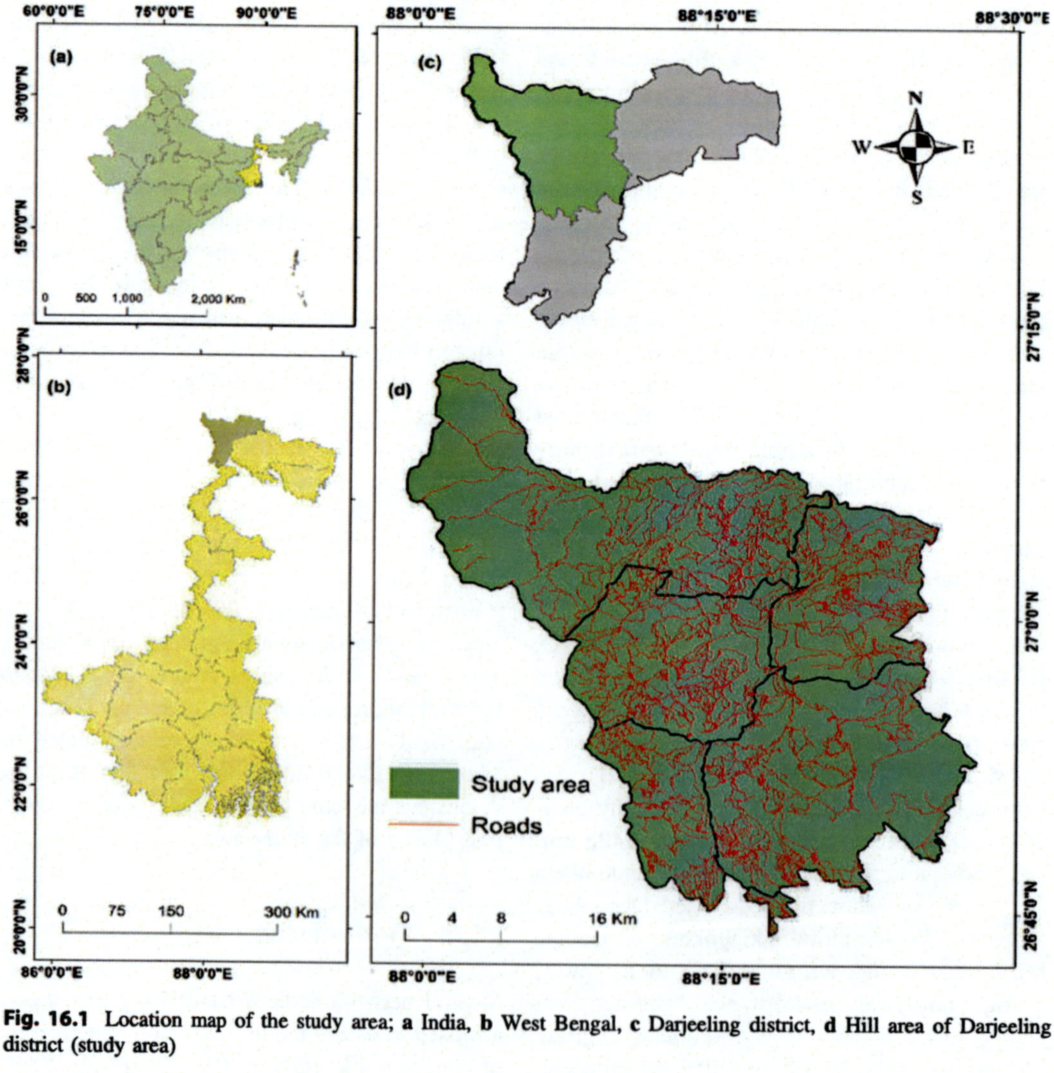

Fig. 16.1 Location map of the study area; **a** India, **b** West Bengal, **c** Darjeeling district, **d** Hill area of Darjeeling district (study area)

16.5 Application of Network Indices Based on Graph Theory

16.5.1 The Alpha Index

A connectivity metric compares the number of cycles in a graph to the maximum number of cycles. A network is more linked when the alpha index is higher. In other words, the value of zero represents simple networks, while a network that is fully linked is represented by a value of 1. Additionally, this index is also presented in percentiles. The alpha index (Fig. 16.2a) is determined using the formula: Alpha index $(\alpha) = (e - v + p)/(2v - 5)$.

16.5.2 Beta Index

The beta index expresses the connection between the number of links (e) over the number of nodes, which assesses the degree of connectedness in a graph. This index's value ranges from 0 to 1, where 1 denotes complete connection, and 0 denotes incomplete connectivity between the

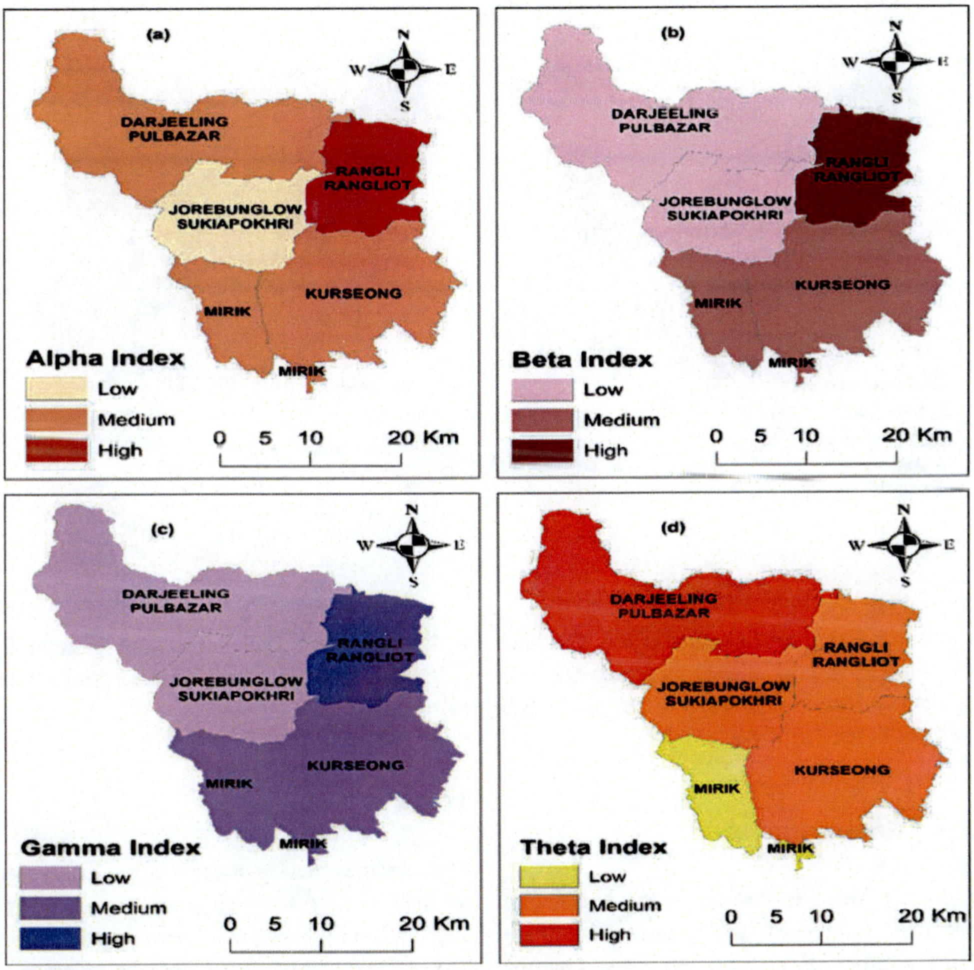

Fig. 16.2 Road network connectivity index **a** alpha index, **b** beta index, **c** gamma index, **d** theta index **e** cyclomatic index, **f** road density, **g** detour index, **h** eta index

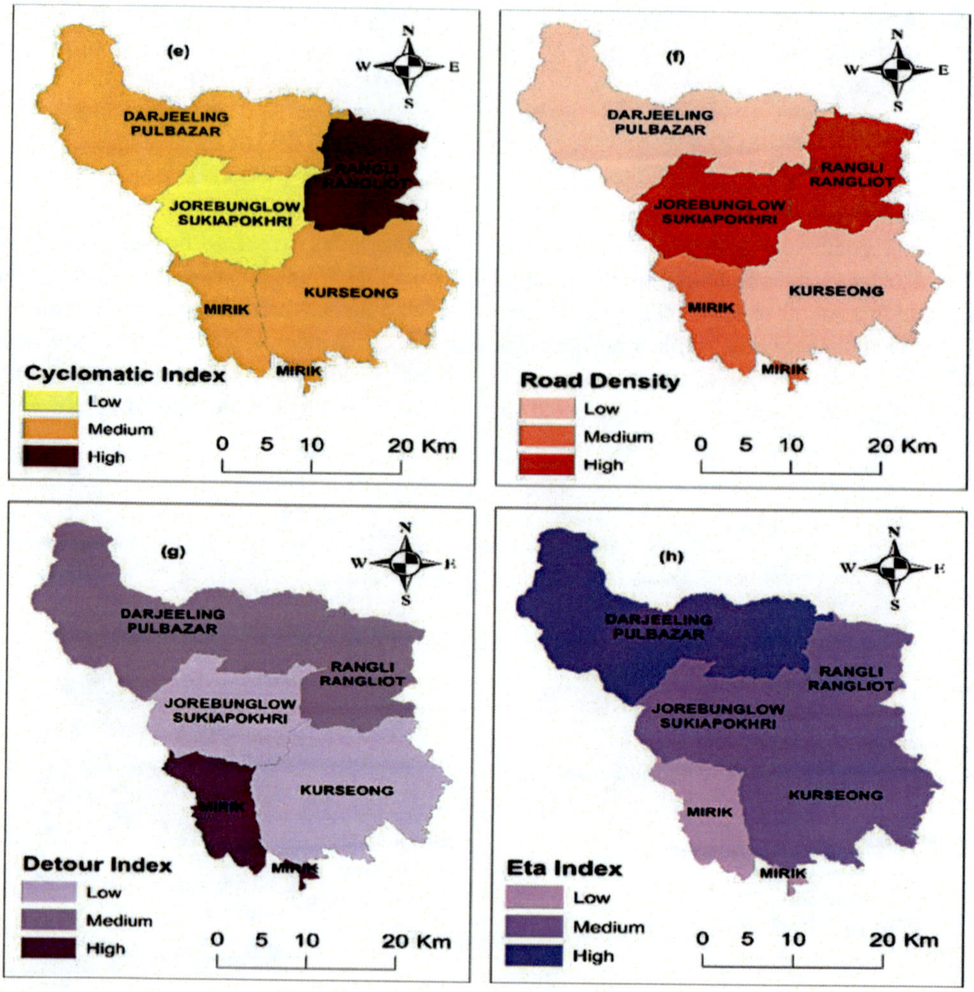

Fig. 16.2 (continued)

roads. When the graph is complicated, the beta index value is greater than 1. The equation to compute the beta index (Fig. 16.2b) is as follows: Beta index (β) = e/v.

16.5.3 Gamma Index

The gamma index is known as the correlation between the number of actual links in a graph and the number of possible linkages. Gamma index values range from 0 to 1, as well. Greater connectedness is associated with greater values and vice versa. Calculating the gamma index (Fig. 16.2c) is done as follows: Gamma index (γ) = e / 3 (v-2).

16.5.4 Theta Index

The average quantity of traffic at each intersection, as measured by the theta index, serves as a node. The load on the network increases with theta value. The metric, which indicates the average load per connection, can also represent the number of links (or edges). The formula expressed as theta

Table 16.1 Basic information of the road network analysis

Blocks	Area (in km^2)	Road length (in km)	Shortest length (in km)	Node (v)*	Edge (e)	Sub-graph (p)
Darjeeling Pulbazar	426.57	602.21	384	399	406	12
Kurseong	377.35	414.68	241.7	369	390	4
Mirik	125.66	235.01	167.8	324	341	4
Jorebunglow Sukiapokhri	222.12	518.66	305	472	465	19
Rangli-Rangliot	272.99	420.36	274	357	439	3

* Where, e is the number of edges, v is the number of nodes, and p is the number of sub-graphs
Source Computed by authors

index = $\frac{\text{Total network distance}}{\text{number of nodes}}$ (Fig. 16.2d and Table 16.1).

16.5.5 Cyclomatic Number

Another crucial metric for network connectedness is the cyclomatic number (l). Higher the worth of it, the more connectivity there is. This formula is used to compute the research area's cyclomatic number (Fig. 16.2e):

$$\text{Cyclomatic Number } (\mu) = e - v + p.$$

16.5.6 Road Density

The number of roads per unit of a geographical area determines how connected and accessible the road network system is, and this metric is known as the road network density (Fig. 16.2f).

$$\text{Expressed as}: \frac{\text{Road length in km}}{\text{Area in km}^2}.$$

16.5.7 The Detour Index

The detour index is a way to evaluate how effectively a transportation system overcomes obstacles like distance or friction caused by distance (Fig. 16.2g). The network is more spatially efficient the closer the detour index is 0 to 1. Rarely, if ever, do networks have a detour index of 1, and most networks would fit on an asymptotic curve that approaches one but never reaches it. It is expressed as DI = $\frac{\text{Shortest distance}}{\text{travel distance}}$ (Flitter et al. 2016; Sarkar et al. 2021).

16.5.8 Eta Index

The eta index denotes the average length of edges per link. The average length of a link gets shorter when more nodes are added, which causes the eta value to fall. Low eta values frequently occur in complex networks (Fig. 16.2h).

$$\text{Expressed as}: \frac{\text{Total network distance}}{\text{Number of edges}}.$$

16.5.9 Aggregate Transport Score

To develop a static transport assignment model that uses smart forms of aggregation to maintain accuracy as much as possible, Mukherjee (2012) established the aggregate transportation score (ATS), which is simply the sum of the ratio and non-ratio indices used in this article and is calculated as follows (Table 16.2):

$$\text{ATS} = \sum \alpha + \beta + \gamma + \mu + \theta \ldots + \text{DI}.$$

Table 16.2 Calculation table for the index of road network connectivity

Blocks	Alpha index (%)	Beta index (%)	Gamma index (%)	Eta index (%)	Theta index (%)	Road density (km/km^2)	Detour index (%)
Darjeeling Pulbazar	2.4	101.75	34.09	148.32	150.93	1.44	63.76
Kurseong	3.41	105.69	35.42	106.33	112.38	1.24	58.28
Mirik	3.27	105.25	35.3	68.91	72.53	1.96	71.40
Jorebunglow Sukiapokhri	1.28	98.52	32.98	111.54	109.88	2.44	58.80
Rangli-Rangliot	11.99	122.97	41.22	95.75	117.74	2.40	65.18

Source Computed by authors

16.6 Indices for Agricultural Characteristics

16.6.1 Gross Cropped Area

The gross cropped area (GCA) is the cumulative area sown once and more than once in a given year is the gross cropped area (GCA). On the other hand, gross cropped area refers to the total area sown once and more than once in a particular year. In this study area, about 3920 hectares are occupied by cultivable land out of the total area (Fig. 16.3a). Crop production in this region depends on the intensity of irrigation as well as the road network. In this region, the nature of cropping is mainly dominated by horticulture and floriculture. Rice, potatoes, pulses, and vegetables are the less dominant cultivated crops, and fruits are also substantial production.

16.6.2 Irrigation Status

Irrigation is the artificial way of pumping water from a river, groundwater, etc. The crop production of the study area largely depends on the existing irrigation facility, as the rainfall is seasonally concentrated and unreliable. On the other hand, most of the rivers are non-perennial, so the irrigation system of the study area is mainly based on shallow tube wells, deep tube wells, etc. The total irrigated area of the district in the year 2010–2011 was 3560 ha; it is observed that Darjeeling-Pulbazar is the highest irrigated block and the lowest irrigation observed in Mirik, Kurseong, and Jorebunglow-Sukiapokhri blocks (Fig. 16.3b).

16.6.3 Total Plantation Area

Tea has been grown in the Darjeeling district for as long as the town of Darjeeling has existed. The district's first commercial tea plantation was established during the first year of the establishment of British colonies in the 1830s and was promoted by British officers. According to the geographical indicators of product regulations, Darjeeling tea was the first commodity in India to have a geographical indication (GI) tag in 2003 (Registration and Protection Act 1999). According to the definition, Darjeeling tea can only be used to describe tea produced in gardens in a select few hilly district sections. In the district (Fig. 16.3c), 87 tea gardens are GI-tagged. A 17,820-ha area in the hills is included in the Darjeeling tea plantation, with an annual production of 8.9 million kg, according to a report by the Indian Tea Board (District Statistical Handbook 2010; 11).

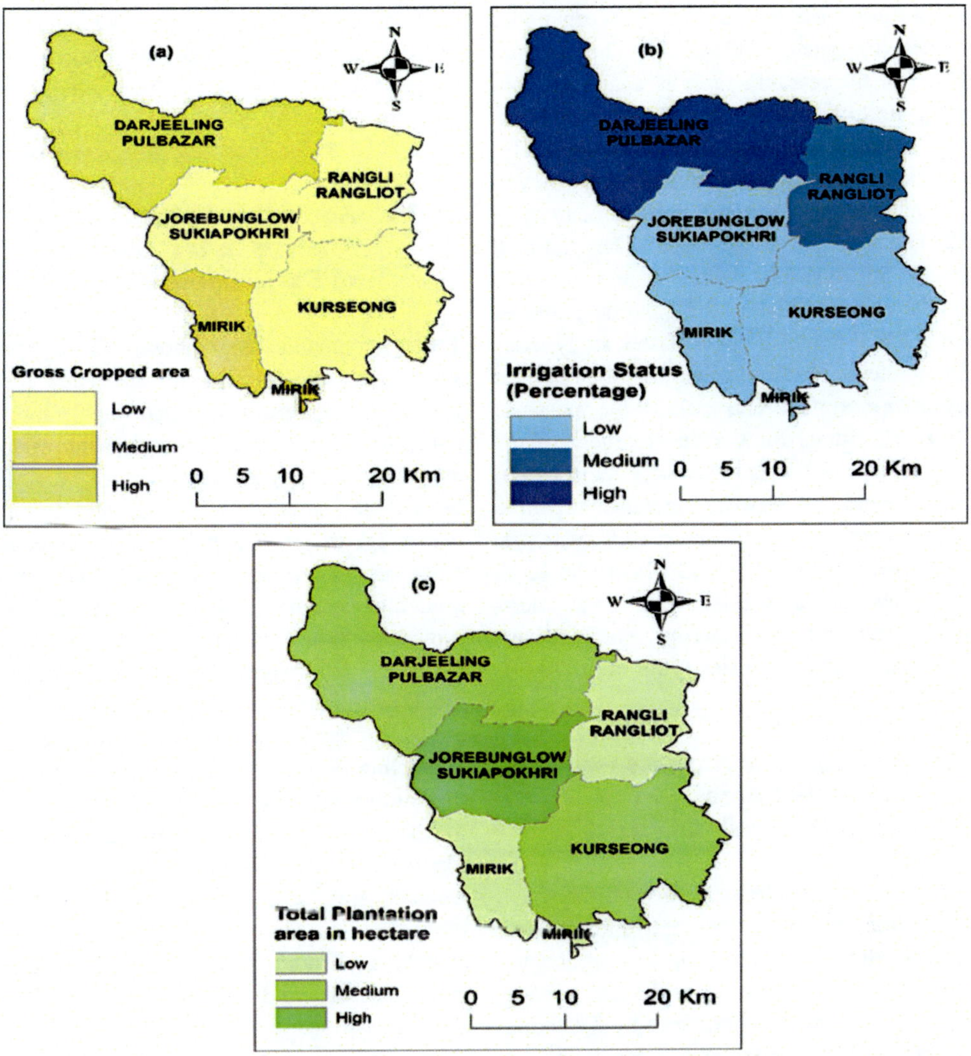

Fig. 16.3 Selected agricultural parameters; **a** gross cropped area, **b** irrigation status, and **c** total plantation area

16.6.4 Agriculture Composite Score

Small groups of data points with a strong conceptual and statistical relationship to one another are represented by composite scores. Combining and presenting these factors as a single score diminishes the risk of information overload. To determine the agricultural characteristics of each block in the study area, the agriculture composite score has been calculated by adding the values of factors, i.e., gross cropped area, irrigation status, and plantation area, and divided by the number of factors.

16.7 Results

The network analysis is crucial for determining the accessibility and connectedness of any area since it assesses the transportation system while also examining the connections between nodes and lines (Reggiani et al. 2011). The network structure of the research topic is analyzed, considering the graph theory notion. The neighboring villages of the Darjeeling District's Hill Areas gained and benefited from highly developed multimodal transportation infrastructure inside and throughout the state. The National Highway 55 (NH-55), which connects North Bengal's plains with its hill regions, passes through the different blocks of the study area. Building new roads, expanding the current transportation system, and making the most use of it are all crucial for the development of the studied region. Network analysis is widely used to comprehend the network infrastructure of a certain location. A branch or network system is created when nodes are joined (Sreelekha et al. 2016). These systems establish a local transportation network by moving commodities, people, and other accessories (Cliff et al. 1979).

16.7.1 Connectivity Analysis in the Hill Area of the Darjeeling District

The most basic characteristics for examining the transportation network are network indices. Rangli-Rangliot Block communities, which are the most connected according to the alpha index, are followed in order of importance by Kurseong, Mirik, etc. The beta index of five blocks has an average value of 1.68%, indicating a complicated transport network. The very low value of the gamma index (35.80) indicates that the main urban cities and villages need to be better connected. The theta index and eta index for the study blocks are quite high in three blocks: Kurseong (1.06, 1.12), Jorebunglow-Sukiapokhri (1.11, 1.09), and Darjeeling-Pulbazar (1.48, 1.51), which indicates that a complex and well-connected network covers the study area as opposed to the other two blocks: Mirik (0.68, 0.72) and Rangli-Rangliot (0.86, 0.92). The GTP index value represents the chessboard network pattern run in the Darjeeling district's hilly regions. Given that the degree of connectivity in the total blocks is 11.55, there is a moderate level of connectivity. The cyclomatic number value suggests low to moderate connectivity in the different blocks of the study area.

16.7.2 Agricultural Characteristics Analysis in the Hill Area of Darjeeling District

Three parameters were chosen to analyze the study area's agricultural characteristics: the irrigation status, gross cropped area, and total plantation area (as tea is the dominant product of the study area). According to the analysis of blocks, the Darjeeling-Pulbazar block has a high percentage of gross cultivated area, whereas all other blocks—aside from Mirik—fall into the moderate category. Due to the research area's proximity to a hilly region, irrigation status is extremely low. Again, the Darjeeling-Pulbazar block was placed first due to the higher frequency of agricultural practices there. Since the Darjeeling district has a 200-year history of tea plantations due to its high terrain, the total area of the plantations has been used as a criterion to identify the agricultural characteristics of the research area. The majority of the well-known tea estates in the world, including Goodrick, Burnesbeg, Boom, Gopaldhara, and others, are situated in this study area.

16.8 Discussion

16.8.1 Impact of the Road Network on Agricultural Development

Following the chapter on road connectivity using various network indices, it is evident that road network connectivity is complicated and poorly connected to major urban areas. Hilly terrain and sloping mountains are undoubtedly contributing factors, but the government should emphasize constructing more rural roads even as access to remote villages improves. In this study, Darjeeling-Pulbazar and Rangli-Rangliot have a

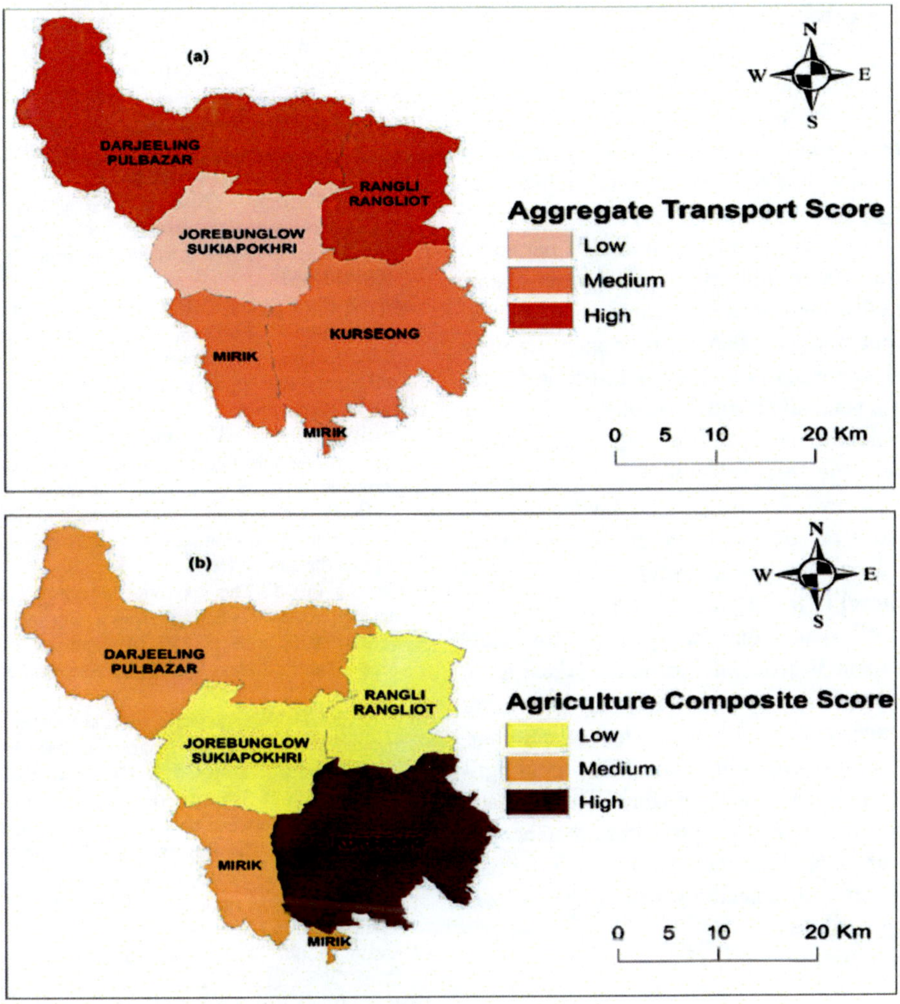

Fig. 16.4 Final output figures **a** aggregate transport score, and **b** agriculture composite score

high connectivity score, while Kurseong and Mirik blocks have a medium connectivity score, and Jorebunglow-Sukiapokhri falls into the low connectivity category. Kurseong Block, Darjeeling-Pulbazar Block, and Mirik Block have the highest agriculture composite scores due to their high gross cropped area and irrigational facilities, respectively. As the majority of the block is controlled by the government or a private firm for a tea plantation, Jorebunglow-Sukiapokhri and Rangli-Rangliot have low values for their agriculture composite score. A comparison between the aggregate transport score and the agricultural composite score was made in this research, and it is obvious that the study area lacks the favorable effects of road connectivity on the development of agriculture (Fig. 16.4a and b). Even if there is little subsistence farming visible, most people rely on plain regions for their grain needs because the study area is located in a sloppy mountain environment with a high value for ruggedness. In the study area, ginger, cardamom, squash, oranges, and orchids are produced on a large scale. Because people grow diverse crops locally to meet their requirements, road connectivity only provides little for the development of agriculture.

16.9 Conclusion

A region's growth and development are generally attributed to the transportation system. Roadways that are well connected and linked provide for easy location linking. As a result, to fulfill the growing population's demands, the study area's blocks have constructed transportation networks. Different network indices based on graph theory indicate that these villages' connectivity is quite good, but not all villages experience seamless network connectivity and accessibility, although there is a generally positive relationship between road network connectivity and accessibility. However, due to adverse conditions of several physical factors, such as climate, landscape, soil, and water, the development of different crops cannot grow that much (except plantation and horticulture) in the study area, which is why the impact of road connectivity on agricultural development is relatively restricted. Except for tea, most grain crops can only be grown for subsistence, making the expansion of commercial agriculture unfeasible. However, local agricultural products like ginger, squash, and cardamom can be exported to other urban centers and major parts of the state if every small village can connect with a nearby urban center by constructing rural roads.

References

Arif M, Gupta K (2020) Application of graph-based model for the quantification of the transport network in the peri-urban interface of Burdwan City, India. Spat Inform Res 28(4):447–457

Annual Reports (2000–2011) Indian Patent Office (IPO). New Delhi, India

Binswanger HP, Khandker SR, Rosenzweig M (1993) How infrastructure and financial institutions affect agricultural output and investment in India. J Dev Econ 41:331–336

Cliff A, Haggett P, Ord K (1979) Graph theory and geography. Appl Graph Theor 293:326

Demšar U, Špatenková O, Virrantaus K (2008) Identifying critical locations in a spatial network with graph theory. Trans GIS 12(1):61–82

Derrible S, Kennedy C (2009) Network analysis of world subway systems using updated graph theory. Transp Res Rec 2112(1):17–25

Derrible S, Kennedy C (2011) Applications of graph theory and network science to transit network design. Transp Rev 31(4):495–519

District Statistical Handbook (2010–2011) Darjeeling district. Handbook. http://wbpspm.gov.in

District Statistical Handbook (2013) Darjeeling District. Handbook. http://www.wbpspm.gov.in/

Dorosh P, Thurlow J (2018) Beyond agriculture versus non-agriculture: decomposing sectoral growth–poverty linkages in five African countries. World Dev 109:440–451

Elmqvist T, Fragkias M, Goodness J, Güneralp B, Marcotullio PJ, McDonald RI, Wilkinson et al (2013) Urbanization, biodiversity, and ecosystem services: challenges and opportunities: a global assessment

Flitter H, Grossmann T (2016) Accessibility (network analysis). Retrieved 2 May 2020, from http://www.gitta.info/Accessibility/en/text/Accessibiliti.pdf

Government of India (2017) State of Indian Agriculture 2016–17. Department of Agriculture and Cooperation, Directorate of Economics and statistics, Ministry of Agriculture, New Delhi

Hafner JA (1971) The economic development of Thai agriculture. By TH Silcock. Ithaca: Cornell University Press, 1970, xiii, p 250. Appendices References, Maps, Tables, Index. $11.50. J Asian Stud 30(4):920–922

Holl A (2007) Twenty years of accessibility improvements. The case of the Spanish motorway building programme. J Transp Geogr 15(4):286–297

Kansky KJ (1963) Structure of transportation networks: relationships between network geometry and regional characteristics. The University of Chicago

Kim H, Song Y (2018) An integrated measure of accessibility and reliability of mass transit systems. Transportation 45(4):1075–1100

Levinson D (2012) Network structure and city size. PLoS ONE 7(1):e29721

Levy H (1996) Kingdom of Morocco: impact evaluation report, socioeconomic influence of rural roads. Operation Valuation Department, World Bank, Washington, DC

Maity B, Mallick SK, Rudra S (2021) Integration of urban expansion with hybrid road transport network development within Haldia Municipality, West Bengal. Egyptian J Remote Sens Space Sci 24(3):471–483

Masters WA, Djurfeldt AA, De Haan C, Hazell P, Jayne T, Jirström M, Reardon T (2013) Urbanization and farm size in Asia and Africa: implications for food security and agricultural research. Glob Food Sec 2 (3):156–165

McGranahan G, Satterthwaite D, Tacoli C (2004) Rural-urban change, boundary problems and environmental burdens. Int Inst Environ Develop

Mukherjee S (2012) Statistical analysis of the road network of India. Pramana 79:483–491

Nagendra H, Munroe DK, Southworth J (2004) From pattern to process: landscape fragmentation and the analysis of land use/land cover change. Agr Ecosyst Environ 101(2–3):111–115

Nagne AD, Gawali BW (2013) Transportation network analysis by using remote sensing and GIS a review. Int J Eng Res Appl 3(3):70–76

Páez A, Scott DM, Morency C (2012) Measuring accessibility: positive and normative implementations of various accessibility indicators. J Transp Geogr 25:141–153

Patarasuk R (2013) Road network connectivity and land-cover dynamics in Lop Buri province, Thailand. J Transp Geogr 28:111–123

Reggiani A, Bucci P, Russo G (2011) Accessibility and impedance forms: empirical applications to the German commuting network. Int Reg Sci Rev 34(2):230–252

Sahitya KS, Prasad CSRK (2020) Evaluation of opportunity based urban road network accessibility using GIS. Spat Inf Res 28(4):487–493

Samanta PK (2015) Development of rural road infrastructure in India. Pacific Business Rev Int 7(11):86–93

Sarkar D (2013) Structural analysis of existing road networks of Cooch Behar district, West Bengal, India: a transport geographical appraisal. Ethiopian J Environ Stud Manage 6(1):74–81

Sarkar T, Sarkar D, Mondal P (2021) Road network accessibility analysis using graph theory and GIS technology: a study of the villages of English Bazar Block, India. Spatial Inform Res 29(3):405–415

Seto KC, Güneralp B, Hutyra LR (2012) Global forecasts of urban expansion to 2030 and direct impacts on biodiversity and carbon pools. Proc Natl Acad Sci 109 (40):16083–16088

Sharma S, Ram S (2023) Investigation of road network connectivity and accessibility in less accessible airport regions: the case of India. Sustainability 15(7):5747

Sreelekha MG, Krishnamurthy K, Anjaneyulu MVLR (2016) Assessment of topological pattern of urban road transport system of Calicut city. Transp Res Proc 17:253–262

Umoren V, Ikurekong EE, Emmanuel A, & Udida AA (2009) Development of road infrastructure as a tool of transforming Ibiono Ibom local government area. Glob J Soc Sci 8(2)

Verburg PH, Ellis EC, Letourneau A (2011) A global assessment of market accessibility and market influence for global environmental change studies. Environ Res Lett 6(3):034019

World Bank (1997) Rural development: vision to action: a sector strategy. The World Bank, Washington DC

17 Landslide and Its Impact on Agriculture in Kottiyoor Panchayath, Kannur District, Kerala

R. Nirmala

Abstract

Kottiyoor panchayath in Kannur district, the birthplace of the Bavali River, faced a severe worst landslide due to heavy rains from August 12 to 16, 2017 (around eight landslides). Kerala received about 116% of the normal rainfall, with 310 mm of rainfall in the state in the first 48 h. According to the Kerala disaster management authority, the 2017 landslide in Kottiyoor panchayath was among the deadliest. The study area is undulating and densely forested with very steep slopes, and excess runoff from the upper hilly areas leads to landslides, leaving the site at a degraded level. Also, many houses and crops were damaged due to the flood, about 3 acres of agricultural land in the Ambayathode area of Kottiyoor panchayath permanently vanished, and 15 acres of cropland were damaged. However, a landslide was repeated in the same area after four years on August 02, 2022, due to heavy rains. As per the study, this terrain mainly belongs to the early stage of landform development and tectonic zone, especially agricultural activities in steep slopes and loose sedimentary soils are prone to erosion and induced slope failure in this region. The only way to prevent landslides in this area is to avoid human activities on the slope of these areas.

Keywords

Agriculture · Kottiyoor panchayath · Landslide · Rainfall

17.1 Introduction

Landslide is considered the most important and common geological process, and hazards occur when they have direct contact with the human habitational zone. The above said natural hazard occurs when part of a natural slope cannot support its weight (Intrieri et al. 2019). The Western Ghats and Nilgiri Hills, which are the stable domains of south India, witness frequent landslide events, mainly due to rains during the South-West and North-East monsoons causing misery to people (Bhasin et al. 2002; Kumar and Sanoujam 2007; Petley et al. 2007; Avasty et al. 2009; Ganapathy et al. 2010; Mayavan and Sundaram 2012; Gurugnanam 2013). Kerala faced one of the worst landslides in 2017, with almost all parts of the state affected by this natural phenomenon, including the Kannur district. Unplanned development and encroachment of lands in the river vicinity have made the Kannur district prone to landslides in hilly areas and have led to floods. However, the relationship between landslides and

R. Nirmala (✉)
Department of Marine Geology, Mangalore University, Konaje 574199, India
e-mail: geonirmala19@gmail.com

J. Das and S. Halder (eds.), *Advancement of GI-Science and Sustainable Agriculture*, GIScience and Geo-environmental Modelling, https://doi.org/10.1007/978-3-031-36825-7_17

rainfall occurrence is very complex. There is no generally accepted method for rainfall-induced landslide prediction. A total of eight landslides occurred in the present study area of Kottiyoor panchayath, followed by five landslides in Ayyankunnu, Eruvessy, and Iritty. Landslide countermeasures are a formidable problem; hence, there is a crucial need for enhanced and modern techniques to analyze slopes and landslide susceptibility (Anbalagan and Parida 2013). The present chapter mainly aims to understand the impact of landslides on agricultural land and their natural and anthropogenic influence.

17.2 Study Area

Kannur district of Kerala has a coastal line on the western side and mountains on the eastern side; its topographical features have a natural slope toward the west. Kottiyoor panchayath, lying between 75° 50′ 00″–75° 55′ 00″ E and 11° 50′ 00″–11° 55′ 00″ N, is a hilly, undulating, and steeply sloping terrain and the extreme eastern part borders Karnataka state with forests at an altitude of 1689 mt. (Figs. 17.1, 17.2 and 17.3a). The study area is drained by river Bavali, a tributary of the Valapattanam River, and it has fourth stream order (Fig. 17.3c). The area is mainly underlain by metamorphic gneiss with patches of schist and amphibole (Fig. 17.3b). Most of the area is covered by dense forest, including evergreen forest and, deciduous forest, grassland, resulting in very few habitats and more agricultural activities around the Bavali River (Fig. 17.3c).

17.3 Data and Methods

The study area was first analyzed from the secondary data source to understand the current status. The LANDSAT 8 OLI/TIRS C2 L1, 2019 satellite

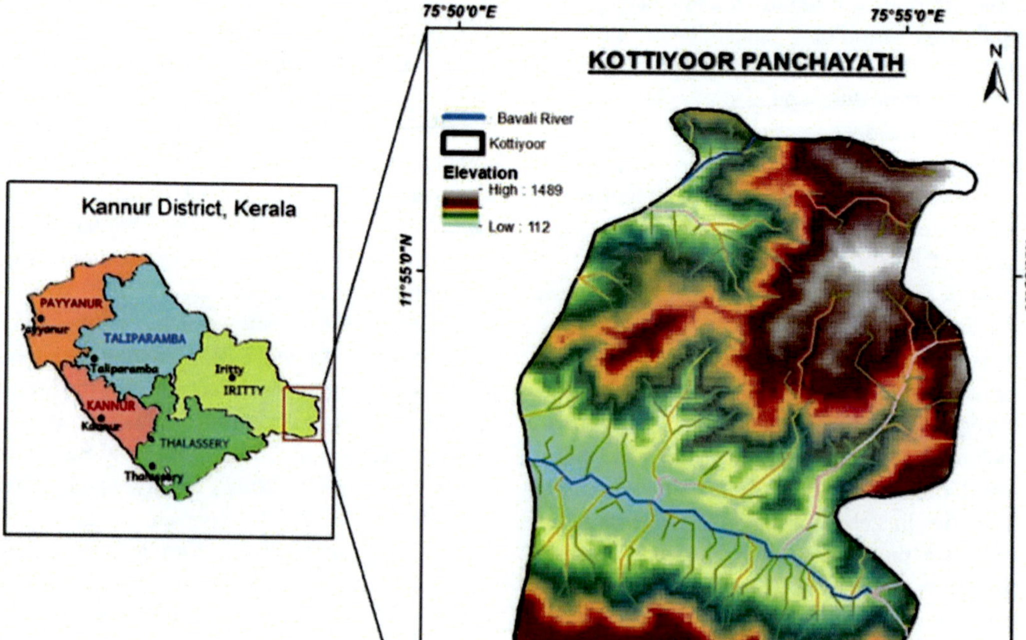

Fig. 17.1 Location map of the study area

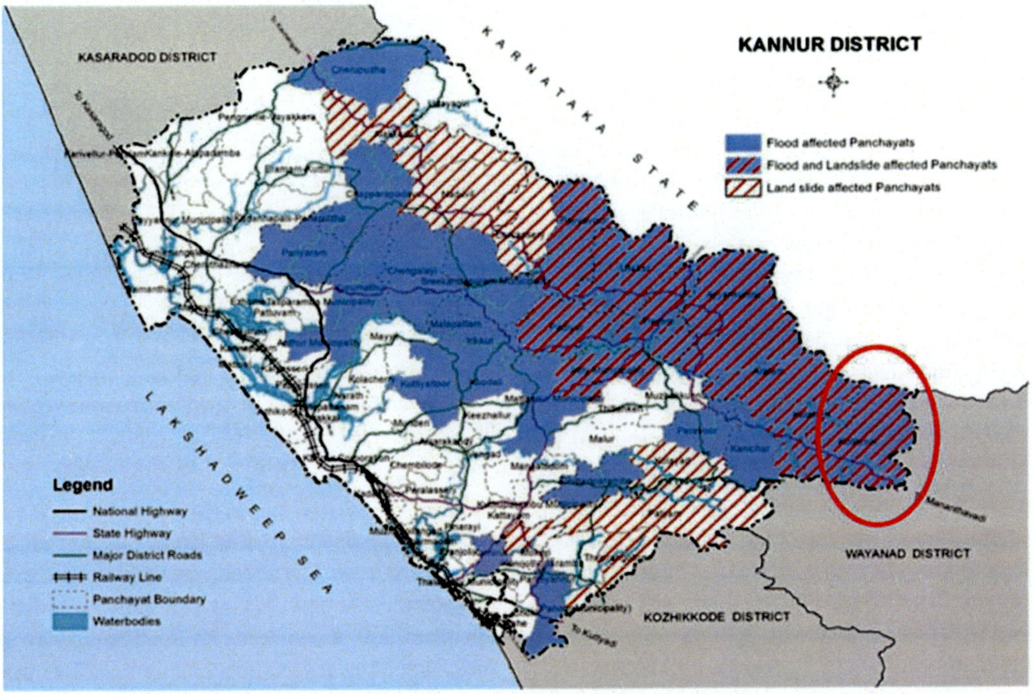

Fig. 17.2 Landslide and flood-affected panchayaths in Kannur district (*Source* GSI, Kerala)

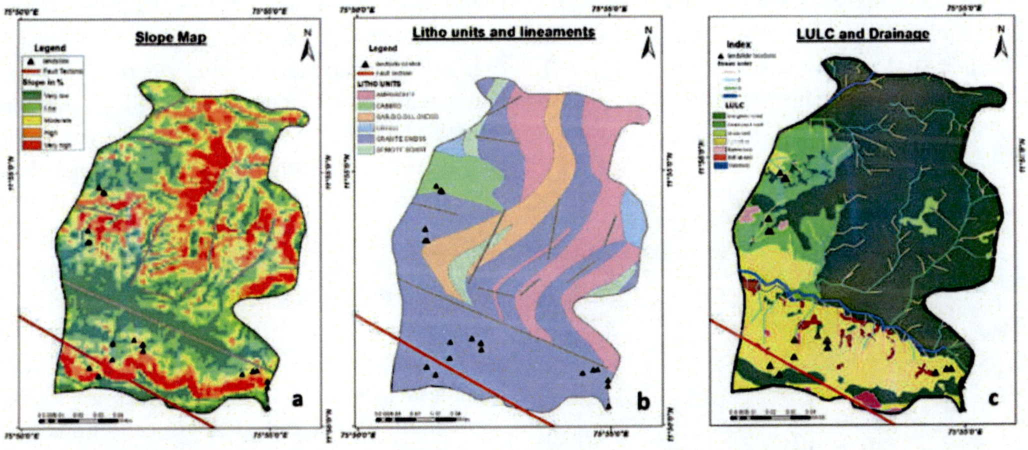

Fig. 17.3 Detailed mapping of the study area; **a** slope, **b** rock units and lineaments with landslide locations, **c** land use and land cover with drainage order

image has been used, and a classification pattern was established for the study area. Geology, soil, geomorphology, and lineaments help understand the cause of landslips (Bhukosh 2019). SRTM DEM image is also used for terrain analysis, such as elevation, slope, and drainage mapping.

17.4 Results

In the last three decades, more than 471 major disasters have occurred in India, and an estimated 1 lakh people have lost their lives. These disasters not only cost lives but also have an irreparable impact on the nation's economy, environment, infrastructure, and other assets (Sooraj et al. 2020). Small creeks, canals, rivers, and water bodies get submerged during floods, and landslides are the main reason behind this. Agriculture land is irreparably damaged.

Kottiyoor, the origin of Bavali River and Panchayath in Kannur district, experienced one of the worst landslides on August 16, 2017, with a total of eight landslides due to heavy rains (Figs. 17.4, 17.5 and 17.6), about 116% more than the usual rainfall in Kerala. In the first 48 h of rain, the state received 310 mm (12 in) of rain (www.deccanchronicle.com). The study terrain is undulating and densely forested with very steep slopes. According to a report by Kasthoori Rangan and Gadgil.

Sooraj et al. (2020), Kottiyoor is the 23rd Wildlife Sanctuary in Kerala, which belongs to the ecologically sensitive units and the landslide-susceptible zone (Fig. 17.2). The Kerala disaster management authority reported the 2017 landslide in Kottiyoor panchayath as one of the most dangerous landslides that ever happened. The surplus water flow from the upstream hill areas leads to landslips, leaving the place in a devastating condition. After four years, the same area again experienced a landslide due to heavy rainfall (> 204.5 mm) on August 02, 2022 (Fig. 17.7). This is mainly because of the youth stage of landform development, tectonic zone along major lineaments, agriculture activities at the steep slope, and saturated lateritic sedimentary soil at the top; shrink and swell weathering leads to vulnerable for erosion and seepage water emanating from the slope (Xu et al. 2020). According to the Central Meteorological Department, on August 02, 2022, strong winds of 40–50 km per hour were gusted along the Kerala-Lakshadweep coast.

Fig. 17.5 Agricultural land has been destroyed by landslides in Kottiyoor (Screengrab/manoramanews Tuesday, August 02, 2022—15:12)

Fig. 17.4 On August 16, 2017, Thursday, a landslide occurred at Ambayathode in Kottiyoor (www.thehindu.com)

As a result of this, it is suggested that irrigation loss from landslides represents a growing agricultural threat and geologic conditions with certain climatic (Duan et al. 2014). Furthermore, recorded landslides worldwide indicate that heavy rainfall causes a collapse in the high effective permeability of soils and high saturation of soil masses (Xie et al. 2021; Jin and Dai 2007). In other words, the more unstable subsurface geologic layers are when saturated due to percolating water from irrigation projects, the more prone to landslides due to a rise in water table level (Slack et al. 1996).

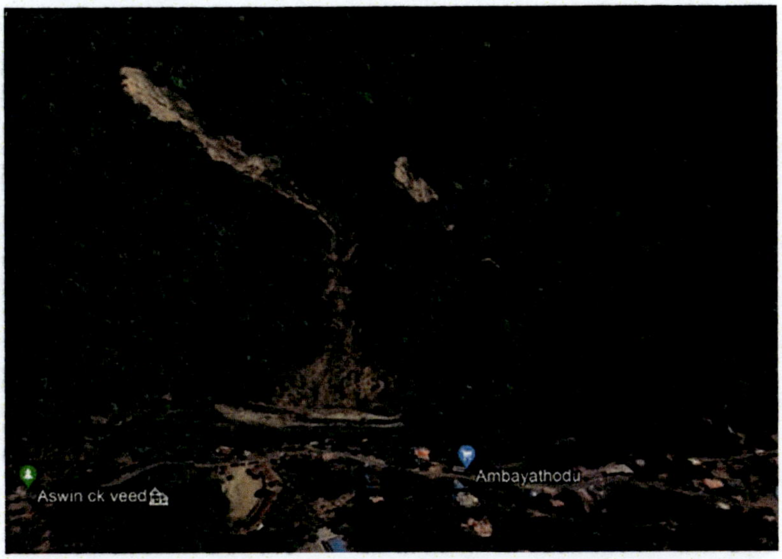

Fig. 17.6 Google earth image of landslide occurred at Ambayathode on August 16, 2017

India Meteorological Department's District Rainfall Forecast
Issue Date 02.08.2022; Time: 10:00 hrs.

District	02.08.2022	03.08.2022	04.08.2022	05.08.2022	06.08.2022
Thiruvananthapuram	ISOL 'H to VH	ISOL 'H to VH	ISOL H	L to M	L to M
Kollam	ISOL 'H to VH	ISOL 'H to VH	ISOL H	L to M	L to M
Pathanamthitta	ISOL 'H to VH	ISOL 'H to VH	ISOL 'H to VH	L to M	L to M
Alappuzha	X H	X H	ISOL 'H to VH	L to M	L to M
Kottayam	X H	X H	ISOL 'H to VH	ISOL H	L to M
Ernakulam	X H	X H	X H	ISOL H	L to M
Idukki	X H	X H	X H	ISOL H	L to M
Thrissur	X H	X H	X H	ISOL H	L to M
Palakkad	X H	X H	X H	ISOL H	L to M
Malappuram	X H	X H	X H	ISOL H	L to M
Kozhikode	X H	X H	X H	ISOL 'H to VH	ISOL H
Wayanad	X H	X H	X H	ISOL 'H to VH	ISOL H
Kannur	X H	X H	X H	ISOL 'H to VH	ISOL H
Kasaragod	ISOL 'H to VH	ISOL 'H to VH	X H	ISOL 'H to VH	ISOL H

Fig. 17.7 Rainfall data of Kannur district on August 02, 2022

Several natural factors are likely to occur simultaneously. In some cases, one risk triggers another (Sooraj et al. 2020). Sudden inundation of rainwater from the hilly areas of Kottiyoor has caused flooding and simultaneous landslides in the lower catchment of the Bavali River, an area occupied by prime agriculture and plantation lands (Fig. 17.3c). About three acres of agricultural land and plantation crops in the study area have been permanently wiped out, and about 15 acres of cropland have been damaged (Fig. 17.5). Accumulation of silt, stones, muddy soil, etc., in agricultural fields, increased soil acidification, loss of topsoil fertility, and mass mortality of soil organisms such as earthworms (Xu et al. 2020; Gu et al. 2019). This natural calamity results in a reduction of latex production in rubber plantations and an increase in crop diseases such as bark rotting, bacterial leaf blight in paddy, and fungal disease in pepper defoliation (Kennedy et al. 2015). The only way to prevent landslides in this area is slope reduction and avoiding construction on steep areas.

17.5 Conclusion

The chapter focuses on the impact of landslides on agricultural land in villages of Kottiyoor panchayath. The time span, selected is 2017 and 2022. Causes of landslides in Kottiyoor panchayath are changes in natural slope by blocking natural grooves and streamlets for cultivation; heavy rains for a week leading to build-up of excess pore water pressure; reducing the cohesive strength of soil particles due to a high degree of saturation, eventually leading to floods and landslips. Although based on land use patterns and geological factors, this area belongs to landslide-prone areas. Proper planning should be done to reduce the severity of disasters caused by natural calamities in the future and help in the sustainable development of hilly areas. Aside these, for better preparedness especially in the hilly landslide-prone (small) regions, rational management (in the shape of pre-disaster and post-disaster management) is earnestly needed. Additionally, under the giant task of regional disaster management, participation of local people and grassroots institutions is also necessary. In this way, we can go forward not only for sustainable development as well as sustainable agricultural avenue.

References

Anbalagan R, Parida S (2013) Geoenvironmental problems due to harmony landslide in Garhwal Himalaya, Uttarakhand, India. Int J Emerg Technol Adv Eng 3 (3):553–559

Avasthy RK, Kumar H (2009) Landslide hazard zonation mapping along Chamba-Bharmaur Road, Chamba District, Himachal Pradesh. Indian Landslides 2(1). http://www.indianlandslide.info/images/v2_2.pdf

Bhasin R, Grimstad E, Larsen JO (2002) Landslide hazards and mitigation measures at Gangtok, Sikkim Himalaya. Eng Geol 64(4):351–368

Bhukosh, GSI (2019) https://bhukosh.gsi.gov.in. Bhuvan Indian Geo-Platform of ISRO: bhuvan.nrsc.gov.in. Central Meteorological Department, Kerala 2019. Deccanchronicle.com/nation/current-affairs/170817

Duan Z, He ZG, Lin HZ (2014) Stability analysis of loess landslides induced by irrigation. Appl Mech Mater 716:395–399

Ganapathy GP, Mahendran K, Sekar SK (2010) Need and urgency of landslide risk planning for Nilgiri District, Tamil Nadu State, India. Int J Geom Geosci 1(10):17–24

Sooraj G, Kumara HS, Naveen K (2020) Sustainable habitat plan for conservation of hilly areas: a case study of Kannur district, Kerala state. J Xi'an Univ Arch Technol 12(7)

Gurugnanam B (2013) GIS data base generation on landslides by tracing the historical landslide locations in Nilgiri District, South India. Int J Remote Sens Geosci 2(6). http://ijrsg.com/Files/fe6e6f08-9b91-4912-ab57-65094cf653dd_IJRSG_10_3.pdf

Hemasinghea H, Rangali RSS, Deshapriyac NL, Samarakoon L (2017) Landslide susceptibility mapping using logistic regression model: A case study in Badulla District, Sri Lanka. In: 7th International conference on building resilience, ICBR2017, 27–29, Bangkok, Thailand. www.sciencedirect.com

Intrieri E, Carla T, Gigli G (2019) Forecasting the time of failure of landslides at slope-scale: a literature review. Earth-Sci Rev 193:333–349

Kennedy ITR, Petley DN, Williams R, Murray V (2015) PLoS currents. https://ncbi.nlm.nih.gov

Kumar A, Sanoujam M (2007) Landslide studies along the national highway (NH 39) in Manipur. Nat Hazards 40(3):603–614

Mayavan N, Sundaram A (2012) An approach for remote sensing and GIS based landslide hazard zonation mapping in Sirumalai Hill, Tamil Nadu. Elixir Remote Sens 51:10829–10833

Petley DN, Hearn G.J, Hart A, Rosser NJ, Dunning SA, Oven K, Mitchell WA (2007) Trends in landslide occurrence in Nepal. Nat Hazards 43(1):23–44

Slack D, Martin E, Sheta A, Fox F Jr, Clark L, Ashley R (1996) Crop coefficients normalized for climatic variability with growing-degree-days. In: Proceedings of the international conference on evapotranspiration and irrigation scheduling, San Antonio, TX, USA, pp 3–6, 892–898

Gu T, Zhang M, Wang J, Wang C, Xu Y, Wang X (2019) The effect of irrigation on slope stability in the Heifangtai Platform, Gansu Province, China. Eng Geol 248:346–356. https://doi.org/10.1016/j.enggeo.2018.10.026

Xie W, Guo Q, Wu JY, Li P, Yang H, Zhang M (2021) Analysis of loess landslide mechanism and numerical

simulation stabilization on the Loess Plateau in Central China. Nat Hazards 106:805–827. https://doi.org/10.1007/s11069-020-04492-w
Jin YL, Dai FC (2007) Mechanism of irrigation-induced landslides of loess. Chinese J Geotech Eng 29(10): 1493–1499
Xu Y, Wang J, Gu T, Kong J (2020) Geological hazards in Loess induced by agricultural irrigation in arid and semiarid regions of China. Adv Civ Eng 11. https://doi.org/10.1155/2020/8859166

18 Agricultural Land Use Change and Its Impact on the Farmers' Livelihood Assets of Maldah District, West Bengal, India

Tapash Mandal and Snehasish Saha

Abstract

Agriculture is an important sector of the Indian economy and plays a pivotal role in the process of economic development of a region. However, the pattern of agricultural land use is not static; it is dynamic. It changes with time and space due to the variation in the quality of land and availability of water resources as well as irrigation facilities. Such agricultural land use changes significantly affect the regional economy and the socio-economic conditions of the farmers. Therefore, this study has been considered to investigate the changing agricultural land use pattern and its impact on the farmer's livelihood assets in Maldah district. Farmer's perceptions have been taken into consideration for this study. The data have been collected from the 420 sample farmers (considering marginal, small, and large) from 15 community development blocks of Maldah district. The stratified random sampling technique has been used for collecting the primary data (Roy, Farmers' perception of the effect of IPM for sustainable crop production. M.S. Thesis, Department of Agricultural Extension Education, Bangladesh Agricultural University, Mymensingh, 2009). Perception index (PI) method has been used to identify the significant impact of agricultural land use change on the farmer's livelihood assets. The results showed that the impact of agricultural land use change significantly varied from farm size, i.e., marginal to large farmers. Such a study will help the policymaker to decide on regional agricultural planning and rural development.

Keywords

Agriculture · Economic development · Livelihood assets · Agricultural planning · Rural development

18.1 Introduction

Land use has been considered a particular area of the earth's surface and probably the most critical asset of the population used by a human from the earliest time to satisfy several needs (Verburg et al. 2004). It is the anticipated service of land management strategy positioned on the land cover by land managers or human agents to feat and reproduce human activities such as residential zones, industrial zones, agricultural fields, grazing, mining, and so on (Zubair 2006). The existing practice of land use in any region has been developed due to the interaction of plentiful

T. Mandal (✉) · S. Saha
Department of Geography and Applied Geography, University of North Bengal, PO. North Bengal University, Darjeeling 734013, India

J. Das and S. Halder (eds.), *Advancement of GI-Science and Sustainable Agriculture*, GIScience and Geo-environmental Modelling, https://doi.org/10.1007/978-3-031-36825-7_18

factors such as the physical characteristics of land, the institutional background, and the arrangement and availability of some other resources, i.e., labor, capital, location of the region, and concerning features of economic development (Mitsuda and Ito 2011). Man uses massive tracts of land to cultivate the crop, and this cropping tract belongs to a rural area and is fundamentally a pastoral activity. The term cropping pattern or agricultural land use symbolizes the gross cropped area computation under different crops during the agricultural year. It is a significant indicator of the appropriate and inappropriate use of land. The district's agricultural land use pattern is typical due to dryland and irrigated culture, which is directly governed by some other geographical aspects and revised by favorable social and economic conditions (Lubowski et al. 2006; Mandal and Saha 2021). Moreover, lack of implementation of new technology, essential machinery, high yielding varieties (HYVs) of seeds, pesticides, fertilizers, and availability of irrigation facilities prevails in the agricultural land use pattern of the district.

The farmers must decide what type of crop and cropping system must be accepted within the existing environment, and the knowledge about the cultivation technique was transferred from one generation to another (Woods et al. 2017). As expected, there was no change in the cultivation procedures or cropping pattern for a very long period. However, most parts of India have recently faced challenges to the increasing demand for food crops and the unequal development of agriculture. On the other hand, farmers' engagement with different crop practices and changing cropping patterns is impacting their livelihood assets (Pervez and Rahman 2017). The livelihood approach is a technique of thinking about the aims and objectives, scope, purposes, and priorities for development. It can be used in various ways, comprising the assets, capabilities, and activities indispensable for a living standard (Carswell 1997). However, the impact of ALU change on the farmer's livelihood assets is not equal in the blocks because of the economic condition of the farmers with the attachment of the various social groups, castes, creeds, and traditions, and they live in various geographical areas, and their attitude to agriculture is quite different. In general, the impact of ALU changes on farmers' livelihood assets varied with the marginal, small, and large farmers. According to the sustainable livelihood framework, a farmer's livelihood assets contain natural, physical, human, financial, and social assets (Chambers 1987; Scoones 1998).

To identify and analyze the farmer's livelihood assets related to ALU change, the present researcher has consulted various literature and gathered some significant assets influenced by ALU change (Chambers 1987; Scoones 1998; Makate et al. 2016; Kuang et al. 2020). These livelihood assets are categorized into four groups —A. physical and cultural assets (soil fertility, crop productivity, area under cultivation, water consumption, pressure on irrigation, pressure on livestock, housing condition, agricultural machinery, use of transportation, availability of labor and household amenities like T.V., and other electronic equipment); B. human assets (knowledge of production, skill of production, working capacity, educational level, family health, punctuality in working, and innovation thinking); C. social assets (interaction with fellow farmers, interaction with farmers outside the village, linkage with the research and development institutions, interaction with agricultural labor, involvement in social activities, interaction with agro-based industries, interaction with other organization, labor migration, and attitude of farmer); and D. financial assets (cash income, credit availability, on-farm income, off-farm income, production cost, fund assistance, and savings). A questionnaire survey assembled the farmer's perceptions to determine the significant impact of ALU changes on farmer's livelihood, which is an extensively helpful technique for identifying the driving forces, influences or impacts, trends, and ALU patterns (Devkota et al. 2017; Zaehringer et al. 2018).

18.2 Study Area

Maldah district has always been a significant basis of power in West Bengal due to its historic location and geographic elements. The district starts just north of the Ganga River, which divides the entire State of West Bengal into two parts, i.e., North Bengal and South Bengal. Currently, the district comprises two subdivisions, i.e., Maldah Sadar and Chanchal, 15 community development blocks, and 16 police stations; among them, one is for women. It lies between latitudes of 24° 40′ 20″ N–25° 32′ 08″ N and longitudes of 87° 45′ 50″ E–88° 28′ 10″ E (Fig. 18.1). The district is encompassed by Murshidabad district in the south, Uttar and Dakshin Dinajpur in the north, Bangladesh in the east, the State Jharkhand and Bihar in the west, and the southwestern limit is restricted by the river Ganga. Based on relief characteristics, the district has been categorized into three regions a. *Tal* is situated in the northern part of Kalindri and the north-western part of Mahananda river; and spread over Harishchandrapur I and Harishchandrapur II, Chanchal I and Chanchal II, Ratua I and Ratua II blocks b. *Diara*, situated in the southern part of Kalindri and the southwestern part of Mahananda river, covers Kaliachak I, Kaliachak II, Kaliachak III, English Bazar, and Manikchak blocks; and c. *Barind*, located in the eastern part of the Mahananda river, contains Old Maldah, Habibpur, Bamangola, and Gazole blocks. According to the NABARD (2020), out of the district's total geographical area of about 373,300 ha, 259,921 ha is the net shown area, and 152,719 ha is the net irrigated area; however, 112,481 ha and 21,145 ha have been considered as the drought-prone and flood-prone area, respectively. The gross cropped area of the district is 474,701 ha, and thus, the cropping intensity is 183. Cereal crops of the district represented five percent of the state's total production, and rice is the main food crop among all food grains (MKVK 2014).

18.3 Methodology

The interview technique has been performed to gather the farmer's perceptions regarding the impact of ALU changes on the farmer's livelihood assets during 1995–96 to 2020–21. A well-structured questionnaire was designed to acquire information on the effects of ALU change on the farmer's livelihood assets across the district. The household survey (HHS) was usually conducted on the head of the family of each household (preferably the oldest person) at the time of the field survey as they can reconstruct the past agricultural activities, recent changes, and impact on livelihood assets suitably. Thirty-four variables were compared in favor and disfavor against the five-point Likert scale. The variables were randomly arranged under four headings, i.e., physical and cultural, human, social, and financial. Each farmer was asked to indicate the level of disagreement or agreement against each variable along a five-point scale, i.e., strongly disagree, disagree, neutral, agree, and strongly agree. The allotted weight to these replies was 1, 2, 3, 4, and 5 in favor and 5, 4, 3, 2, and 1 for disfavor options. The respondent's total score was ascertained by summating the weights for replies against all 34 variables. Therefore, a farmer's perception score ranges from 34 to 170. On the other hand, the perception score of each farmer was computed using perception index (PI) technique (Roy 2009) using the following formula:

$$\mathrm{PI} = (5 \times \mathrm{SA}) + (4 \times A) + (3 \times N) + (2 \times \mathrm{DA}) + (1 \times \mathrm{SDA}) \quad \text{(in favor)} \tag{18.1}$$

$$\mathrm{PI} = (1 \times \mathrm{SA}) + (2 \times A) + (3 \times N) + (4 \times \mathrm{DA}) + (5 \times \mathrm{SDA}) \quad \text{(in disfavor)} \tag{18.2}$$

where PI refers to perception index; SA is the total no. of respondents asserting their perception

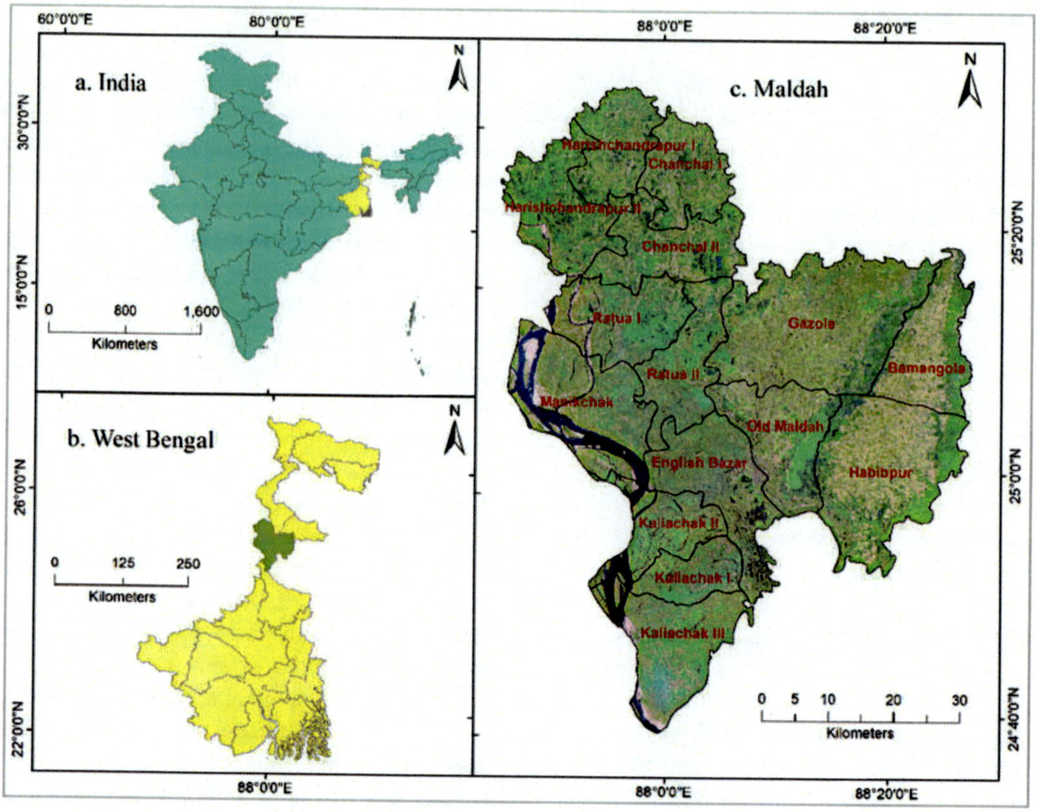

Fig. 18.1 Location map of the study area; **a** India, **b** West Bengal, and **c** Maldah

'strongly agree' for the statements; *A* is the total no. of respondents describing their perception 'agree' for the statements; *N* denotes the total no. of respondents asserting their perception 'neutral' for the statements; DA refers to the no. of respondents describing their perception 'disagree' for the statements; and SDA is the total no. of respondents expressing their perception 'strongly disagree' for the statements.

The vector normalization technique has also been performed to obtain the variables' rank. For those variables which have a direct relationship with the ALU change, the normalization is performed using the following formula (Lotfi and Fallahnejad 2010; Ildoromi et al. 2019):

$$V(x_{ij}) = \frac{x_{ij} - \min\{x_{ij}\}}{\max\{x_{ij}\} - \min\{x_{ij}\}} \quad (18.3)$$

Few variables have an indirect relationship with ALU change; for such cases, the following formula of normalization is used (Lotfi and Fallahnejad 2010; Ildoromi et al. 2019):

$$V(x_{ij}) = \frac{\max\{x_{ij}\} - x_{ij}}{\max\{x_{ij}\} - \min\{x_{ij}\}} \quad (18.4)$$

Before analyzing the data, the reliability of the datasets was tested by Cronbach's alpha using Statistical Packages for Social Sciences (SPSS) V26. Lee Cronbach proposed Cronbach's alpha in 1951 to measure the consistency or reliability of the data. The following formula has been used for calculating Cronbach's alpha:

$$\alpha = \frac{n}{n-1}\left(1 - \frac{\Sigma V_i}{V_i}\right) \quad (18.5)$$

where

n = Item's number,

V_i = Total score's variance,

V_t = Item score's variance,

α = Cronbach's alpha.

The Cronbach's alpha coefficient results show that the datasets are reliable and can be used for further analysis (Table 18.1 and Fig. 18.2).

On the other hand, relative changes area of the crops was obtained using the following formula:

$$\text{RD} = \frac{A(2020-21) - B(1995-96)}{B(1995-96)} \times 100 \quad (18.6)$$

where

A (2020–21) is the area of the specified crop for the year 2020–21

B (1995–96) is the area of the specified crop for the year 1995–96.

18.4 Results and Discussion

18.4.1 Changes in Agricultural Land Use

The growth and spatial variation of the area of the crops have been analyzed under four headings, i.e., A. cereals, B. pulses, C. oil seeds, and D. cash crops.

18.4.1.1 Spatio-Temporal Changes of Cereals

Rice, wheat, and maize are the major cereals in the study area. Among these three, rice and wheat are the principal crops grown in the highly fertile land of the district. The area under cereals in the district accounts for 78.26% of the total cropped area in 2020–2021 and 74.74% in 1995–96, showing significant change during the study period. The spatial variation of the cereals is evaluated using the data of 1995–96 and 2020–21 due to the unavailability of block-wise reliable data before 1995–96. The spatial variation of cereals with their relative change in percent in the different blocks is shown in Fig. 18.3. The variation is categorized into three equal zones based on their areal coverage and relative deviation (RD). Figure 18.3 clearly indicates a decrease in the area of autumn rice (Aus) in all the blocks of the study area except Harishchandrapur I and Harishchandrapur II during 1995–96 to 2020–21, and the rate of decrease is quite high (RD = > −66.67%). At the same time, Harishchandrapur I and Harishchandrapur II blocks showed a comparatively low decreasing rate (RD = < −33.33%). In these blocks, the cultivation of maize and summer rice (boro) has increased significantly. So, it may be summed up that the farmers switched their cultivation from autumn rice (Aus) to maize and boro in these blocks. For winter rice (Aman), decreasing and increasing areas are observed. Decreasing area (RD = −92.79–0.00) is mainly observed in the Chanchal I, Chanchal II, English Bazar, Gazole, Harishchandrapur I, Harishchandrapur II, Manikchak, Kaliachak I, and Kaliachak II blocks. In these blocks, a tendency of land conversion from cropping fields to mango orchards has been observed more. The abovesaid situation is the main cause of these blocks' decreased area under winter rice (Aman). The increasing winter rice (Aman) area is observed in the rest of the blocks. A comparatively low rate (RD = 0.00–800%) of increase is noticed in the Bamangola, Habibpur, Old Malda, Ratua I, and Ratua II blocks; and a high increasing rate (800%) is found in the

Table 18.1 Cronbach alpha's level of reliability

Level of reliability	Cronbach's alpha score
Less reliable	0.00–0.20
Rather reliable	0.21–0.40
Quite reliable	0.41–0.60
Reliable	0.61–0.80
Very reliable	0.81–1.00

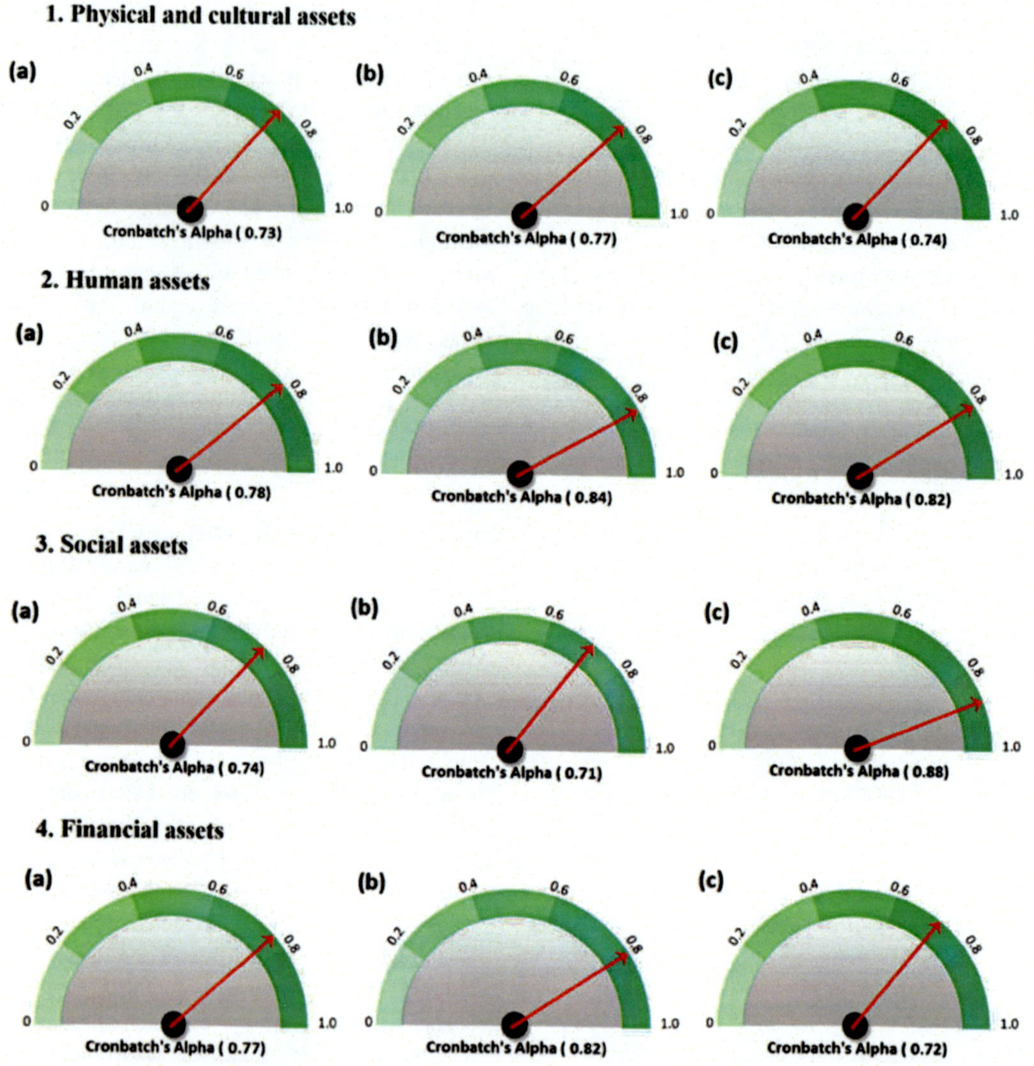

Fig. 18.2 Cronbach's alpha coefficient of different livelihood assets influenced by ALU change for **a** marginal, **b** small, and **c** large farmers

Kaliachak II block (Fig. 18.3). In these blocks, the area of autumn rice (Aus) has converted to the area under winter rice (aman). The area of the summer rice (Boro) also displayed negative and positive changes during the study period. Nine blocks (Chanchal I, Harishchandrapur I, Harishchandrapur II, English Bazar, Manikchak, Kaliachak I, Kaliachak II, Kaliachak III, and Ratua II) out of fifteen are caught with negative RD (−92.79–0.00%). In these blocks, the cultivation of maize has increased remarkably, and the summer rice (Boro) area has been converted to maize. On the other hand, Chanchal II, Gazole, and Ratua I have low positive changes (RD = 0.00–70.00%). At the same time, Bamangola, Habibpur, and Old Maldah are observed with high RD (>70.00%) in the district (Fig. 18.3).

Wheat cultivation areas also showed a similar tendency to summer rice (boro). Eleven blocks (Chanchal I, Chanchal II, Habibpur, Harishchandrapur I, Harishchandrapur II, Manikchak, Kaliachak I, Kaliachak III, Old Maldah,

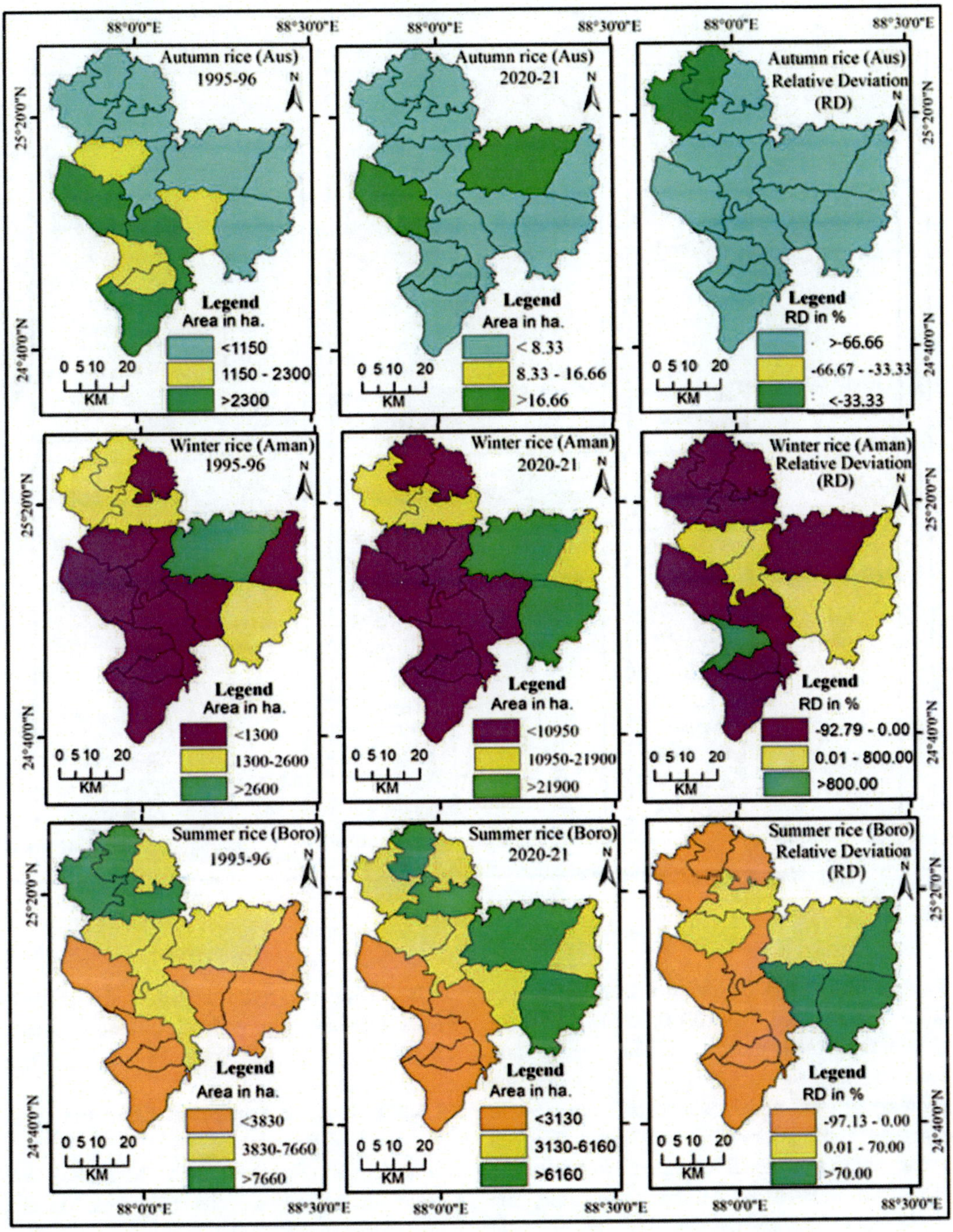

Fig. 18.3 Spatial variation and relative changes of an area of cereals (1995–1996 and 2020–2021)

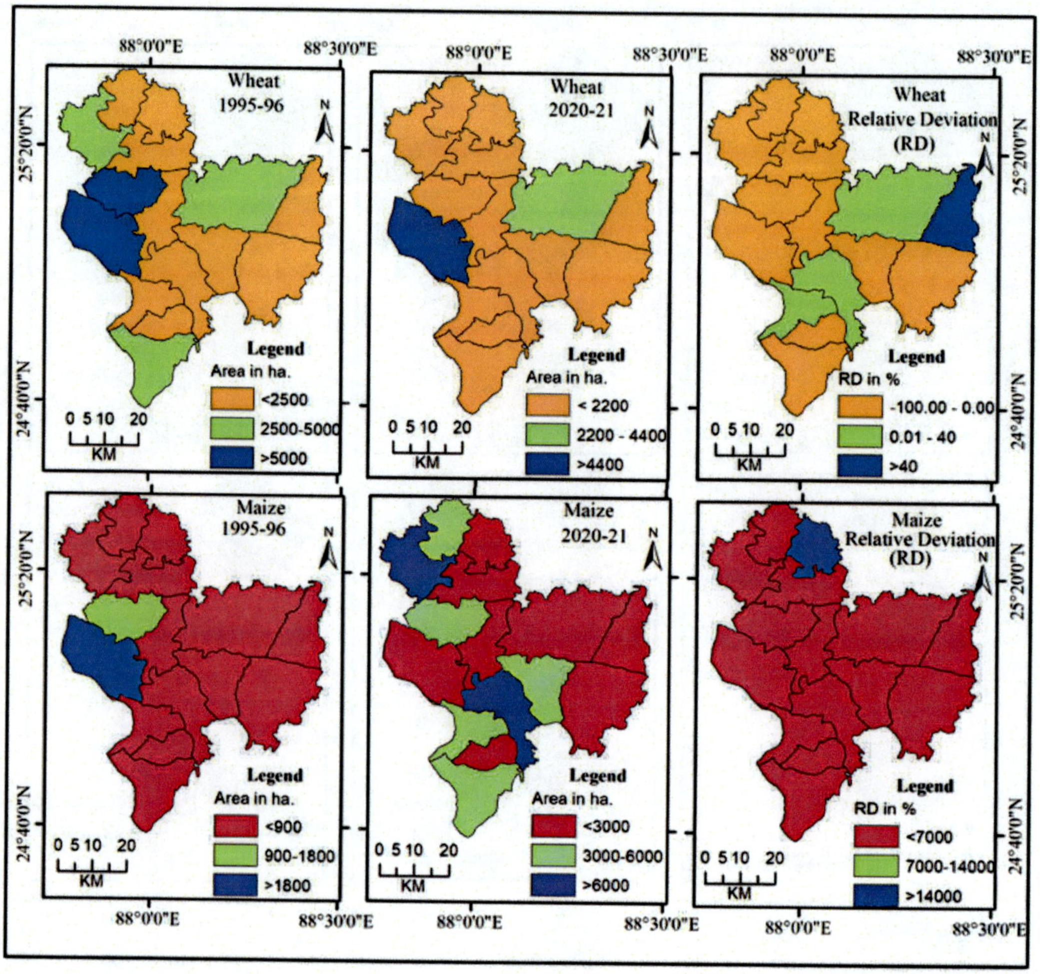

Fig. 18.3 (continued)

Ratua I, and Ratua II) out of fifteen are caught with negative RD (−100–0.00%). In these blocks, the farmers mostly cultivate rapeseed and mustard instead of wheat; therefore, the decreasing area under wheat is converted to rapeseed and mustard. While Gazole, English Bazar, and Kaliachak II have positive RD of the area with 0.01–40%. At the same time, the Bamangola block has the maximum positive RD (>40.00%). The area under maize has also increased positively from 1995–96 to 2020–21, and the increment magnitude is so high in all the blocks. All the blocks except Chanchal I showed an increase of < 7000; for Chanchal I, it was > 14,000 (Fig. 18.3). So maize is the district's most significant crop, which increased its area at a higher rate in all the blocks.

18.4.1.2 Spatio-Temporal Changes of Pulses

Gram, lentils (Masoor), urad (Maskalai), and Khesari are the pulses of the study area. The area under pulses in the district accounts for 6.76% of the total cropped area in 2020–2021 and 11.42% in 1995–96, showing significant change during the study period. The spatial variation of pulses with their relative change in percent in different blocks is shown in Fig. 18.4. Figure 18.4 shows

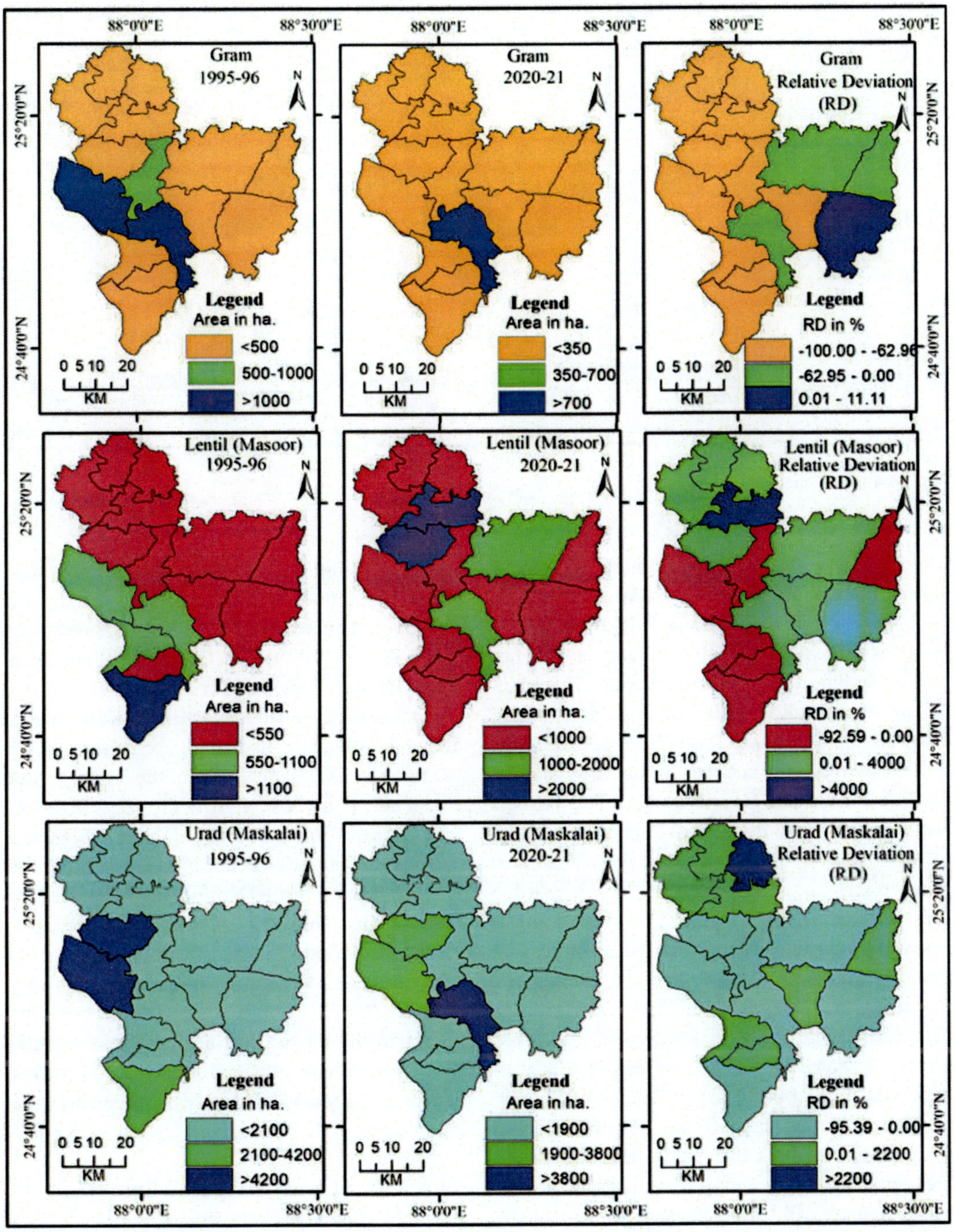

Fig. 18.4 Spatial variation and relative changes of an area of pulses during 1995–96 and 2020–21 (for Khesari 2003–2004 and 2020–2021)

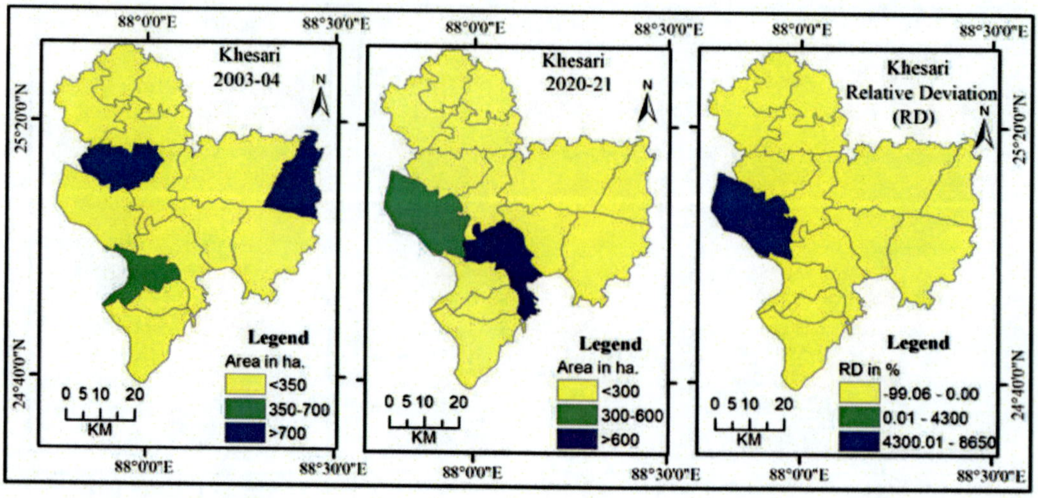

Fig. 18.4 (continued)

a decrease in the area of a gram in all the blocks except Bamangola during 1995–96 to 2020–21. High negative changes (RD = −100.00–62.96%) are observed in Chanchal I, Chanchal II, Harishchandrapur I, Harishchandrapur II, Manikchak, Kaliachak I, Kaliachak II, Kaliachak III, Old Malda, Ratua I, and Ratua II. Comparatively, a low decrease is noticed in the Bamangola, English Bazar, and Gazole blocks. The farmers switched their land from gram to urad (Maskalai) in these blocks. However, Bamangola showed positive RD from 1995–96 to 2020–21, and the increase rate is 0.01–11.11%. The area of lentils (masoor) showed decreasing and increasing trends during the study period, though the increasing area dominated the decreasing area. In nine blocks, the area of lentils (Masoor) has increased; however, the rate of increase is very high (RD = > 4000%) in the Chanchal II block.

Comparatively, high (0.01–4000%) increase is observed in the Chanchal I, English Bazar, Habibpur, Harishchandrapur I, Harishchandrapur II, Gazole, Old Maldah, and Ratua I block. While negative changes (RD = −92.59–0.00%) are observed in Bamangola, Manikchak, Kaliachak I, Kaliachak II, Kaliachak III, and Ratua II blocks (Fig. 18.4). The area of urad (Maskalai) also displayed negative and positive changes during the study period. Seven blocks (English Bazar, Gazole, Habibpur, Manikchak, Kaliachak III, Ratua I, and Ratua II) are caught with negative changes (RD = −95.39–0.00%) from 1995–96 to 2020–21. On the other hand, Bamangola, Chanchal II, Harishchandrapur I, Harishchandrapur II, Kaliachak I, Kaliachak II, and Old Maldah had high positive changes (RD = 0.01–2200.00%). At the same time, Chanchal I is observed with very high RD (>2200.00%) of the area of urad (Maskalai) cultivation (Fig. 18.4). The cultivation of khesari showed decreasing RD in all the blocks except Manikchak. The RD of the area decreasing blocks is −99.06–0.00%. In comparison, the RD rate of the Manikchak block is very high (RD = 4300.01–8650%). So, the khesari cultivation in the district has declined continuously. The farmers have changed their ALU from khesari to urad (Maskalai) and lentil (masoor) in the study area.

18.4.1.3 Spatio-Temporal Changes of Oilseeds and Cash Crops

Rapeseed and Mustard are the only oilseed, and jute, sugarcane, and potato are important cash crops in the study area. The area under rapeseed and mustard in the district accounts for 4.95% of the total cropped area in 2020–2021, which was

3.17% in 1980–1981, showing change during the study period. The area under cash crops is 10.02% of the total cropped area in 2020–2021 and 10.66% in 1980–81. The spatial variation of oilseed and cash crops with their relative change in the different blocks is shown in Fig. 18.5. From Fig. 18.5, it is found that the area under rapeseed and mustard has decreased in three blocks (Harishchandrapur I, Kaliachak I, and Kaliachak III) and increased in twelve blocks (Bamangola, Chanchal I, Chanchal II, English Bazar, Gazole, Habibpur, Harishchandrapur II, Manikchak, Kaliachak I, Old Maldah, Ratua I, and Ratua II). The increasing rate is maximum (>2000%) in Old Maldah and Ratua II blocks (Fig. 18.5). Medium to high positive changes (RD = 0.01–2000%) are observed in the Bamangola, Chanchal I, Chanchal II, English Bazar, Gazole, Habibpur, Harishchandrapur II, Manikchak, Kaliachak I, and Ratua I. While decreasing amount is −64.48–0.00% in the Harishchandrapur I, Kaliachak I, and Kaliachak III blocks. The farmers switched their cropland from rapeseed and mustard to maize cultivation in Harishchandrapur I, Kaliachak I, and Kaliachak III. The area under jute showed only positive changes from 1995–96 to 2020–21, and the magnitude of the increase is high in all the blocks. Comparatively, low high (<2950%) increasing rate of the area is observed in the Chanchal I, Chanchal II, Gazole, Harishchandrapur I, Harishchandrapur II, and Ratua I blocks. Medium–high (2950–5900%) increasing rate is observed in the English Bazar, Habibpur, Manikchak, Kaliachak I, Kaliachak II, Kaliachak III, Old Maldah, and Ratua II blocks. In contrast, a very high increasing rate (>5900%) is observed only in the Bamangola block. So, jute is another significant crop in the district, which increased its area at a higher rate in all the blocks.

The area of sugarcane showed both a decrease and an increase during the study period; however, the number of blocks with decreasing rates dominated the district. The district's sugarcane area has decreased to eleven blocks: Bamangola, Chanchal I, Chanchal II, Habibpur, Harishchandrapur I, Harishchandrapur II, Kaliachak II, Kaliachak III, Old Maldah, Ratua I, and Ratua II. In these eleven blocks, the decreasing amount is −100–0.00%. The farmers switched their cropland from sugarcane to maize and jute cultivation in these blocks. On the other hand, two blocks (English Bazar and Gazole) showed a very high increasing amount (RD = > 500%), and two blocks (Manikchak and Kaliachak I) showed a comparatively low increasing amount (RD = 0.01–500%) during the study period (Fig. 18.5). The area of potato also displayed negative and positive changes during the study period. Eight blocks (English Bazar, Chanchal I, Habibpur, Harishchandrapur I, Manikchak, Kaliachak I, Kaliachak III, and Old Maldah) out of fifteen are caught with negative changes (RD = −78.26–0.00%) from 1995–96 to 2020–21. In these blocks, the farmers changed their cropland from potato to rapeseed and mustard and maize cultivation. On the other hand, Bamangola, Chanchal II, Gazole, Kaliachak II, Ratua I, and Ratua II blocks areas observed high positive changes (RD = 0.01–500%). At the same time, Harishchandrapur II is observed with very high positive changes (RD = > 500%) in the area of potato cultivation (Fig. 18.5). The patterns of agriculture within the blocks have been changed over time and from place to place because of the socio-economic conditions of the farmers.

18.4.2 Farmer's Perception: Impact of ALU Change on Livelihood Assets

Many studies have employed farmer's perceptions through interviews to assess the status, trends, driving forces, impacts, and adaptation measures related to climate and changing agricultural land use patterns (Mertz et al. 2009; Devkota et al. 2017; Zaehringer et al. 2018). The HHS results showed that ALU change significantly impacts the farmers' livelihood assets in the study area. The summarized results reveal that the ALU changes affected mostly the physical and cultural assets (rank 1st), followed by financial (rank 2nd), social (rank 3rd), and human assets (rank 4th) of the marginal farmer (Table 18.2). At the same time, the financial

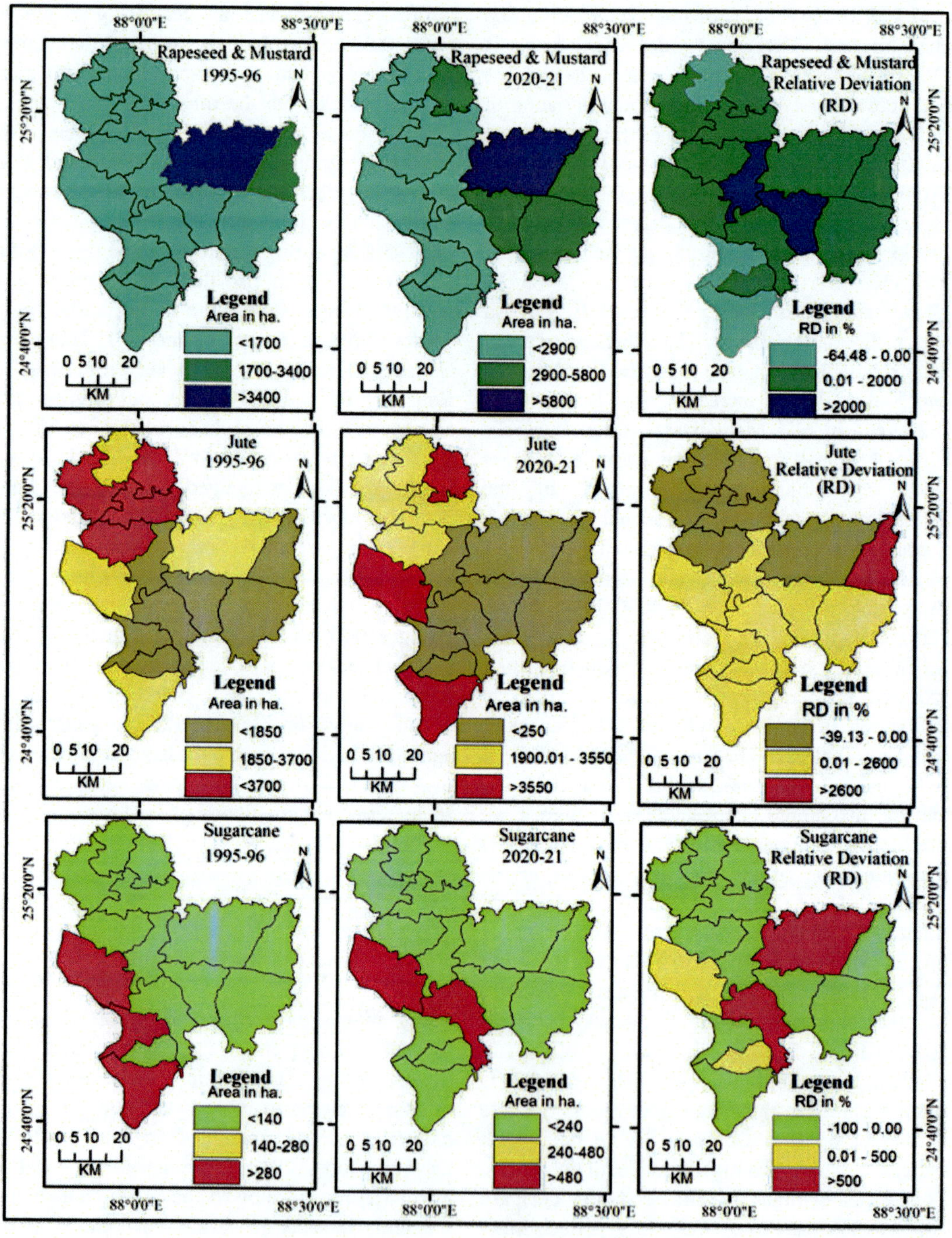

Fig. 18.5 Spatial variation and relative changes of an area of oilseeds and cash crops (1995–96 to 2020–21)

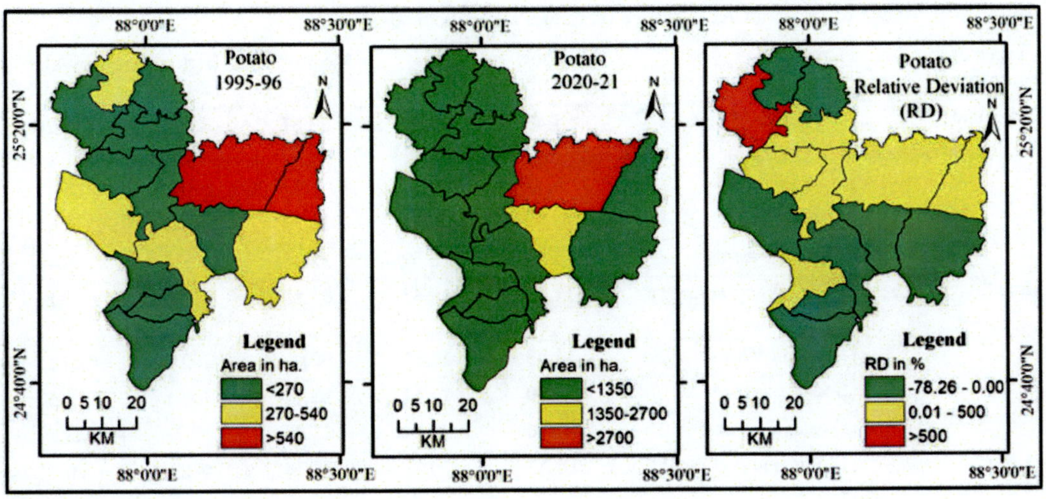

Fig. 18.5 (continued)

assets ranked first in the impact assessment of ALU change for small and large farmers. According to small and large farmers, the physical and cultural assets were also affected by the ALU change and ranked 2nd among the four major components. In contrast, social and human assets ranked 3rd and 4th for small farmers for the impact assessment of ALU change in the study area. For large farmers, human assets were positioned 3rd, and social assets were placed 4th in terms of the impact of ALU change (Table 18.2).

18.4.3 Nature of the Impact of ALU Changes

Table 18.3 shows the nature or quantification of the impact of ALU changes on the farmer's livelihood assets. Most marginal farmers (45.71%) revealed that the ALU changes have highly impacted physical and cultural assets. While 33.57% of the marginal farmers replied that ALU changes have a medium impact on physical and cultural assets. However, 14.29, 5.00, and 1.43% of the marginal farmers

Table 18.2 Farmer's responses about the rank of the impacts of ALU change on their different livelihood assets

Sl. no.	Parameter	Marginal farmers		Small farmers		Large farmers	
		Mean response	Rank	Mean response	Rank	Mean response	Rank
1	Physical and cultural assets	2.31	**1**	2.35	2	2.53	2
2	Human assets	3.00	4	3.22	4	2.71	3
3	Social assets	2.35	3	2.44	3	2.73	4
4	Financial assets	2.34	2	2.09	**1**	2.01	**1**

Table 18.3 Nature of impact of ALU change on farmer's livelihood assets

Types of the farmer	Criteria	Very low (%)	Low (%)	Medium (%)	High (%)	Very high (%)
Marginal	Physical and cultural assets	1.43	5.00	33.57	**45.71**	14.29
	Human assets	15.71	24.29	3.57	**30.00**	26.43
	Social assets	10.71	28.57	15.71	**31.43**	13.57
	Financial assets	14.29	8.57	**32.86**	27.86	16.43
Small	Physical and cultural assets	11.43	16.43	**37.86**	25.00	9.29
	Human assets	11.43	17.86	**30.00**	20.00	20.71
	Social assets	5.00	17.14	**33.57**	30.71	13.57
	Financial assets	27.86	7.14	21.43	10.71	**32.86**
Large	Physical and cultural assets	1.43	20.71	37.14	**40.00**	0.71
	Human assets	**25.71**	17.14	23.57	11.43	22.14
	Social assets	15.71	23.57	**27.86**	12.14	20.71
	Financial assets	10.00	13.57	10.71	**37.86**	27.86

responded that ALU changes impacted very high, low, and very low physical and cultural assets, respectively (Table 18.3). In the case of the effects on human assets, 30% of marginal farmers reported that the ALU changes have a high impact on this variable, followed by the very high (26.43%), low (24.29%), very low (15.71%), and medium (3.57%) impact, respectively. At the same time, 31.43% of the marginal farmers replied that the ALU changes highly affected the social assets, followed by low (28.58%), medium (15.71%), very high (13.57%), and very low (10.71%), respectively. The other essential livelihood assets of the marginal farmers are financial assets which were also influenced by the ALU changes. The majority (32.43%) of the marginal farmers said that the impact was medium, followed by high (27.86%), very high (16.43%), very low (14.29%), and low (8.57%), respectively.

The replies obtained from the majority of the small farmer showed that ALU changes had impacted medium on physical and cultural assets (37.86%), human assets (30%), social assets (33.57%), and very high on financial assets (32.86%). In contrast, the high impact of ALU changes was assessed on the physical and cultural (25%) and social assets (30.71%) by the small farmers (Table 18.3).

On the other hand, the majority of large farmers observed the high impact of ALU changes on physical and cultural assets (40%) and financial assets (37.86%). At the same time, medium and low impact of ALU changes were noticed by the majority of the large farmers on social assets (27.86%) and human assets (25.71%), respectively (Table 18.3). Conversely, a comparatively less significant percentage of influence was noticed on physical and cultural (37.14%) and human assets (23.57%), termed as medium impact and low impact on social assets (23.57%) and very high impact on financial assets (27.86%).

18.4.4 Ranking of the Impact Variables of ALU Change

Table 18.4 shows the rank of the thirty-four impact variables under the main component. The rank has been prepared based on the value obtained from perception index (PI) and the vector normalization. Here, a composite rank has been prepared considering the four main

Table 18.4 Rank of the impact variable based on their importance obtained from the marginal farmers' perception

Major component	Effects (impacts)	Perception index (PI)	Vector normalization	Rank
Physical and cultural assets	Decreased soil fertility	432	0.617	25
	Increased productivity	537	0.767	2
	Increased cultivation area	534	0.763	3
	Increased water consumption	412	0.589	27
	Increased pressure on irrigation	504	0.720	18
	Decreased pressure on livestock	513	0.733	14
	Improved housing condition	533	0.761	4
	Increased use of agricultural machinery	522	0.746	8
	Increased use of transportation means	520	0.743	10
	Increased availability of labor	526	0.751	7
	Increased household amenities like T.V. and other electronic equipment	539	0.770	1
Human assets	Increased knowledge of production	482	0.689	22
	Enhanced skill in production	516	0.737	13
	Increased working capacity	506	0.723	17
	Improved educational level	532	0.76	5
	Improved family health	509	0.727	16
	Enhanced punctuality in working	502	0.717	20
	Enhanced innovative thinking	418	0.597	26
Social assets	Increased interaction with fellow farmers	521	0.744	9
	Increased interaction with farmers outside of the village	529	0.756	6
	Built linkage with research and development institution	390	0.557	31
	Increased interaction with agricultural labor	494	0.706	21
	Increased involvement in social activities	333	0.476	34
	Increased interaction with agro-based industries	392	0.560	30
	Increased interaction with other organizations	503	0.719	19
	Decreased migration of labor	450	0.643	24
	Increased farmers attitude	404	0.577	29
Financial assets	Increased cash income	519	0.741	11
	Increased credit availability	511	0.730	15
	Increased on-farm income	406	0.580	28
	Increased of farm income	357	0.510	33
	Decreased production cost	517	0.739	12
	Increased fund assistance	470	0.671	23
	Increased savings	366	0.523	32

Table 18.5 Rank of the impact variable based on their importance obtained from the small farmers' perception

Major component	Effects (impacts)	Perception index (PI)	Vector normalization	Rank
Physical and cultural assets	Decreased soil fertility	450	0.643	14
	Increased productivity	482	0.689	1
	Increased cultivation area	336	0.48	34
	Increased water consumption	382	0.546	30
	Increased pressure on irrigation	466	0.666	8
	Decreased pressure on livestock	363	0.519	33
	Improved housing condition	452	0.674	5
	Increased use of agricultural machinery	474	0.677	3
	Increased use of transportation means	444	0.634	16
	Increased availability of labor	436	0.623	18
	Increased household amenities like T.V. and other electronic equipment	473	0.676	4
Human assets	Increased knowledge of production	470	0.671	6
	Enhanced skill in production	438	0.626	17
	Increased working capacity	430	0.614	20
	Improved educational level	459	0.656	10
	Improved family health	429	0.613	21
	Enhanced punctuality in working	426	0.609	23
	Enhanced innovative thinking	456	0.651	12
Social assets	Increased interaction with fellow farmers	409	0.584	28
	Increased interaction with farmers outside of the village	469	0.670	7
	Built linkage with research and development institution	411	0.587	27
	Increased interaction with agricultural labor	460	0.657	9
	Increased involvement in social activities	401	0.573	29
	Increased interaction with agro-based industries	422	0.603	25
	Increased interaction with other organizations	367	0.524	32
	Decreased migration of labor	472	0.646	13
	Increased farmers attitude	427	0.610	22
Financial assets	Increased cash income	447	0.639	15
	Increased credit availability	413	0.59	26
	Increased on-farm income	432	0.617	19
	Increased of farm income	380	0.543	31
	Decreased production cost	478	0.683	2
	Increased fund assistance	424	0.606	24
	Increased savings	458	0.654	11

components, i.e., physical and cultural, human, social, and financial assets. According to the value of PI, the ALU changes have significantly impacted the marginal farmer's household amenities like T.V. and other electronic equipment as it ranked 1st out of the 34 variables (Table 18.4). The marginal farmer's household amenities and electronic equipment increased from 1995–1996 to 2020–2021. The productivity of the crops and cultivation area of the marginal farmers has also been raised during the study period. These two variables ranked 2nd and 3rd in the impact assessment of ALU change (Table 18.4). Additionally, it significantly impacted the marginal farmer's housing conditions and educational level. According to the PI value, the marginal farmer's housing conditions and educational level improved during the study period due to ALU changes. These two indices ranked 4th and 5th among the total variables. The other essential variables affected by ALU change were increased interaction with farmers outside the village (rank 6th), increased availability of labor (rank 7th), increased use of agricultural machinery (rank 8th), increased interaction with fellow farmers (rank 9th), and increased use of transportation means (rank 10th) (Table 18.4). As per the marginal farmer's opinion, the rest of the variables, which ranked 11th–36th, were also impacted by the ALU change; however, these were comparatively less affected by the ALU change.

The rank of the thirty-four variables affected by ALU change in the study area based on the PI of small farmers has shown in Table 18.5. Also, a composite rank has been prepared considering the four main components associated with farmers' livelihood. As per the perception of the small farmer, the rank of the different variables was more or less similar to the perception of the marginal farmer. The PI and normalized vector indices clearly showed that the variable's dominant rank has significantly impacted by ALU change in the district. The productivity of crops was the most influential index among all the variables affected by the ALU changes. The crop productivity of the district has increased because of ALU change during 1995–96 to 2020–21, ranking 1st among all the variables. The 2nd and 3rd-ranking remarkable variables affected by the ALU changes were the production cost and the use of agricultural machinery. The production costs have decreased, and the use of agricultural machinery in the farming sector increased during the 25 years study period (Table 18.5). Also, from the PI and normalized indices results, it was found that the ALU changes significantly impacted the household amenities like T.V. and other electrical equipment and the housing conditions of the small farmers. The household amenities like T.V. and other electrical equipment and the housing conditions have increased or improved during the study period. These two indices ranked 4th and 5th among the total variables. The other variables affected by ALU change and ranked 6th–10th are increased knowledge of production, interaction with farmers outside the village, increased pressure on irrigation, interaction with agricultural labor, and improved educational level, respectively (Table 18.5).

According to the large farmer, the ALU changes have dramatically impacted the housing conditions, savings of the farmers, cash income, household amenities like T.V. and other electronic equipment, and pressure on irrigation. These five variables have increased or improved during the study period (1995–96 to 2020–21) and ranked 1st–5th, respectively (Table 18.6). Due to the ALU change, the attitude to agriculture has also changed and probably increased and is ranked 7th among all variables. The ALU change has also influenced the farmers' crop productivity, use of agricultural machinery, knowledge of agricultural production, and use of agricultural transportation, which ranked 7th, 8th, 9th, and 10th, respectively (Table 18.6). Additionally, the ALU changes significantly impacted the production cost, fund assistance, on-farm income, credit availability, and innovative thinking of the large farmer. According to the PI and the value obtained from vector normalization, these indices ranked 11th–15th among all the variables. The results also revealed that the variables increased during the study period. The other variables which were affected

Table 18.6 Rank of the impact variable based on their importance obtained from the large farmers' perception

Major component	Effects (impacts)	Perception index (PI)	Vector normalization	Rank
Physical and cultural assets	Decreased soil fertility	414	0.591	23
	Increased in productivity	478	0.683	7
	Increased cultivation area	370	0.529	32
	Increased water consumption	288	0.411	34
	Increased pressure on irrigation	486	0.694	5
	Decreased pressure on livestock	456	0.651	16
	Improved housing condition	536	0.766	1
	Increased use of agricultural machinery	476	0.680	8
	Increased use of transportation means	470	0.671	10
	Increased availability of labor	378	0.540	28
	Increased household amenities like T.V. and other electronic equipment	492	0.703	4
Human assets	Increased knowledge of production	472	0.674	9
	Enhanced skill in production	448	0.646	18
	Increased working capacity	447	0.64	19
	Improved educational level	422	0.603	22
	Improved family health	412	0.589	24
	Enhanced punctuality in working	453	0.647	17
	Enhanced innovative thinking	458	0.654	15
Social assets	Increased interaction with fellow farmers	380	0.543	27
	Increased interaction with farmers outside of the village	376	0.537	29
	Built linkage with research and development institution	374	0.534	30
	Increased interaction with agricultural labor	402	0.574	25
	Increased involvement in social activities	446	0.637	20
	Increased interaction with agro-based industries	436	0.623	21
	Increased interaction with other organizations	372	0.531	31
	Decreased migration of labor	382	0.546	26
	Increased farmers attitude	482	0.689	6
Financial assets	Increased cash income	504	0.72	3
	Increased credit availability	460	0.657	14
	Increased on-farm income	462	0.66	13
	Increased of farm income	367	0.524	33
	Decreased production cost	468	0.669	11
	Increased fund assistance	466	0.666	12
	Increased savings	518	0.740	2

by ALU change were decreased pressure on livestock (rank 16th), enhanced punctuality in working (rank 17th), enhanced skill of production (rank 18th), increased working capacity (rank 19th), and increased involvement in social activities (rank 20th) (Table 18.6).

18.5 Conclusion

The present study has assessed the ALU changes in Maldah district and its impact on farmers' livelihood assets based on the farmer's perceptions. The impact assessment using PI significantly affected farmers' livelihood assets from 1995–96 to 2020–21. However, the impacts varied with the farm's size, i.e., marginal, small, and large farmers. The significant rank obtained from PI and normalized vector was changed with farm size based on the farmer's perception; however, the crucial ranks alternate among marginal, small, and large farmers. As a result, the ALU changes most affected the variables that rank up to tenth. As per marginal farmers, the physical and cultural assets were more influenced by the ALU changes, and for the small and large farmers, financial assets were mostly affected. Additionally, the ALU changes have significantly impacted the marginal farmers' increasing household amenities like T.V. and other electronic equipment (agreed 68.6% and strongly agree 8.6%). From per small farmer's view, it significantly impacted the production cost (agreed 67.10% of the sample farmers), which increased during the study period. In contrast, the maximum agreed responses have come for increased cash income (agreed 79.20% of the sample farmers) for the large farmer.

References

Carswell G (1997) Agricultural intensification and rural sustainable livelihoods: a 'think piece'. IDS Working paper 64. https://opendocs.ids.ac.uk/opendocs/handle/20.500.12413/3368

Chambers R (1987) Sustainable livelihoods, environment and development: putting poor rural people first. Agriculture and rural problem. Institute of Development Studies

Cronbach LJ (1951) Coefficient alpha and the internal structure of tests. Psychometrika 16(3):297–334

Devkota RP, Pandey VP, Bhattarai U, Shrestha H, Adhikari S, Dulal KN (2017) Climate change and adaptation strategies in Budi Gandaki River basin, Nepal: a perception-based analysis. Clim Change 140 (2):195–208. https://doi.org/10.1007/s10584-016-1836-5

Ildoromi AR, Sepehri M, Malekinezhad H, Kiani-Harchegani M, Ghahramani A, Hosseini SZ, Artimani MM (2019) Application of multi-criteria decision making and GIS for check dam layout in the Ilanlu Basin, Northwest of Hamadan Province, Iran. Phys Chem Earth Parts a/b/c 114:102803. https://doi.org/10.1016/j.pce.2019.10.002

Kuang F, Jin J, He R, Ning J, Wan X (2020) Farmers' livelihood risks, livelihood assets and adaptation strategies in Rugao City, China. J Environ Manage 264:110463. https://doi.org/10.1016/j.jenvman.2020.110463

Lotfi FH, Fallahnejad R (2010) Imprecise Shannon's entropy and multi attribute decision making. Entropy 12(1):53–62. https://doi.org/10.3390/e12010053

Lubowski RN, Bucholtz S, Claassen R, Roberts MJ, Cooper JC, Gueorguieva A, Johansson RC (2006) Environmental effects of agricultural land-use change: the role of economics and policy. https://doi.org/10.22004/ag.econ.33591

Malda Krishi Vigyan Kendra (MKVK) (2014) Accessed date 05.03.2021 https://maldakvk.ubkv.ac.in

Makate C, Wang R, Makate M, Mango N (2016) Crop diversification and livelihoods of smallholder farmers in Zimbabwe: adaptive management for environmental change. Springer Plus 5(1):1–18. https://doi.org/10.1186/s40064-016-2802-4

Mandal T, Saha S (2021) Spatio-temporal changes of cropping intensity in Maldah district, West Bengal, India. Shodh Sanchar 11(41):54–59

Mertz O, Mbow C, Reenberg A, Diouf A (2009) Farmers' perceptions of climate change and agricultural adaptation strategies in rural Sahel. Environ Manage 43(5):804–816. https://doi.org/10.1007/s00267-008-9197-0

Mitsuda Y, Ito S (2011) A review of spatial-explicit factors determining spatial distribution of land use/land-use change. Landscape Ecol Eng 117–25. https://doi.org/10.1007/s11355-010-0113-4

NABARD (2020) Annual Report 2019–20, National Bank for Agriculture and Rural Development, Head Office, Mumbai

Pervez AS, Rahman S, Al-Amin AK (2017) Change in cropping patterns and its impacts on farmers 'livelihood in some selected areas of Mymensingh district. Bangladesh J Agric Econom 36(454–2017–716):1–3

Roy BS (2009) Farmers' Perception of the effect of IPM for sustainable crop production. M.S. Thesis, Department of Agricultural Extension Education, Bangladesh Agricultural University, Mymensingh

Scoones I (1998) Sustainable rural livelihoods: a framework for analysis

Verburg PH, Schot PP, Dijst MJ, Veldkamp A (2004) Land use change modelling: current practice and research priorities. Geo J 61:309–324. https://doi.org/10.1007/s10708-004-4946-y

Woods BA, Nielsen HØ, Pedersen AB, Kristofersson D (2017) Farmers' perceptions of climate change and their likely responses in Danish agriculture. Land Use Policy 65:109–20. https://doi.org/10.1016/j.landusepol.2017.04.007

Zaehringer JG, Wambugu G, Kiteme B, Eckert S (2018) How do large-scale agricultural investments affect land use and the environment on the western slopes of Mount Kenya? Empirical evidence based on small-scale farmers' perceptions and remote sensing. J Environ Manage 213:79–89. https://doi.org/10.1016/j.jenvman.2018.02.019

Zubair AO (2006) Change detection in land use and land cover using remote sensing data and GIS (a case study of Ilorin and its environs in Kwara State). University of Ibadan, Department of Geography, p 176

19 Assessment of Drought in Meteorological Data Using SPI and SPEI Indicators for Sustaining Agricultural Productivity in the Agra Division of Uttar Pradesh, India

Shekhar Singh, Anil Kumar, and Sonali Kumara

Abstract

The 'drought' phenomenon is regarded as a matter of concern as it negatively impacts agricultural productivity, the health of numerous water bodies, and the long-term survival of the earth's inhabitants. It is one of the oddities that cause the most serious corollaries in terms of security, especially in places that depend on agriculture. In this current chapter, the Standardized Precipitation Index (SPI) and Standardized Precipitation Evapotranspiration Index (SPEI) were employed to analyze meteorological drought events over 70 years, from 1951 to 2020, in four districts of State Uttar Pradesh, namely Agra, Firozabad, Mathura, and Mainpuri. The above-said methods (SPI and SPEI) were computed on various time scales of 1, 3, 6, 9, 12, and 24 months utilizing data on monthly climatic parameters (rainfall and temperature). According to the SPI24 findings, all stations experienced consecutive wet years from 1955 to 1965, whereas Agra and Mathura stations had consecutive drought years from 2000 to 2011. However, periods of consecutive drought were noted in Firozabad and Mainpuri from 2014 to 2018 and 2011 to 2018, respectively. The outcomes of this study revealed that, in most circumstances, the results provided by SPI and SPEI were identical. The findings of this study can also be used to create a mitigation strategy for drought in the study districts, which would involve reserving excess water during wet periods and using it for domestic and agricultural needs during dry periods.

Keywords

SPEI · SPI · Meteorological drought · Agra division · Uttar Pradesh

19.1 Introduction

Recently, 'drought' is a continuous recurring event of dryness and water scarcity in various regions on the surface of the earth, also classified as one of the vulnerable climatic events (Alam et al. 2023). Typically, drought can be described as an extended absence or noticeable deficiency of precipitation. A precipitation deficiency causes a water scarcity for some function of some group,

S. Singh · A. Kumar
Department of Soil and Water Conservation Engineering, G.B. Pant University of Agriculture and Technology, Pantnagar, Udham Singh Nagar 263145, Uttarakhand, India
e-mail: 52783@gbpuat.ac.in

S. Kumara (✉)
Department of Soil Conservation and Watershed Development, Office of Project Director, Watersheds, Balangir 767001, Odisha, India
e-mail: sonalikumura05@gmail.com

J. Das and S. Halder (eds.), *Advancement of GI-Science and Sustainable Agriculture*, GIScience and Geo-environmental Modelling, https://doi.org/10.1007/978-3-031-36825-7_19

or a period of remarkably dry weather sufficiently prolonged to establish an unexpected hydrological imbalance (Trenberth et al. 2014). It is also regarded as an adverse peculiarity that has critically influenced the financial, agricultural, and ecological circles and meant to be a longstanding average state of equilibrium between precipitation and evapotranspiration. It is an extended, recurring event of aridity in diverse regions of the earth system (atmosphere, lithosphere, and hydrosphere) (Mahmoudi et al. 2019) and responsible for unwanted damage (approximately one-fifth) occurred due to natural disasters worldwide (Srinivasarao et al. 2020). It is among the most unsafe oddities causing serious corollary, particularly in the areas dependent on agriculture. Despite the reality, it is revealed as a slow process to grasp the attention of the community on global scale; its impact continues long after the event has ended (Zhang et al. 2021). Farming methods have a greater impact on farm vulnerability, leading to drought. Some researchers and advocates believe sustainable agriculture systems are far less vulnerable to climate change than conventional agriculture (Kim and Jehanzaib 2020).Extreme drought episodes have occurred throughout human history (Uddin et al. 2020).

In India, 11 states, i.e., Karnataka, Odisha, Uttar Pradesh, Telangana, Jharkhand, Gujarat, Maharashtra, Madhya Pradesh, Chhattisgarh, and Rajasthan, were declared drought-affected areas in the year 2015–2016. Environmental changes are becoming riskier every year, putting humans and a wide variety of other life forms in danger. Both substantial flooding and drought affect states in India like Odisha, Jharkhand, and parts of the North East. In general, it creates the impression of scarcity as a consequence of inadequate precipitation, high evapotranspiration, misuse of groundwater sources, or the cumulative effect of these factors (Mahmoudi et al. 2019) and also had far-reaching consequences on ecosystems and the environment. Water availability in reservoirs, the quantity of rainfall, moisture contained soil, and depths of groundwater are all environmental indicators of drought (Kwon and Sung 2019; Cammalleri et al. 2022). In the arid and semi-arid regions, climatic variation has increased the harshness and interval of meteorological droughts. Drought will likely increase in the upcoming decades between 2020 and 2050 (Kulkarni et al. 2016). The fact is that the drought effects are unparalleled.

According to a study, 600,000 people were killed in a single region of northern India in 1792 by a severe drought (Thompson et al. 2000). Drought has far-reaching consequences on ecosystems and the environment. Water availability in reservoirs, temperature inversions, soil moisture, and groundwater depths are all environmental indicators of drought. According to Dracup et al. (1980), the major factors responsible for drought are precipitation (meteorological), soil moisture (agriculture), and streamflow (hydrological). According to White and Walcott (2009), droughts are of four kinds: (a) 'meteorological drought,' (b) 'hydrological drought,' (c) 'socio-economic drought,' and (d) 'agricultural drought' (Mishra and Singh 2011).

Meteorological drought is a situation where there is a decline in rainfall over a specific period of less than a specified amount, meaning that the actual rainfall in an area is much lower than the area's climatologically mean rainfall. Its analysis depends on the rainfall data. After the Meteorological Department of India (IMD) it can be said that, *it* (MD) takes place when the average annual rainfall is <75% of the normal rainfall (Das et al. 2020). The water shortage causes hydrological drought, it is the state that occurs when the water level in various reservoirs or storages, such as lakes and aquifers, falls below the limit. Also, its analysis depends on streamflow data and reservoir data/lake data.

Agricultural drought arises when the level of soil moisture is inadequate to support plant growth. Irregular rainfall patterns and insufficient soil moisture bring on crop failures. Cropping patterns are a reason for worry because they have a bad effect on a country's economy, company, and income. Socio-economic drought occurs when a lack of water affects the natural ecosystem's ability to produce food, causing environmental harm such as cattle, wildlife, and forest tree extinctions. Meteorological drought, hydrological drought, and agricultural drought are environmental indicators, and socio-economic drought is regarded as a water resources indicator. According to the Meteorological Department of India (IMD), meteorological drought occurs when precipitation in the form of rainfall during a calendar year or the Southwest Monsoon Season (June to September) is <25% of its average annual rainfall over the area. For proper water resources management, it is imperative to conduct a scientific inquiry and study the rate of recurrence, severity, return period, and extreme drought episodes (Bhunia et al. 2020). Recently, several global studies on the assessment and forecasting of droughts have been carried out, such as Rezaeian-Zadeh and Tabari (2012), Mishra and Desai (2006), Shirmohammadi et al. (2013), Belayneh and Adamowski (2012), Morid et al. (2007), Ozger et al. (2012), and Mehr et al. (2014), and few are discussed below.

Kar et al. (2007) studied the drought situation of western Odisha using SPI and RDI indices and computed at an interval of one month, two months, and three months time scale. The findings indicated that the district named Balangir was highly affected by drought. Log Pearson distribution and Weibull's formulae were used to find the probability of moderate to severe drought occurrence. Lorenzo-Lacruz et al. (2010) compared SPI and SPEI for three, twelve, twenty-four, and forty-eight months for three hydrological variables parameters, i.e., river flow, storage reservoir, and outflow from the reservoir and determined that the two drought indices (SPI and SPEI) efficiently recognized water shortages for multiple time scales and found the utmost correlations with inflow reservoirs observed at time scales of four–twelve months for both the SPI and the SPEI. Almedeij (2014) studied the historical precipitation data of Kuwait to analyze hydrological drought using SPI. The result showed a non-systematic trend. The result was also used for determining the drought event and evaluating the risk for drought. Also, the result helped for adopting the measures for drought. Deo and Sahin (2015) investigated the data of monthly precipitation from 1915 to 2012 using Standardized Precipitation Index (SPI) and Standardized Precipitation Evapotranspiration Index (SPEI) and examined the potential of artificial neural networks (ANN) for drought forecasting in Eastern Australia. They asserted that employing ANN models, monthly SPI, and SPEI forecasting in the research region could be accomplished satisfactorily.

Basak et al. (2022) identified the meteorological drought using Prophet, Support Vector Regression (SVR), and Multiple Linear Regression (MLR) in India based on multi-scalar SPI and suggested that these methods can be adopted for studying 'drought forecasting.' Abbasi et al. (2019) researched the drought condition over the Urmia Lake area in Iran and determined SPI and SPEI indices at different time scales, and the study outcome showed that the performance of the forecasting amplified with the upsurge in drought time scale, i.e., from one month to forty-eight months. There are numerous techniques and indices for analyzing droughts, and they measure various drought-causative and drought-responsive characteristics, identifying and categorizing drought in accordance with their results. However, a correlation between different types of drought is particularly challenging because these characteristics do not have linear relationships (Stagge et al. 2017).

In addition, different indices were developed for quantifying drought traits (meteorological drought and hydrological drought) to assess drought parameters, intensity, and its effect. A monthly time scale is typically employed and considered more appropriate for monitoring the situation while studying the drought characteristic. Droughts, for example, are thought to be the primary source of agroecosystem instability in recent years, and they have a significant impact on sustainable agriculture because they are becoming more frequent, intense, unpredictable, and vulnerable to climate change (Kar et al. 2007; Stocker et al. 2013; Liu et al. 2015; Zhang et al. 2021). The methods *like* SPI and SPEI have a benefit in that they use previous cumulative climatic conditions, which produce more valuable results, to identify the period in which previous values influence the proportional value of the present index (Vicente-Serrano et al. 2012; Abbasi et al. 2019). These two drought indicators are crucial for measuring productivity of agriculture, calculating the frequency of wildfires, examining water levels, and identifying precipitation accrual, among other things, as SPI is grounded on precipitation data and SPEI is grounded on precipitation and evaporation data (Deo and Sahin 2015; Malik and Kumar 2020; Cammalleri et al. 2022). Studies based on drought are crucial for populations that depend on groundwater and farmland supplied by rain (Singh et al. 2021; Gururani et al. 2023). It is essential to classify drought events according to their duration, grimness, and intensity to support proactive policy and planning. Also, drought has several negative effects, but its impact on agriculture is the most serious and immediate, which poses a threat to the country's food safety and stability (Dai 2011; Malik and Kumar 2020; Yacoub and Tayfur 2020; Fanadzo et al. 2021; Meza et al. 2021). Consequently, addressing the consequences of drought on sustainable agriculture and figuring out how to mitigate them have grown to be top scientific priorities. In order to measure the severity of drought, the World Meteorological Organization (WMO) suggested the SPI and SPEI indices. Rainfall and temperature are the primary determinants of water availability because they control evapotranspiration rates (Guntu et al. 2023). These parameters can be used to detect drought. Thereafter, the above-mentioned indicators are transformed into drought indices depicting the various characteristics.

Based on the above literature cited, a thorough investigation into the occurrence of meteorological drought during a 70-year period (1951–2020) was done using Standardized Precipitation Index (SPI) and Standardized Precipitation Evapotranspiration Index (SPEI) at different districts of Agra division, Uttar Pradesh, i.e., Agra, Firozabad, Mathura, and Mainpuri.

19.2 Study Area and Data Acquisition

Uttar Pradesh is regarded as one of the most important places for the political and economic strength of the nation. Basically, the economy of the state like Uttar Pradesh is agrarian. The state experiences three seasons: winter (October to February), summer (March to June), and monsoon (July to September). While assessing the growth rate of the state, agriculture and its allied sectors, including horticulture, animal husbandry, dairying, and fisheries, played a significant role and have diverse agro-climatic zones. The average holding size in Uttar Pradesh is 0.83 ha, while the per capita land area is 0.14 ha. The state is divided into four zones: the eastern zone, the western zone, the central zone, and the Bundelkhand zone. The western zone includes twenty-seven districts, the central zone includes ten districts, the Bundelkhand zone includes seven districts, and the eastern zone consists of twenty-seven districts. The eastern plain is heavily waterlogged, while Bundelkhand has a climate that makes it susceptible to droughts. Out of all districts, Agra and Jhansi experienced more heat in the summer. The state can sustain the production of a wide variety of crops because of its enormous alluvial soil resource and diverse agro-climatic conditions. The districts taken for drought analysis are Agra, Firozabad, Mathura, and Mainpuri, using Standardized Precipitation Index (SPI) and Standardized Precipitation Evapotranspiration Index (SPEI). The properties of drought on various water resources required by various policymakers are reflected in the timelines of SPI and SPEI. Streamflow, reservoirs, and groundwater respond to longer-term precipitation anomalies for the order of 6 months to 24 months or longer, whereas meteorological and soil moisture conditions (agriculture) respond to precipitation anomalies on relatively short time scales, for instance, one to six months (World Meteorological Organization 2012). In spite of the SPI index, the SPEI was also considered because temperature and precipitation were also employed. The benefit of utilizing SPEI is that two crucial meteorological factors (precipitation and temperature) were also prioritized (Bazrafshan 2017).

The gridded rainfall and temperature data for 70 years, i.e., from 1951 to 2020, of different metrological stations, were obtained from IMD,

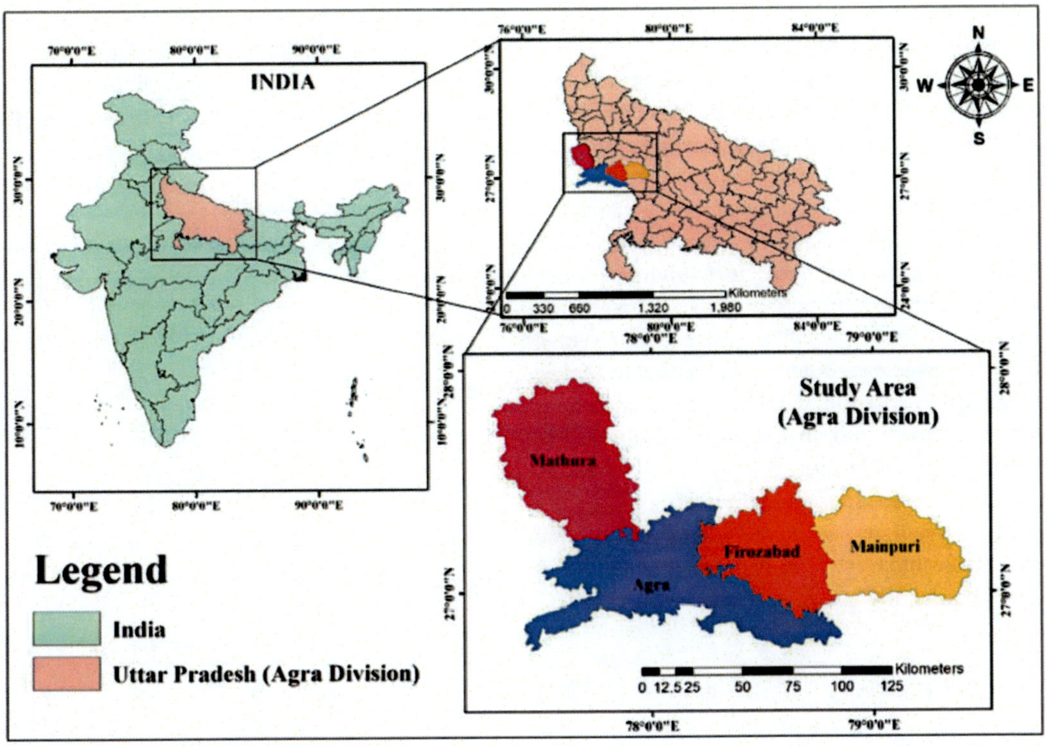

Fig. 19.1 Location of the study area

Pune. The detailed map of the Agra division is shown in Fig. 19.1, in order to present the most accurate model (a combination of several lag times) for predicting 'drought' in each of selected climatic regions mentioned earlier.

19.3 Material and Methods

Many indices have been established for assessing drought severity and characteristics, including the Palmer Drought Severity Index (PDSI), Standardized Precipitation Index (SPI), Standardized Precipitation Evapotranspiration Index (SPEI), and Soil Moisture Deficit Index (SMDI). Each of these indices focuses on one or more environmental parameters, like annual precipitation, temperature, streamflow, and soil moisture.

19.3.1 Standardized Precipitation Index (SPI)

Drought is usually classified as a multi-scalar event due to the wide temporal gap between the arrival of water supplies and the resource availability. Drought indices associated with a particular timeframe are used to monitor and manage various water supplies (White and Walcott 2009; Vicente-Serrano et al. 2012; Almedeij 2014). An effective indicator for describing meteorological drought over a diversity of periods is the Standardized Precipitation Index (SPI). As a standard deviation from a predefined probability distribution function that repeats the precipitation data, it expresses the amount of observed precipitation (Wu et al. 2017). The Standardized Precipitation Index (SPI) values were calculated and reported by McKee et al. in 1993 using a variety of time scales, including one, three, six, nine, twelve, twenty-four, and forty-eight months.

The SPI's usefulness in assessing changes caused by climate change has been criticized because it overlooks changes in evapotranspiration. With this approach, the total rainfall for various regions is added up across a variety of time frames, including one, three, six, nine, twelve, twenty-four, and forty-eight months. Positive values represented above-average precipitation, while negative values represented below-average precipitation. According to Mckee et al. (1993) criterion, these consistent anomalies are frequently used to categories droughts, as illustrated in Table 19.1.

The cumulative probability of a particular rainfall event is then calculated using the obtained parameters. The formula for the cumulative probability is:

$$G(x) = \frac{1}{r(x)} \int_{0}^{x} t^{z-1} e^{-t} \mathrm{d}t \tag{19.1}$$

Rainfall data may have zero values, the gamma function is undefeated for $x = 0$, and the cumulative probability function for the gamma distribution is calculated as,

$$H(x) = q + (1 - q)G(x) \tag{19.2}$$

where q is the likelihood of zero rainfall and is determined by the proportion of dataset zeros to the total number of years with records. Abramowitz and Stegun (1965) developed an approximation approach that translates a cumulative probability to the standard mal random variable, making it simple to calculate the SPI values, also known as Z values (Z).

$$\text{For } 0 < H(x) \leq 0.5$$

$$Z = \mathrm{SPI} = -\left(t - \frac{c_2 + c_2 t + c_2 t^2}{1 + d_2 t + t_2 t^2 + d_2 t^1}\right) \tag{19.3}$$

where

$$t = \sqrt{\left(\ln \frac{1}{(H(x))^2}\right)} \tag{19.4}$$

For $0.5 < H(x) \leq 1.00$

$$Z = \mathrm{SPI} = +\left(t - \frac{c_0 + c_2 t + c_2 t^2}{1 + d_2 t + d_2 t^2 + u_4 t^4}\right) \tag{19.5}$$

where

$$t = \sqrt{\left(\mathrm{in} \frac{1}{(1 - H(x))^2}\right)} \tag{19.6}$$

d the coefficient is given as;

$$\begin{array}{lll} c_1 = 2515517 & c_2 = 0.802853 & c_2 = 0.010328 \\ d_1 = 1.432788 & d_2 = 0.189269 & d_2 = 0.001308 \end{array}.$$

Table 19.1 Drought classification of SPI and SPEI values

SPI and SPEI values	Drought classification
Extremely wet	≥2.0
Severe wet	1.50–1.99
Moderate wet	1.0–1.49
Mild wet	0.50–0.99
Normal	0.49 to −0.49
Mild drought	−0.50 to −0.99
Moderate drought	−1.00 to −1.49
Severe drought	−1.50 to −1.99
Extremely drought	≤−2.00

19.3.2 Standardized Precipitation Evapotranspiration Index (SPEI)

Like SPI, the Standardized Precipitation Evapotranspiration Index (SPEI) combines rainfall with the average minimum and maximum temperatures. The Hargreaves method was employed in this chapter, and some stations lacked data to satisfy the requirements of the FAO-Penman–Monteith approach. The Hargreaves method requires rainfall, the highest and lowest temperature, latitude, and extraterrestrial radiation (R_a). The SPEI is calibrated in the same manner as the SPI. The SPEI, on the other hand, uses the difference between precipitation and reference evapotranspiration as the input rather than precipitation. The SPEI compares the maximum possible evapotranspiration (or evaporative demand by the atmosphere) to the current availability of water (Vicente-Serrano et al. 2012).

SPEI is based on a climatic water balance that is determined by the difference between the monthly potential evapotranspiration (PET) and precipitation (P) for that month and is represented as:

$$\text{PET} = 0.0023 \times (T_{\text{mean}} + 17.8) \times \left(\sqrt{T_{\text{max}} - T_{\text{min}}}\right) \times R_{\text{a}} \quad (19.7)$$

where PET = potential evapotranspiration (mm/day), T_{max} and T_{min}= maximum and minimum temperatures (°C), R_a = extraterrestrial radiation (mm/day), and T_{emin}= mean temperature (°C).

PET is determined by the difference between precipitation and PET (D_i).

$$D_i = P_i - PET_i \quad (19.8)$$

PET_i = monthly potential evapotranspiration (mm), D_i = climatic water balance (CWB), and P_i = monthly precipitation (mm). The total deviation between PET and P over various time intervals can be expressed as:

$$D_n^k == \sum_{i=0}^{k-1} (P_{n-i} - \text{PET}_{n-i}) \quad (19.9)$$

where k is distinct time scale and n is the total number of computations. For evaluating the SPEI series in normalized D with a mean of 0 and a standard deviation of 1, the formula is as follows:

$$f(x) = \frac{\beta}{\alpha}\left(\frac{x-\gamma}{\alpha}\right)^{\beta-1}\left[1 + \frac{x-\gamma}{\alpha}\right]^{-2} \quad (19.10)$$

The parameters β, α, and γ are shape, scale, and location parameters, respectively, for D values in the range $(\gamma < D < \infty)$. Thus, the probability distribution function can be stated as:

$$F(x) = \frac{\beta}{\alpha}\left[1 - \frac{\alpha}{x-\gamma}\right]^{-1} \quad (19.11)$$

It is simple to identify the SPEI as the standardized $F(x)$ with an estimate of

$$\text{SPEI} = W_i - \frac{2.515517 + 0.8022853W_i + 0.010328W_i^2}{1 - 1.432788W_i + 0.189269W_i^2 + 0.001308W_i^3} \quad (19.12)$$

Regardless of whether p is more than 0.5, less than 0.5, or equal to 0.5, the equation holds true. The formula is as follows:

$$W_i = \begin{cases} \sqrt{-2\ln(p)} & \text{for } p \le 0.5 \\ \sqrt{-2\ln(1-p)} & \text{for } p > 0.5 \end{cases} \quad (19.13)$$

where W_i = probability-weighted moment

The benefits of SPEI and SPI are that they only require climatological information, which is often available from multiple sources; they do not require any assumptions; and, most importantly, they compute climatological irregularity for exact length periods (Belayneh and Adamowski 2012; Vicente-Serrano et al. 2012; Kim and Jehanzaib 2020).

19.4 Results and Discussion

The meteorological drought was examined using SPI and SPEI at different time scales, including one month, three months, six months, nine months, twelve months, and twenty-four months, for each of four specified stations. Additionally, an investigation was conducted to discover the correlation between SPI and SPEI at the same scale within the Agra division (Agra, Firozabad, Mathura, and Mainpuri) of Uttar Pradesh state, India, spanning from 1951 to 2020.

19.4.1 Results of SPI and SPEI over Selected Stations

Figures 19.2a–f display time series for the SPI values (SPI_1, SPI_3, SPI_6, SPI_9, SPI_12, and SPI_24) for Mainpuri station, and the maximum and minimum values range from 3.26 to 3.60, 2.75 to −3.77, 2.88 to −3.59, 2.89 to −3.26, 2.30 to −2.91, and 2.61 to −3.85, respectively. Similarly, Figs. 19.3a–f display time series for the SPI values (SPI_1, SPI_3, SPI_6, SPI_9, SPI_12, and SPI_24) calculated for Mathura station, and the maximum and minimum values range between 2.94–3.71, 2.39–4.05, 2.67–4.01, 2.22–3.26, 2.01–3.31, and 2.23–2.78, respectively. Figures 19.4a–f display time series for the SPI values (SPI_1, SPI_3, SPI_6, SPI_9, SPI_12, and SPI_24) calculated for Agra station, and the maximum and minimum values range from 3.41 to −3.47, 2.69 to −3.05, 2.49 to −3.14, 2.45 to −3.47, 2.24 to −3.22, and 2.15 to −2.62. Figures 19.5a–f display time series for the SPI values calculated for Firozabad station for the periods of 3, 6, 12, and 24 months, and the maximum and minimum values range from 3.37 to −3.92, 2.57 to −4.12, 2.67 to −3.61, 2.54 to −3.37, 2.40 to −2.77, and 2.96 to −3.05, respectively.

While Figs. 19.6a–f show SPEI_1, SPEI_3, SPEI_6, SPEI_9, SPEI_12, and SPEI_24 values for Mainpuri range, the values range from 2.79 to −3.94, 2.55 to −2.14, 2.44 to −2.20, 2.51 to −2.25, 2.27 to −2.19, and 2.78 to −2.36, respectively. For Mathura, as demonstrated in Figs. 19.7a–f, SPEI_1, SPEI_3, SPEI_6, SPEI_9, SPEI_12, and SPEI_24 have maximum and minimum values ranging from 2.58 to −3.39, 2.27 to −3.20, 2.35 to −2.19, 2.19 to −2.05, 2.07 to −2.34, and 2.01 to −2.26, respectively.

As shown in Figs. 19.8a–f, the maximum and minimum values of SPEI_1, SPEI_3, SPEI_6, SPEI_9, SPEI_12, and SPEI_24 for Agra range from 2.70 to −2.99, 2.36 to −2.62, 2.40 to −2.26, 2.32 to −2.19, 2.32 to −2.18, and 2.30 to −2.16, respectively. Similarly, Figs. 19.9a–f show the maximum and minimum values of SPEI_1, SPEI_3, SPEI_6, SPEI_9, SPEI_12, and SPEI_24 for Firozabad range, and the values range from 2.70 to −5.67, 2.41 to −2.28, 2.46 to −2.21, 2.50 to –2.31, 2.28 to –23.5, and 2.65 to –2.48, respectively. In light of the above findings, SPI and SPEI showed a tendency to detect consecutive drought or wet events over a longer time scale (Malik and Kumar 2020; Bera et al. 2021). Its frequency, however, decreases as time scales increase.

The maximum and minimum values for each SPI value are also reported for each station and time period. These values represent the range of the SPI values for the given time period and station, and they can provide information about the severity of drought conditions during that period. The range of values can be used to assess the extent of the drought conditions and their potential impact on agriculture, water resources, and other sectors.

The most significant feature of the drought was the variation in drought frequency as the time scale changed. For a shorter time scale, i.e., one month, three months, and six months, drought frequency increased; however, its length was reduced. In other words, drought happens more frequently but continued for a smaller duration on shorter time scales, which also showed that the SPI responded quickly to wet and dry periods. The index reacted more slowly as the time scale lengthens. On the other hand, as the time scale lengthens, each new month had less effect on the overall amount and showed fewer longer-lasting droughts. For different water resources, the two cases (Case 1: fewer droughts but longer duration, Case 2: more droughts but shorter duration) might be inferred differently. It was noteworthy to observe that as the time scale raises, the frequency of drought increases. At the longer time scale, drought frequency decreases,

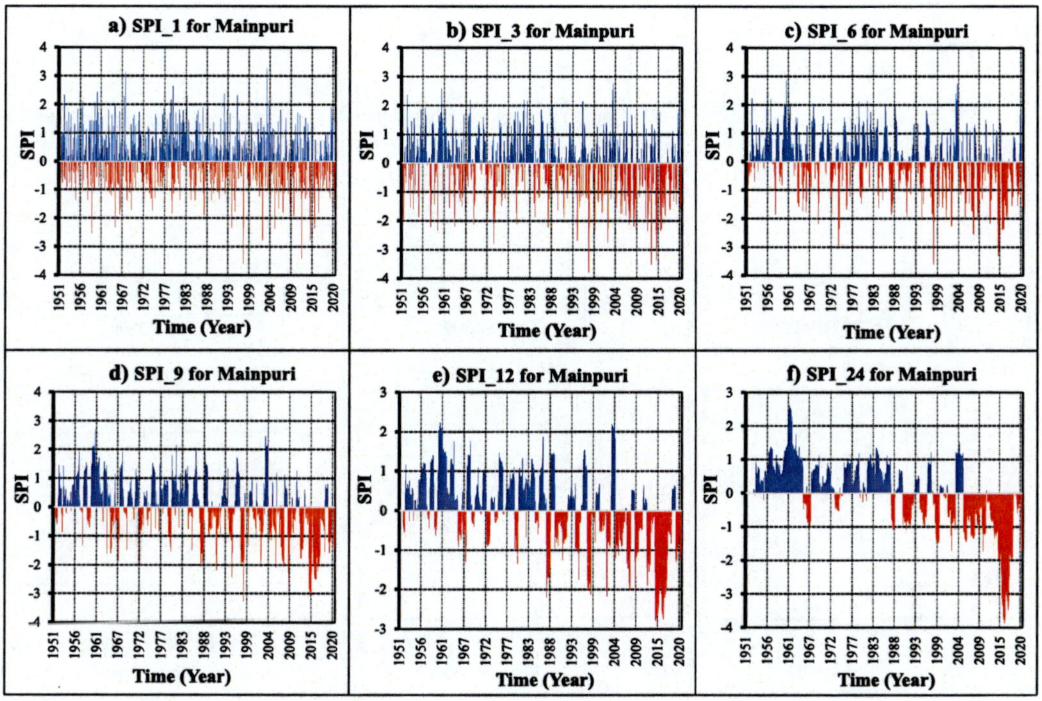

Fig. 19.2 Results of SPI for Mainpuri station

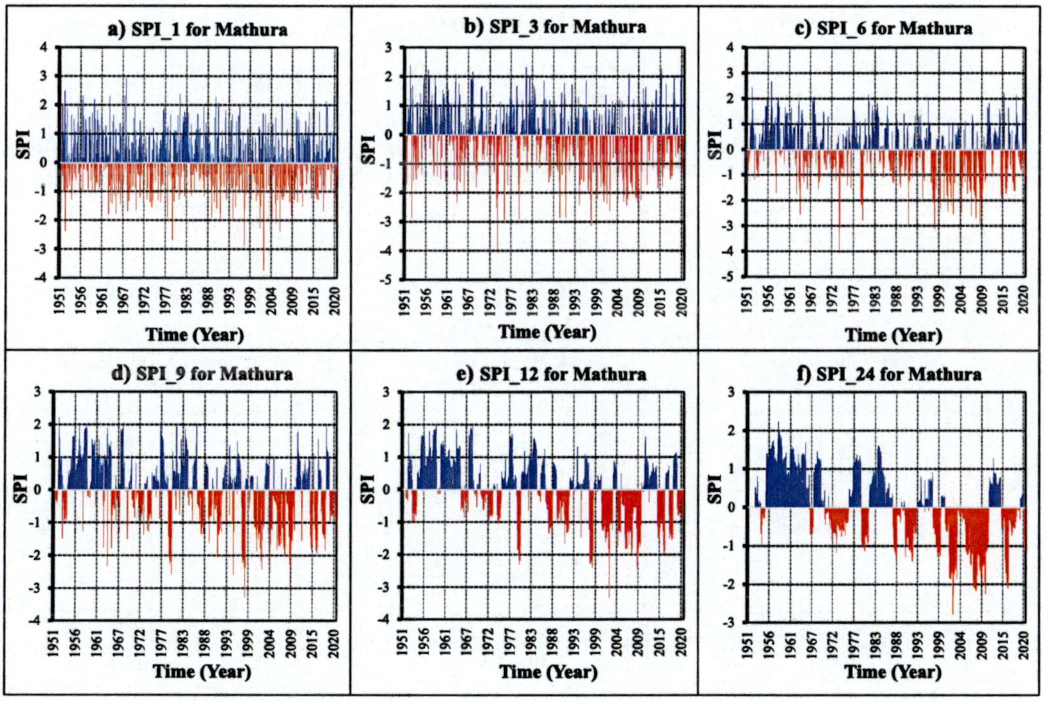

Fig. 19.3 Results of SPI for Mathura station

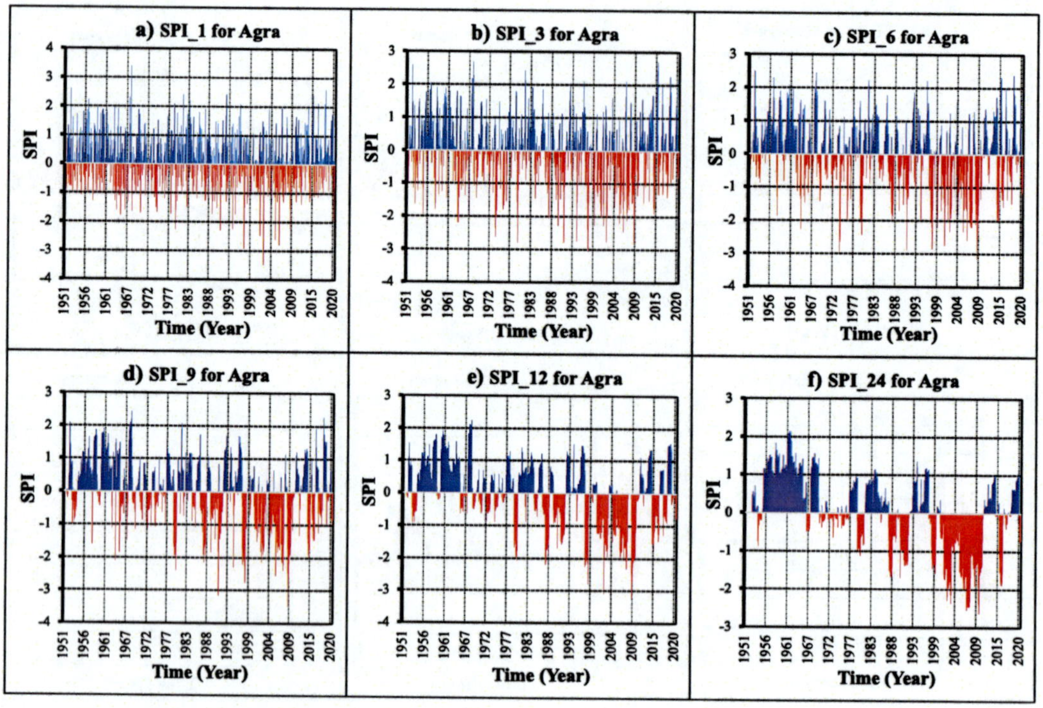

Fig. 19.4 Results of SPI for Agra station

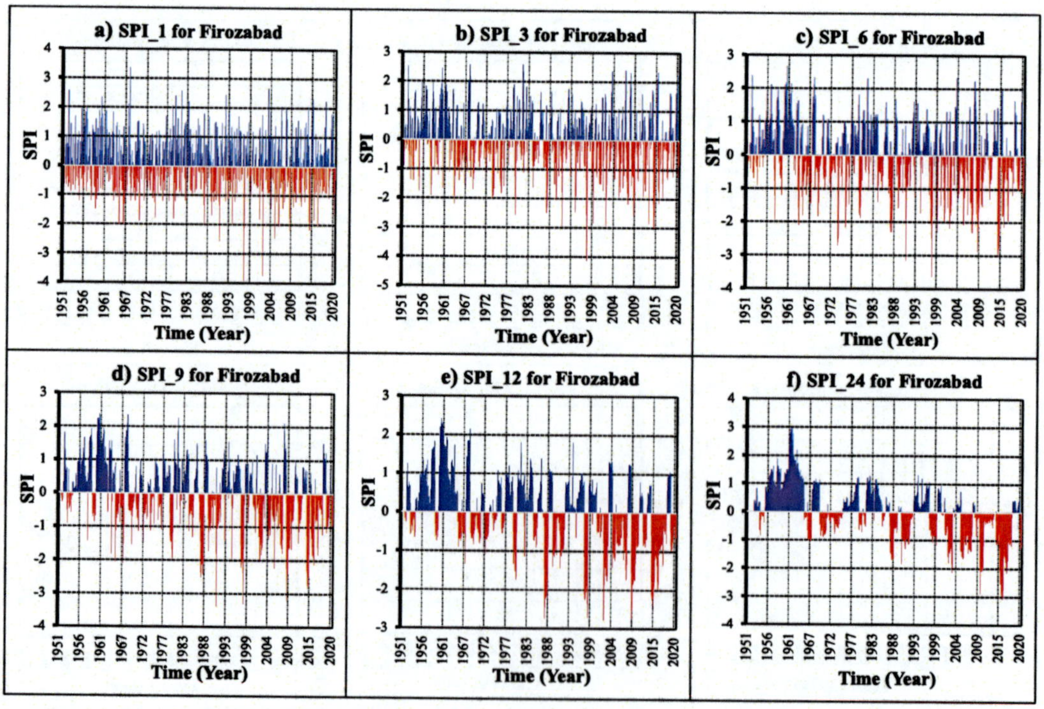

Fig. 19.5 Results of SPI for Firozabad station

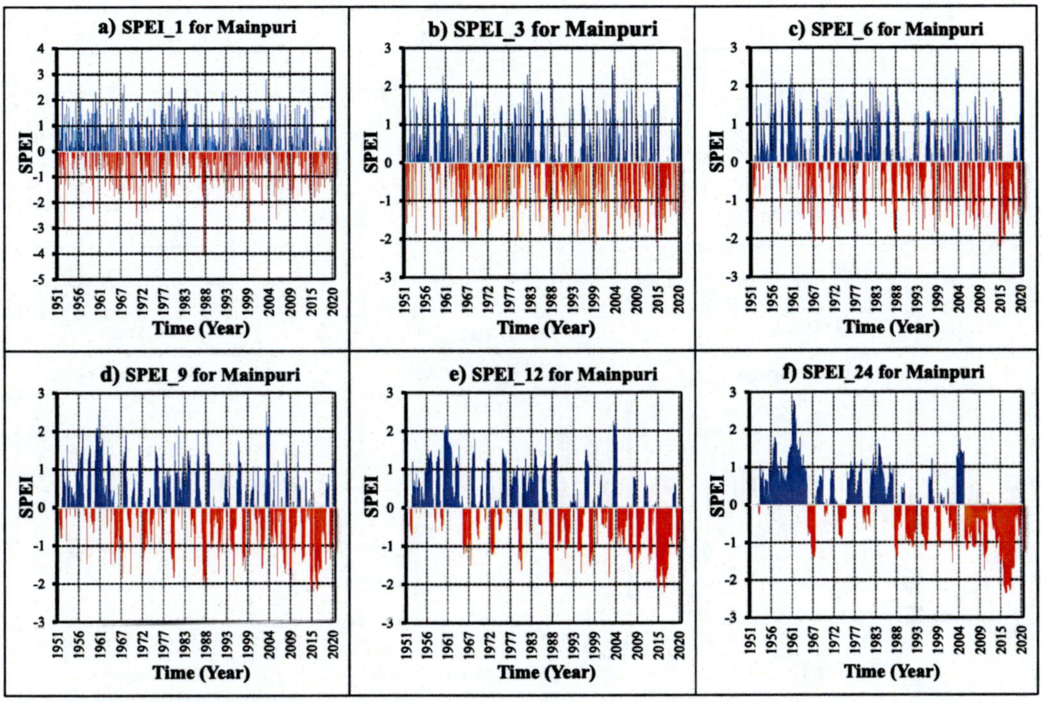

Fig. 19.6 Results of SPEI for Mainpuri station

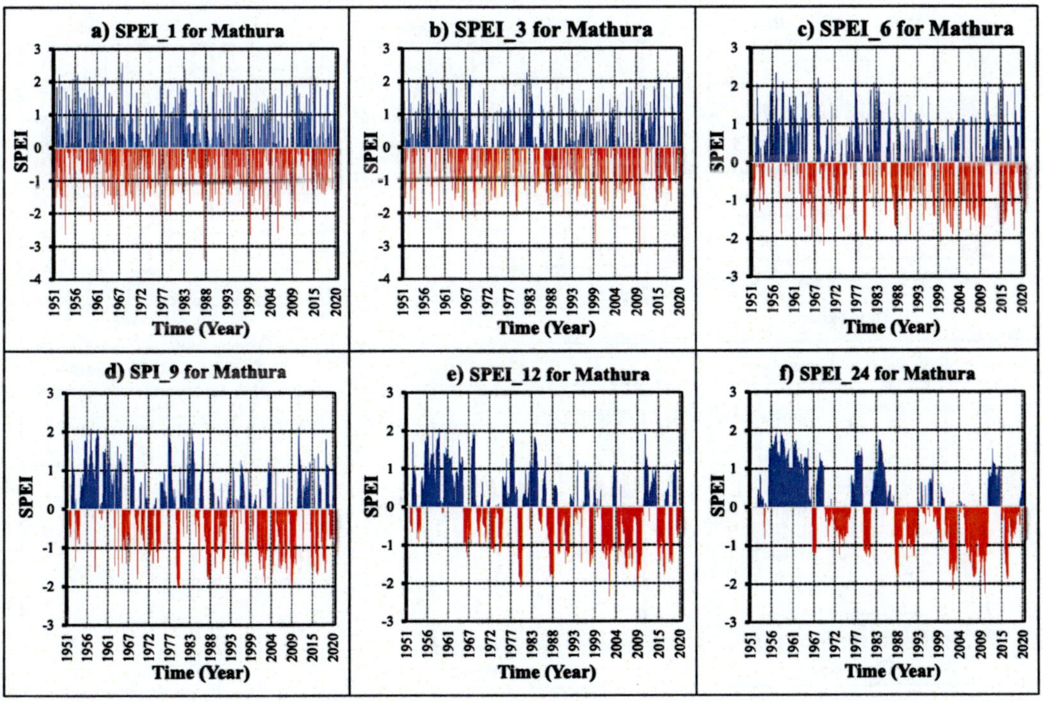

Fig. 19.7 Results of SPEI for Mathura station

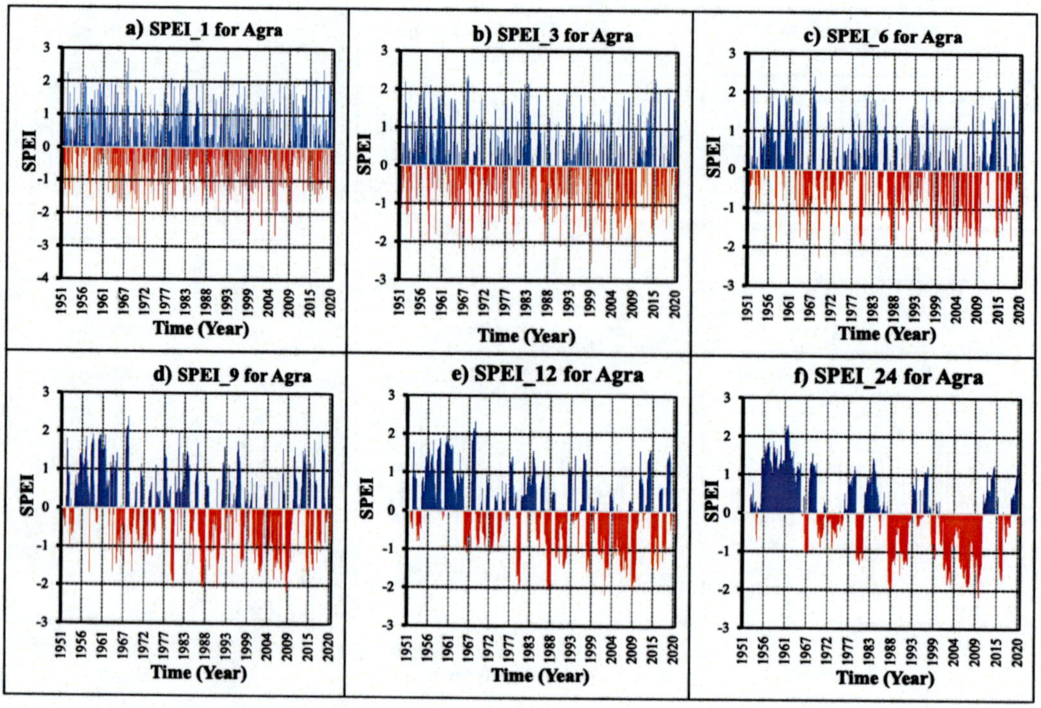

Fig. 19.8 Results of SPEI for Agra station

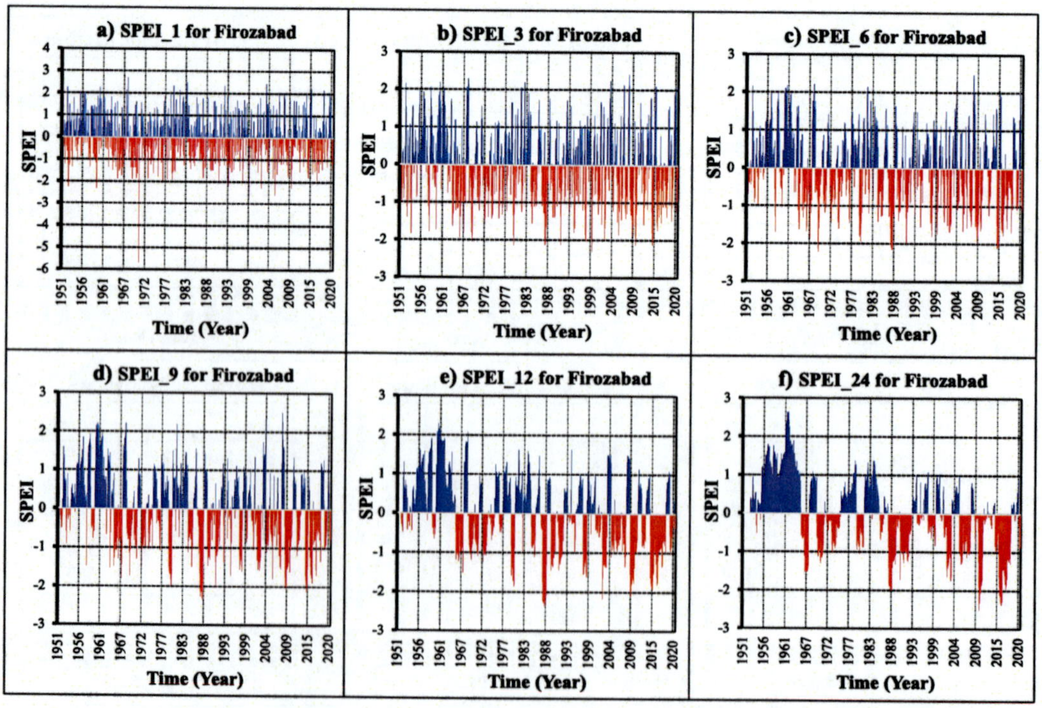

Fig. 19.9 Results of SPEI for Firozabad station

but its duration increases, leading to the identification of consecutive drought and consecutive wet events.

The monthly differentiation characteristics of the SPI and SPEI were perceptible at different timeframes, representing the change in the magnitude of dry and wet degrees of each month in the current study area, especially after 1999. The dry degree of some months grew as well. Following 1999, the two indices depicted comparable monthly dry and wet temperatures. The SPI and SPEI's representation of drought periods can be regarded to be typically consistent. Due to the factor described above of rising temperatures, the SPEI's portrayal of the severity of the drought in some months was higher with the lengthening of the time series than the SPI's. Most studies also found variations in regional drought between the SPI and SPEI. When evaluated from the standpoint of a time series, the SPI and SPEI variations at each period were relatively similar. The highest differences throughout the short period were seen in the two indices that varied the most frequently. Although there were slight variations in the drought intensity, the SPI and SPEI fluctuated less often, and their differences widened at the long time scale. The results also provided evidence in favor of the theory that variations in the SPI and SPEI values and their significant regional variability were responsible for drought.

19.4.2 Correlation Between SPI and SPEI

A high correaltion between SPI and SPEI values was observed at different time scale. However, a comparison of the long-term scale (at 9, 12, and 24 months) and the short-term scale (at 1, 3, and 6 months) revealed that the prolonged time scale recorded a strong correlation, while the short-term time scale recorded weaker correlation values between SPI and SPEI across all stations as depicted in Table 19.2.

19.4.3 Percentage of SPI and SPEI Values Over Different Stations

19.4.3.1 Percentage of SPI and SPEI Values of Mainpuri Station

Figure 19.10 depicted the relationship between the percentage of SPI and SPEI values and the time scale of Mainpuri. It was found that the maximum range was seen for normal, mild drought, moderate drought, mild drought, and moderate drought. The normal drought value for SPI_1 is 42.38, SPI_3 is 37.62, SPI_6 is 39.40, SPI_9 is 38.93, SPI_12 is 35.24, and SPI_24 is 32.02. The mild drought value for SPI_1 is 15.36, SPI_3 is 12.02, SPI_6 is 11.79, SPI_9 is 12.62, SPI_12 is 16.43, and SPI_24 is 14.17. The moderate drought value for SPI_1 is 7.74, SPI_3 is 9.17, SPI_6 is 7.02, SPI_9 is 5.60, SPI_12 is 5.95, and SPI_24 is 8.21. The mild wet value for SPI_1 is 15.36, SPI_3 is 14.76, SPI_6 is 14.29, SPI_9 is 16.67, SPI_12 is 15.83, and SPI_24 is 20.26. The moderate wet value for SPI_1 is 9.64, SPI_3 is 11.90, SPI_6 is 13.10, SPI_9 is 12.02, SPI_12 is 12.98, and SPI_24 is 12.38. The normal drought value for SPEI_1 is 36.79, SPEI_3 is 33.93, SPEI_6 is 34.64, SPEI_9 is 33.93,

Table 19.2 Correlation between SPI and SPEI at different time scales

Time scale (month)	Stations				Correlation
	Agra	Mathura	Mainpuri	Firozabad	
1	0.90	0.90	0.89	0.89	r
3	0.93	0.92	0.94	0.93	r
6	0.95	0.94	0.96	0.96	r
9	0.97	0.97	0.97	0.97	r
12	0.97	0.97	0.97	0.97	r
24	0.97	0.97	0.97	0.97	r

SPEI_12 is 32.62, and SPEI_24 is 29.17. The mild drought value for SPEI_1 is 15.95, SPEI_3 is 15.00, SPEI_6 is 14.76, SPEI_9 is15.83, SPEI_12 is 15.36, and SPEI_24 is 17.86. The moderate drought value for SPEI_1 is 10.36, SPEI_3 is 12.02, SPEI_6 is 11.90, SPEI_9 is 10.60, SPEI_12 is 11.31, and SPEI_24 is 10.36. The mild wet value for SPEI_1 is 12.74, SPEI_3 is 14.29, SPEI_6 is 12.98, SPEI_9 is 15.24, SPEI_12 is 14.40, and SPEI_24 is 19.64. The moderate wet value for SPEI_1 is 9.88, SPEI_3 is 11.43, SPEI_6 is 12.26, SPEI_9 is 11.67, SPEI_12 is 13.21, and SPEI_24 is 9.64. In Mainpuri, the SPI and SPEI values range from 0.24 to 42.38% and 0.36 to 36.79%, respectively. The above-discussed study outcome indicates that most Mainpuri is within a normal range, i.e., neither a drought nor a wet period.

19.4.3.2 Percentage of SPI and SPEI Values of Mathura Station

Figure 19.11 represents the relationship between the percentage of SPI and SPEI values and the time scale of Mathura. It was found that the maximum range was seen for normal, mild drought, moderate drought, mild drought, and moderate drought. The normal drought value for SPI_1 is 44.40, SPI_3 is 37.14, SPI_6 is 35.83, SPI_9 is 36.43, SPI_12 is 32.26, and SPI_24 is 31.67. The mild drought value for SPI_1 is 14.88, SPI_3 is 11.79, SPI_6 is 13.10, SPI_9 is 12.26, SPI_12 is 13.10, and SPI_24 is 14.17. The moderate drought value for SPI_1 is 14.88, SPI_3 is 11.79, SPI_6 is 13.10, SPI_9 is 12.26, SPI_12 is 13.10, and SPI_24 is 15.36. The mild wet value for SPI_1 is 15.12, SPI_3 is 16.90, SPI_6 is 17.98, SPI_9 is 17.86, SPI_12 is 18.69, and SPI_24 is 14.29. The moderate wet value for SPI_1 is 6.07, SPI_3 is 8.69, SPI_6 is 8.10, SPI_9 is 9.29, SPI_12 is 11.79, and SPI_24 is 10.00. The normal drought value for SPEI_1 is 36.79, SPEI_3 is 33.21, SPEI_6 is 32.74, SPEI_9 is 30.60, SPEI_12 is 30.48, and SPEI_24 is 29.76. The mild drought value for SPEI_1 is 16.07, SPEI_3 is 16.31, SPEI_6 is 14.88, SPEI_9 is 14.05, SPEI_12 is 13.33, and SPEI_24 is 16.43. The moderate drought value for SPEI_1 is 11.67, SPEI_3 is 12.14, SPEI_6 is 12.02, SPEI_9 is 14.17, SPEI_12 is 15.12, and SPEI_24 is 11.90. The mild wet value for SPEI_1 is 12.86, SPEI_3 is 14.05, SPEI_6 is 15.60, SPEI_9 is 16.07, SPEI_12 is 15.60, and SPEI_24 is 11.90. The moderate wet value for SPEI_1 is 8.45, SPEI_3 is 10.83, SPEI_6 is 9.64, SPEI_9 is 10.95, SPEI_12 is 11.43, and SPEI_24 is 15.12. The SPI and SPEI values for a station like Mathura vary from 0.12 to 44.40% and 0.12 to 36.79%, respectively. The derived result shows that the highest percentage of severe extreme events is exceptionally rare, indicating that water demand may be readily satisfied.

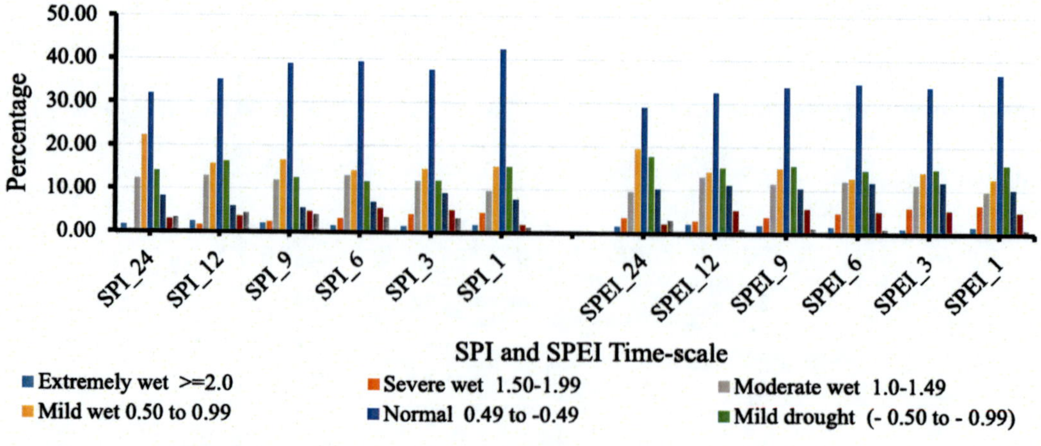

Fig. 19.10 Percentage of SPI and SPEI values of Mainpuri station

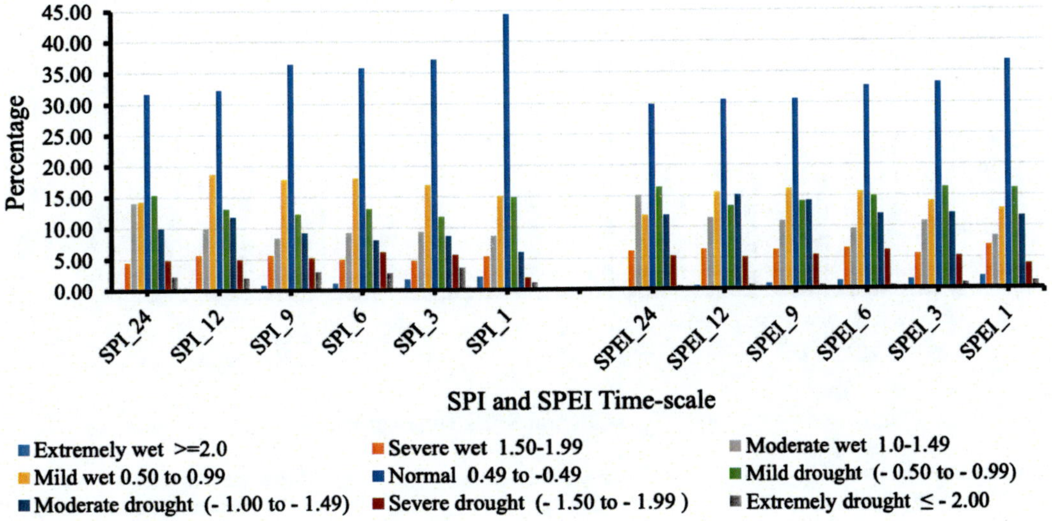

Fig. 19.11 Percentage of SPI and SPEI values of Mathura station

19.4.3.3 Percentage of SPI and SPEI Values of Firozabad Station

Figure 19.12 displays the relationship between the percentage of SPI and SPEI values and the time scale of Firozabad. It was found that the maximum range was seen for normal, mild drought, moderate drought, mild drought, and moderate drought. The normal drought value for SPI_1 is 43.45, SPI_3 is 40.48, SPI_6 is 38.69, SPI_9 is 34.76, SPI_12 is 30.71, and SPI_24 is 40.12. The mild drought value for SPI_1 is 16.67, SPI_3 is 12.38, SPI_6 is 13.10, SPI_9 is 15.00, SPI_12 is 19.52, and SPI_24 is 13.45. The moderate drought value for SPI_1 is 10.60, SPI_3 is 11.31, SPI_6 is 10.83, SPI_9 is 10.12, SPI_12 is 10.60, and SPI_24 is 10.95. The mild wet value for SPI_1 is 15.60, SPI_3 is 14.40, SPI_6 is 15.60, SPI_9 is 19.05, SPI_12 is 18.10, and SPI_24 is 11.55. The moderate wet value for SPI_1 is 5.12, SPI_3 is 8.21, SPI_6 is 6.79, SPI_9 is 7.02, SPI_12 is 6.79, and SPI_24 is 9.05. The normal drought value for SPEI_1 is 38.69, SPEI_3 is 35.24, SPEI_6 is 33.81, SPEI_9 is 30.24, SPEI_12 is 29.17, and SPEI_24 is 35.12. The mild drought value for SPEI_1 is 15.12, SPEI_3 is 15.95, SPEI_6 is 15.24, SPEI_9 is 18.21, SPEI_12 is 17.74, and SPEI_24 is 13.33. The moderate drought value for SPEI_1 is 10.60, SPEI_3 is 11.31, SPEI_6 is 10.83, SPEI_9 is 14.17, SPEI_12 is 15.12, and SPEI_24 is 11.90. The mild wet value for SPEI_1 is 11.55, SPEI_3 is 12.98, SPEI_6 is 14.76, SPEI_9 is 17.26, SPEI_12 is 17.98, and SPEI_24 is 15.24. The moderate wet value for SPEI_1 is 10.00, SPEI_3 is 11.07, SPEI_6 is 11.67, SPEI_9 is 10.24, SPEI_12 is 10.83, and SPEI_24 is 10.00. The SPI and SPEI values range from 1.31 to 43.45% and 1.07 to 38.69% for Firozabad, respectively, showing that SPEI can realistically simulate information about drought events (Zhang et al. 2009; Vicente-Serrano et al. 2012; Kim and Jehanzaib 2020).

19.4.3.4 Percentage of SPI and SPEI Values of Agra Station

Figure 19.13 showcases the relationship between the percentage of SPI and SPEI values and its time scale of Agra. It was found that the maximum range was seen for normal, mild drought, moderate drought, mild drought, and moderate drought. The normal drought value for SPI_1 is 42.98, SPI_3 is 38.57, SPI_6 is 39.05, SPI_9 is 35.60, SPI_12 is 33.93, and SPI_24 is 37.86. The mild drought value for SPI_1 is 15.71, SPI_3 is 12.50, SPI_6 is 11.90, SPI_9 is 12.26, SPI_12 is 12.74, and SPI_24 is 9.88. The moderate drought value for SPI_1 is 6.55, SPI_3 is 9.17, SPI_6 is

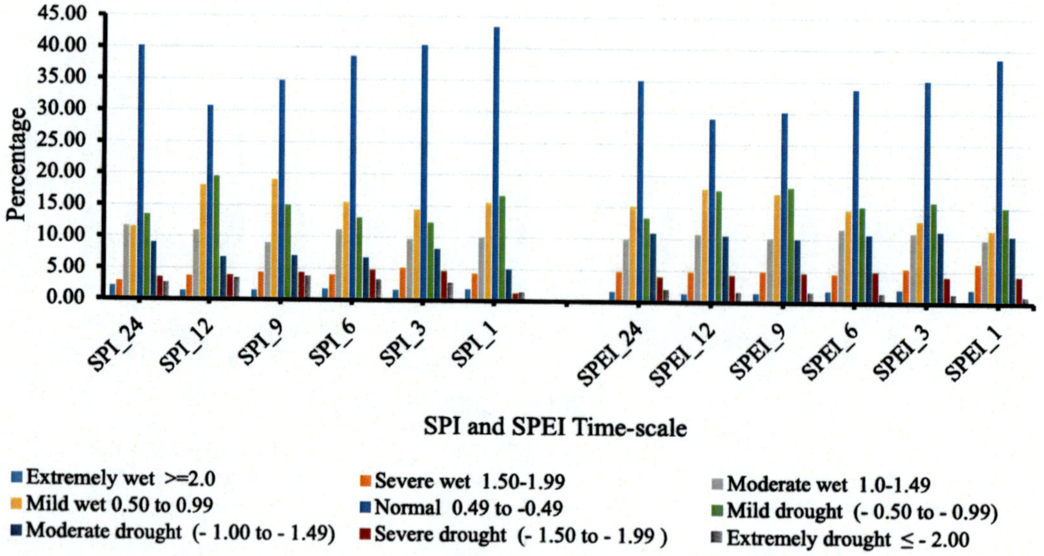

Fig. 19.12 Percentage of SPI and SPEI values of Firozabad station

7.38, SPI_9 is 7.14, SPI_12 is 7.74, and SPI_24 is 7.86. The mild wet value for SPI_1 is 15.95, SPI_3 is 15.36, SPI_6 is 15.60, SPI_9 is 17.62, SPI_12 is 17.26, and SPI_24 is 13.10. The moderate wet value for SPI_1 is 9.52, SPI_3 is 10.24, SPI_6 is 10.00, SPI_9 is 10.71, SPI_12 is 13.10, and SPI_24 is 15.83. The normal drought value for SPEI_1 is 37.38, SPEI_3 is 33.33, SPEI_6 is 33.57, SPEI_9 is 31.43, SPEI_12 is 31.07, and SPEI_24 is 33.45. The mild drought value for SPEI_1 is 15.48, SPEI_3 is 17.38, SPEI_6 is 15.60, SPEI_9 is 16.90, SPEI_12 is 17.02, and SPEI_24 is 11.67. The moderate drought value for SPEI_1 is 11.79, SPEI_3 is 10.95, SPEI_6 is 10.95, SPEI_9 is 10.95, SPEI_12 is 11.43, and SPEI_24 is 13.10. The mild wet value for SPEI_1 is 12.38, SPEI_3 is 15.00, SPEI_6 is 14.52, SPEI_9 is 15.00, SPEI_12 is 15.12, and SPEI_24 is 12.14. The moderate wet value for SPEI_1 is 9.64, SPEI_3

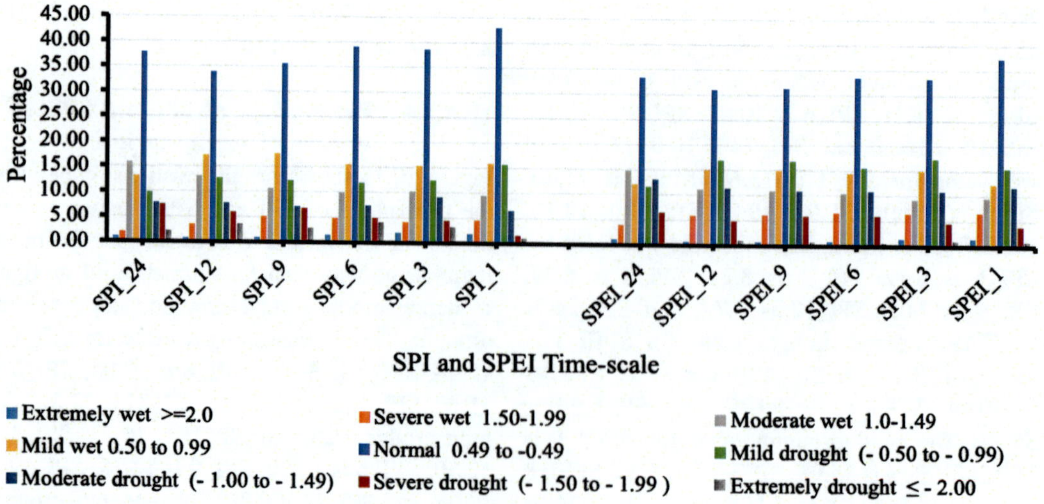

Fig. 19.13 Percentage of SPI and SPEI value of Agra station

is 9.29, SPEI_6 is 10.48, SPEI_9 is 10.95, SPEI_12 is 11.19, and SPEI_24 is 14.88. In addition, the values of SPI and SPEI for Agra range from 0.83 to 42.98% and 0.24 to 37.38%, respectively. In short, SPEI has been able to record precise values of the drought percentage due to consideration of the temperature and rainfall, while SPI recorded a relatively high percentage (Belayneh and Adamowski 2012; Vicente-Serrano et al. 2012; Joetzjer et al. 2013; Malik and Kumar 2020; Bera et al. 2021).

19.5 Conclusions

The present chapter was carried out to illustrate the metrological drought and wet events using the Standardized Precipitation Index (SPI) and Standardized Precipitation Evapotranspiration Index (SPEI) at multiple time scales of 1, 3, 6, 9, 12, and 24 months for the Agra division (Agra, Firozabad, Mathura, and Mainpuri) which is situated in the state of Uttar Pradesh, India. The gridded rainfall and temperature data for 70 years (1951–2020) were used in this study. The correlation between SPI and SPEI was also investigated. According to the findings of this chapter, the highest percentages of drought and wet events were concentrated in the normal to moderate category, and all the stations experienced consecutive wet years from 1955 to 1965, whereas Agra and Mathura stations had consecutive drought years from 2000 to 2011. However, Firozabad and Mainpuri had successive droughts from 2014–2018 to 2011–2018, respectively. The SPI and SPEI indices were also correlated by a correlation test, which showed that as the time scale increased, the correlation between SPI and SPEI substantially increased, which highlighted that these indices performed better when used together on a longer time scale (9–24 months). The outcome of the chapter suggested that the SPI and SPEI had similar outcomes in most stations, which further indicates their applicability in the characterization of drought and wet events.

Additionally, SPEI has measured the precise value of the drought percentage due to consideration of rainfall and temperature, making it an effective tool for assessing drought events. The end products of this research can help national and regional policymakers and agricultural producers develop drought mitigation strategies in the chapter districts, which entail storing excess water during wet seasons and utilizing it for agricultural purposes during dry seasons. This investigation would be useful for better agricultural water resources planning and its management in drought-hit areas for ensuring the sustainability of agricultural productivity and adaptation strategies in the study areas.

References

Abbasi A, Khalili K, Behmanesh J et al (2019) Drought monitoring and prediction using SPEI index and gene expression programming model in the west of Urmia Lake. Theor Appl Climatol 138(1):553–567

Abramowitz M, Stegun A (1965) Handbook of mathematical formulas, graphs, and mathematical tables. Dover Publ New York 470

Alam J, Saha P, Mitra R, Das J (2023) Investigation of spatio-temporal variability of meteorological drought in the Luni River Basin, Rajasthan, India. Arabian J Geosciences 16(3):201. https://doi.org/10.1007/s12517-023-11290-8

Almedeij J (2014) Drought analysis for Kuwait using standardized precipitation index. The Scientific World Jornal

Basak A, Rahman ATMS, Das J, Hosonod T, Kisi O (2022) Drought forecasting using the Prophet Model in semi-arid climate region of western India. Hydrol Sci J 67(9):1397–1417. https://doi.org/10.1080/02626667.2022.2082876

Bazrafshan J (2017) Effect of air temperature on historical trend of long-term droughts in different climates of Iran. Sustain Water Resour Manag 31(14):4683–4698

Belayneh A, Adamowski J (2012) Standard precipitation index drought forecasting using neural networks, wavelet neural networks, and support vector regression. Appl Comput Intell Soft Comput.https://doi.org/10.1155/2012/794061

Bera B, Shit PK, Sengupta N, Saha et al (2021) Trends and variability of drought in the extended part of Chhota Nagpur plateau (Singbhum Protocontinent), India applying SPI and SPEI indices. Environ Chall 5:100310.

Bhunia P, Das P, Maiti R (2020) Meteorological drought study through SPI in three drought prone districts of West Bengal, India. Earth Syst Environ 4(1):43–55

Cammalleri C, Spinoni J, Barbosa P et al (2022) The effects of non-stationarity on SPI for operational

drought monitoring in Europe. Theor Appl Climat 142 (6):3418–3430
Dai A (2011) Drought under global warming: a review. Wiley Interdisciplinary Reviews: Clim Change 2 (1):45–65
Das J, Gayen A, Saha P, Bhattacharya SK (2020) Meteorological drought analysis using Standardized Precipitation Index over Luni River Basin in Rajasthan, India. SN Applied Sciences 2(9):1–17. https://doi.org/10.1007/s42452-020-03321-w
Deo RC, Sahin M (2015) Application of the artificial neural network model for prediction of monthly standardized precipitation and evapotranspiration index using hydrometeorological parameters and climate indices in eastern Australia. Atmos Res 161: 65–81
Dracup JA, Fee KS, Paulson EG Jr (1980) On identification of drought. Wat resour Res 16(2):297–302
Fanadzo M, Ncube B, French A et al (2021) Smallholder farmer coping and adaptation strategies during the 2015–18 drought in the Western Cape, South Africa. Phys Chem Earthhttps://doi.org/10.1016/j.pce.2021.102986
Guntu RK, Merz B, Agarwal A (2023) Increased likelihood of compound dry and hot extremes in India. Atmos Res 290:106789
Gururani DM, Singh S, Joshi H, Kumar Y, Kumar A, Bohra MS, Mehta P (2023) Assessment of groundwater prospects zones using RS, GIS, and MIF Methods. In: Surface and Groundwater Resources Development and Management in Semi-arid Region: Strategies and Solutions for Sustainable Water Management. Cham: Springer International Publishing. pp 317–335
Joetzjer E, Douville H, Delire C et al (2013) Hydrologic benchmarking of meteorological drought indices at interannual to climate change timescales: a case study over the Amazon and Mississippi river basins. Hydrol Earth Syst Sci 17(12):4885–4895
Kar G, James BK, Singh R, Mahapatra IC (2007) Agroclimate and extreme weather analysis for successful crop production in Orissa Water Technology centre for Eastern Region, Bhubaneswar, Orissa. India Res Bull 22(2007):1–76
Kim TW, Jehanzaib M (2020) Drought risk analysis, forecasting and assessment under climate change. Water 12(7):1862
Kulkarni A, Gadgil S, Patwardhan S (2016) Monsoon variability, The 2015 Marathwada drought and rainfed agriculture. Curr Sci 111(7):1182–1193
Kwon M, Sung JH (2019) Changes in future drought with Had GEM2-AO projections. Water 11(2):312
Liu X, Zhu X, Pan Y et al (2015) Spatiotemporal changes of cold surges in Inner Mongolia between 1960 and 2012. J Geogr Sci 25(3):259–273
Lorenzo-Lacruz J, Vicente-Serrano SM, López-Moreno et al (2010) The impact of droughts and water management on various hydrological systems in the headwaters of the Tagus River (central Spain), J Hydrol 386:13–26
Mahmoudi P, Rigi A, MiriKamak M (2019) A comparative study of precipitation-based drought indices with the aim of selecting the best index for drought monitoring in Iran. TheorApplClimatol 137(3):3123–3138
Malik A, Kumar A (2020) Meteorological drought prediction using heuristic approaches based on effective drought index: a case study in Uttarakhand. Arab J Geosci 13(6):1–17. https://doi.org/10.1007/S12517-020-5239-6
Mckee TB, Doesken NJ, leist J (1993) The relationship of drought frequency and duration to time scales. Preprints 8th Conference on Appl Clim 17:179–184
Mehr AD, Kahya E, Ozger M (2014) A gene–wavelet model for long lead time drought forecasting. J Hydrol 517:691–699
Meza I, EyshiRezaei E, Siebert S et al (2021) Drought risk for agricultural systems in South Africa: drivers, spatial patterns, and implications for drought risk management. Sci. Total Environ
Mishra AK, Desai VR (2006) Drought forecasting using feed-forward recursive neural network. Ecol Model 198:127–138
Mishra AK, Singh VP (2011) Drought modelling-a review. J Hydrol 403:157–175
Morid S, Smakhtin V, Bagherzadeh K (2007) Drought forecasting using artificial neural networks and time series of drought indices. Int J Climatol 27:2103–2111
Ozger M, Mishra AK, Singh VP (2012) Long lead time drought forecasting using a wavelet and fuzzy logic combination model: a case study in Texas. J Hydrometeorol 13:229–284
Rezaeian-Zadeh M, Tabari H (2012) MLP-based drought forecasting in different climatic regions. Theor Appl Climatol 109(3–4):407–414
Shirmohammadi B, Moradi H, Moosavi V et al (2013) Forecasting of meteorological drought using wavelet ANFIS hybrid model for different time steps (case study: south-eastern part of East Azerbaijan province, Iran). https://doi.org/10.1007/s11069-013-0716-9.
Singh S, Kumara S, Kumar V (2021) Analysis of groundwater quality of haridwar region by application of nemerow pollution index method. Indian J Ecol 48 (4):1149–1154
Srinivasarao C, Rao KV, Gopinath KA et al (2020) Agriculture contingency plans for managing weather aberrations and extreme climatic events: development, implementation and impacts in India. Adv Agron 159:35–91. Academic Press Inc. https://doi.org/10.1016/bs.agron.2019.08.002
Stagge JH, Kingston DG, Tallaksen LM et al (2017) Observed drought indices show increasing divergence across Europe. Sci Rep 7(1):1–10
Stocker TF, Qin D, Plattner GK et al (2013) IPCC 2013, climate change: the physical science basis. Contribution of Working Group I to the Fifth Assessment Report of the Intergovernmental Panel on Climate Change
Thompson LG, Yao T, Mosley Thompson EPN et al (2000) A high-resolution millennial record of the

South Asian monsoon from Himalayan ice cores. Science 289(5486):1916–1919

Trenberth KE, Dai A, van der Schrier et al (2014) Global warming and changes in drought. Nat Clim Change 4 (1):17–22). https://doi.org/10.1038/nclimate2067

Uddin M, Hu J, Islam ARM et al (2020) A comprehensive statistical assessment of drought indices to monitor drought status in Bangladesh. Arab J Geosci 13(9):1–10

Vicente-Serrano SM, Begueria S, Lorenzo-Lacruz J et al (2012) Performance of drought indices for ecological, agricultural, and hydrological applications. Earth Interact 16:10

White DH, Walcott JJ (2009) The role of seasonal indices in monitoring and assessing agricultural and other drought: a review. Crop Pasture Sci 60:599–616

Wilhite DA (2003) Drought in JR Holton (Ed.). J Atmos Sci. Academic Press, pp 650–658

World Meteorological Organization (2012) Standardized precipitation index user guide, Svoboda M, Hayes M, Wood M (eds), WMO-No. 1090, Geneva

Wu J, Chen X, Yao H et al (2017) Non-linear relationship of hydrological drought responding to meteorological drought and the impact of a large reservoir. J Hydrol 551:495–507

Yacoub E, Tayfur G (2020) Spatial and temporal of variation of meteorological drought and precipitation trend analysis over whole Mauritania. J Afr Earth Sci. https://doi.org/10.1016/j.jafrearsci.2020.103761

Zhang H, Ding J, Wang Y et al (2021) Investigation about the correlation and propagation among meteorological, agricultural and groundwater droughts over humid and arid/semi-arid basins in China. J Hydrol 603:127007

Zhang JY, Su YS, Song DS et al (2009) The historical droughts in China during 1949–2000

20

Understanding the Utility of Print Media for Dissemination of Insect Pest Management Tactics to Rice Farmers at Hooghly, West Bengal, India

Eureka Mondal and Kaushik Chakraborty

Abstract

Information on crop protection in print media as a conception that incarcerates how can be flawlessly put the advice on agroecology to the service of crop production after adopting integrated pest management (IPM) as the prime motto in rice eco-ecosystem has been debated for a long time. Print media plays an important role in communicating agricultural information among literate farmers on improved agricultural practices and informing the public in general. The role of newspapers in disseminating agricultural knowledge among the farmers to adopt better crop management strategies is essential in the present scenario. Depending on the habit of reading agriculture-related information in newspapers, 120 selected farmers in the block Tarakeswar covering two adjacent villages of the district Hooghly, West Bengal, India, were broadly categorized into two groups, viz. contact farmers-CF (read newspapers 2–3 h/day) and non-contact farmers-NCF (no habit of newspaper reading). Depending on the age of the farmers, each group was again subdivided into three sub-groups (3 sub-groups for CF + 3 sub-groups for NCF), viz. CF-1 and NCF-1 (20–29 years), CF-2 and NCF-2 (30–39 years), and CF-3 and NCF-3 (40–49 years), respectively. Each sub-group comprised 60 heads. A questionnaire having 20 questions, each with two alternative options, was prepared. Questions encompass prior knowledge and outline information on rice insect and pest management practices. Then, farmers were interrogated independently to the prepared questionnaire and they responded and were categorized accordingly. Grossly, contact farmers (CF) responded more accurately to the actual rice insect pest infestation level. The maximum percentage of identification of adult yellow stem borer (YSB) was successfully made by 90% of CF B male farmers and 77.5% of CF-B female farmers, whereas maximum identification of YSB larva was made by 75% of male respondents of CF-B and 69.2% female respondents of CF-A. Identification of rice crop damage due to YSB larva was also made successfully by 95% of CF-B male farmers and 85% of CF-B female farmers. Thus, the habit of reading agriculture-related news on a regular basis was found indispensable to maximize the benefit of the integrated pest management program for rice crops.

Keywords

Sustainable agriculture · Newspaper · Farmers · Article · Data responses · Age group · Yellow stem borer · Insect pest management

E. Mondal (✉) · K. Chakraborty
Department of Zoology, Raiganj University, Raiganj, Uttar Dinajpur, West Bengal, India
e-mail: eurekazoology10@gmail.com

J. Das and S. Halder (eds.), *Advancement of GI-Science and Sustainable Agriculture*, GIScience and Geo-environmental Modelling, https://doi.org/10.1007/978-3-031-36825-7_20

20.1 Introduction

Sustainable agriculture ropes sufficiently safe healthy food production (Das 1991). Agricultural information is the main apprehension of agricultural extension agencies to culminate sustainable agriculture. Agro-information is the accumulation of all the necessary components responsible for changing the farmers' outlook (Bentley et al. 2015). Such education concerns to give them the latest knowledge of crop cultivation and protection. Different agencies work to catalyze agricultural innovation and subsequent diffusion to the farmers. However, agricultural publications in daily newspapers are considered an effective tool for disseminating agricultural information among the farming community (Anderson 1985). The achievement of agencies for agricultural output in developing countries largely depends on the type and dimension of the use of mass media to mobilize farm people for development (Marek 2012). Meanwhile, radio and television have been acclaimed as the most effective media for diffusing scientific knowledge to the masses but have proven cost-effective (Gill and Sandhu 1986). In this context, the print media gains momentum. Newspapers mainly address the main factors influencing agricultural production, technology, information regarding crop varieties, and insect pest management (IPM) which are also faintly covered. Print media can cope with future agricultural problems and lend them a hand in the solution of their problems (Ananta and Tauffiqu 2016).

Investing agro-information in extension services is recommended to increase awareness and adoption of IPM. Moreover, modifying the current cultivation approach by targeting not just the primary farmers but also the members of their families can help in the adoption of rice-IPM (Mengech et al. 1995). Effective monitoring of insect pests as a part of IPM is a prerequisite for any successful plant protection program. The decision on whether and when to follow control measures is based on the information available on the pest population at a particular time. Mass media can be classified as print media and electronic media (Zijp 2002). Print media include words, pictures, publications, brochures, posters, and other types of printed materials that are physical items. Identifying the determinants of IPM and disseminating information by mass media at the grass root level is crucial for promoting the use of more ecologically benign pest control tactics in the agricultural sector (Levy and Windahl 1984). Among the mass media, newspapers are vital in communicating agricultural information among literate farmers. In contrast, the escalating rate of literacy in the country offers new promises and projections for pertaining print media as a means of communication. The print media broaden the range of communication. It is cheap, and people can afford to buy and read them at their expediency Shuwa et al. (2014). It is a permanent medium as the messages are imprinted permanently with high storage value, which makes them suitable for reference and research.

The success of agricultural development programs and successful implementation of rice-IPM in developing countries like India largely depends on the nature and extent of effective use of mass media to mobilize people and disseminate newly evolved agricultural technologies (Birkenholtz and Maricle 1991). Among the mass media, newspapers and farm magazines are very important. Print media is low in cost and has widened the scope of visual communication (Olowu and Oyedokun 2000). Farm magazines and newspapers with agro-techniques are commonly used to teach farmers about rice-IPM. It offers agricultural information to literate farmers to improve their knowledge level (Purushothaman 2003). Multi-stage sampling methods were used to select study participants. In the first phase, two villages were chosen. In the final phase, 120 rice growers from three distinct age groups and from two villages who directly participated in agricultural activities were selected for interview against the prepared questionnaire, which was made with appropriate information.

Agricultural publications play a very pivotal role in the diffusion of innovation. These publications help to change the primitive agricultural methods in the land forming. The present chapter will ascertain why the farmers are disposed to agricultural information in newspapers. The chapter will also reveal that the age, gender, and educational level of the farmers influence the perception of agricultural information in newspapers. The role of the newspaper as a forerunner for awareness creation and dissemination of agricultural information to improve wider reach within and outside the farming population will also be evaluated.

20.2 Material and Methods

20.2.1 Study Area

The place of study was Tarakeswar, Hooghly, West Bengal. Tarakeswar is located at 22.89°N 88.02°E. It has an average elevation of 18 m (59 feet). It is under the Chandannagore subdivision of the Burdwan Division in West Bengal (Fig. 20.1).

The town is 48 km from Chinsurah, the district headquarters and 45 km from Chandannagore. The district Hooghly encompasses 239,500 agricultural families, 88,536 small-range farmers, 135,827 marginal farmers, and 242,564 agricultural laborers (District Annual plan on Agriculture, Hooghly, Govt of West Bengal, 2002–2003). The present work contemplates farmers' perception of rice-IPM and the importance of newspapers distributing agricultural-related information among the farmers to check rice insect pests, especially rice yellow stem borer (YSB).

20.2.2 Methodology

Because of the increased national literacy level of 52.11% during 1991, print media has acquired greater momentum in disseminating information relating to improved agricultural practices to farming communities. Grossly, 558 registered newspapers in India, at about 35% of the total,

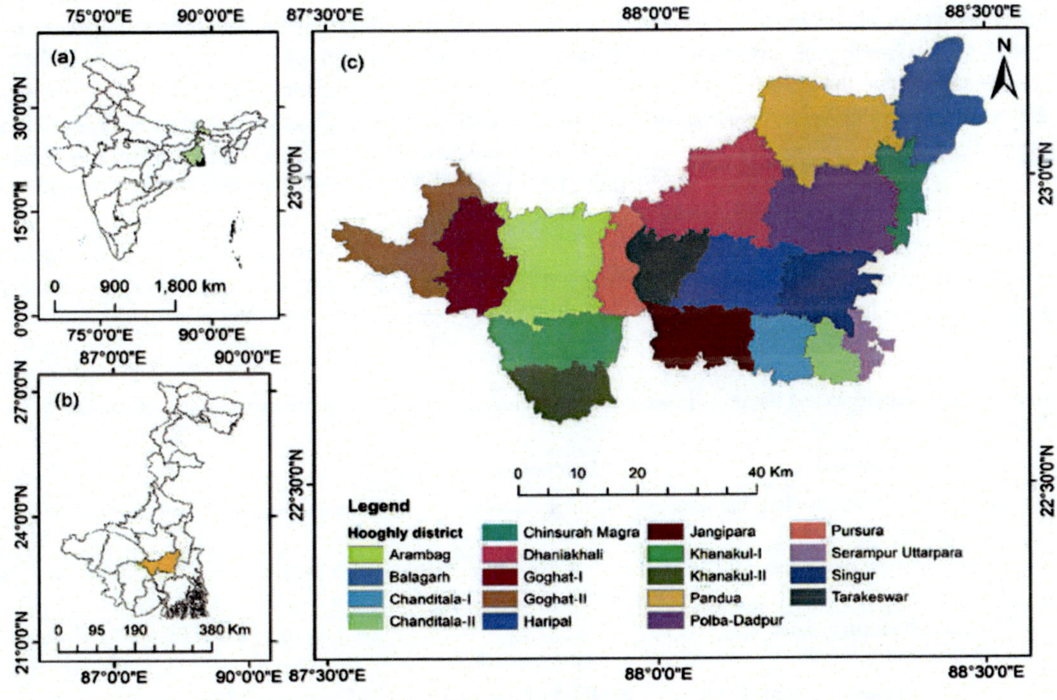

Fig. 20.1 Location of the study area (marked by *) at Hooghly, West Bengal, India

are being circulated in the district Hooghly, which is available in West Bengal. Further, about 11 weekly, 12 fortnightly, two monthly, and four other newspapers circulated in this region (district information and culture office, Hooghly, West Bengal, 2002–2003). The issues of the available daily newspaper in regional languages, especially in Bengali, published between 2020 and 2021 and had agricultural-related information, were purposively selected for the study. The selected papers analyze the cover page, content, format, and illustration components. Those articles were purposively selected for the study and were related to agricultural production, especially insect pest control encompassing scientific and cultural practices, proper insecticide application, field management, etc.

In general, farmers of the Hooghly district cultivate a fairly large area with only a small number of workers. They cannot spend all their time farming. Sometimes, they cannot sow all their crop properly; other times, they weed some crops poorly or too late; often, they have very little information about rice insect pests. Farmers use only huge quantities of insecticides and fertilizer, if any at all, at regular intervals to check pest menace. Grossly, the farmers that are selected for the study have the criteria: (i) the majority of the farmers were literate and were acquainted with farm literature; (ii) all were progressive farmers undertaking agriculture, livestock production and other allied enterprises; (iii) farmers were aware of the basic norms of farm journalism.

The interview schedule consisted of structural and open-ended questions in the Bengali language for data collection, and the respondents were individually interviewed. The questionnaire consisted of mainly three parts: (i) demographic characteristics of farmers, (ii) questions about identification, damage, economic threshold level (ETL), and non-target pest outbreaks in the rice fields, and (iii) farmers' knowledge and perception about the management of insect pests in rice fields (Table 20.1).

Adopted a bunch of varietal techniques for collecting data (Damalas and Khan 2016). Two adjacent villages (Bhanjipur and Dhalyan, respectively) were primarily chosen randomly from the administrative block Tarakeswar, West Bengal, for the survey. At the same time, a group of 120 farmers (60 male and 60 female) with a regular habit of reading newspapers from three distinct age groups mostly related to rice farming from each of the two villages were selected. Farmers were successfully interviewed against the prepared questionnaire. Farm magazine effectiveness index (FMEI) was assessed against nine major components as developed by Nataraju and Perumal (1995). Information obtained with questionnaires was enlisted, and the data was checked through SPSS to correct entry errors, including typographical, and transcriptional errors. The statistical analysis was mainly descriptive. The Chi-square tests were adopted to assess the linearity between the ability of farmers to identify insects and the selected socioeconomic variables of farmers correctly. Similarly, the independent t-test was used to assess whether significant differences existed between farmers who correctly identified the rice insect pests.

Table 20.1 Components and the description of the questionnaire to assess the importance of agro-information to the farmers

Types	Components	Description of questions
Structural	Farmer's bio-data and farm details	Marital status, age, educational level, land tenure, farm holding size
Functional	About newspaper reading	Language of newspapers they preferred, regularity, causes of choice, the information they collected
	About integrated pest management	Identification of the yellow stem borer in the field or by pictorial representation, damage symptoms, and economic threshold levels
	Commonly adopted pest management through pesticides	Dose of insecticide, time of application, times, dose, mode of application, insecticide trends, effectiveness

20.2.3 Data Source

The data during the survey was obtained in two different ways, primarily by observing what was going on in the fields or by interviewing the farmers. Information gathered by interviewing is likely less accurate than that collected by observation because the farmers may need to remember or wish to tell. On the other hand, the quantity of information gathered by interviews is much greater, so interviewing is much cheaper per item of information obtained. During the study, data obtained from interviews was combined with observations and accordingly tabulated. As per the work requirement, we set up a regular interview schedule with the farmers. If there is an unnecessary delay, a farmer is more likely to forget some information or make a mistake (Fig. 20.2).

20.3 Results and Discussion

20.3.1 Component-Wise Discussions

Item 1: Farmer and Farm Characteristics

The majority of the respondents (83%) were males. Among male farmers, 50% had no formal education, 38% had basic education, 9% had secondary education, and 3% had only tertiary education. In contrast, among female respondents, 57% had no formal education, 34% had basic education, 7% had secondary education, and 1% had only tertiary education, respectively. Depending on the habit of newspaper reading, randomly selected 120 farmers from 2 nearby villages of Tarakeswar block under district Hooghly, West Bengal, were broadly categorized into two groups, such as contact farmers (CF) and non-contact farmers (NCF). CF reads newspapers 3–4 h a day, while NCF has no such habit. Depending on the age of the farmers, each group was again subdivided into three sub-groups, viz. CF-A and NCF-A (18–27 years), CF-B and NCF-B (28–37 years), and CF- C and NCF-C (38–47 years). So collectively, six sub-groups were formed, each sub-group comprised 20 heads (Fig. 20.3).

Item 2: Observation on Farmers' Habit of Newspaper Reading

Farm publications offer new agricultural practices and proper techniques for new crops, which is a blessing for farming families to acquaint them with the new cultivation strategy.

In Village-1 (Bhanjipur), 40.83% of male and 25.83% of female farmers read articles with a duration of 30–59 min daily, whereas 42.50% of male and 17.50% of female farmers utilized 60–89 min. About 8.33% of male and 3.33% of

Fig. 20.2 a Collection of data at field level, b interaction with the farmers

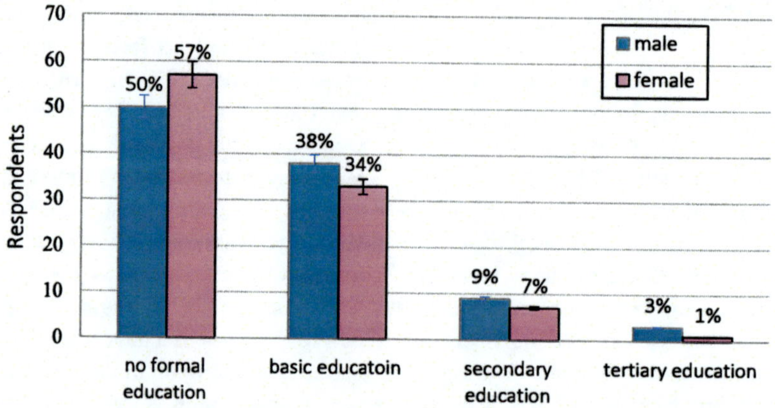

Fig. 20.3 Demographic representation of the educational status of the respondents depending on the gender of the farmers

female farmers had an average of 90–119 min of reading habits. Only 5.83% of male and 2.50% of female farmers showed daily 120–149 min of reading habits. On the other hand, the scenario was almost the same in Village-2 (Dhalyan), 22.50% of male and 15% of female farmers used to read the articles for a duration of 30–59 min daily, 7.50% of male and 17.50% of female farmers showed 60–89 min of reading habit, 4.17% of male and 3.33% of female farmers showed 90–119 min of reading habit, and 3.33% of male and 1.67% of female farmers showed 120–149 min daily reading habit, respectively (Fig. 20.4).

Item 3: Title of the Newspaper

The title is the most imperative part of a newspaper to give agro-information which should be agreeable to read. The title must be descriptive, direct, accurate, appropriate, interesting, concise, precise, and unique. The newspaper title was perceived to be satisfactory by 100% CF-A, 75% CF-B, and 65% CF-C among the male farmers. For 85% of female respondents of the CF-A group, it was perceived to be more satisfactory. Whereas 55% of female respondents of CF-B had opined that the title of the newspaper was important to disseminate the view. For 60% of female farmers of the CF-C group, the

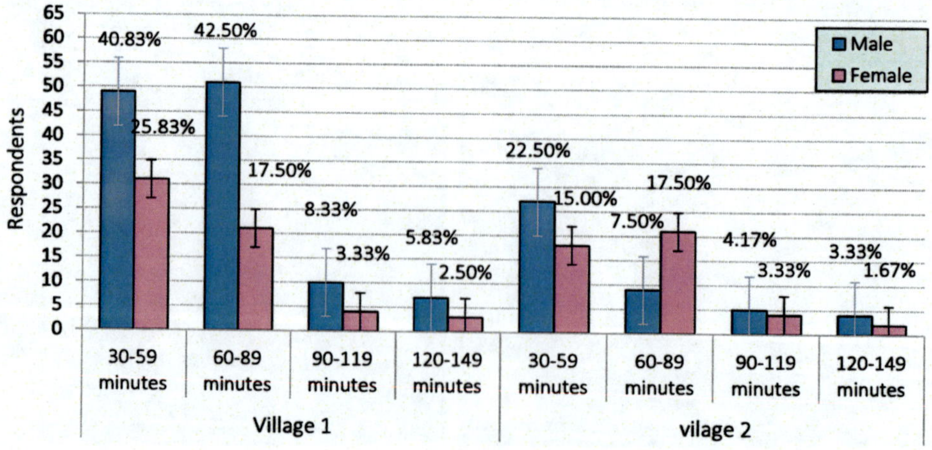

Fig. 20.4 Demographic representation of the number of farmers with a habit of newspaper reading in Village-1 (Bhanjipur) and Village-2 (Dhalyan), respectively

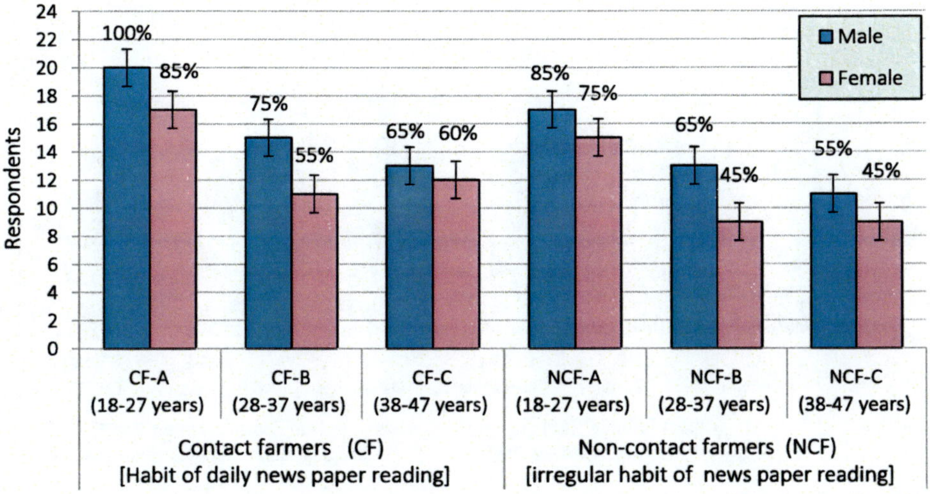

Fig. 20.5 Acceptance of farm-related information in newspapers depending on the title of the newspaper, the habit of the newspaper reading, age, and gender of the farmers

importance of the title of the newspaper is crucial. In contrast, 85% of the NCF-A male farmers perceived the newspaper title to be satisfactory. Additionally, it was satisfactory to 65% of NCF-B and 55% of NCF-C among the male farmers or 75% of female respondents of the NCF-A group. It was perceived to be more satisfactory. 45% of female respondents of both NCF-B and NCF-C groups had opined that the newspaper's title is important to disseminate the view (Fig. 20.5).

Item 4: Readability of the Cover Page Write-Up
The way a text looks matters to a reader, so it should matter to a writer. Letters and reports are more than words on a page or a screen. Low or inferior ideas are arranged and delivered in physical form, whether electronically or on paper, making reading seem intimidating, confusing, or downright unfriendly, even if the content itself is perfect. Simultaneously, the cover page write-up's readability was perceived to be satisfactory by 90% of CF-A male farmers. Additionally, it was satisfactory to the 65% CF-B and 65% CF-C male farmers. For 75% of female respondents of the CF-A group, it was perceived to be more satisfactory. 50% of female respondents of CF-B had opined that the readability of the cover page write-up was important to disseminate the view. For 5% of female farmers of the CF-C group, the importance of the magazine's title was crucial. The readability of the cover page write-up was perceived to be satisfactory by 85% of NCF-A male farmers. Additionally, it was satisfactory to 60% NCF-B and 50% NCF-C male farmers, respectively. It was professed to be more satisfactory for 75% of the NCF-A group female respondents. 45% of female respondents of NCF-B had opined that the readability of the cover page write-up was important to disseminate the view. For 40% of female respondents of the NCF-C group, the importance of readability of the cover page write-up was crucial (Fig. 20.6).

Item 5: Design of the Magazine
One of the important basics of news is the design of the magazine. The most esthetically pleasing magazines always have great covers. An attention-grabbing magazine cover design is vital for selling news to readers and inviting them to delve deeper into the publication. 60% of CF-A male farmers perceived the magazine design to be satisfactory. Additionally, it was satisfactory to 55% CF-B and 55% CF-C male group of farmers. For 50% of female respondents of the CF-A group, it was perceived to be more satisfactory. 45% of female respondents of CF-B had

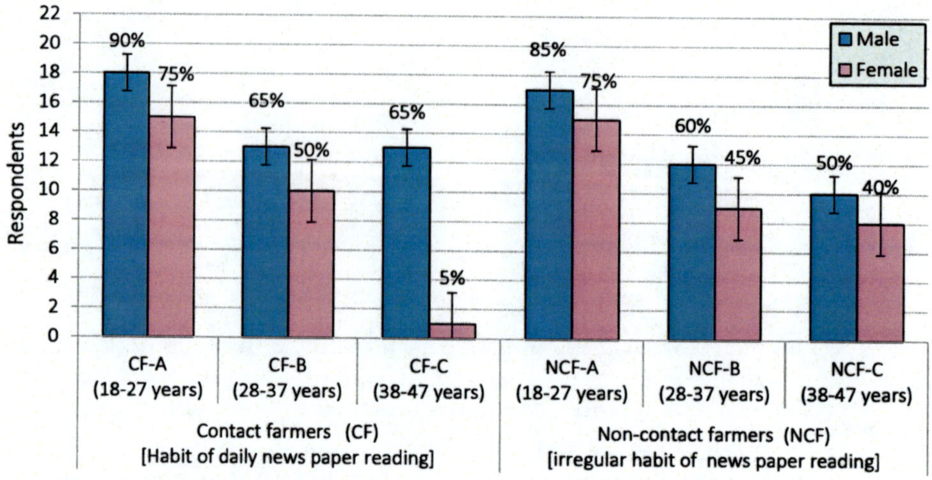

Fig. 20.6 Acceptance of farm-related information in newspapers depending on the readability of the cover page write-up of the magazine, the habit of the newspaper reading, age, and gender of the farmers

opined that the magazine's design was important to disseminate the view. For 45% of female farmers of the CF-C group, the importance of the magazine's design was crucial. In contrast, the magazine design was perceived to be satisfactory by 45% of NCF-A, 45% of NCF-B, and 40% of NCF-C of male farmers. Whereas 35% of female respondents of the NCF-A group, 35% of female respondents of NCF-B and 35% of female farmers of the NCF-C group had opined that the magazine's design was important to disseminate the view (Fig. 20.7).

Item 6: Length of the Article

Subject-specific and elimination of unnecessary sentences and inclusion of very basic information are obvious to the target audience. In a text, there should be statements, especially at the beginning of the introduction and discussion, which will be catchy, give the audience very obvious

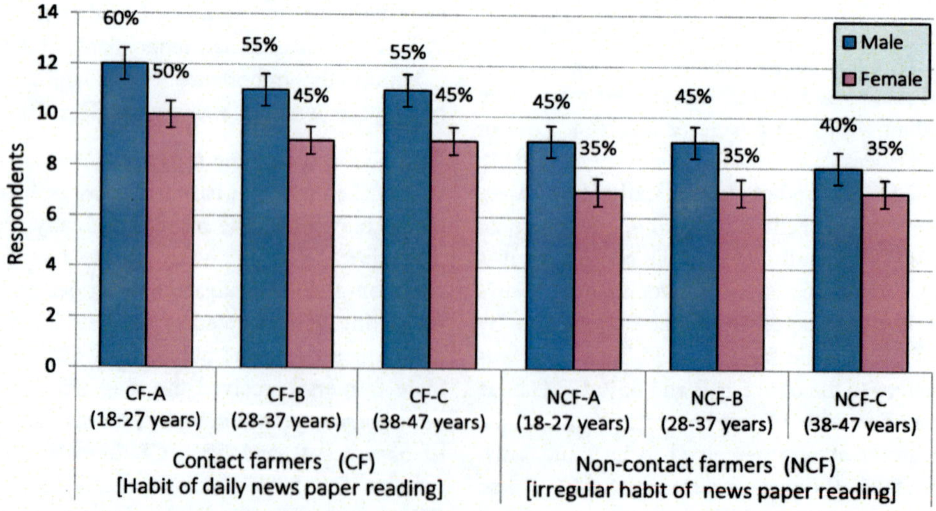

Fig. 20.7 Acceptance of farm-related information in newspapers depending on the design of the magazine, the habit of the newspaper reading, age, and gender of the farmers

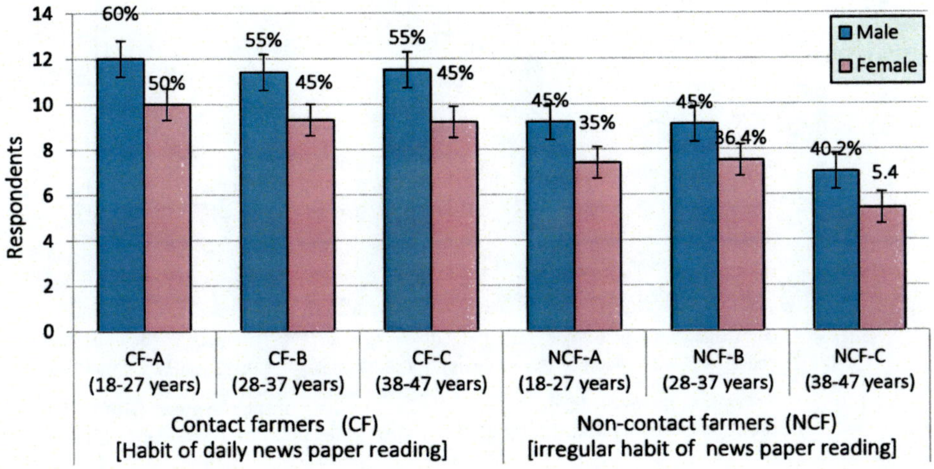

Fig. 20.8 Acceptance of farm-related information in newspapers depending on the length of the article of the magazine, the habit of the newspaper reading, age, and gender of the farmers

information, and patronize the text. The perceived length of the chapter was to be satisfactory to 60% of CF-A male farmers. Additionally, it was satisfactory to 55% of CF-B and 55% CF-C male group of farmers. For 50% of female respondents of the CF-A group, it was perceived to be more satisfactory. 45% of female respondents of CF-B had opined that the article's length was important to disseminate the view, and 45% of female farmers of the CF-C group the importance of the chapter's length was crucial. In contrast, the chapter's length was satisfactory by 45% of NCF-A, 45% of NCF-B, and 40.2% of NCF-C of male farmers. For 35% of female respondents of the NCF-A group, it was perceived to be more satisfactory. Whereas 36.4% of female respondents of the NCF-B and 5.4% of female farmers of the NCF-C group had opined that the length of the article was crucial to disseminate the view (Fig. 20.8).

Item 7: Timeliness of the Article

To create timeliness in scientific work, information should be regular. The abovesaid situation increases the chance that results from the research are included in policymaking. The timeliness of research is important for evidence-based policymaking. In order to create timeliness in research, the interaction between researchers and policymakers is important. The timeliness of the article was perceived to be satisfactory by 75% of CF-A male farmers. Additionally, it was satisfactory to 70% of CF-B and 85% of the CF-C male group of farmers or 65% of female respondents of the CF-A group. It was perceived to be more satisfactory. 60% of female respondents of CF-B had opined that the timeliness of the article was important to disseminate the view. In contrast, for 80% of female farmers of the CF-C group, the importance of timeliness of the article is crucial. In contrast, the timeliness of the article was perceived to be satisfactory by 45% of NCF-A, 45% of NCF-B, and 65% of NCF-C among the male farmers. Whereas 35% of female respondents of the NCF-A group, 37.5% of female respondents of the NCF-B group, and 15% of female farmers of the NCF-C group had opined that the timeliness of the article was important to disseminate the view (Fig. 20.9).

Item 8: Relevancy of the Message to the Season and Region

The article should be addressed to the audience. The chapter will give more information on the type of audience. Focus on the main message will keep the attention of the readers. Subject-specific writing helps in effective communication. Perceived relevance of the message to the season and region to be satisfactory by 80% of CF-A male farmers. Additionally, it was satisfactory to the 70% CF-B and 60% CF-C male

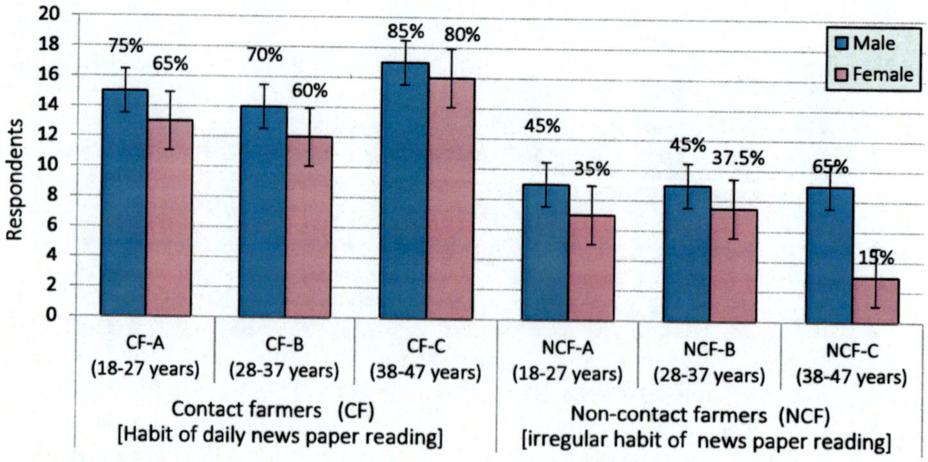

Fig. 20.9 Acceptance of farm-related information in newspapers depending on the timeliness of the article in the magazine, the habit of the newspaper reading, the age and gender of the farmers

group of farmers. For 60% female respondents of CF-A group, it was perceived to be more satisfactory. It was perceived to be more satisfactory. 60% of female respondents of CF-B had opined that the relevancy of the message to the season and region was important to disseminate the view. For 55% of female farmers of the CF-C group, the importance relevancy of the message to the season and region is crucial. Perceived relevancy of the message to the season and region was to be satisfactory by 35% of CF-A male farmers. Additionally, it was satisfactory to the 40% CF-B and 45% CF-C male group of farmers. For 30% female respondents of the CF-A group, it was perceived to be more satisfactory. 37.5% of female respondents of CF-B had opined that the magazine's title was important to disseminate the view and for 15% of female farmers of the CF-C group, the importance of the relevancy of the message to the season and region was crucial (Fig. 20.10).

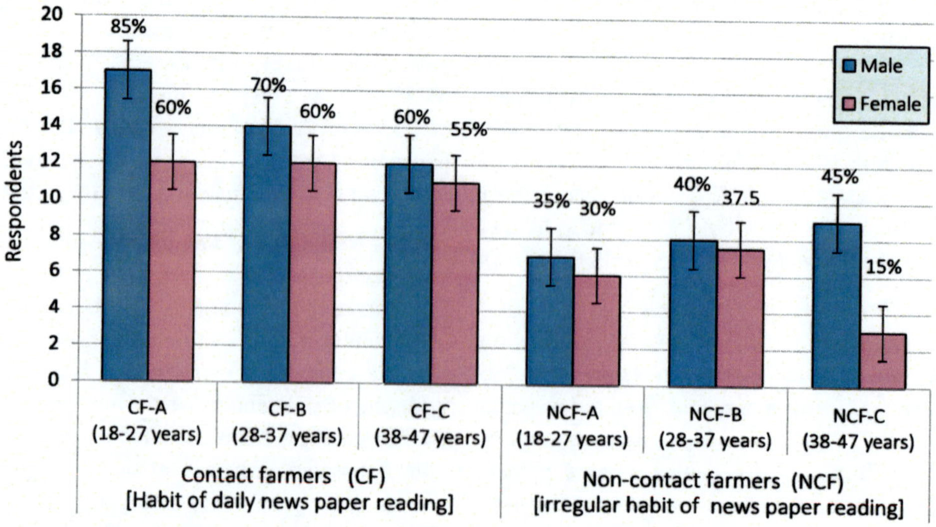

Fig. 20.10 Acceptance of farm-related information in newspapers depending on the relevancy of the message to the season and region, the habit of the newspaper reading, age, and gender of the farmers

Item 9: Number of Pages per Article
A limited page number of an article is essential to deliberate information specifically. The number of pages per article was perceived to be satisfactory by 90% of CF-A male farmers. Additionally, it was satisfactory to the 90% CF-B and 80% CF-C male group of farmers, or 85% female respondents of the CF-A group; it was perceived to be more satisfactory. 75% of female respondents of CF-B had opined that the numbers of pages are important to disseminate the view. For 80% of female farmers of the CF-C group, the importance of the number of pages per article was crucial. In contrast, the number of pages per article was perceived to be satisfactory by 45% of NCF-A, 45% of NCF-B, and 45% of NCF-C male farmers. It was perceived to be more satisfactory for 35% of female respondents of the NCF-A group. 37.5% of female respondents of NCF-B had opined that the magazine title is important to disseminate the view and for 15% of female farmers of the NCF-C group, the importance of the number of pages per article was crucial (Fig. 20.11).

Item 10: Quality and Clarity of Illustrations
An illustration can explain an idea without the text nearby. The illustration is a very interesting area that simultaneously contains a lot of creativity. The quality and clarity of illustrations were perceived to be satisfactory by 90% of CF-A male farmers. Additionally, it was satisfactory to 85% CF-B and 70% CF-C male group of farmers. For 75% female respondents of the CF-A group. It was perceived to be more satisfactory. 60% of female respondents of CF-B had opined that the title of the quality and clarity of illustrations disseminate the view. For 55% of female farmers of the CF-C group, the importance of quality and clarity of illustrations was crucial. In contrast, the quality and clarity of illustrations were perceived to be satisfactory by 35% of NCF-A male farmers. It is also satisfactory to 40% NCF-B and 40% NCF-C male farmers. It was perceived to be more satisfactory for 30% of female respondents of the NCF-A group. 37.5% of female respondents of NCF-B and 25% of female farmers of the NCF-C had opined that quality and clarity of illustrations are important to disseminate the view (Fig. 20.12).

20.3.2 Observation on the Identification of Major Pests of Paddy

20.3.2.1 Identification of Major Insect Pest Problem

One hundred twenty farmers, irrespective of age and gender, were shown pictures of rice insect pests and then asked to identify the insect pests

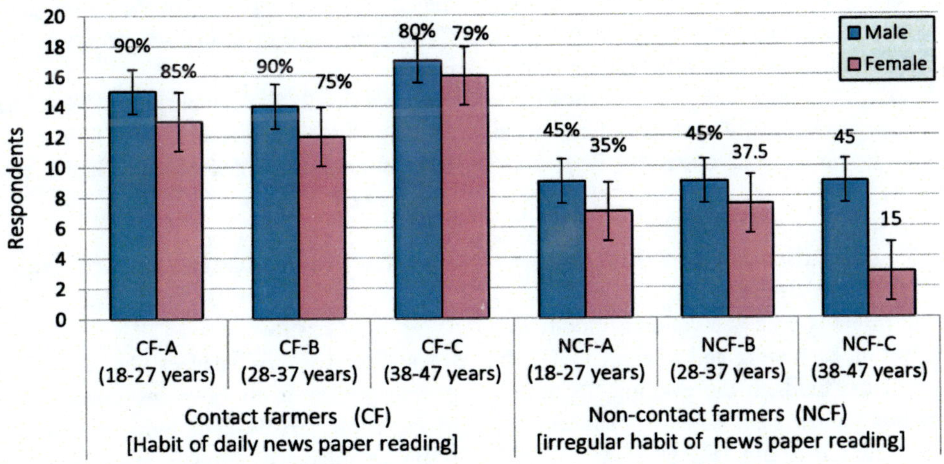

Fig. 20.11 Acceptance of farm-related information in newspapers depending on the number of pages per article, the habit of the newspaper reading, age, and gender of the farmers

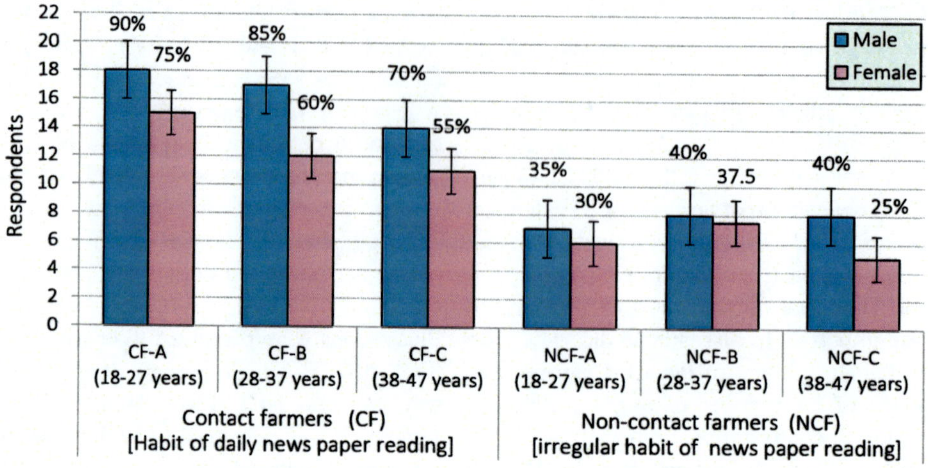

Fig. 20.12 Acceptance of farm-related information in newspapers depending on the quality and clarity of illustrations of the article, the habit of the newspaper reading, age, and gender of the farmers

and to correlate the pest with damage symptoms. The responses in percentage were taken into consideration for this executive research. Farmers' knowledge and perception of rice insect pests were assessed accordingly. Brown plant hopper (51.96%) and yellow stem borer (41.36%) ranked first and second as economically important insect pests. Then, it was followed by rice grasshoppers (34.7%), mole cricket (18.3%), Gandhi bug (15.86%), rice hispa (10.33%), rice caseworm (8.3%), gall midge (8.2%), and white fly (5.53%) in decreasing order as per the opinion of the farmers. Non-contact farmers who have little accustomed to newspaper reading have similar opinions about enlisting the important insect pests in the rice field. However, in such a case, probably due to a low knowledge level, a comparatively low number of farmers can recognize the insect pests correctly (Figs. 20.13 and 20.14).

20.3.2.2 Identification of Adult Yellow Stem Borer

The adult insect is usually a yellowish moth with a black dot on each wing and a pointed apex. he adults visit the fields in the evening. Identification of adult yellow stem borer was successfully made by 75% CF-A and 45% NCF-A, 90% CF-B and 35% NCF-B, 80% CF-C, and 35% NCF-C of male farmers. In contrast, the same identification was successfully made by 60% CF-A and 30% NCF-A, 77.5% CF-B and 37.5% NCF-B, and 60% CF-C and 25% NCF-C of female farmers (Figs. 20.15 and 20.16).

20.3.2.3 Identification of YSB Larva

The young larvae are found inside the stem and are off-white with a red head and no visible legs. Identification of YSB larva after splitting the rice stem was satisfactorily made by 65% CF-A, 75% CF-B, and 70% CF-C male group of farmers. At the same time, 69.2% of female respondents of the CF-A group and 35% of the CF-B group and CF-C group had identified the yellow stem borer larvae. The same identification was made successfully by 25% of male and 15% of female NCF-A, 30% of male and 15% of female NCF-B, 20% of male and 14.5% of male NCF-C group of farmers (Fig. 20.17).

20.3.2.4 Identification of Rice Crop Damage Due to YSB Larva

Infestation by YSB causes the drying of the central panicle of the growing plant and results in the characteristic symptoms of a dead heart (DH) at the early growth stage and a white head (WH) at the late growth stages.

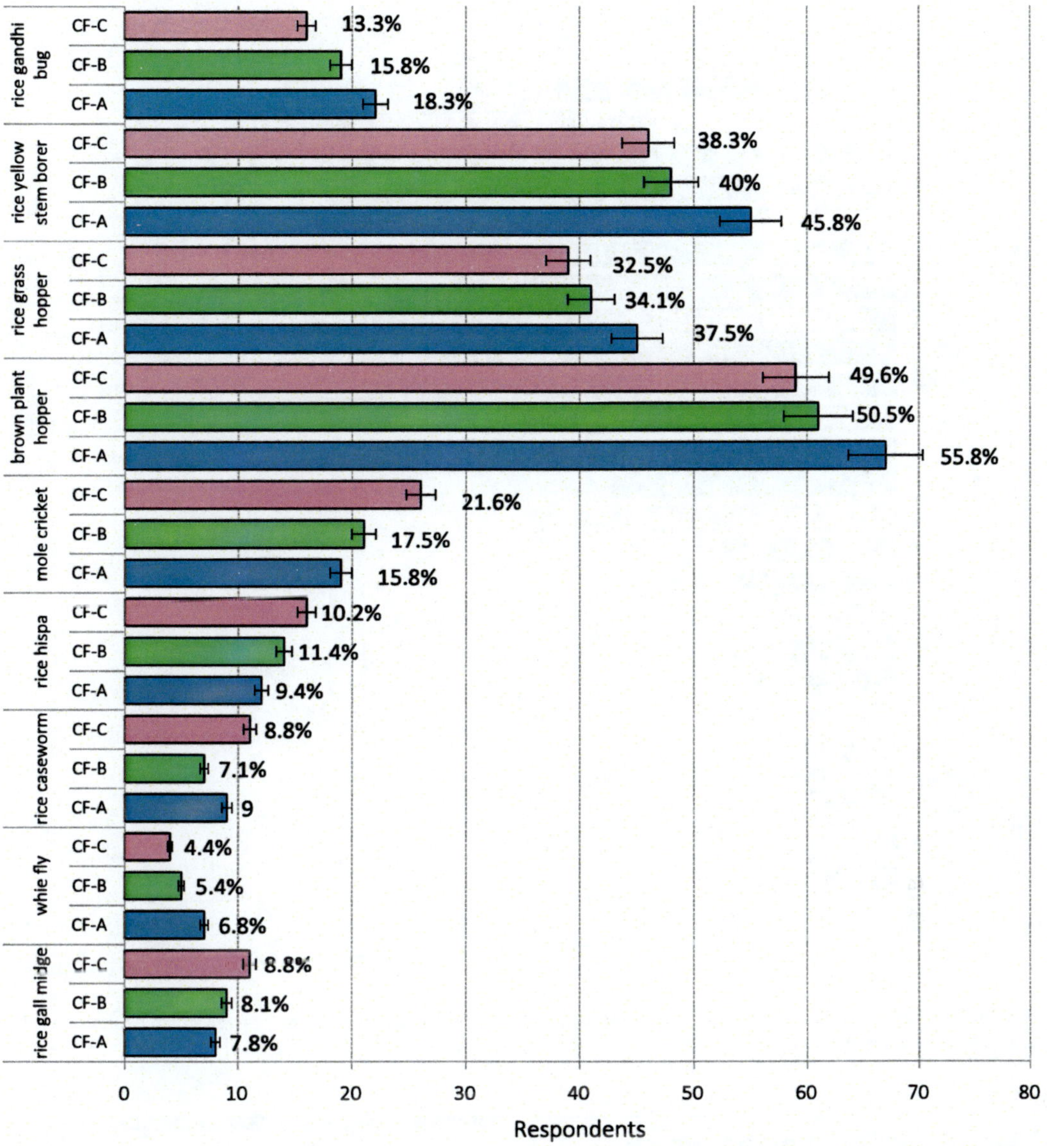

Fig. 20.13 Identification of the major insect pest in the locality by the farmers having a regular reading of newspapers with age groups

Identification of rice crop damage due to YSB larva was made successfully by 85% CF-A, 95% CF-B and 90% CF-C male group of farmers. In comparison, 75% of the CF-A group, 85% of the CF-B, and 75% of the CF-C female farmers had identified the rice crop damage due to YSB larva. On the other hand, identification of rice crop damage due to YSB larva was made successfully by 45% NCF-A, 37.5% NCF-B, and 35% NCF-C male farmers. In contrast, 30% of the CF-A group, 37.5% of the CF-B, and 27% of CF-C female farmers had identified the rice crop damage due to YSB larva (Fig. 20.18).

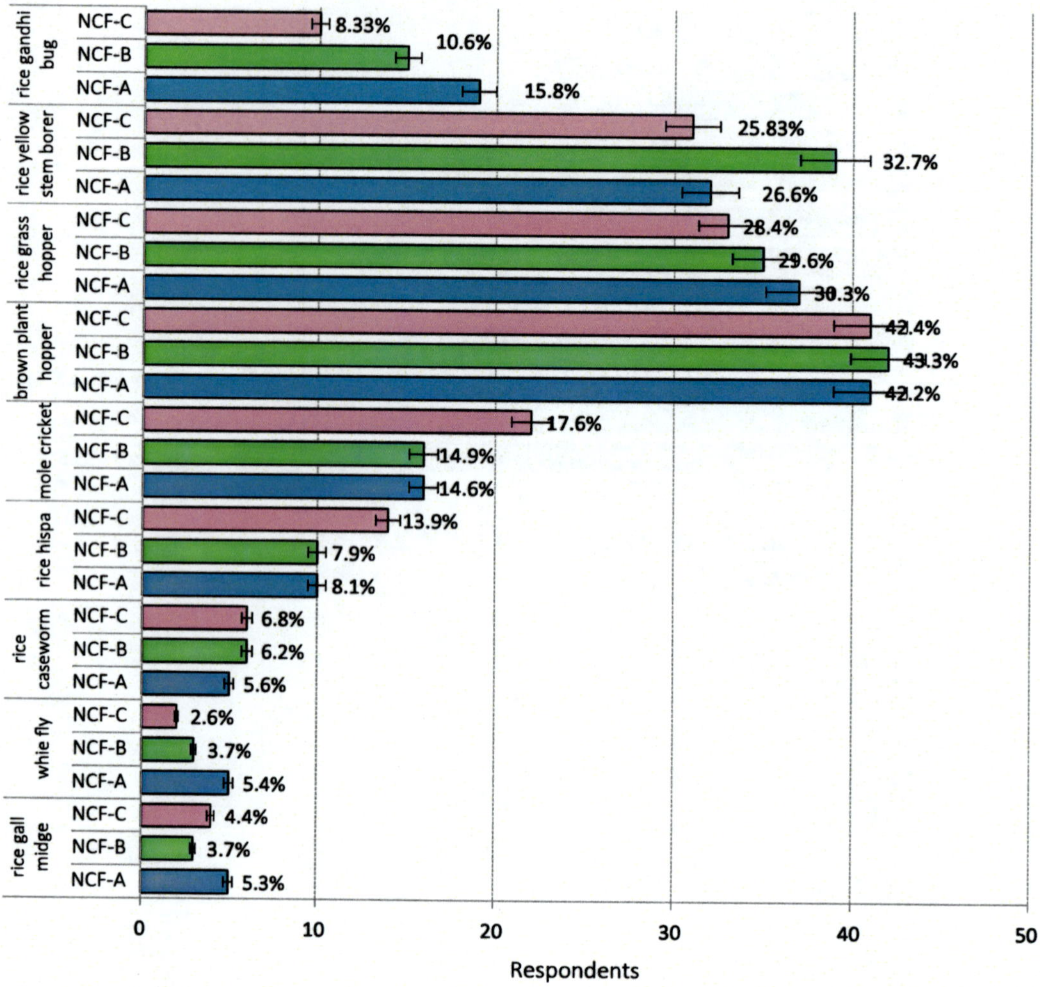

Fig. 20.14 Identification of the major insect pest in the locality by the farmers having irregular habits of reading newspapers with age groups

20.4 Policy Recommendations

Information by print media can effectively be used to address the needs and interests of the target rice growers, offer options and facilitate decision-making, encourage the adaptation of technology to a local situation, provide a more explicit treatment of sustainability concerning the technical content, and give information on the economic and financial implications of any recommended technologies, including the uncertainties and risks involved at the district Hooghly, West Bengal. Agricultural extension workers can use printed materials along with other communication channels to reinforce the learning process of farmers. Therefore, it is important to develop media policies that specifically target and design to meet the needs of agricultural communities, which will ultimately lead to the development of agro-sustainability, which farmers can guarantee by considering the local environment and favored language. The dissemination of agricultural information would be lucrative, catchy, and conversion motivated to a specific crop to ensure farmers' attraction to the news or this special attention to be given to the cover paper of the newspaper, length of the news, style of

Fig. 20.15 Yellow stem borer and consequences of damage **a** adult female yellow stem borer, **b** adult male yellow stem borer, **c** fifth instar larvae within paddy stem, **d** a solitary fifth instar larva, **e** entry point of larvae within paddy stem, **f** the dead heart (DH) with white panicle, **g** white head (WH) within paddy field, **h** farmers are identifying white heads (WH) from the field

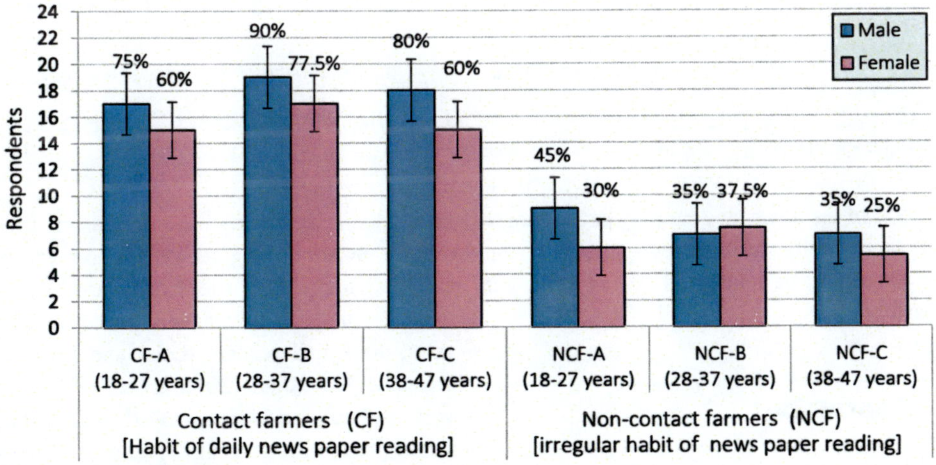

Fig. 20.16 Number of respondents who can correctly identify the adult of yellow stem borer moth depending on the habit of the newspaper reading

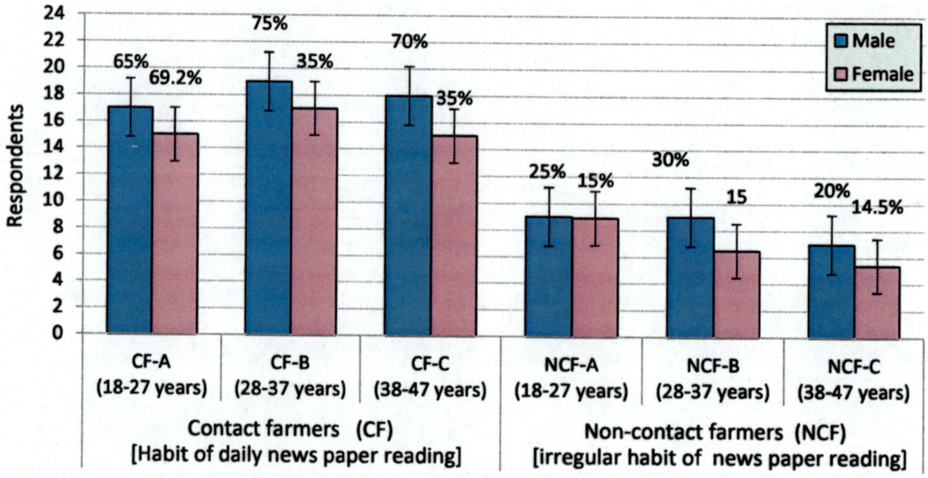

Fig. 20.17 Number of respondents who can correctly identify the larva of yellow stem borer moth depending on the habit of the newspaper reading

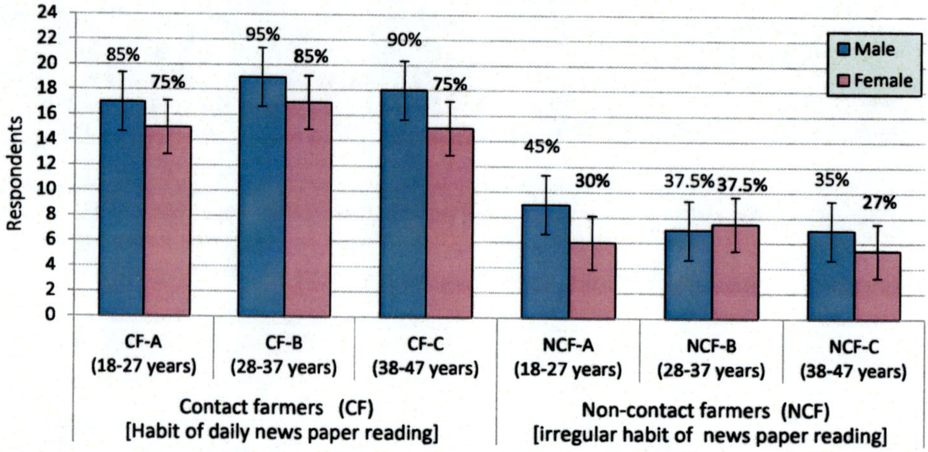

Fig. 20.18 Number of respondentswho can correctly identify the damage symptoms (DH and WH) of yellow stem borer moth depending on the habit of the newspaper reading

deliberation of the news, and lucidity of language. Further accessibility of the newspapers to the farmers is also to be addressed so they can get them easily.

20.5 Conclusion

In the current information age, communication services, particularly the print media, have been intended to sustain national development in an association for progress (Halakatti et al. 2010). Print media had premeditated to assist the community in improving its production capabilities and overall quality of life (Norton and Way 1990). Resourceful print media with attractive and gorgeous look call readers (Grist and Lever 1981). Damalas and Khan (2016) observed that it is a very difficult and complicated task to control insect pests with poor agricultural knowledge by most of the farmers. The necessity to improve agricultural output by adopting agricultural technologies using print media is very overriding in supporting agricultural invention (Shekara et al.

2016). The resource-poor farmers may need more funds to collect newspapers and therefore need to be encouraged and supported. Allahyari et al. (2017) have stated that the scarcity of information about smallholder farmers' knowledge of major crop pests is a prime limitation for effective pest management recommendations. Hashemi et al. (2009) stated that rice Gandhi bug infestation was followed by yellow stem borer in standing crops, while grain moths followed rice grain weevil to be major in storage. From an experiment, Okonya and Kroschel (2015) reported 64% and 60% of the respondents could correctly identify insect pests when they saw them in pictures. Only 48% correctly identified a yellow stem borer when by picture. Consequently, about lack of ability of farmers to identify major insect pests on their crops due to poor knowledge was also reported in Pakistan (Damalas and Khan 2016), sweet potato farmers in Tanzania (Nataraju and Perumal 1995), and beans farmers in Rwanda (Trutmaun et al. 1993). Meijer et al. (2015) reported that improving farmers' knowledge had taken paramount importance in the decision-making process for insect pest management on their farms. The above finding highlights the need for education to enable farmers to understand the biology and behavior of key insect pests and assist them in their effective pest management. In addition, a participatory research approach can increase farmers' knowledge of pests and enhance their pest control (Soniia and Christopher 2011). Educated farmers and those with many years of farming experience are expected to easily identify major pests as a reflection of their knowledge and experience (Taghdisi et al. 2018). It has been found that role of extension workers as a source of information on these matters could have been improved (Shekara 2016). Several-related information like banned pesticides, color symbols, read labeled instructions, and diagnosis symptoms are known from the print media (Smith and Haverkamp 1977).

Improved educational programs focused on pest control are observed in developing countries (Koul et al. 2004). Plant clinics designed to educate farmers about pest biology and appropriate controls have gained momentum in recent years. Accurate pest identification, reasonable estimates of potential damage, and determination of appropriate control have been given in the 'toolbox' of alternatives (CAST 2004). But its success mostly depends on the effective communication the newspapers may serve. Farmers' decisions about crop protection strategy may depend, among other factors, on their knowledge and experience with pests and the damage inflicted on cultivated plants, as well as their awareness of the existence of natural enemies (Segura et al. 2004). It is widely accepted that pest management can be more meaningful when farmers' perceptions and practices are considered (Segura et al. 2004). It is, therefore, very important to understand farmers' perceptions and knowledge of pest identity (Nurzaman et al. 2000). Thus, research on their knowledge, perceptions, and pest control practices may be useful in developing integrated pest management (IPM) programs that more farmers can quickly adopt as an alternative to broad-spectrum pesticide application.

References

Allahyari MS, Damalas CA, Ebadattalab M (2017) farmers' technical knowledge about integrated pest management (IPM) in olive production. Agriculture 7:101

Ananta N, Tauffiqu A (2016) Role of media in accelerating women empowerment. Int J Adv Educ Res 1 (1):16–19

Anderson GA (1985). Future young and adult farmer programs. Agricult Educ Magazine 58(6):14–15

Bentley J, VanMele P, Harun-ar-Rashid M, Timothy JK (2015) Distributing and showing farmer learning videos in Bangladesh. J Agril Edu and Ext. https://doi.org/10.1080/1389224 X.2015.1026365

Birkenholtz RJ, Maricle GL (1991) Adult education in agriculture: a national survey. J of Agril Edu 32 (4):19–24

CAST (2004) Management of pest resistance: strategies using crop management. Biotechnology and Pesticides. Council of Agricultural Science and Technology, Washington, DC

Damalas CA, Khan M (2016) Farmers' attitudes towards pesticide labels: Implications for personal and environmental safety. Int J Pest Manage 62:319–325

Das B (1991) A study on adoption of improved cultivation practices of rapeseed and mustard by the farmers of Berpeta district. M.Sc. (Agric.) Thesis, Assam Agricultural University, Jorhat

Gill LS, Sandhu NS (1986) Measurement of readability ease in farm magazines. PAU Res J 21(4):597–599
Grist DH, Lever RJ (1981) Pests of rice. Global Pattern Biodiver 5(1):220–227
Halakatti SV, Gowda DS, Natikar KV (2010) Role of mass media in transfer of agricultural technologies. Res J Agric Sci 1:290–113
Hashemi SM, Hosseini SM, Damalas CA (2009) Farmers' competence and training needs on pest management practices: participation in extension workshops. Crop Prot 28:934–940
Koul O, Dhaliwal GS, Cuperus GS (2004) Integrated pest management: potential constraints and challenges. AB international, Wallingford, UK, 336. https://doi.org/10.1002/ps.1021
Levy MR, Windahl S (1984) Audience activity and gratification-a conceptual clarifications and exploration. Comm Res 11:51–78
Marek R (2012) Technology transfer and innovation indicators best practices and policy issues, universities. Third Mission Indication and Good Practices 2–3, Wood Way, Ireland
Meijer SS, Catacutan D, Ajayi OC (2015) The role of knowledge, attitudes and perceptions in the uptake of agricultural and agroforestry innovations among smallholder farmers in sub-Saharan Africa. Int J Agril Sustain 13(1):40–54
Mengech AN, Saxena AN, Gopalan HNB (1995) Integrated pest management in tropics: Current status and future prospects. Wiley, New York
Nataraju MS, Perumal G (1995) Effectiveness of farm magazines. Communicator 30(1):23–28
Norton GA, Way J (1990) Rice pest management systems-past and future. In: Pest management in rice Grayson BT, Green MB, Copping LG (eds), pp 1–14
Nurzaman M, Islam MN, Ahmed S (2000) Practice of integrated pest management by FFS and non-FFS farmers. Bangladesh J Train and Develop 3(1–2):219–227
Okonya JS, Kroschel J (2015) A cross-sectional study of pesticide use and knowledge of smallholder potato farmers in Uganda IoMed Res Int, 759049, 9. https://doi.org/10.1155/2015/759049
Olowu TA, Oyedokun OA (2000) Farmer accessibility of agricultural marketing information. Bayo Publishers, Ado-Ekiti
Purushothaman C (2003) Role of mass media in agriculture. Global Communication Research Association. Centre for International Communication, 1–3
Segura HR, Barrera JF, Morales H, Nazar A (2004) Farmers' perceptions, knowledge, and management of coffee pests and diseases and their natural enemies in Chiapas, Mexico. J Econom Entomol 97(5):1491–1499
Shekara PC (2016) A holistic perspective of scientific agriculture: a joint initiative to impart farmers with technical knowledge on basic agriculture. Nat Inst Agricult Extension Management (MANAGE), Hyderabad, p 1–154
Shuwa MI, Shettima L, Makinta BG, Kyari A (2014) Impact of mass media on farmers agricultural production, case study of Borno State, Agricultural Development Programme. Acad J of Scien Res 3(1):008–014
Smith RM, Haverkamp KK (1977) Toward a theory of learning how to learn. J Agril Edu 28:3–21
Soniia D, Christopher A (2011) farmer knowledge as an early indicator of ipm adoption: a case study from cocoa farmer field schools in Ghana. J Sustain Dev in Africa 13(4):123
Taghdisi MH, Besheli BA, Dehdari T, Cherati JZ, Khalili F, Toorani AH (2018) Factors influencing the intention of behavior the use of pesticides among farmers in Mazandaran Province: an application of the theory of planned behavior. Revista Publicando 5 (3):488–499
Trutmaun P, Voss J, Falrhead J (1993) Management of common bean diseases by farmers in the central African highlands. Int J of Pest Manage 39:334–342
Zijp W (2002) Changing the way World Bank thinks about and supports agricultural extension. World Bank, Washington, DC

Check for updates

21 Faulty Waste Water Usage Versus Agricultural Sustainability: An Assessment of East Kolkata Wetlands

Divyadyuti Banerjee, Sweta Sinha, and Kathakali Bandopadhyay

Abstract

In recent times, excessive population growth and increasing demand for habitation continue to drive rapid urbanization, illegal construction, and modification of land cover in the surrounding areas of East Kolkata Wetlands. These expansions lead to the disruption of basic physiochemical and hydrological functioning, excessive silting, pollution and heavy metal concentration, overexploitation of fragile ecosystems, loss of biodiversity coupled with the change of soil properties and drop in both quality and quantity of sewage indirectly affecting both agriculture and pisciculture. This study aimed to identify the linkage of urban expansion and wastewater usage as non-sustainable agricultural practices in East Kolkata Wetlands. The data from different fields of geospatial and earth observatory technology (satellites and Google View), laboratory testing (from field samples) along with empirical observations have been used. Here, the boundary of EKW for 1968, 1977, 1990, and 1998 has been drawn based on the same visualizing characteristics of the EKW and considering seven parameters: tone, texture, size, shape, pattern, shadow, and site situation. This area was a marshy land around 30–40 years ago, which gradually transformed into agricultural areas and later resulted in an urban scape due to illegal construction. The food and feeding habits of the fishes are one of the detrimental factors regarding bioaccumulation of heavy metals. A decrease in organic carbon concentration indicates an interruption in the biogeochemical cycle. Leaf area index value has been analyzed to identify the light energy interception. The gross and net primary productivity values are examined for the survival level of crops. The results obtained show unsustainable wastewater usage in agricultural practice in the East Kolkata Wetlands area, thus adversely influencing human food grain. This study further suggests more influences of empirical research on the ground of scientific correlation between the water quality parameters and adverse human health impact.

Keywords

Agriculture · Pisciculture · Wetlands · Vegetation · Productivity · Sustainability

D. Banerjee (✉)
Jadavpur University, Makardaha, Howrah, West Bengal, India
e-mail: divyadyuti30@gmail.com

S. Sinha
North 24 Paraganas, Mothijheel, West Bengal, India

K. Bandopadhyay
Vidyasagar University, Kadamtala, Howrah, West Bengal, India

J. Das and S. Halder (eds.), *Advancement of GI-Science and Sustainable Agriculture*, GIScience and Geo-environmental Modelling, https://doi.org/10.1007/978-3-031-36825-7_21

21.1 Introduction

A wetland is often considered an unwanted swampy area, but sometimes as history states, it has been of great use in maintaining the homeostasis of an ecosystem. East Kolkata Wetlands, a distinct wetland of West Bengal, has varied uses for Kolkata and its adjoining areas. This wetland was the epitome of pisciculture and often acted as an agricultural hub for the adjacent areas. However, the faulty water usage methods have led to havoc, creating less opportunity for sustainable agriculture. This paper tries to investigate this core problem with the help of an analytical study. Various eminent hydrologists and environmentalists have discussed this issue in their working papers. Various authors have evaluated this problem, extensively studied it, and tried to provide an empirical conclusion of the abovementioned situation.

Roy et al. (2016a), in their study, entitled 'Water Quality Index (WQI) of East Kolkata Wetlands using dissolved oxygen as proxy' has tried to investigate the reason for declining WQI by considering various factors and also has tried to provide a concrete solution regarding the issue. In contrast, Roy et al. (2016b) have tried to identify the causes of spatiotemporal variations of surface water temperature, pH, dissolved oxygen, nitrate, phosphate, and silicate on chlorophyll concentration in three water bodies meant for fish culture (locally known as Bheries) in East Kolkata Wetlands. Chattopadhyay et al. (2002) have extensively studied the toxic concentration of metal in the East Calcutta Wetlands. Haque (2020) also has tried to evaluate the spatiotemporal changes in the wetland's water. Similarly, Ghosh and Das (2019) stated about the recent anthropogenic interference, which acted as an obstacle and a threat to wetlands conservation.

Thus, a wetland denotes a distinct ecosystem, flooded by water at regular intervals (permanently or temporarily), where oxygen-free processes are common. East Kolkata Wetlands (henceforth EKWs) is no exception. The wetlands acted as a sewage treatment area in Kolkata and the nutrients harnessed from there helped to withstand agriculture. The preliminary hydrological functioning of the canals and fishing ponds has been severely affected by a huge mass of siltation, along with lower levels of sewage quality that the wetlands used to receive, thus wholly affecting the production of crops. The circumstances described above pose a serious threat to the water treatment systems. So it is a prerequisite to study the changes over time to ameliorate the ill effects until it gets too late. The studies regarding single-parametric analysis of agricultural sustainability in East Kolkata Wetlands are many years old. Still, it is necessary to carry out researches on multi-parametric analysis based on non-sustainable agricultural practices. Hence, this particular problem is the prime target in the selected study area. Thus, this research has been of great interest as there is very less evidence of such researches in the past. This study is a small step toward the above lacunae. Moreover, this research's prime objective is to identify the extremely vulnerable zones regarding agricultural sustainability.

In this chapter, the methodology is planned with the identification of certain selected parameters evaluating their purpose and the techniques used in such evaluation. An empirical study has been conducted in East Calcutta Wetlands to identify the various changes like chlorophyll concentration, the concentration of carotenoids, leaf area index, inability to perform photosynthesis, low energy, and gas exchange rate. Moreover, the above analysis concludes along with the identification of ecologically vulnerable zones.

21.1.1 Relevance

East Kolkata Wetlands is the most important wetland and has received an extensive response as various environmentalists, geologists, hydrologists, and geographers have studied in detail this particular wetland. However, the constant implementation of faulty techniques of wastewater conservation has created an inverse effect on the various types of agricultural practices and their allied activities in the EKW. The paper tries to evaluate the exact reason behind such an anomaly which has resulted in gradual disruption and has prevented from using sustainable agricultural techniques. This paper is significant as no such initiative has been taken so far to examine the details and the existing relationship between the usage of wastewater and agricultural practices that are being carried out in East Calcutta Wetlands.

21.2 Data and Methodology

21.2.1 Data Source

As per the prime objective of this chapter, the data from different fields of geospatial and earth observatory technology (satellites and Google View), along with the empirical observations, have been used to understand the extremely vulnerable zones in terms of agricultural sustainability. Datasheet details have been furnished below (Table 21.1). Here, the boundary of EKW for 1968, 1977, 1990, and 1998 has been drawn based on the same visualizing characteristics of the EKW and considering seven parameters: tone, texture, size, shape, pattern, shadow, and site situation.

21.2.2 Methodology

A literature review from different articles has been prepared along with the collection of satellite images from different sources; the sample is prepared in the laboratory for digestion. Tests have been prepared using different instruments, and a pilot survey has been conducted to have a clear glimpse of the area. Samples have been collected from Dhapa, Topsia DPS, Chowbaga DPS, Tannery site, Palmer Bazar DPS, and Kulti lock gate) for ground truth verification, and locations have been detected by the trilateration method with GPS. Testing of samples using different methods (Walkley–black acid digestion method, spectrophotometer method, ascorbic acid method, electrical conductivity method, gravimetric analysis, iodometric method; chlorophyll extraction method and fluorometric technique), calibration of ground truth with satellite for data analysis have also been carried out. Findings and solutions have been provided in the following Table 21.2.

21.2.3 Study Area

East Kolkata Wetlands (22°25′N to 22°40′N; 88° 20′E to 88°35′E) is located in the districts of North and South 24 Parganas of the Indian state of West Bengal, occupying a total surface area of 12,500 hectares. It consists of salt marshes, agricultural fields, sewage farms, and settling ponds and also serves as a natural sewage treatment plant for Kolkata. The name East Kolkata Wetlands (EKW) was propounded by Dhrubajyoti Ghosh (Special Advisory, Commission on Ecosystem Management, IUCN). The EKW is a kind of human-induced ecosystem involved in resource recovery practices. It has transformed itself from 'nona-bheris' (salty swamps) to a sound sewage-fed fish farm in 250 years, leading to a typical livelihood pattern as per a set of interdependent vocations in the area. EKW is a peri-urban wetland supplying a multifaceted perspective in terms of both fish criteria and waterfowl habitats. It also hosts a large variety of flora and fauna. The aquatic vegetation mainly comprises of floating microphytes. There are also about 100 plant species in EKW. It also serves as a home to various indigenous animal species and snakes. Several fish species are farmed in sewage-fed ponds called 'bheris' (Fig. 21.1).

21.3 Results and Discussion

21.3.1 Lack of Chlorophyll Concentration

Poor drainage, damaged roots, high alkalinity [deficiency of iron or manganese, both of which are present but unavailable in high pH soils ($pH > 7.2$)], and nutrient deficiencies in the plant have caused a huge amount of chlorophyll destruction (Baker 2008) in the northeastern and northwestern part of the wetlands region from 1977 to 2020.

Table 21.1 Catalogue of different types of data sources

Data type	Sub-type		Nature	Source	Purpose
Earth observatory images (secondary data)	Sensor	MSS	$r = 148$, $p = 45$ December 1975	https://earthexplorer.usgs.gov/	Image classification for land cover change detection, NDBI, NDMI, LST, BSI, NDVI, NPCRI, habitat fragmentation, and disturbance mapping
		TM	$r = 138$, $p = 44$ November 1990		
		OLI-1	$r = 128$, $p = 45$ February 2020		
		MODIS	MOD17A2H V-6	NASA LPDAAC collection	Determining average evapotranspiration value
		MODIS	MOD16A2 V-6		Determining GPP value
		ALOS PALSAR	$r = 138$, $p = 45$ 2009 and 2020	https://earthdata.nas a.gov/	For preparing a 3d urban model and to detect topographic wetness
Maps (secondary data)	Toposheets	79B/6 on 1:50,000 (surveyed in 1958-59) toposheet on inch map (surveyed on 1922–1924)		SOI-1973 SOI-1968	The base map for assessing the previous land resources, infrastructures, and settlements
	Other thematic maps	Gross biomass, surplus biomass, and bioenergy potential map		ISRO Bhuvan portal	To visualize gross biomass, surplus biomass, and bioenergy potential
		Dry matter productivity map, FAIR, COVER, leaf area index, soil water index, surface soil moisture map, VCI, VPI		Copernicus Global Land Service	For determining dry matter productivity rate, spatial variation of photosynthesis rate, fraction of green cover, leaf area, presence of surface water, surface soil moisture content, vegetation condition and vegetation productivity

In 1977, through an archival study, comparatively dense vegetation cover was found in almost all the wetlands regions like Gharal, Tardaha Kapasati, Samukpota, Tardaha, Pratapnagar, Tihuria, Nayabad, Kantipota, Ranabhutia, Atghara, northern part of Dhapa Manpur, northern part of Panchuria, Hadia, Jagatipota, and Bhagabanpur, while low-dense vegetation cover was found in Boinchtala, Dhalenda, Paschim Chaubaga, Nonadanga, the northern part of Dhapa, the eastern part of Beonta, the northeastern part of Tardaha Kapasati, and very low vegetation cover was found in the middle part of Dhapa.

In 1990, very high dense vegetation cover was observed in the eastern part of Boinchtala, Dhapa, the middle eastern part of Dhapa Manpur, the western part of Hadia, Beonta, Karimpur, the eastern part of Bhagabanpur, the western part of Deara, a small part of Hatgacha and middle part of Tardaha Kapasati. On the contrary, high dense vegetation cover was found in Ranabhutia, Nayabad, Gharal, Samukpota, Bhagabanpur,

Table 21.2 Detail description of methodology

Parameters			Purpose	Methods
Primary productivity	GPP	Net	To find the food-web	(Data collected from NASA LPDAAC Collection)
		Nano		
		Total		
	NPP	Net		
		Nano		
		Total		
	TP	Net		
		Nano		
		Total		
Normalized difference vegetation index			To detect the Chlorophyll concentration	(B4-B3)/(B4 + B3) (LANDSAT-5) (B5-B4)/(B5 + B4) (LANDSAT-8)
Normalized pigment chlorophyll ratio index			To identify the status of plant stability	(B3-B1)/(B3 + B1) (LANDSAT-5) (B4-B2)/(B4 + B2) (LANDSAT-8)
Leaf area index			To measure the amount of light energy interception	Copernicus Global Land Service
Fraction of absorbed photosynthetically active radiation index			To measure the amount of photosynthesis and to detect the presence of dead leaves	Copernicus Global Land Service
Amount of top of canopy reflectance index			To detect the state of ecophysiological condition	Copernicus Global Land Service
Vegetation productivity index			To measure the respiration rate and growth rate and detect vegetation state	Copernicus Global Land Service
Amount of dry matter productivity			To measure the stability of the base of trophic structure, energy flow, and bioenergy potentiality	Copernicus Global Land Service
Amount of average evapotranspiration			To measure the rate of transpiration	NASA LPDAAC collections
Amount of gross biomass			To measure the stability of the base of the trophic structure, the energy flow	ISRO Bhuvan portal
Amount of surplus biomass			To measure the stability of the base of the trophic structure, the energy flow	ISRO Bhuvan portal
Amount of bioenergy potential			To measure the bioenergy potentiality	ISRO Bhuvan portal

Jagatipota, Atghara, Karimpur, the northeastern part of Tardaha Kapasati, Tardaha, Tihuria, and Beonta. Bongtala, Dhalenda, Paschim Chaubaga, Nonadanga, North Dhapa Manpur, North Hatgacha, Eastern Beonta, North Panchuria, South Tardaha Kapasati, and Hatgacha had a low vegetation cover. Whereas, very low vegetation cover was found in the middle part of Dhapa Manpur, the western and eastern parts of Hadia, the southern part of Dhapa, the western part of Khodhati, the middle and northwestern part of Tardaha Kapasati, the southwestern part of Tihuria, Deara, and the southeastern and northeastern part of Bhagabanpur. In 2020, very dense vegetation cover is absent in any part of the wetlands. In the meantime, high vegetation cover

Fig. 21.1 View of surveyed wetland, **a** inlet of wetland, **b** outlet of wetland

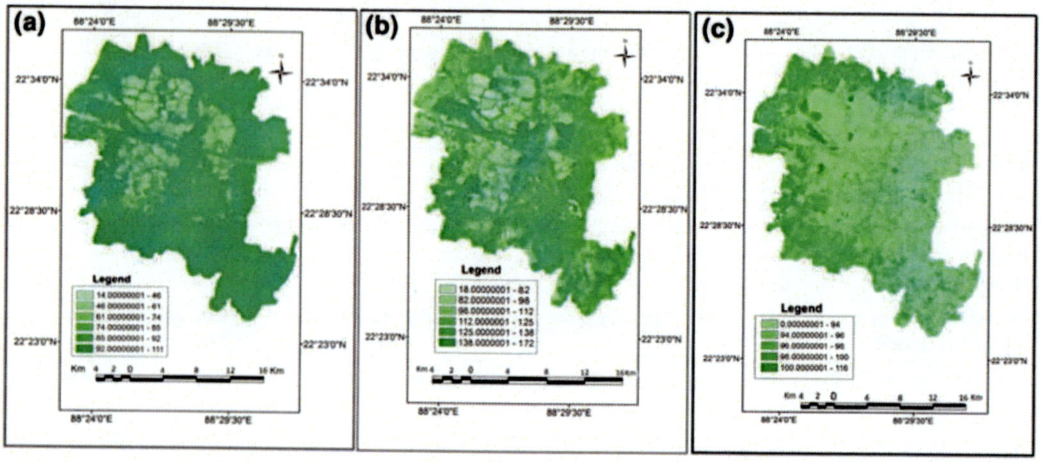

Fig. 21.2 Normalized difference vegetation index **a** 1977, **b** 1990, and **c** 2020

is limited to some blocks like West Dhapa Manpur, northeast Dhapa, southwest Dhapa, Paschim Chaubaga, Nonadanga, west Chaubaga, and west Dhalenda. Medium and dense vegetation cover prevails in Boinchtala, north Dhapa Manpur, north Hatgacha, north Nanchuria, north Kulberia, Karimpur, Bhagabanpur, Jagatipota, Algarah, Ranabhutia, Kantipota, Nayabad, Tihuria, middle and western Tardaha, a small part of Samukpota, and a small patch in south Tardaha Kapasati, and low vegetation cover in the middle and southern part of Dhapa Manpur, Hadia, Beonta, south Pachuria, and the southwestern Kulberia, middle and southeastern part of Dhapa, Tardaha Kapasati, the eastern part of Tardaha, Deara, Kheadaha, Kumar Pukuria, Goalpara, and south Chaubagha (Fig. 21.2).

21.3.2 Lack of Amount of Carotenoids

On the western side, metal pollution affects the transfer of carotenoids (Sillanpää, et al., 2008) across the trophic levels. Although the plant's growth rate is high there due to the presence of sufficient amount of nitrogen, the stability is low due to the presence of a few carotenoids. There is

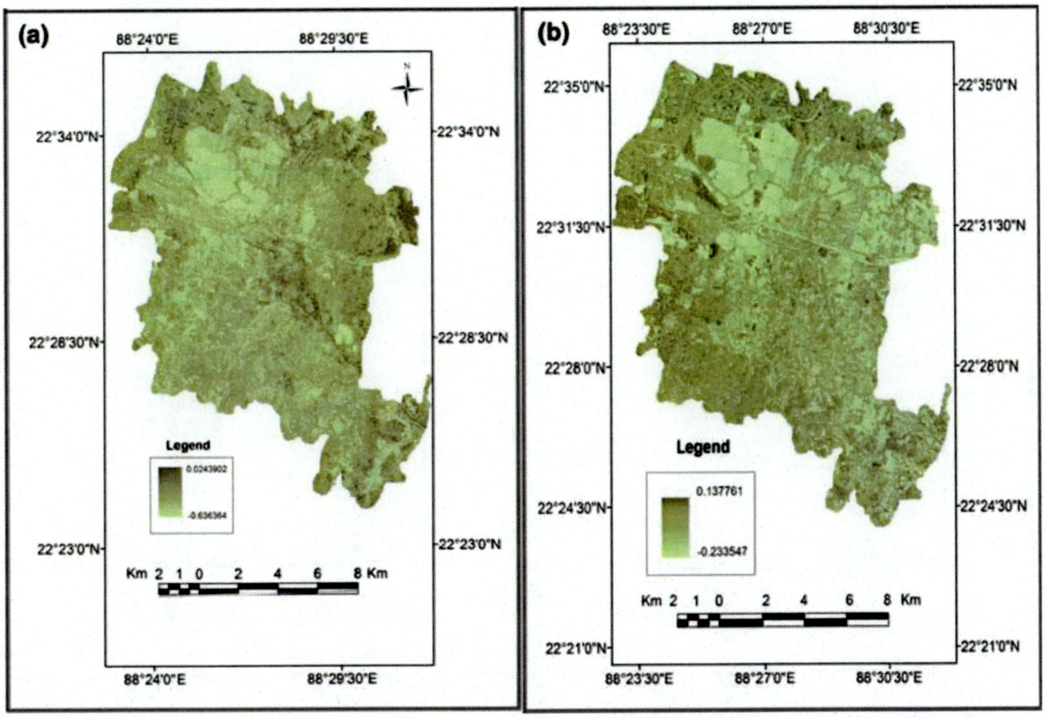

Fig. 21.3 Normalized Pigment Chlorophyll Ratio index **a** 1990, and **b** 2020

a low survival response of plants due to non-photo-protection and low energy absorption, which is used in photosynthesis. From Normalized Pigment Chlorophyll Ratio Index, it is evident that stable ecosystem conditions have shifted from the north and eastern sites to the southern and northern sites in the wetlands region from 1990 to 2020. In 1990, a very high value was found in East Beonta (0.0243902), and a high value was seen in North Dhapa, Chaubaga, Nonadanga, east Khodhati, a small patch near North Goalpara, northwestern part of Tardaha Kapasati and a very small patch near South Ghara, South Samukpota, and southeast Tardaha Kapasati, indicating more stable ecosystem because the large survival response of plants due to photo-protection via non-photochemical quenching and high amount of energy absorption which is used in photosynthesis. Despite the low presence of nitrogen, large amounts of carotenoids are responsible for a long stability (Swapnil, 2021). It has been found that very low value was found in the middle and southern part of Dhapa Manpur, Dhapa, Hatgacha, south Bonchtala, south Panchuria, west Beonta, Hadia, Kharki, Kaarimpur, Kalikapur, Deara, Bhagabanpur, Jagatipata, Atghara, Ranabhutia, Kantpota, Nayabad, west Khodhati, Kheadaha, Tihuria, Tardaha, and the middle part of Tardaha Kapasati. From the Normalized Pigment Chlorophyll Ratio Index, it is clear that a stable ecosystem has shifted from the north and eastern sites to the western, southern, and northern sites in the wetlands region from 1990 to 2020 (Fig. 21.3).

21.3.3 Increase of the Shortage of Leaf Area

Climatic factors (mainly temperature), light intensity, nutrient availability, soil moisture, N supply, and the air humidity regime can influence leaf size and shape. Due to the lack of those ideal conditions, a very low leaf area index value

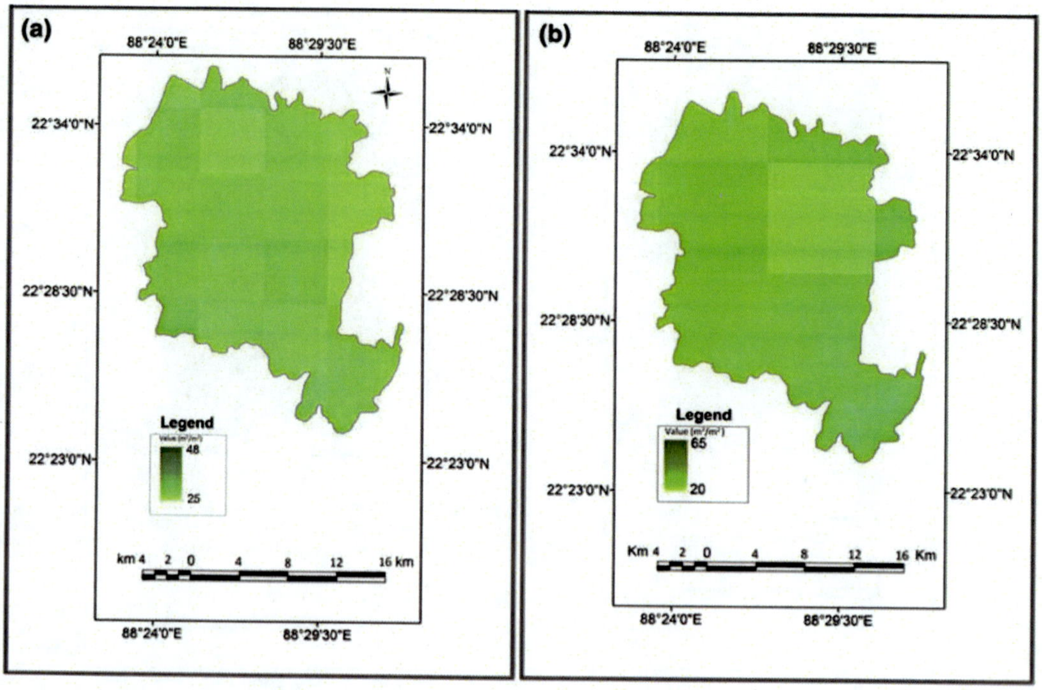

Fig. 21.4 Leaf area index **a** 1999, and **b** 2020

(almost $20m^2/m^2$) is found in Hadia, east Beonta, north Khodhati, south Panchuria, south Hatgacha, southeast Dhapa, Manpur which indicates a small amount of light energy interception. As a result, the primary production is also very negligible (due to the small leaf area per unit of ground area) (Srinivasan et al. 2017). Primary production and light energy absorption have changed in the wetlands region's northern, southern, and western sites from 1990 to 2020. Gross Primary Productivity is very high (33,000 g C m^{-2} $year^{-1}$) in the middle and northwestern part of Dhapa Manpur, north Dhapa, Bonchtala, Dhalenda, Paschim Chaubagha, Nonadanga, northwest Chaubagha, south Chaubagha, west Kalikapur, northwest Beonta, and south Dhapa. In comparison, maximum Gross Primary Productivity regions are located in the northwestern and middle part of the wetlands, indicating a high metabolism rate, cellular respiration, and high rate of tissue building. Very low Gross Primary Productivity (10 g cm-$year^{-2}$ $year^{-1}$) is found in Pratapnagar, Gharal, Samukpota, Tardaha, Tihuria, Nayabad, Kantipota, Beonta, Khodhati, Goalpara, Kumar Pukuria, Kheadaha, Deara, Bhagabanpur, Kharki which points out low metabolism rate, low cellular respiration and low rate of tissue building (Fig. 21.4).

21.3.4 Lack of Photosynthesis and Increase in the Number of Dead Leaves

Leaf spectral properties are the main analyzing factor for light composition within the plant communities. The photosynthetic pigments absorb sunlight in the upper part of the plant (top spectrum). The leaves often function as a filter for blue and red wavelengths. The resultant transmitted wavelengths mainly consist of green and far red (bottom spectrum). The greener tissue in the plant sometimes reflects far red, thus reducing the R/FR ratio. In a forested area, a high reflection of far red light is noticed, especially from the

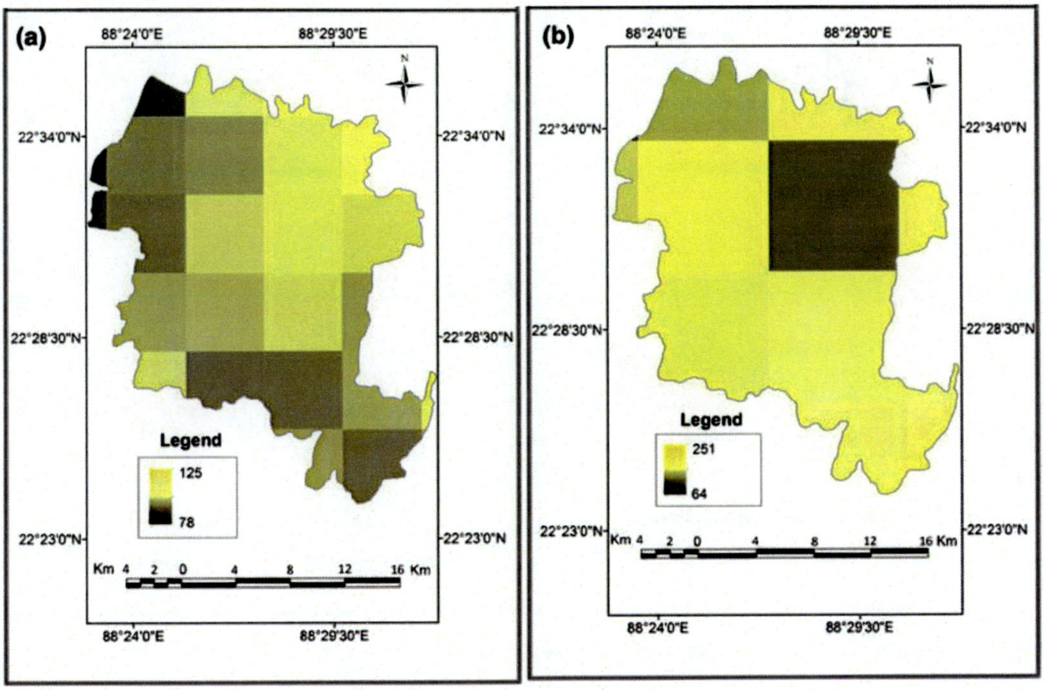

Fig. 21.5 Fraction of absorbed photosynthetic active radiation **a** 1990, and **b** 2020

neighboring plants, to indicate a strong competitive environment under an evergreen canopy; light is filtered by the presence of leaves at a higher level and lower parts that are the understoey forests receives a low light intensity. Under a canopy, light is strongly filtered by high tree leaves, and the understory receives a much lower light intensity, characterized by low UV-B, low photosynthetic active radiation, and low R/FR. FAPAR value becomes very low (1990–2020) in the northeastern and northwestern parts, which indicates not only a low amount of photosynthesis but also the presence of dead leaves (Fig. 21.5).

21.3.5 Very Low Energy and Gas Exchange Rate

Due to water-logging (rainy season), high temperatures, and inadequate mineral content of the soil, the presence of phytotoxic compounds with high reflectance is found in the northwestern, southeastern, northern, and western parts, which indicates less presence of pigments (including chlorophyll a & b) and canopy water content. The above said phenomenon shows poor ecophysiological conditions (where CO_2 consumption, O_2 release, and the energy exchange rate is improper). Low reflectance is found in East Kharki, Deara, southeast Bhagabanpur, Goalpota, west Kheadaha, north Nayabad, and north Tihuria, which indicates more presence of pigments (including chlorophyll a & b) and canopy water content (Fig. 21.6). The above-stated situation is a sign of better ecophysiological condition (where CO_2 consumption, O_2 release, and the energy exchange rate is proper).

21.3.6 Decrease in the Amount of Vegetation Productivity

Despite the high value of Gross Primary Productivity, low Net Primary Productivity value is found in northwestern and western parts due to

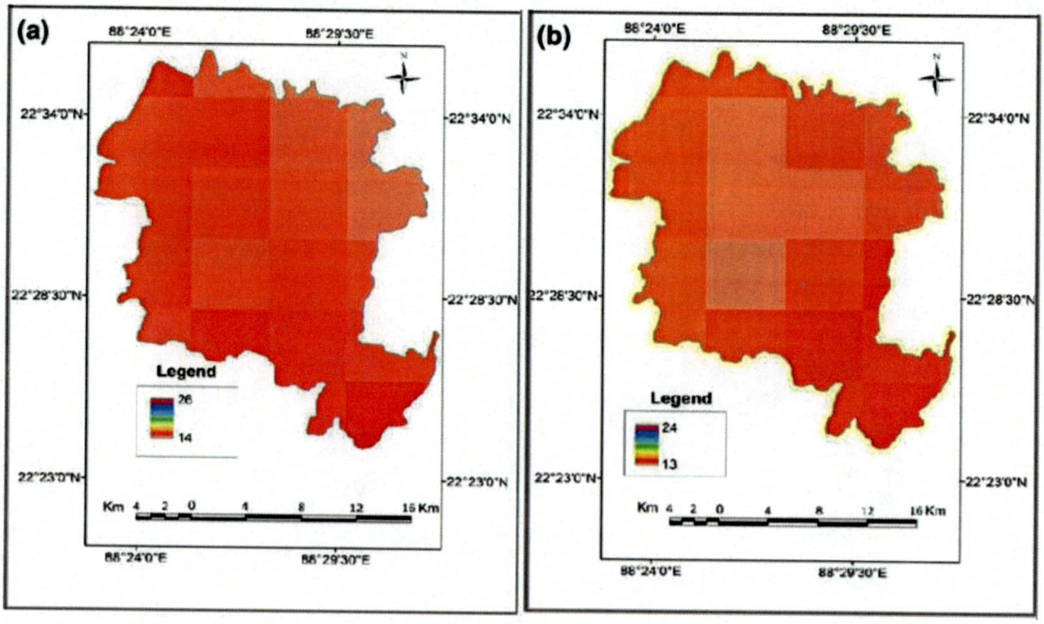

Fig. 21.6 Amount of top of canopy reflectance **a** 1999 and **b** 2005

huge anaerobic respiration, which is essential for plants' cellular activities, growth and maintenance of all plant tissues, carbon balance of individual cells, and to get the energy to stay alive and creating the ability to survive by fighting various diseases (Ward et al. 2019) It indicates that the worst vegetation state has a low growth rate. The average evapotranspiration rate is also very high (32,766 mm) there. High Net Primary Productivity in other sites indicates a more or less good vegetation state (Figs. 21.7 and 21.8).

21.3.7 Spatial Variation of Amount of Gross, Surplus Biomass, and Bioenergy Potential

Low Gross Biomass (<0.15 Kilo tons Km^{-2}) in the middle part of the wetlands reflect poor ecological process (less control of plants on nutrients, water, and solar resources), and the base of the trophic structure is very weak (Table 21.3). Low surplus biomass (<0.02 kilo tons Km^{-2}) in the southeastern site indicates interrupted energy flow through different trophic levels (Rapport et al. 1985) in an ecosystem. Low bioenergy potentiality (0–0.4 Giga Joule Km^{-2}) is found in the middle part of the wetlands, but the northwestern part has a huge bioenergy potentiality (Figs. 21.9, 21.10, 21.11, 21.12, and 21.13).

21.3.8 Identification of Ecologically Vulnerable Zone

This study indicates that the various 'mouza' (demarcated areas under a community development block) in the vicinity of Kolkata megacity and Salt Lake City are at high risk and under a higher vulnerable zone, and the southern part that is including the Sonarpur municipality area is also at a high-risk zone. On the other hand, as the distance increases from the Kolkata municipality

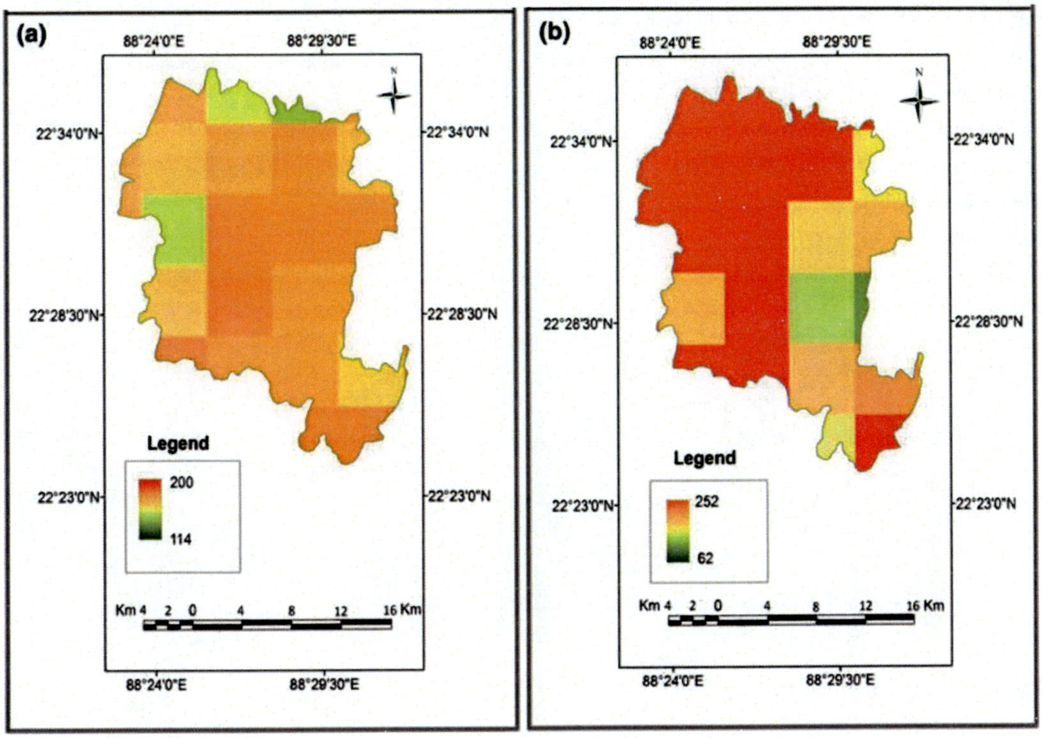

Fig. 21.7 Vegetation productivity index a 2013, and b 2020

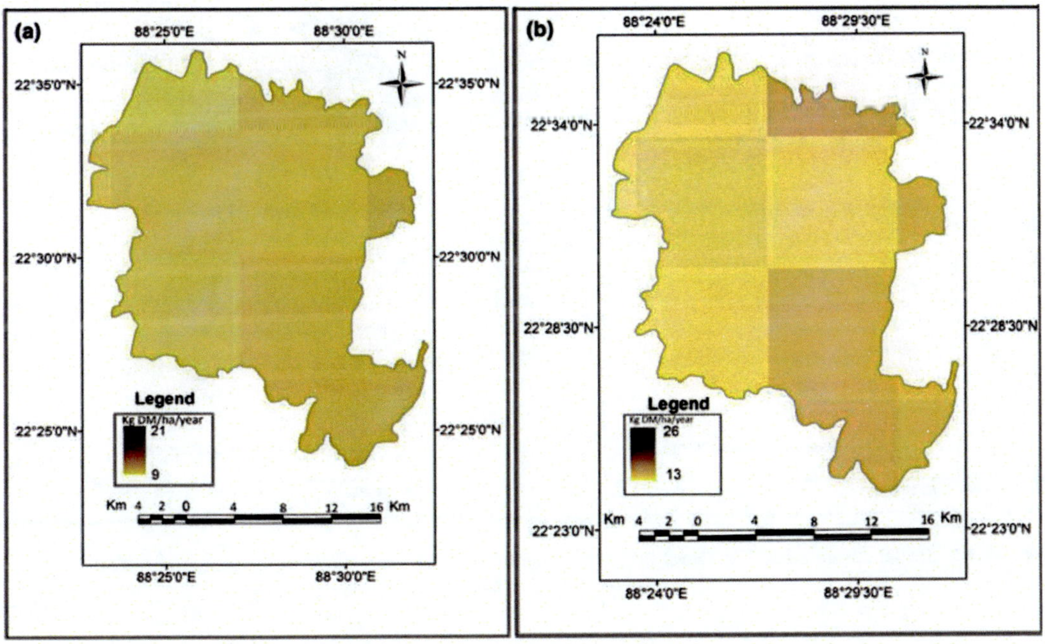

Fig. 21.8 Amount of dry matter productivity a 2014 and b 2020

Table 21.3 Biomass data of different crops, 2020

Biomass	Kharif rice	Rabi rice	Wheat	Cotton	Sugarcane	Total
Gross biomass	0.37	11.54	0	0	0	–
Surplus biomass	0.02	0.57	0	0	0	–
Bioenergy potential	0.28	8.86	0	0	0	9.14

Source ISRO Bhuvan portal

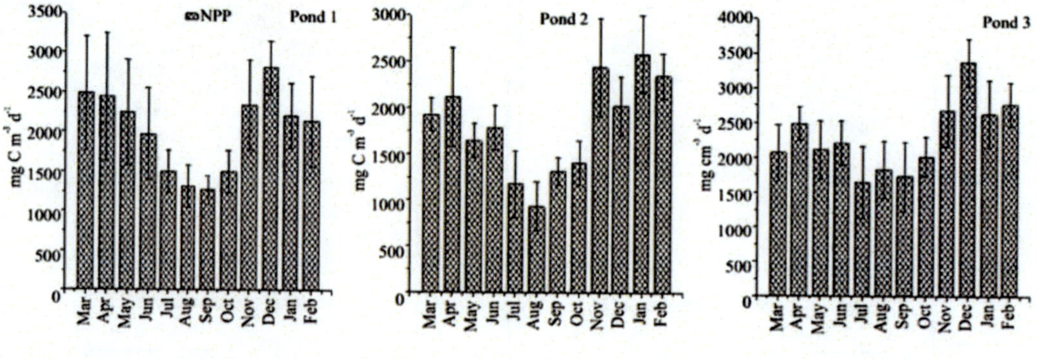

Fig. 21.9 Net primary productivity

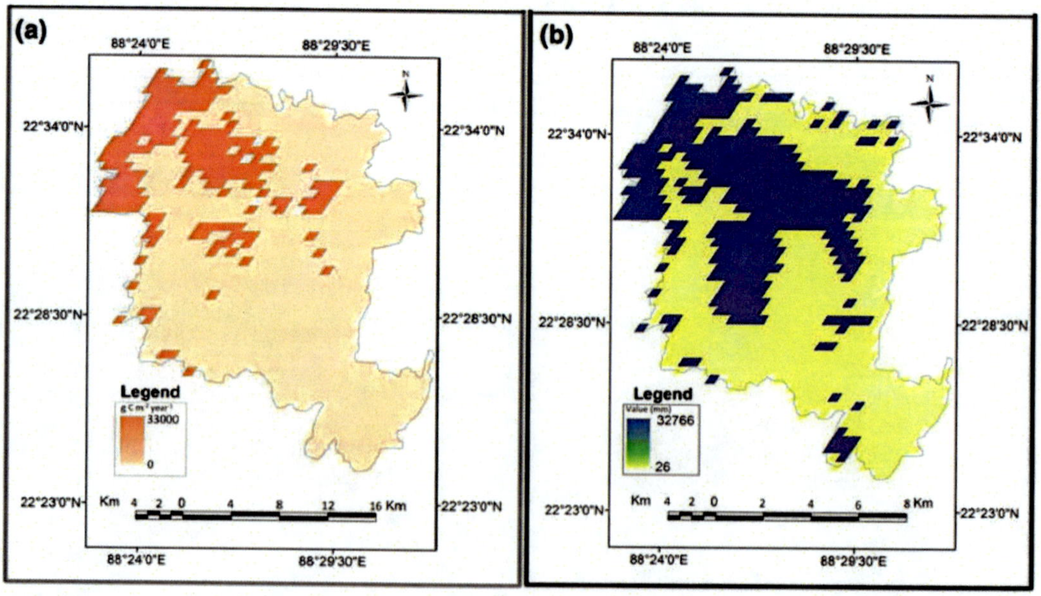

Fig. 21.10 Spatial scenario EKW **a** Gross Primary Productivity (2020) and **b** average evaporation (2020)

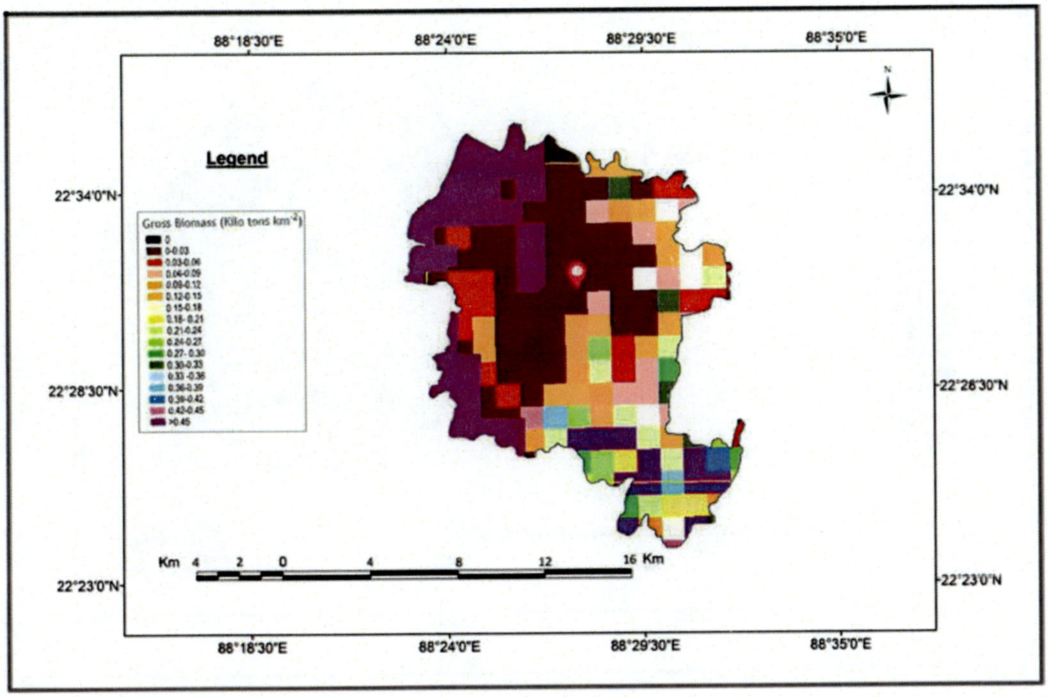

Fig. 21.11 Spatial distribution of gross biomass in EKW (2020)

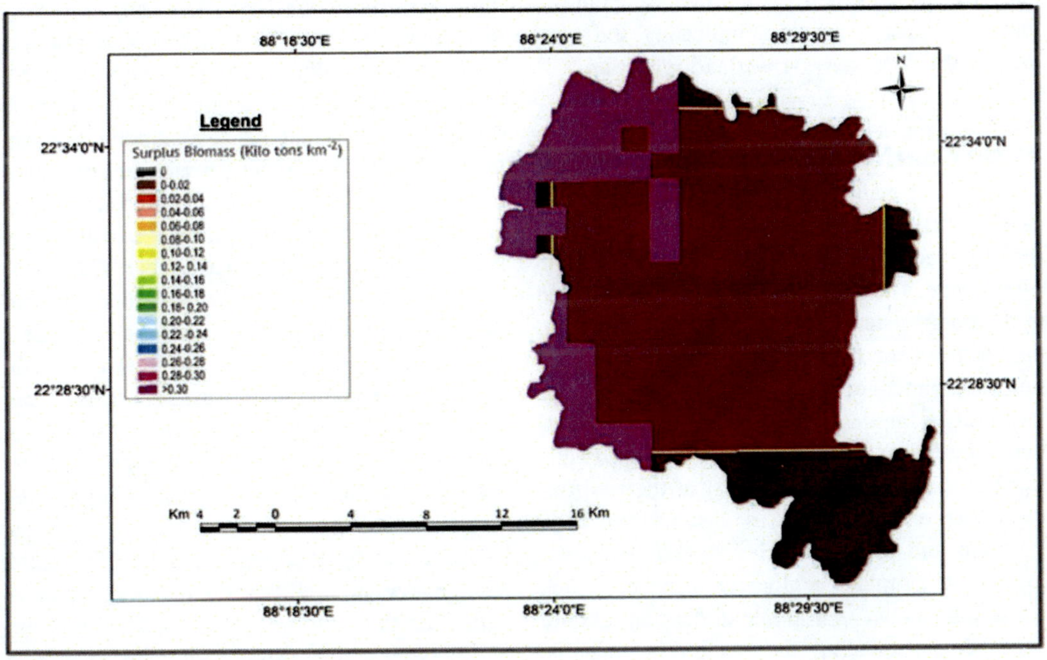

Fig. 21.12 Spatial distribution of surplus biomass in EKW (2020)

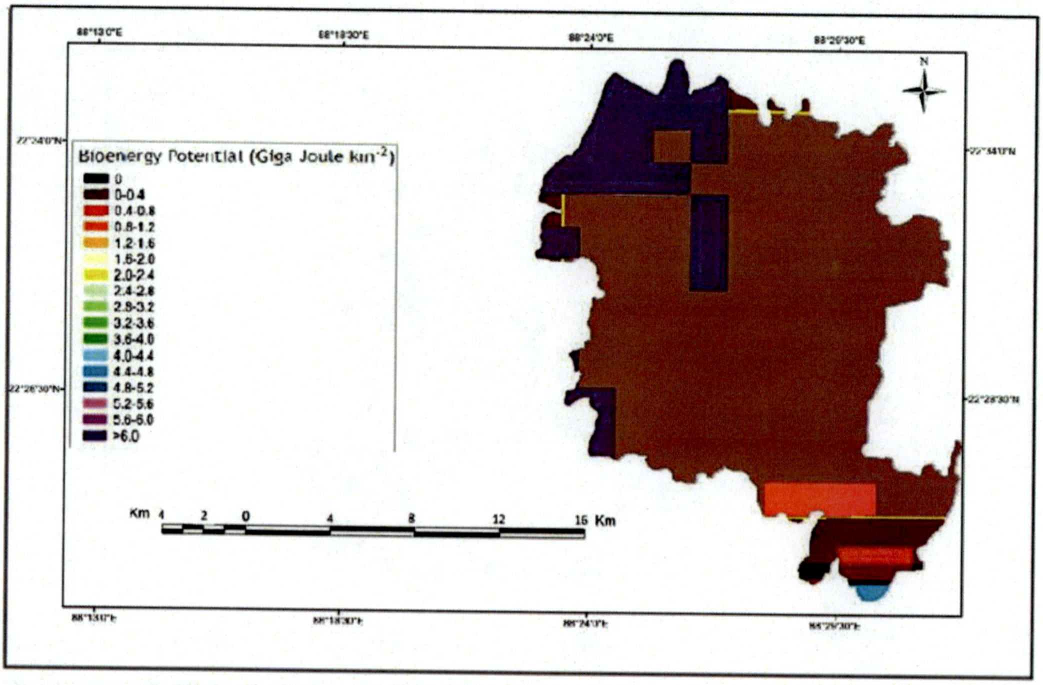

Fig. 21.13 Spatial distribution of bioenergy potential in EKW (2020)

area and other urban areas, the vulnerability magnitude also decreases, indicating the high impact of urban areas on wetland reduction.

21.4 Conclusion

The western side of East Kolkata Wetlands (EKW) is ecologically more sustainable than the eastern side. The ecosystem was most sustainable in the northwestern, and the most fragile was in the southeastern side. Recently, rapid urbanization has started on the western side, which is responsible for destroying this healthy ecosystem. Therefore, it is imperative to control this rapid urbanization in the interest of the environment and retain Kolkata as a subsidized city. Relevant and proper policies should be implemented to maintain ecological balances which are as follows: an appropriate buffer zone should be made surrounding the small but valuable wetlands, with the help of guideline provided by the East Kolkata Wetlands Authority. Secondly, to maintain the health of these concerned wetlands of east Kolkata there is an earnest need to allow biological movements, and interconnections (through small cannels) among the wetlands. Thirdly, to restrict the adverse impacts of urbanization on the wetlands and its micro-ecology, specific management is necessary to ban littering, vegetation destruction, etc. Other than these, for making the aforementioned recommendations fruitful for the purpose of wetland conservation, for its future agricultural sustainability, stimulating public awareness is a pre-requisite. Nothing is possible without bottom-up approach. Sustainability and ecological conservation of the area can be restored through public awareness about the problems faced, and it can be mitigated by teaching the daily life attitude and behavioral modification of local residents, supplying them with necessary articles and providing them with assistance for enjoying a better living.

References

Baker NR (2008) Chlorophyll fluorescence: a probe of photosynthesis in vivo. Annu Rev Plant Biol 59:89–113. https://doi.org/10.1146/annurev.arplant.59.032607.092759

Bhattacharyya A, Sen S, Roy PK, Mazumdar A (2008) A critical study on status of East Kolkata Wetlands with special emphasis on water birds as bio-indicator. In: Proceedings of Taal 2007: The 12th World Lake Conference, Vol 1561, p 1570

Bhattacharya S, Ganguli A, Bose S, Mukhopadhyay A (2012) Biodiversity, traditional practices and sustainability issues of East Kolkata Wetlands: a significance Ramsar site of West Bengal, (India). Res Rev Biosci 6 (11):340–347

Bose PC (1944) Calcutta sewage and fish culture. Proc Natl Inst Sci India 10:443–450

Chattopadhyay B, Chatterjee A, Mukhopadhyay SK (2002) Bioaccumulation of metals in the East Calcutta wetland ecosystem. Aquat Ecosyst Health Manage 5(2):191–203

CPCB (1997) Standards for liquid effluents, gaseous emissions, automobile exhaust, noise and ambient air quality.Central Pollution Control Board, Parivesh Bhawan, Delhi

Ghosh S, Das A (2019) Urban expansion induced vulnerability assessment of East Kolkata Wetland using Fuzzy MCDM method. Remote Sensing Appl: Soc Environ 13:191–203

Haque M (2020). Assessing infrastructural encroachment and fragmentation in the East Kolkata Wetlands. In: Water management in South Asia. Springer, Cham, pp 233–257

Hussan A (2016) Threats to fish diversity of East Kolkata wetlands and conservation needs. Aquaculture times 2(6):10–15

Rapport DJ, Regeir HA, Hutchinson TC (1985) Ecosystem behavior under Stress. Am Nat 125(5):617–640

Roy A, Amin G, Agarwal SK, Pramanick P, Banerjee A, Roy M, Mitra A (2016a) Water Quality Index (WQI) of East Kolkata Wetland using dissolved oxygen as proxy. Int Adv Res J Sci, Eng Technol 3(2): 22–24

Roy A, Roy R, Sarkar D, Roy M, Mitra A, Lake Campus S (2016b) Hydrological parameters of east Kolkata wetlands: time series analysis. Int J Life Sci Scient Res 2(6):692–699

Roy AS, Gorain PC, Paul I, Sengupta S, Mondal PK, Pal R (2018) Phytoplankton nutrient dynamics and flow cytometry based population study of a eutrophic wetland habitat in eastern India, a Ramsar site. RSC Adv 8(17):9530–9545

Srinivasan V, Kumar P, Long SP (2017) Decreasing, not increasing, leaf area will raise crop yields under global atmospheric change. Glob Change Biol 23(4):1626–1635

Swapnil P, Meena M, Singh SK, Dhuldhaj UP, Marwal A (2021) Vital roles of carotenoids in plants and humans to deteriorate stress with its structure, biosynthesis, metabolic engineering and functional aspects. Current Plant Biol 26:100203

Ward LM, Rasmussen B, Fischer WW (2019) Primary productivity was limited by electron donors prior to the advent of oxygenic photosynthesis. J Geophys Res: Biogeosciences 124:2. https://doi.org/10.1029/2018JG004679

Index

A

Abiotic stress, 155–158, 167, 172, 173
Accuracy assessment, 225, 229, 233, 234, 237
Active tectonics, 257, 260
Age group, 306, 308
Agricultural drought, 41, 42, 44, 47, 49, 55, 156, 286, 287
Agricultural suitability, 19
Agriculture, 3–9, 13–17, 19, 20, 25, 41, 43, 44, 55, 60, 61, 81, 86, 96, 100, 109, 113–115, 135–137, 149–152, 156, 157, 169, 187, 189, 199, 201, 212, 219, 223, 225, 244, 245, 251, 253, 254, 260, 261, 265, 266, 275, 281, 285–288, 292, 305–308, 323, 324
Agriculture 4.0, 149
Agriculture industry, 4, 13, 149, 150, 152
Agroclimatic zones, 5, 17, 105, 288
Agro-information, 306, 308, 310
Agronomic adaptation strategies, 60, 301
Agronomic practice, 116
Agrotourism, 8
Amount of carotenoids, 328
Amount of gross biomass, 327, 332
Amount of surplus biomass, 327, 332
Analytic Hierarchy Process Change, 25, 29, 86, 95
ArcGIS, 41, 43, 46, 65, 84, 103, 113, 116, 118, 119, 137, 139, 202–204, 229, 245
Aridity, 286
Article, 4–6, 14, 187, 249, 308–310, 312, 313, 315, 316, 325, 336
Ascorbate peroxidase (APX), 167, 169–171

B

Beneficial and non-beneficial criteria, 139, 204
Biate village, 136
Bioenergy potential, 327, 332, 334, 336
Biogenesis of Nanoparticles, 168
Biogenic nanomaterials, 155, 157
Bioreduction, 159, 161, 165
Biosorption, 159, 161
Biotechnology, 5

C

Cartographic techniques, 4
Catalase (CAT), 158, 167, 169–171
Contact Farmer (CF), 305, 309
Climate change, 14, 15, 20, 42, 60, 78, 82, 99–102, 107, 108, 114, 150, 152, 156, 167, 171, 218, 286, 287, 290
Combined compromise solution, 135, 138
Competitive farming, 149
Contingent valuation method, 188
Correlation, 25, 29, 64, 68, 118, 120–122, 130, 131, 234, 235, 248, 287, 297, 301, 323
Criteria importance through inter criteria, 25, 29
Crop suitability analysis, 101
Crop suitability assessment, 201, 208–210
Crop yield prediction, 113, 114

D

Data responses, 305
Dead Heart (DH), 316, 319
Dinhata sub-division, 7, 8, 19, 199, 201–205, 207–213
Disease-free appropriate environment, 189
Disruption of basic hydrological functioning, 323
Disruption of basic physio-chemical functioning, 323
Diversification of sericulture production, 197
Drought, 16, 18, 41–45, 47, 49, 51, 54–56, 59, 61, 78, 100, 104, 155–157, 168, 169, 171, 267, 285–290, 292, 297–299, 301

E

ECOCROP model, 99, 101, 104, 107
Eco-friendly, 8, 14, 155, 156, 158, 159, 166, 167, 171, 173
Ecological, 5, 6, 8, 17, 100, 108, 120, 217, 286, 332, 336
Ecologically vulnerable zone, 324, 332
Economic prospects of sericulture, 188
Economic Threshold Level (ETL), 308
Education, 186, 227, 244, 306, 309, 321
Efficient cropping zone, 99–101, 103, 106, 107, 109, 110

J. Das and S. Halder (eds.), *Advancement of GI-Science and Sustainable Agriculture*, GIScience and Geo-environmental Modelling, https://doi.org/10.1007/978-3-031-36825-7

Employment opportunities, 186, 195
Entropy Weighting Method, 199–202, 213
Environmentalist, 3, 324
Error matrix, confusion matrix, or kappa analysis, 233
Evading import dependency, 197
Evaporation, 26, 156, 226, 230, 237, 288, 334
Evapotranspiration, 25–27, 31, 32, 45, 156, 236, 285, 287–289, 291, 301, 327, 332
Excessive silting, 323

F
FAO Penman-Monteith, 25, 26, 28, 39
Farm Magazine Effectiveness Index (FMEI), 308
Farmers, 13, 15–17, 20, 43, 55, 61, 76, 78, 81, 83, 100, 101, 108, 109, 114, 135, 149–152, 185–193, 197, 199, 212, 223, 244, 265, 266, 269, 270, 272, 274, 275, 277–283, 305–321
Field level, 189, 309
Financial constraints, 195
Finger Millet, 99–101, 103–110
Flood prone area, 16, 17, 201, 267
Forest, 44, 83, 113, 114, 119, 199, 225–228, 230, 231, 233–235, 237, 239, 258, 286, 331
Functional chain mapping, 188
Fuzzy analytical hierarchy process, 19, 25, 81, 86, 95

G
Geographical Information System, 4, 13, 14, 19
Geomatics, 4
Geospatial data, 3–9, 17
Geo spatial modelling, 3, 100
Geospatial science, 5, 6, 9
GIScience, 3–8
Global smart agriculture market, 150
Global supply chains, 185
Global value chain, 186
Green chemicals, 158
Green synthesis, 155, 157, 158, 160–162, 164, 166, 171–173
Group, 64, 87, 137, 139, 159, 164–166, 188, 221, 227, 251, 266, 285, 305, 308–318

H
Heavy metal, 155, 156, 158, 159, 163, 169–171, 323
Heavy rainfall and high permeability of soil, 260
High yield variety, 266
Hilly areas, 227, 244, 245, 257, 261, 262
Hot spot analysis, 119, 123
Humanists, 3, 9

I
Impact of LULC change, 225
Information technology, 7, 149
Insect pest management, 306, 321
Integrate Farming, 149
IoT, 131, 149–152
IoT Sustainable, 152
Inter governmental Panel on Climate Change (IPCC), 60, 99, 100, 108
IPM Integrated Pest Management, 305, 306, 321
Iril watershed, 225, 228–230, 232–238

J
Jhum cycles, 227
Jhum land, 135, 137, 144, 145, 234

K
Kappa statistics, 144, 233, 237
Kendall's tau,, 59, 60, 64, 67, 71
Kharif season, 59, 68, 69, 73, 74, 76, 78, 130, 213
Knowledge, 4, 13, 55, 101, 139, 152, 157, 201, 243, 266, 279–282, 305, 306, 308, 316, 320, 321
Kriging interpolation, 118, 119

L
Land degradation, 82, 100
Landsat, 41, 44, 45, 49, 228, 229, 233, 237, 258, 327
Landslide impact on agriculture, 257, 258, 262, 288
Land suitability assessment, 19, 83, 200
Land surface temperature, 18, 44, 45, 49, 52, 53
Landuse and landcover, 46, 47
Land use land cover classification, 233
Leaf area index, 114, 117, 324, 327, 329, 330
Linear regression, 64, 67, 70, 73, 114, 115, 287
Linkage of urban expansion, 323
Loss of biodiversity, 323

M
Machine learning, 5, 19, 114, 131, 213
Magazine, 306, 308, 311–315
Maize, 18, 82, 83, 101, 114, 115, 167, 199, 201–203, 205, 207–213, 269, 270, 272, 275
Manipur, 225–229, 234, 237
Mann-Kendall test, 59, 61, 221
Meteorological stations, 61–63
Model building, model calibration and validation, 231
Model simulation, 225, 236, 238
Mulberry leaves, 185, 189–191, 197
Multi-Criteria Decision Analysis, 8, 25, 26, 29, 30, 33
Multi-Criteria Decision-Making, 17, 19, 39, 82, 86, 90, 114, 135, 136, 138, 145, 199–201, 204, 213
Mycosynthesis, 165

N
Nanoparticles, 156–158, 165, 168, 170
Nanotechnology, 155–157, 166, 173
Newspaper, 305–321
Ngopa, 136
Non-Contact Farmer (NCF), 305, 309, 311–313, 315, 316
Non-sustainable agricultural practice, 323, 324
Normalized Difference Vegetation Index, 18, 41, 42, 44, 47, 48, 117, 328

Normalized matrix, 139, 140, 204
North East India, 227, 286

O
Objective model, 14, 27, 95, 213
Operational technology, 149
Organic carbon concentration, 210, 323
Overexploitation of fragile ecosystem, 323
Oxidative Stress, 158, 167, 169–171

P
Pairwise comparison matrix, 30, 31
Parambikulam Aliyar Basin, 99, 101, 106
Photogrammetry, 4
Photosynthesis, 167, 168, 170, 324, 327, 329–331
Pollution and heavy metal concentration, 323
Poly house, 25, 27, 29, 31, 32, 34, 36–39
Polyphenol oxidase (POX), 168
Potato cultivation, 82, 84, 86–89, 92–95, 275
Poverty reduction, 185, 186, 243, 244
Precision Agriculture, 113, 114
Pre-Kharif season, 59, 68, 73, 78
Protected cultivation structures, 25–27, 29, 31, 32, 34, 39

Q
Questionnaire, 187, 266, 267, 305, 306, 308

R
Rabi season, 59, 67–69, 71, 73, 101, 115, 130
Rainfall, 17, 18, 25, 27, 29, 31, 33, 38, 42, 43, 47, 49, 59–64, 66–70, 74–78, 81–84, 86, 88–90, 93–95, 101, 102, 115, 136–140, 142, 144, 145, 168, 199, 200, 202, 204–209, 213, 219, 221, 223, 226, 230–232, 250, 257, 258, 260, 261, 285–288, 290, 291, 301
Reactive Oxygen Species, 156
Reeling, 185, 187, 189, 191–193, 196
Remote sensing, 4, 6, 7, 14, 16–18, 41–45, 55, 82, 113–115, 123, 226
Remote sensing and GIS techniques, 16, 82
Remote sensing indices, 41, 42, 55
Respondents, 187, 267, 268, 305, 308–311, 313–316, 320, 321
Rice and potato crop, 113, 115, 125, 130, 131
Rice crop, 115, 120, 122, 124, 126, 128, 130, 305, 316, 317

S
Semivariogram analysis, 113, 118, 131
Sensitivity analysis, 19, 25, 31, 33, 36, 131, 226
Sen's slope estimator, 59, 60, 76
Sentinel-2, 8, 41, 118, 130, 131, 202, 203
Sericulture industry, 185, 186, 189, 190, 197
Sericulture value chain, 188, 191, 193
Shifting cultivation, 225, 227, 234
Silkworm rearing, 186, 190, 191, 193, 196, 197
Silk yarn, 185, 189, 190, 192, 193, 196, 197
Skilled human resources, 193
Smart agriculture, 19, 150, 151
Soil and water assessment tool, 225, 226
Spatial data, 4, 100, 230
Spatial distribution, 65, 67–72, 201, 335, 336
Spearman's rho test, 59, 65, 68
Spherical Model, 113, 118, 122, 123, 131
Superoxide Dismutase (SOD), 158, 167, 169, 170
Supply chain, 108, 185, 186, 189
Sustainability, 6, 152, 157, 173, 227
Sustainable agriculture, 3–9, 100, 109, 199, 219, 286–288, 306, 324

T
Temperature, 16–18, 25–29, 31, 32, 39, 41, 42, 44, 45, 49, 54, 59–63, 65, 66, 70–74, 76–78, 81–84, 88–90, 93, 100–102, 115, 136–142, 144, 145, 152, 155, 156, 158, 163–166, 168, 171, 199, 200, 202–206, 209, 210, 213, 219, 226, 230, 231, 232, 285, 286, 288, 289, 291, 297, 301, 324, 329, 331
Temperature Condition Index, 45, 55
Terrain analysis, 259
Top soil nitrogen content and, 137, 170
Top soil organic carbon content, 92–95, 138, 139, 141, 145, 203, 209
Transpiration, 26, 42, 170, 237, 286, 290, 327

V
Validation, 82, 95, 96, 113, 118–120, 125, 130, 131, 135, 145, 225, 226, 231, 232, 236, 238, 245
Value chain analysis, 185, 186, 190
Value chain development, 185, 186
Value chain process, 186
Vegetation Condition Index, 41, 42, 44, 45, 50, 51
Vegetation Health Index, 41, 42, 45, 56, 57
Vegetation productivity, 327, 331, 333

W
Waste water usage, 323
Weight Aggregated Sum Product Assessment, 25, 30
Weight assignment, 25, 30
Weighted Product Model, 30, 33, 138
Weighted Sum Model, 30, 138
Western Ghat, 101, 257
White Head (WH), 316, 319
WordClim Data Sets, 137
Worst landslide, 257, 260

Y
Yellow Stem Borer (YSB), 45, 122, 305, 307, 308, 311, 316, 319–321